2nd Aceh International Symposium on Civil Engineering (AISCE 2019)

IOP Conference Series: Materials Science and Engineering
Volume 933

Banda Aceh, Indonesia
18-19 September 2019

ISBN: 978-1-7138-4769-4
ISSN: 1757-8981

Printed from e-media with permission by:

Curran Associates, Inc.
57 Morehouse Lane
Red Hook, NY 12571

Some format issues inherent in the e-media version may also appear in this print version.

Institute of Physics
Dirac House, Temple Back
Bristol BS1 6BE UK

Phone: 44 1 17 929 7481
Fax: 44 1 17 920 0979

techtracking@iop.org

Additional copies of this publication are available from:

Curran Associates, Inc.
57 Morehouse Lane
Red Hook, NY 12571 USA
Phone: 845-758-0400
Fax: 845-758-2633
Email: curran@proceedings.com
Web: www.proceedings.com

TABLE OF CONTENTS

IOP Conf. Series: Materials Science and Engineering **933** (2020) 011002 doi:10.1088/1757-899X/933/1/011002

Peer review declaration

All papers published in this volume of *IOP Conference Series: Materials Science and Engineering* have been peer reviewed through processes administered by the Editors. Reviews were conducted by expert referees to the professional and scientific standards expected of a proceedings journal published by IOP Publishing.

- **Type of peer review:** Double-blind (involving internal and external reviewers)
- **Conference submission management system:** EDAS Conference and Journal Management System
- **Number of submissions received:** 66 papers
- **Number of submissions sent for review:** 61 papers
- **Number of submissions accepted:** 59 papers
- **Acceptance Rate (Number of Submissions Accepted / Number of Submissions Received X 100):** 89.39%
- **Average number of reviews per paper:** 2
- **Total number of reviewers involved:** 31 reviewers
- **Any additional info on review process:**

  The organizer received draft papers and distributed the papers to reviewers through the EDAS system. Each paper was sent to two reviewers. The papers were reviewed based on the following aspects: Clarity, Suitability, Originality, Methodology Appropriateness, Clarity of finding, and Scientific.
  If reviewers accepted the papers, then the organizer sent the papers back to authors so the authors can address all comments from the reviewers.
  The authors sent the final manuscripts back to the organizer through the EDAS system. Later, the AISCE 2019 editors finalized the final manuscripts by proofreading and checking the format of manuscripts based on IOP Publishing guidelines.

- **Contact person for queries:** Teuku Budi Aulia (aulia@unsyiah.ac.id)

IOP Conf. Series: Materials Science and Engineering **933** (2020) 012001 doi:10.1088/1757-899X/933/1/012001

Identification of fire protection system availability in faculty of engineering, University of Teuku Umar

I Y Salena[1], M Safriani[1] and M A R Amin[1]

[1] Civil Engineering Department, Engineering Faculty, Universitas Teuku Umar, Meulaboh, Aceh Barat, Indonesia

Corresponding e-mail: inseunsalena@utu.ac.id

Abstract. The objective of this study is to identify the availability and implementation of fire protection systems in the Engineering Faculty of Teuku Umar University. The fire protection components identified were the access and water supply component, the mean of egress, passive and active fire protection systems, by conducting direct observations to the field using a checklist form. The building identification refers to the Minister of Public Works Regulation No. 26/PRT/M/2008, and the collected data was analyzed using the Likert scale method, which shows the level of conformity to the regulations based on the generated data. The result shows that the access and water supply components obtain 2.75 of the Likert scale. The mean of egress obtains an average value of 2.24 of the Likert scale. The values mean these facilities do not fulfill the regulation. The passive protection system obtains 2.48 of the Likert scale; it means the facilities do not fulfill the regulation. Active protection systems get an average value of 1.4 of the Likert scale; it means the facilities do not fulfill the regulation. The results of the study show that the fire protection system in the Faculty Of Engineering Building did not meet regulation No. 26/PRT/M/2008 overall so that further improvement is needed to fulfill the requirement of fire protection systems to overcame the danger of fire.

1. Introduction

The Southwest Aceh Regency area often experiences fire disasters. Throughout 2018, cases of fires have engulfed several buildings, including residential houses, Islamic Boarding Schools, and the West Aceh Government Office. The fire, which broke out in Sawang Teube Village, Kaway XVI Subdistrict, Aceh Barat District, burned three houses and one car. Based on information from the West Aceh BPBD (2017), material losses reached 370 million rupiahs. A massive fire also occurred at the Nur Yaqdhah Islamic Boarding School in Labuhan Haji Barat District, South Aceh Regency. Fifty-four units of the students' rooms burned. The warehouse fire also occurred at the location of the West Aceh Government Office Complex.

Based on the background that put forward, the identification of the availability of fire protection systems based on regulations needs to be implemented. The number of fire cases shows the fire protection system and the readiness to deal with fire has not optimally applied. The potential fire hazard is large, can not be considered easy. Government regulations regarding the implementation of fire protection for buildings already exist, but supervision and control of the application of the fire protection system do not perform optimally.

Teuku Umar University (UTU) was the first state university in the South West Aceh region; the university was inaugurated on April 2, 2014. UTU consists of 6 faculties, one of those is the Faculty of

Engineering. The Faculty of Engineering consists of a 2-level building that has a building area of 720 m^2 containing 8 rooms, which are separated by brick walls. The other parts of the building are equipped with room partitions. The appearance of the Faculty of Engineering Building is shown in figure 1.

Figure 1. Engineering faculty.

Based on the background above, it is essential to know the current conditions of the availability and application of fire safety systems in the UTU Faculty of Engineering Building. The purpose of this study is to identify the availability of fire protection systems at the Faculty of Engineering of Teuku Umar University.

2. Methodology

Field observation and inspection of protection components were performed to identify the availability and implementation of the fire protection system. According to [1], the fire protection system was divided into four components, each of those was divided into sub-components, as shown in table 1.

Table 1. Fire protection systems [1].

No	Components	Subcomponent
1	Access and water supply	Water sources
		Access road
		Distance between buildings
		Private hydrant
2	Means of egress	Exits
		Means of egress reliability
		Doors
		Enclosure and protected stairs
		Number of means of egress
		Arrangement of means of egress
		Discharge from exit
		Illumination of means of egress
		Emergency lighting
		Marking of means of egress
3	Passive protection system	Fire resistive or noncombustible construction
		Availability of fire windows and fire doors
		Interior finish materials
		Fire barriers
		Smoke barrier partitions
		Smoke barriers
4	Active protection system	Automatic sprinkler systems

	Fire pumps
	Water supplies
	Fire extinguisher
	Detection, alarm, and communications systems

2.1. Data collections

Data collection is carried out by conducting direct observations in the field using the checklist form shown in table 2. The checklist form consists of components and sub-components of fire protection systems, the availability of the subcomponent based on the direct observation, and description of the information data collected from the existing building.

Table 2. Checklist form of field observation [5].

No.	Component	Condition		Information
		available	not available	

2.2. Data analysis

Data collection result was analyzed using the Likert Scale. Likert scale is a tool to measure or collect data by measuring and weighing the questions [5]. The rating on the Likert scale is shown in table 3, which consists of the suitability of the available fire protection system with the applicable regulations accordance with Ministry regulation No. 26/PRT/M/2008. The rating starts from 1, as the lowest rate which does not fulfill the regulation, and 5, as the highest rating which fulfills the regulation.

Tabel 3. Rating in likert scale [5].

No.	Description	Likert Scale
1	Very accordance with regulation	5
2	Accordance with regulation	4
3	Enough accordance with regulation	3
4	Less accordance with regulation	2
5	Not accordance with regulation	1

The results of the assessment of the system components will be classified under certain criteria. To get the final results of each component, the below formula was used:

$$\text{Average} = \frac{\text{Amount of Values}}{\text{Data Collect X}}$$

$$X = \frac{x_1 + x_2 + \ldots + x_n}{n} \quad (1)$$

The higher the value, the closer it is to the maximum value of a set of numerical values. The relationship can be formulated as a percentage value [5].

3. Results and discussion

After the survey was conducted, the results of observations are set in a descriptive form. Then, they were analyzed using the Likert scale method. The following conditions show the fire safety component that observed. Table 4 shows the results of the identification of access and supply of water.

Table 4. Assessment of access and water supply.

No	Fire safety component	Criteria accordance to PerMen PU No:26/PRT/M/2008	Existing condition	Likert scale
1	2	3	4	5
1	Water sources	Available with a capacity that meets the minimal requirements to the function of the building	The only available water sources are the drill well. There is no fire well and water reservoir	2
2	Access road	Fire department access road must have a width of not less than 6 m, and a minimum length of 15m. The road must be paved and has the minimum width of the entrance of 4 m. The road must not be obstructed	Paving block pavement. Width is 12 m, and the length is 49 m. The entrance width is 4 m	3
3	Distance between buildings	The Minimum distance from the nearest building (m) based on the hight of the building. 1) < 8 m, distance 3 m 2) > 8 -14 m, distance > 3 - 6 m 3) >14-40 m, distance > 6 - 8 m 4) > 40 m, distance > 8 m	The height of the Faculty of Engineering Building is about 12 m. The distance between the buildings next to it is 23 m	5
4	Private hydrant	1) Private hydrants must be installed at an accessible area	Not available	1
		2) Completely functioning and complete	Not available	1
		3) Water supply is 38 l/sec; 35 Bar for pressure	Not available	1

Identification for Component of Means of Egress can be seen in table 5 below.

Table 5. Assessment for means of egress.

No	Fire safety component	Criteria accordance to PerMen PU No:26/PRT/M/2008	Existing condition	Likert scale
1	2	3	4	5
1	Exit access corridor/exit	1) Exit corridor and exits must be separated from other parts of the building	On the 1st floor, the exits are separated from the corridor. However, on the 2nd floor, the exits are not separated from the corridor.	2
		2) Construction of the separator must have a fire-resistance level of at least 1 hour in case the exits connect three floors or less	- On the 1st floor: construction of the separator is a partition of glass and aluminum - On the 2nd floor: there is no dividing construction. The available construction is a corridor that directly leads to the exit, which consists of plywood partitions and brick walls	2

2	Means of egress reliability	1) Furniture, decoration, or other objects should not be placed carelessly that they could disturb the exits, the access to it, the way to get out of there, or could disturb the view.	- On the 1st floor: exits access corridor is blocked by a banner located on the right side of the corridor - On the 2nd floor: there are no obstacles	3
		2) There must be no disturbances due to railing, barriers, or doors of the track lane	- On the 1st floor: there are doors at both ends of the corridor before heading exits - On the 2nd floor: corridor goes directly to exits	2
		3) Mirrors should not be installed inside or near any exits	There is no mirror attached on the 1st and the 2nd floors	5
		4) Every door and the main entrance to the exits must be visible and directly from the road to the exits	- On the 1st floor: the door and entrance go directly to the path for an exit - On the 2nd floor: the door and entrance go directly to the exit stairs that lead to the exit lane	4
3	Doors	1) Every door on the outlet must be of the type of side hinges or swing doors	At 1st and 2nd floors: exit doors are made from the type of side hinges	3
		2) The door must open towards the outside lane	At 1st and 2nd floors: the door opens inward	1
4	Enclosure and protected stairs	1) All stairs inside a building, which serves an exit or component, must be enclosed	The stairs are on the outside corridor connecting between 1st and 2nd floors. The stairs lead to the exit.	4
		2) Spaces used in protected exits are prohibited, including under stairs that have the potential to interfere with the road to the outside	Under the stairs, there are tables and the waiting chairs that are used by students. The things block the way out	2
		3) Stairs must be provided with special identification; the marking must indicate the level of the floor, the top, and the bottom end of the staircase.	There is no identification and marking of the stairway	1
		4) The marking must be in a protected room placed close to 1.5 m above the landing and must be easily visible	There is no marking of the stairway	1
		5) Markings must be painted or written on the wall	There is no marking of the stairway	1
5	Number of means of egress	The minimum number of facilities to escape from each balcony, mezzanine, the floor, is two	The exit is difficult to achieve	3
6	Arrangement of means of egress	Minimum number of facilities to escape from each balcony, mezzanine, and floor is two	The exit is difficult to achieve	3

7	Discharge from exit	1) The exit access must be arranged so that the exit is easily achieved at all times	The exit width only serves 3 (three) exits	2
		2) Access to the exits must not be through kitchens, warehouses, break rooms, workspaces, closets, bedrooms, or other spaces and places that may be locked	Access to the exit is through the waiting room and lecturer hall	1
		3) Hangers or curtains must not be installed above the exit door because it makes the exit hidden or unclear	There are no curtains installed	5
8	Illumination means of egress	1) All exits must end directly on public roads or the exterior of the release exits	All exits end on a public road or the outside of a release exit	4
		2) The exit releases must be arranged and marked to make clear the directions from the road to the public road	There is no sign of the release exit	1
9	Emergency lighting	Emergency lighting equipment for road exits must be available	No emergency lighting equipment available	1
10	Marking of means of egress	1) Exit signs that can be touched should read: "EXIT" and must be placed on each exit door	No exit sign available	1
		2) Every required mark must be placed in such a size, a real color and easily seen	There is no marking on the way out	1
		3) Direction signs indicating the direction of the track must be placed at each location if the direction of the track reaches the nearest exit is unclear.	No directions sign	1

Identification result for Component of Passive Protection Systems shown in table 6 below:

Table 6. Assessment result for passive protection systems.

No	Fire safety component	Criteria accordance to PerMen PU No:26/PRT/M/2008	Existing condition	Likert scale
1	2	3	4	5
1	Fire resistive or noncombustible construction	The construction that meets the fire resistance requirements based on the type of construction (firewall, fire barrier, fire resistant ceiling, fire retardant partition, roof cover)	Type II construction, in which structural elements are non-flammable. The partitions are made of glass and aluminum. Ceilings are made of PVC	2
2	Availability of fire windows and fire doors	Fireproof doors and windows to protect openings in walls, floors, and ceilings against the spread of fire and smoke inside, inside and outside buildings	Doors are made of wood; glass for partition doors and windows are made of glass	2

3	Interior finish materials	Interior coatings are fire-resistant	Interior coating is painted	4
4	Fire barriers	1) Walls, doors/windows, floors, floor/ceiling mounting constructions, that are constructed as fire barriers, constructed using fire-resistant materials that meet the testing requirements.	Brick walls, wooden doors/windows from wooden and glass frames, PVC ceilings. It is an existing structure that already exists and is not constructed as a fire barrier	2
		2) Equipment or fire stopper systems for cables, cable racks, conduits, piping, tubes, smoke vents and exhaust vents, maintenance and similar equipment to accommodate electrical, mechanical, plumbing and communication systems that pass walls, floors, or floor/ceiling construction partners that are constructed as a fire barrier	There is a portion of fire-stopping equipment, but there is also a part that is exposed	2
		3) Openings in fire barriers for air conditioning ducts or air movement must be protected	There are no openings specifically for air conditioning ducts, there is the air ventilation	3
5	Smoke barrier partitions	1) The partition must be installed stretching from the floor to the bottom of the roof or the roof deck above, passing through hidden spaces such as above the suspended ceiling, and passing through the intermediate spaces for structure and mechanics.	Some partitions on the 1st floor did not reach the bottom of the roof, partitions attached to the ceiling for 2nd floor	2
		2) The door must not have air grilles	The door does not have a grille	4
		3) The door must close alone or close automatically	There are two doors which are broken and cannot close automatically, exit door on L1 can close automatically	2
6	Smoke barriers	1) The doors in the smoke barrier must completely cover the door openings, leaving only a minimum gap for the smooth operation of the door and there should be no openings on the door leaf, air cavities or door grille or grille	There is no special smoke barrier door, there is a wooden door covering the door opening, there is no gap in the door but it still has a large gap.	2
		2) The doors on the smoke barrier must be of a type that can close itself or close itself automatically	The door cannot close by itself	2

Identification result for Component of Active Protection Systems shown in table 7 below.

Table 7. Assessment result for active protection systems.

No	Fire safety component	Criteria accordance to PerMen PU No:26/PRT/M/2008	Existing condition	Likert scale
1	2	3	4	5
1	Automatic sprinkler systems	Automatic sprinklers must be installed and fully ready to operate in the type of dwelling within the technical requirements/standards referred to.	Not available	1
2	Fire pumps	Fire pump available	Not available	1
3	Water supplies	The main pipelines of private fire services must be installed under SNI 03-3989-2000 standards	No main pipe available	1
4	Fire extinguisher	The installation, maintenance, selection and distribution of APAR must comply with SNI 03-3987-1995	1 APAR available for the whole building	3
5	Detection, alarm, and communications systems	Refer to SNI 03-3985-2000 or the latest edition "Procedures for Planning and Installing Fire Detection and Alarm Systems for the Prevention of Fire Dangers in Buildings and Houses".	Not available	1

The evaluation results of fire safety components based on observations in the field and conformity assessment of regulations using the Likert scale is presented in the table below. Table 8 shows the value of the Likert Scale for Water Access and supply, which is 2.75. This value indicates that the requirements for access and water supply for fire safety in the Faculty of Engineering building is not following the regulations. Sub-components identified for the access road, and the distance between buildings meets the requirements, but for water supply and fire wells, reservoir water and hydrant is not available.

Table 8. Assessment result of access and water supply.

No	Components	Likert Scale
1	Water sources	2
2	Access road	3
3	Distance between buildings	5
4	Private hydrant	1
The Average result of access and water supply component		2.75

The identification component of the means of egress is shown in table 9. The average value obtained is 2.24 on the Likert scale. From the ten subcomponents observed, three subcomponents meet compliance with regulations namely Means of Egress Reliability, Number of Means of Egress, and Arrangement of Means of Egress. Seven other subcomponents do not meet the regulation.

Table 9. Assessment result for the means of egress.

No	Components	Likert scale
1	Exit access corridor/exit	2

2	Means of egress reliability	3.5
3	Doors	2
4	Enclosure and protected stairs	1.8
5	Number of means of egress	3
6	Arrangement of means of egress	3
7	Discharge from exit	2.6
8	Illumination means of egress	2.5
9	Emergency lighting	1
10	Marking of means of egress	1
The average result of means of egress		2.24

The result for the passive protection system in table 10 shows the value of the Likert scale of 2.48 which indicates that the requirements for the passive protection system do not meet the regulation. The subcomponent of interior-finishing material is following the regulation but the other subcomponents are not following the regulations.

Table 10. Assessment result for passive protection systems.

No	Components	Likert scale
1	Fire resistive or noncombustible construction	2
2	Availability of fire windows and fire doors	2
3	Interior finish materials	4
4	Fire barriers	2.3
5	Smoke barrier partitions	2.6
6	Smoke barriers	2
The average result of passive protection systems		2.48

Table 11 below shows the average result of the Likert scale for active protection systems, which is 1.4. This value indicates that the requirements for this component in the Faculty of Engineering Building are not following the regulations. The sub-components that fulfill the regulation are fire extinguishers. For the other component, they do not meet the regulation.

Table 11. Assessment result for active protection systems.

No.	Components	Likert scale
1	Automatic sprinkler systems	1
2	Fire pumps	1
3	Water supplies	1
4	Fire extinguisher	3
5	Detection, alarm, and communications systems	1
The average result for active protection systems		1.4

4. Conclusion

Based on the research results that have been obtained from observations and the analysis according to Likert scale, it is concluded that the access and water supply component obtains the value of 2.75 in Likert scale, the means of egress obtains 2.24 in Likert scale, the passive protection system obtains 2.48 in the average value of Likert scale, and active protection system obtains 1.4 averagely in Likert scale. As a result, all fire safety system components that have been assessed do not meet the requirement of the regulation. The components of access and water supply, means of egress, passive and active protection systems show a low value that does not fulfill the applicable regulations. Observations indicate the availability and implementation of building safety system components are still minimal at the Building of Faculty of Engineering of Teuku Umar University. Improvement is needed for fire safety protection systems in the environment of Teuku Umar University Campus to prepare and overcome the danger of fire.

Acknowledgments
The authors want to thank the staff of the Engineering Faculty for enabling the visit to the faculty building to observe. Also, the authors want to thank the Planning Department Staff of Teuku Umar University for their generosity by providing the documentation needed for this research.

References
[1] Peraturan Menteri PU No 26 PRT/M/2008 2008 *Persyaratan Teknis Sistem Proteksi Kebakaran pada Bangunan Gedung dan Lingkungan* (Jakarta: Departemen Pekerjaan Umum)
[2] Napitupulu P and Dulbert B T 2015 *Sistem Proteksi Kebakaran Kawasan Pemukiman dan Perkantoran* (Bandung: P.T. Alumni)
[3] National Fire Protection Association NFPA 220 2018 *Standard on Type Of Building Constructions section 7.2 NFPA 5000®: Building Construction and Safety Code®*
[4] Badan Standarisasi Nasional 2000 *SNI 03-1736-2000 Tata Cara Perencanaan Sistem Proteksi Pasif Untuk Pencegahan Bahaya Kebakaran pada Bangunan Rumah dan Gedung*
[5] Kurniawati E 2012 *Evaluasi Sistem Proteksi Kebakaran pada Bangunan Apartemen Ditinjau dari Sarana Penyelamatan dan Sistem Proteksi Pasif (Studi Kasus di Apartemen Solo Paragon)* (Surakarta: Jurusan Teknik Sipil Universitas Sebelas Maret)
[6] Saptaria E, Mulyanto S and Maryono 2006 *Pemeriksaan Keselamatan Kebakaran Bangunan Gedung* (Jakarta: Badan Litbang PU Departemen Pekerjaan Umum, Pd-T-11-2005-C)
[7] Badan Standarisasi Nasional 2000 *SNI 03-1746-2000. Tata Cara Perencanaan dan Pemasangan Sarana Jalan Keluar Untuk Penyelamatan Terhadap Bahaya Kebakaran pada Bangunan Gedung*
[8] Badan Standarisasi Nasional 2000 *SNI 03-6571-2000 Sistem Pengendalian Asap Kebakaran pada Bangunan Gedung*
[9] Badan Standarisasi Nasional 2000 *SNI 03-6574-2001 Tata Cara Perencanaan Pencahayaan Darurat, Tanda Arah dan Sistem Peringatan Bahaya pada Bangunan Gedung*
[10] Adiwidjaja R 2012 Studi tingkat keandalan distem proteksi kebakaran pada bangunan apartemen (studi kasus apartemen di Surabaya) *Dimensi J. of Archit. and Built Envir.* **39** 1 15-22
[11] Safriandi, Trikomara R and Sebayang M 2013 *Analisa Keandalan Sistem Keselamatan Bangunan Terhadap Bahaya Kebakaran (Studi Kasus Gedung Surya Dumai Group Dan Bank Tabungan Negara Kota Pekanbaru* (Riau: Jurusan Teknik Sipil, Fakultas Teknik, Universitas Riau)
[12] Kurniawan P A, Sugiyarto S and Laksito B 2014 Evaluasi penerapan sistem proteksi kebakaran pada bangunan rumah sakit (studi kasus Rs. Ortopedi Prof. Dr. R. Soeharso Surakarta) *E J. Matriks Teknik Sipil* 824-32.
[13] Pratama R and Trikomara R 2017 Analisis keandalan sistem keselamatan bangunan terhadap bahaya kebakaran (studi kasus: Gedung Rumah Sakit Ibu dan Anak Eria Bunda Kota Pekanbaru) *Jom FTEKNIK* **4** 2 1-6
[14] Stephanie 2017 *Evaluasi Sistem Proteksi Pasif Kebakaran Bangunan (Studi Kasus: Millennium Ict Centre* (Medan: Departemen Arsitektur Fakultas Teknik USU)
[15] Sugiyono 2009 *Metode Penelitian Kuantitatif Kualitatif dan R&D* (Bandung: CVAlfabeta)

IOP Conf. Series: Materials Science and Engineering **933** (2020) 012002 doi:10.1088/1757-899X/933/1/012002

Analysis of NATM tunneling method using CYCLONE modeling and simulation tools

B B Nataadiningrat[1], A W Prabowo[1], I M A B Rasmawan[1], A T Putri[1], M Abduh[2] and R D Wirahadikusumah[3]

[1] Master Student in Civil Engineering, Faculty of Civil and Environmental Engineering, Bandung Institute of Technology, Indonesia
[2] Associate Professor, Faculty of Civil and Environmental Engineering, Bandung Institute of Technology, Indonesia
[3] Professor, Faculty of Civil and Environmental Engineering, Bandung Institute of Technology, Indonesia

Corresponding e-mail: abduh@si.itb.ac.id

Abstract. The NATM or New Austrian Tunneling Method became a popular method for tunneling in Indonesia recently, since at least two strategic and huge construction projects utilized it. It is considered as a new method introduced to the Indonesian construction industry and therefore it is interesting to be studied and anticipated for the implementation. This paper discusses modeling the operation of the NATM and then simulation of the model for finding the productivity, effectiveness, and problems that may come in the implementation. The CYCLONE modeling and three simulation tools, i.e., WebCYCLONE, COST, and Symphony.Net, were utilized for this purpose. Data used for durations of tasks associated with the method were gathered from previous work in the field of different projects and literature. The results of this analysis then compared to other similar analyses but for a different project. A comparison of those two analyses was used to identify the most important tasks that will determine the performance of the method with a safety-first requirement in mind. The sensitivity analysis identified secondary resources such as a concrete wet sprayer; the shotcrete team and the formwork team are considered sensitive to improve the performance of the NATM.

1. Introduction

The development of science and technology in all aspects of the scientific field is a continuous process. Technological developments in the construction sector will contribute directly to the performance of a work operation in the project, where the objectives can be achieved with optimal conditions. The optimal condition means each resource, the existing constraints, and their objectives can be well controlled according to their purpose and are effective. Productivity is important in construction activities, which will have a direct impact on the final cost of project completion. Important control activities in construction operations to achieve expected effectiveness and efficiency, control efforts can be carried out by operating simulation activities designed in such a way as to be able to describe the real conditions in the field.

The object of this study is a tunnel section with a diameter of 12-14 meters, as some segment of the overall project that will be used by the transportation fleet. A mega project of the railway with the total investment in this project amounted above 50 trillion rupiahs, a slight delay would potentially result in a large cost overrun, which need to be avoided. In infrastructure projects in Indonesia, tunneling is still rarely found while in this overall project the portion of tunnel work is 15.23% of the total project length, therefore thorough planning was required. Tunneling is a complex job with high uncertainty such as soil conditions that are at risk of erosion while vertical digging is being done. Although tunneling is complex, the operation can be repetitive so that the modeling simulations can be carried out. The construction method used is the New Austrian Tunneling Method. Cyclone modeling is used for the construction operations modeling in this study and the simulations are conducted with the help of three simulation tools, namely WebCYCLONE, COST, and Simphony.NET.

The purposes of this research are to develop tunneling simulation models in tunnels that are part of the train project, analyze the productivity of the tunneling method used based on the model that has been made, compare the simulation results which is the productivity of tunnel construction operation to the previous research on the construction operation of Cisumdawu toll road tunnel and last but not least is to determine the effect of resource variations on the modeled productivity system.

1.1. New Austrian tunneling method

New Austrian Tunneling Method (NATM) is a method developed to make tunnel construction, especially for tunnels made in soft soil conditions. Soft soil categories are lands that require the use of structural support, either to maintain balance or to limit the displacement of land around the tunnel construction area. The purpose of the NATM method is to provide safe and economical support for tunnels where the type of soil unable to support their own weights - for example, crushed rocks, debris, and even soil. Structural support can be obtained by mobilizing any strength with hard rock equivalent strength. This structural support can be obtained by using surface stabilization such as thin shotcrete layers and other reinforcement for example steel rib, inverted arch lining, and so on [1].

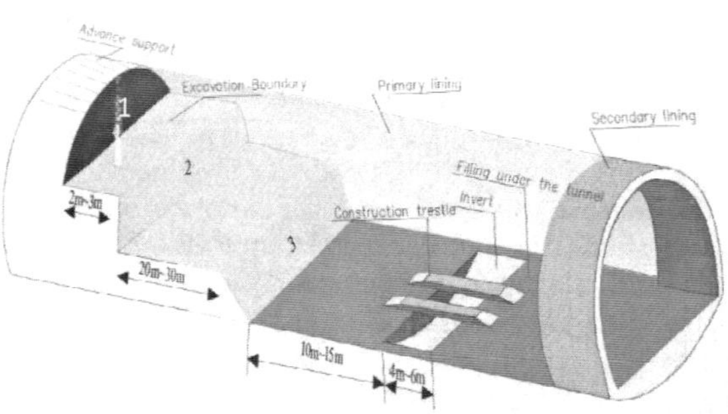

Figure 1. Tunnel excavation.

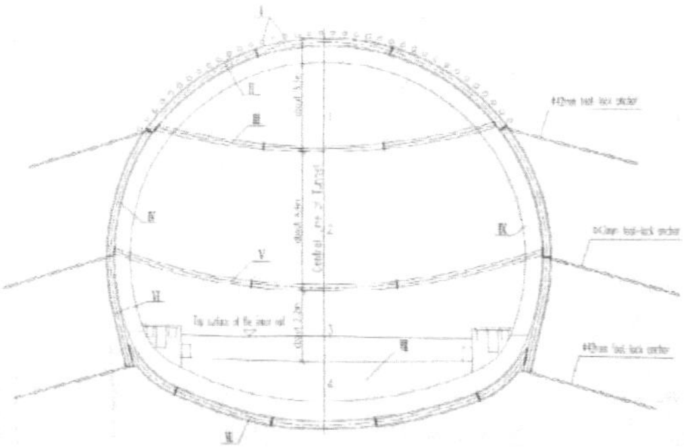

Figure 2. Division of tunnel section.

1.2. Modeling of construction operation

Models are representations of real situations. The purpose of modeling a construction process is to find out more about the actual situation of a construction process occurring in the field by analyzing and investigating. Basics symbols of cyclic operations network systems modeling developed from the [2] as in table 1. The procedure in modeling the construction process involves four basic steps [2], namely :

- Identify the resource flow, at this stage a number of resources are related to the implementation of the construction operation. Related resources include workers, equipment, materials, workspace, and access to information;
- Develop the cycle of resource flow, develop each of the stage's cycle which includes several type and number of resources;
- Integrate the flow unit cycle, each cycle that represents each stage of the work is then integrated into a system that describes the stages of the overall operation;

- Flow unit Initiation, to analyze the model made, each flow unit needs to be named and numbered in the logical sequence.

Table 1. Basic elements of CYCLONE modeling [2].

No	Symbol	Name	Description
1	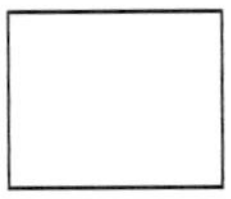	Normal Activity (NORMAL)	The normal work task modeling element, which is unconstrained in its starting logic and indicates active processing of (or by) resource entities
2	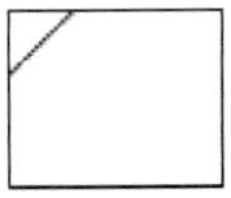	Combination Activity (COMBI)	The constrained work task modeling element, which is logically constrained in its starting logic, otherwise similar to the normal work task modeling element
3		Queue Node (QUEUE)	The idle state of a resource entity symbolically representing a queueing up or waiting for use of passive state of resources
4		Accumulator (COUNTER)	The countermeasures the modeled system's production rate
5		Arcs	The resource entity directional flow modeling element

Besides, the model developed must be modified to provide control for system performance. Simulations in construction operations are needed to provide information on work performance with existing conditions.

1.3. Simulation of construction operation

Simulation is a modeling of a process or system to mimic the original system of events that occur at a certain time [3]. Simulation is done to determine the condition of the system, to determine the system response and the imbalance between resources. This imbalance can have an impact on bottlenecks and inefficiencies that must be avoided or minimized [2]. The steps taken in conducting the simulation are: defining the system, modeling the system, analyzing input and output, as well as validation and verification.

Based on the previous study [4], the development of simulation techniques for repetitive construction operations has been going on for a long time, namely since the 1970s where Daniel W. Halpin developed CYCLONE modeling. Until now, a web-based simulation technique has been developed called WebCYCLONE. Some simulation techniques for repetitive construction operations that have been developed are Insight (Kalk 1980), RESQUE (Chang 1986), UM-CYCLONE (Loannou 1989), MicroCYCLONE (Halpin 1990), COOPS (Liu 1990), CIPROS (Odeh et al. 1992), DISCO (Huang 1994), STROBOSCOPE (Martinez 1996), SIMPHONY (AbouRizk and Mohamed 2000), COST (Cheng et al. 2000), RISim (Chua and Li 2002), and WebCYCLONE (Halpin et al. 2003).

In this study, three soft wares are used in the simulation, namely WebCYCLONE, COST, and Simphony.NET. The use of each software is quite different in terms of input and output data. In WebCYCLONE, the users must input the model in a particular programming language, while in COST, input data is a parameter of each element in the model. Unlike the two soft wares mentioned earlier, Simphony.NET uses input data in the form of images so that the result is easier to see. Apart from being different in terms of inputting data, the output given by these three soft wares is also different, table 2 shows the comparative output of each software.

Table 2. The comparative output of each soft wares.

Output	Web Cyclone	COST	Symphony.NET
Productivity graph	✓	✓	✓
Event list		✓	
Chronological list	✓	✓	
Productivity per time unit	✓	✓	
Trace chart/ Time chart for every que	✓		✓
% que occupied	✓	✓	✓
Waiting time for que	✓	✓	✓
% occupied graph for que		✓	
% busy graph for normal & combi		✓	
Emission report			✓
Cost report		✓	✓
Histogram and CDF for each element			✓

2. Methodology

The methodology begins with conducting a preliminary study of the construction operations in a project. This preliminary process is carried out to see how repetitive operating processes in construction projects can be examined using Cyclone modeling to obtain productivity values with work tasks and available resources. Then a literature study was conducted regarding the use of NATM (New Austrian Tunneling Method) in tunnel construction operations, and identifying the objectives of the research, followed by field observations for tunneling projects and collecting data in the form of interviews with parties who carry out operations in the field. The data is also obtained by analyzing the simulation video of the work made by the contractor. To complete the simulation, secondary data from the previous studies [4] are used to cover the data that could not be obtained from the construction project.

The study was continued by using the data obtained to make the Cyclone model using work tasks that have a significant influence on tunneling operations, along with data on the number of resources and the duration that has been obtained during data collection. The model will be run using Webcyclone, COST, and Symphony.NET software, if there is a failure in running the model, the model will be made to re-check the existing model. The use of three soft wares is intended to obtain results that can be compared between one software with another software.

The results of the simulation will be analyzed regarding the value of productivity which will then be compared with previous studies which will then be analyzed for the causes of the differences. The next step is sensitivity analysis, sensitivity analysis is used to compare several possible uses of resources with different amounts and then will produce a number of different productivity results with the same operating cycle, those results then will be considered as suggestions for the most efficient use of resources. All results of analysis then used as a basis for making conclusions and recommendations.

3. Results and discussion

Construction operation can be defined as a collection of specific work tasks, where work tasks are the basic elements of work, each work task will be related to the technology and tools that support it and the resources involved in it [2]. The simulation model is intended to be able to describe the real conditions, it is expected that by modeling the results of the output values achieved will be close to the real output results in the field. Simulations can describe resources that will move from one work task to another in accordance with the task definition of each resource. Identification of resources, elements of the work task, and how the flow of resources between work tasks is the basic rational modeling that is carried out.

3.1. Work task and duration identification
The work task is a readily identifiable component of a construction process or operation whose description to a crew member implies what is involved in and required [2], thus work task identification in the model is the selection of work groupings that have significant value-added to the progress of the work. The duration of each work task affects the work task productivity value. There are two types of duration distributions in each work task in the study. The first type is a uniform distribution, where each value of the random variable has the same probability to be selected and the type of triangular distribution is a continuous opportunity distribution with 3 parameters, they are minimum value, most likely value, and maximum value, respectively.

Table 3 shows a list of work tasks in tunneling operations that are reviewed and their respective durations, the duration of which comes from a previous study [5,6] regarding tunneling on the Cisumdawu Toll Road project with adjustments. The assumption for the duration is the length of time needed to complete the work until the secondary lining for 10.5 meters has finished, thus defined as one operating cycle.

3.2. Identification of resource needs
Resource requirements used in tunnel work activities are Drilling Machine, Pneumatic Drill, Anchor Pump, Excavator, Loader, Dump Truck, Labor (work team), Concrete Wet Sprayer, Ready Mix, Concrete Pump, Inverse Arc Hydraulic model, Multi-function Laying Trolley-Formwork Platform, Crane, and Traveler Concrete Formwork. Resource requirements for each work task in tunnel construction activities are shown in table 4.

3.3. Cyclone model development
Modeling development based on the work breakdown structure of tunnel work. A model was developed to get the possibility of improvements in terms of productivity. Assume that in modeling, three-bench conditions in the excavation are considered to have occurred, operations are running in series stages where there are no parallel jobs that occur, this may not be accurate when compared to the actual conditions but with limited resources, it needs to be considered.

Figure 3 illustrates how a tunneling operation cycle works, simply explained as follow, assume the three bench condition (figure 1) has been achieved, preparation and reinforcement were conducted first to prevent the upper soil from collapsing, excavation then begins from the top-level then continued with the initial reinforcement immediately, excavation and initial reinforcement of the middle and bottom level then carried out sequentially (top, middle, and bottom level of the tunnel shown by number 1, 2, and 3 respectively in figure 1), the next step is excavation, reinforcement, and lining for the inverted section, the last part is the reinforcement and lining of the tunnel section (above the inverted part which was previously lining). Black dots that are assumed to be the initial condition for resources will move in its path to carry out the task, then return again to its initial condition. The Cycle is counting as one cycle of this operation when the resources pass the counter element, then the cycle time will stop counting as the result of cycle duration.

Table 3. The type of probability distribution of work task duration.

ID	Work Task	Duration (Hours)			
		Type	Parameter		
WT1	Assessment	uniform	0.5		1
WT2	Advanced Support	uniform	1.5		2
WT3	Excavation Phase 1	triangular	9	10	11

WT4	Initial concrete Spray	uniform	20		22
WT5	Anchor Installation + Wire Mesh + H beam + temporary Support	uniform	6		8
WT6	Shotcrete 30 cm	uniform	20		22
WT7	Excavation Phase 2	triangular	9	10	11
WT8	Initial concrete Spray	uniform	20		22
WT9	Anchor Installation + Wire Mesh +H beam + Temporary Arc	uniform	6		8
WT10	Shotcrete	uniform	20		22
WT11	Excavation Phase 3	triangular	9	10	11
WT12	Excavation Phase 4	uniform	7		9
WT13	H beam arc	triangular	5	6	7
WT14	lean Concreting h Beam pouring	uniform	3		6
WT15	Inverted Rebar work + Formwork	uniform	21		23
WT16	Inverted Lining	uniform	18		20
WT17	Instal Geotextile	triangular	22	23	24
WT18	Rebar work + Formwork	triangular	32	40	48
WT19	Concrete Lining	triangular	10	13	16

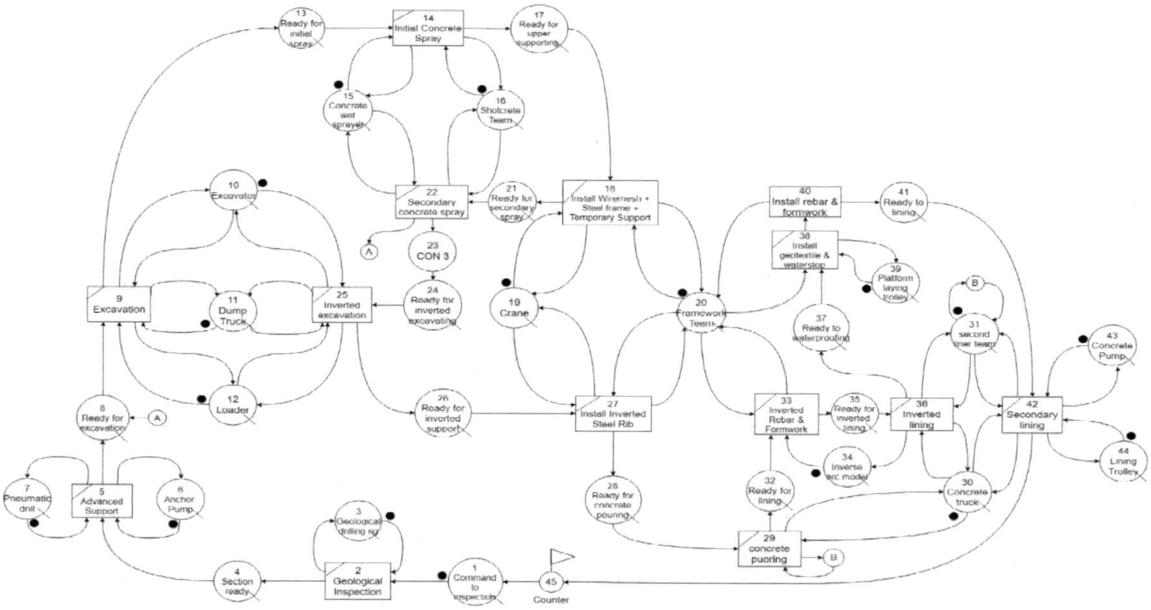

Figure 3. Cyclone model of tunneling.

Table 4. Integration flow units' cycle and initial resource.

Integration Flow units' cycle	Initial Resource		WT 1	WT 2	WT 3	WT 4	WT 5	WT 6	WT 7	WT 8	WT 9	WT 10	WT 11	WT 12	WT 13	WT 14	WT 15	WT 16	WT 17	WT 18	WT 19
Drilling Machine	R1	1 Unit	✓																		
Pneumatic Drill	R2	1 Unit		✓																	

Anchor Pump	R3	2 Unit	✓																	
Excavator	R4	2 Unit		✓				✓				✓	✓							
Loader	R5	2 Unit		✓				✓				✓	✓							
Dump Truck	R6	3 Unit		✓				✓				✓	✓							
Labour	R7	1 Team				✓	✓			✓	✓			✓	✓	✓	✓	✓	✓	✓
Concrete Wet Sprayer	R8	1 Team			✓		✓		✓		✓									
Ready-mix	R9	3 Unit													✓		✓			✓
Concrete Pump	R10	1 Unit																		✓
Inverse Arc Hydraulic model	R11	1 Unit														✓				
Multi-function Platform Laying Trolley-Formwork	R12	1 Unit																✓		
Crane	R13	1 Unit				✓				✓				✓						
Traveller Concrete Formwork	R14	1 Unit																	✓	✓

3.4. Simulation result analysis

The simulation of tunneling operations is carried out for 100 cycles. With this number, the system has reached a stable state, as in figure 4 for simulation using WebCyclone, figure 5 for simulations using Symphony.Net software, and figure 6 for simulations using COST software. The steady states show that the process has reached equilibrium conditions, where changes in time do not affect the value of productivity.

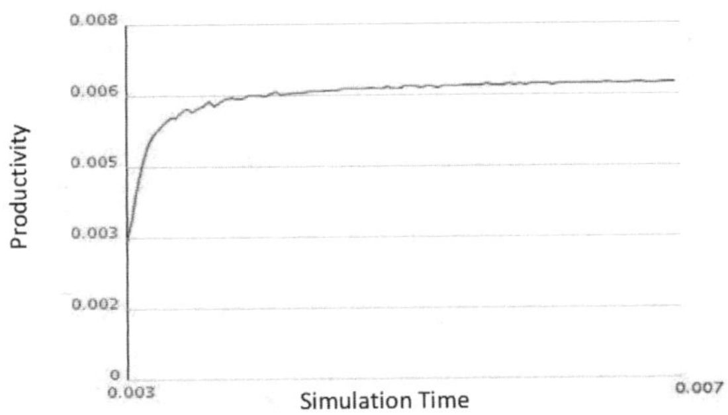

Figure 4. Simulation result using WebCyclone.

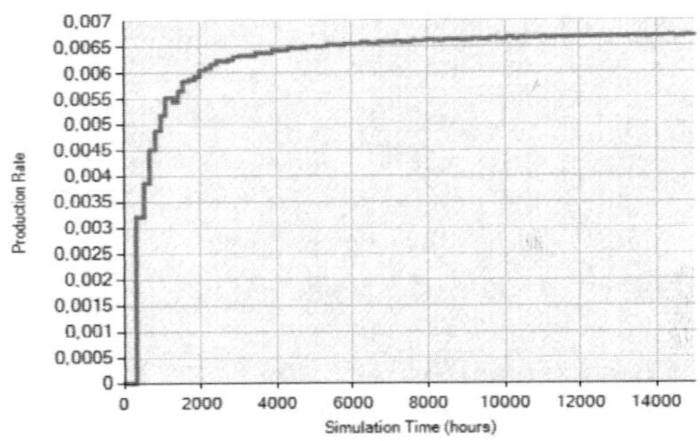

Figure 5. Simulation result using Simphony.NET.

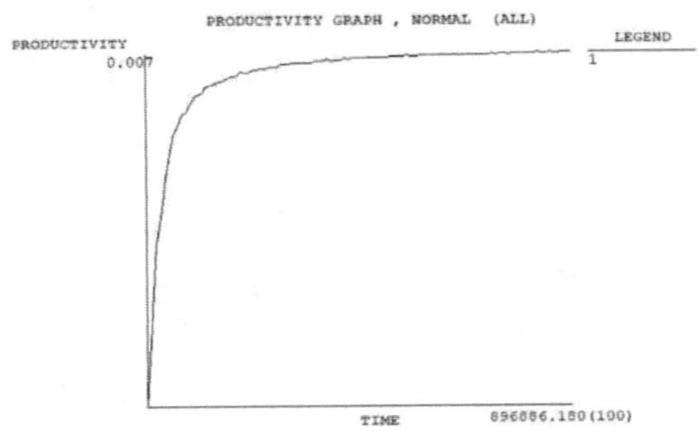

Figure 6. Simulation result using COST.

The operating simulation results produce output values of productivity (or production rate, it's the same just different in the term used in the software) as units produced in unit time (hour in this case). In this simulation, 1 production unit is defined as 10.5 meters of the part that has been done secondary lining, with the 12 hours of working hours in one day, the time needed to produce 10.5 meters of the tunnel can be calculated using equation (1).

$$Dur.\,for\ 1\ unit\ prod.\,(days) = \frac{1}{Working\ hours\ assumption\ \times Productivity\ (unit/hour)} \quad (1)$$

Table 5. Productivity value from simulations.

Productivity	WebCyclone	Symphony.NET	COST
Productivity (Unit/hour)	0.0067	0.006	0.007
Productivity (meter/day)	0.84	0.76	0.88
Duration for 1 unit production (days)	12.4	13.9	11.9

The simulation output in the form of productivity in table 5 shows that to produce 10.5 meters of the tunnel it takes around 12-14 days. There is a gradual change in the value of productivity when running simulations before finally achieving productivity in a steady state.

Table 6. The idleness percentage of resources.

Resources	Percentage of Idleness (%)		
	WebCyclone	Symphony.NET	COST
Anchor Pump (node 5)	99.98%	100.00%	100.00%
Concrete Pump (node 42)	91.36%	91.30%	91.29%
Concrete Truck (node 30)	100.00%	100.00%	100.00%
Crane (node 19)	77.22%	62.40%	77.20%
Dump Truck (node 9)	99.98%	100.00%	100.00%
Excavator (node 11)	95.76%	94.30%	94.38%
Formwork Team (node 18)	20.46%	20.30%	20.11%
Geological Drilling rig (node 2)	97.98%	99.50%	99.51%
Inverse Arch Model (node 33)	84.64%	85.30%	85.23%
Lining Trolley (node 43)	91.36%	91.30%	91.29%
Loader (node 10)	95.76%	94.30%	94.38%
Multifunction Work Platform (node 38)	57.89%	57.80%	57.69%
Pneumatic Drill (node 6)	95.26%	98.80%	98.82%
Second liner Team (node 29)	75.67%	75.50%	75.69%
Shotcrete Team (node 14)	0.82%	1.00%	1.03%
Spray Robot (node 15)	0.82%	1.00%	1.03%

The idleness percentage of most resources in table 6 shows that there is a high waiting time in the system (model), this can occur due to the dependency from one work task to another so that the work sequence in the system is sequential (series), there are too many resources available, or there are too many work tasks that use these resources.

3.5. Sensitivity analysis

The low idleness percentage shows that the resource works continuously. It can occur because the resource is used in many work tasks and /or the amount is too little, therefore the addition of the number of resources (which has a low percentage) is considered to increase system productivity value. Based on the simulation results shown in table 6, resources that have a low percentage and are considered to need additional resources, namely the Formwork Team, Shotcrete Team, and Concrete Wet Sprayer. The

alternative number of configurations is made from these three resources for the sensitivity analysis process.
Sensitivity analysis to find out the number of resources can be done using WebCyclone. For modeling on tunnel construction operations, variations in resources are as follows:

- *Formwork team* : 1 to 3 teams
- *Shotcrete team* : 1 to 3 teams
- *Concrete Wet Sprayer* : 1 to 3 units of equipment

Based on the results of the sensitivity analysis, the system (model) productivity value increases with the addition of certain resources. The addition of each of them to the Shotcrete Team, Concrete Wet Sprayer, and the Formwork Team to be 2, led to almost twice the productivity increase (6.5 days for production of 10.5 meters) but the addition of each of these resources to 3 has not increased productivity significantly. An increase in the number of resources could lead to additional cost, even though cost analysis were not conducted its logical to conclude that the optimal amount of Shotcrete Team, Concrete Wet Sprayer, and Formwork Team are 2 for each resource (the decision ultimately depend on the project needs). Table 7 is a resume of the sensitivity results analysis, alternative 1 with 1 resource each, alternative 2 with 2 resources each, and alternative 3 with 3 resources each. The addition of a number of resources causes the idleness percentage of all three resources to increase, but unlike other resources, the percentage of the idle formwork team increases dramatically.

Table 7. Comparison of the productivity value and the percentage of each alternative condition.

Name	% Idleness		
	Alternative 1	Alternative 2	Alternative 3
SHOTCRETE TEAM	0.78	5.78	8.79
SPRAY ROBOT	0.78	5.78	8.79
FORMWORK TEAM	20.45	39.56	68.63
Productivity (unit/hour)	**0.006709**	**0.012945**	**0.01496**

3.6. Comparison with previous research
Comparisons in the result are made to previous studies [5], The total duration of the tunneling work can be calculated using equation (2), both previous study and this study has a different length of the tunneling work path (608 meters in this study), so that has an impact on total duration. Moreover, in the previous study [5], the working hours in one day were 8 hours, while in this study the working hours in one day were assumed to be 12 hours because the project needs to meet the completion target.

$$Total\ duration\ (day) = \frac{Total\ length\ of\ tunnel\ (m)}{Productivity\ (m/day)} \tag{2}$$

Table 8. The comparison of productivity value.

Description	Previous study [5]	Current study
Tunneling Method	NMT	NATM
Productivity (m/day)	0.554	0.84
Total Duration (day)	653	724

As shown in table 8, the results of each study are displayed based on the construction method, the value of work productivity, and the total duration needed to complete the work. There are differences in the tunnel construction method used as the object of study, in the previous study [5] that simulated tunneling operations on the construction of Cisumdawu toll road, Norwegian Method of Tunneling (NMT) were used in the project, while this study used the New Austrian Tunneling Method (NATM).

4. Conclusion

The conclusion of this study are as follows:

- The tunneling operation in this study can be modeled with Cyclone to figure out the system's condition and the response based on the simulation conducted;
- The productivity value obtained from the simulation results is around 12 days for 10.5 m, assuming the working hours are 12 hours a day. Meanwhile, according to interviews in the field, tunneling work has a productivity of 1.2 meters per day, which means that the work methods in the field are better than the simulated methods. This can be caused by the assumption of working hours used in the model might be different from the project;
- Compared to the previous study of tunneling at Cisumdawu toll road, there are differences in the value of productivity, this is due to differences in tunneling construction methods; and
- Sensitivity analysis results that the resource concrete wet sprayer, shotcrete team, and formwork team are quite sensitive in influencing system productivity, where the addition of the number of those resources influences the increase in productivity, with the optimum number of two resources for each of concrete wet sprayer, shotcrete team, and formwork team.

Some recommendations based on this research were given as follows:

- The duration of modeling used should come from project data that is being reviewed, if possible, use data from field observations to obtain a more reliable duration; and
- The construction operation process used in the modeling assumptions should be adjusted to the actual conditions, namely by observing directly (if possible) and interviewing the construction implementers in the field and validate the model with the expert on the project.

References

[1] Rabcewicz L and Golser J 1973 *Principles of dimensioning the supporting system for the New Austrian Tunnelling Method Water Power*

[2] Halpin D W and Riggs L S 1992 *Planning and Analysis of Construction Operations* (Canada: John Wiley & Sons)

[3] Schriber T J 1987 *The Nature and Role of Sim in the Design of Manufacturing System Society for Computer Simulation*

[4] Abduh M 2007 Perkembangan teknologi simulasi untuk perancangan dan analisa operasi konstruksi berulang *Seminar Nasional Teknik Sipil* (Bandung)

[5] Abduh M et al 2017 Simulation of tunneling construction methods of the Cisumdawu Toll Road *Proc. of the 3rd Int. Conf. on Constr. and Build. Eng. (ICONBUILD)* (Bali).

[6] Maisarah F S C S and Abduh M 2009 Survei tingkat utilisasi simulasi untuk operasi konstruksi berulang *Konferensi Nasional Teknik Sipil 3* (Jakarta: Universitas Pelita Harapan)

IOP Conf. Series: Materials Science and Engineering **933** (2020) 012003 doi:10.1088/1757-899X/933/1/012003

The effect of wood ash and styrofoam on the characteristics of AC-WC mixture

R M Yusputri[1], S M Saleh[2] and M Isya[2]

[1] Civil Engineering Postgraduate Program, Department of Civil Engineering, Universitas Syiah Kuala, Banda Aceh, Indonesia
[2] Department of Civil Engineering, Universitas Syiah Kuala, Banda Aceh, Indonesia

Corresponding E-mail: fikayusputri@gmail.com

Abstract. In this study, two types of substitutions, wood ash, and styrofoam, were used in asphalt concrete. The substitution of wood ash and styrofoam for asphalt concrete's filler is expected to enrich the variations of the asphalt concrete material as well as to reduce the existing industrial waste. This study objective is to determine the effect of the asphalt mixture, which filler was substituted with the wood ash, on the Marshall test results. The initial stage of the research is to examine the physical properties of the asphalt and aggregate. Then, the optimum asphalt content (OAC) of the asphalt mixture was estimated. Then, the asphalt mixture's filler was substituted with the wood ash and the asphalt material was substituted with the styrofoam. The result of this study is the best combination of the wood ash in asphalt mixture, which is 25% of the total filler, and the best combination of the styrofoam in asphalt mixture, which is 9% of the total asphalt content. With these best combinations of the wood ash and the styrofoam, the obtained Marshall stability of the asphalt concrete was 1,249.01 kg, which fulfills the '2018 Bina Marga' requirement.

1. Introduction

Asphalt, as a binding material of paved mixtures in the flexible pavement, has a significant influence on the service life of the pavement layer. Therefore, it is necessary to improve the quality of construction of the transportation infrastructure that is environmentally friendly, inexpensive, and durable. One of the alternatives is the modified asphalt material. Modifying asphalt material can also be an alternative solution for the waste by using the waste polymer materials that are difficult to decompose in the soil. One of the polystyrene wastes is styrofoam that functions as an unusable food container. Putri [1] stated that using styrofoam waste or foamed plastic as an additive for asphalt forms a new binding material for road construction pavement due to its properties, which is thermoplastic solid at room temperature and liquid at higher temperatures. According to Listiani [2], besides the technical advantages, the utilization of polystyrene waste is also beneficial to Indonesia's environment and economy. Also, modified asphalt has high viscosity at low temperatures and low viscosity at high temperatures, which is favorable for construction [3].

Most fillers have a high price. For this reason, innovation is needed by using substitute materials that are more profitable and efficient in an asphalt mixture and also accessible but lessens in the utilization, like

IOP Conf. Series: Materials Science and Engineering **933** (2020) 012003 doi:10.1088/1757-899X/933/1/012003

wood ash. In this study, wood ash was used as a filler substitute for portland cement. Wood ash waste has the potential to be used as a filler substitution for the asphalt concrete wearing course (AC-WC) because this material passed No. 200 filter, which is the requirement of filler material. Hossain [4] stated that wood ash is suitable for polymer asphalt filler because of its cost-effectiveness, low density, high rigidness, high strength, and low price.

2. Literature review

2.1. Aggregate
According to the general specifications [5], the coarse aggregate must retain the No.4 sieve (4.75 mm) and the fine aggregates, which must be composed of sand or the product from crushed stones, must pass the No.4 sieve (4.75 mm) and retain the No. 200 filter (0.075 mm).

2.2. Fillers
The filler is a non-plastic material with a minimum of 75% of the weight passing the No. 200 filter (0.075 mm). Filler lies between the coarse aggregate cavity and fine aggregate to reduce the size of the cavity, increase the density, and increase the stability of the asphalt mixture. Filler can be limestone dust, stone ash, fly ash, or other materials. Filler must be in a state dry with a maximum water content of 1% [6]. The requirements for fillers are shown in the following table 1.

Table 1. Requirements for fillers [5].

Testing	Standard	Value
Pass the No. 200 filter	SNI 03-4428-1997	Min. 75 %
Free of organic ingredients	SNI 03-4428-1997	Maks. 4%
Filler specific gravity	SNI 0013 - 81	2,5 gr/cm^2.

2.2.1. Wood ash. Ash is the leftover material from the wood-burning process. Sulaiman [7] stated that wood ash could contain much silica if it is burned at high temperatures reaching 500-700 °C.

Figure 1. Wood ash.

2.3. Asphalt
Asphalt or bitumen is a brownish-black and viscoelastic material. It will soften and melt if it is heated up and hardens if the temperature decreases. This viscoelastic character makes asphalt envelopes the aggregates and make the aggregates durable [8]. Sukirman [9] stated that asphalt is the material which is solid at room temperature. The amount of asphalt in the pavement mixture ranges from 4-10% by the mixture weight or 10-15% by the volume.

2.3.1. Modified asphalt. Modified asphalt is made by mixing concrete asphalt with additional material. Modified asphalt is often called polymer bitumen (plastomer and elastomer). The percentage of polymer addition into the concrete asphalt mixtures must be determined by laboratory testing because adding sufficient material can improve the asphalt rheological properties such as penetration, thickness, softening point, and concrete asphalt elasticity, while the excessive additions will generate the adverse effect. The improvement of asphalt rheological properties is generally influenced by the content of the polymer, bitumen grade, and a particular type of polymer [10].

2.3.2. Styrofoam. Styrofoam is one of the types of polymer that is thermoplastic: it will soften if it is heated and harden when it becomes cold. According to Billmeyer in Giri [11], styrofoam is a compatible material in terms of mechanism and temperature, but it is somewhat fragile and soft at temperatures below 100 °C. According to Crawford in Giri [11], styrofoam has a density of up to 1,050 kg/m^3, a tensile strength of up to 40 MN/m^2, a flexural modulus of up to 3 GN/m^2, shear modulus of up to 0.99 GN/m^2, and the Poisson ratio of up to 0.33.

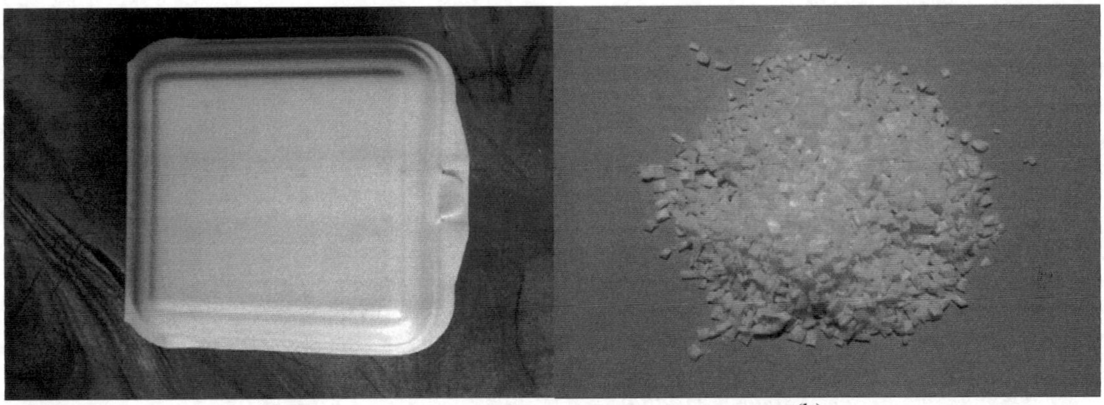

(a) (b)

Figure 2. a) Styrofoam as a food container: b) Styrofoam after being cut into pieces.

3. Methods and analysis

3.1. Aggregate testing
The aggregate is the crushed gravel, which is classified as fine aggregate. It is produced by PT. Dana Dinamika Persada, located in Leupung Baleu, Cot Glie, Aceh Besar, Aceh. The testing on the aggregate physical properties includes testing on asphalt consistency (penetration testing, ductility), testing on temperature sensitivity (softening point), and testing on asphalt specific gravity.

3.2. Preparation of fillers

In this study, the filler was wood ash and portland cement. The wood ash is generally in the form of powder and is the left of material from the wood-burning. The wood ash is the waste wood from the sawmill and the furniture making. Formerly, the wood is burned for approximately 30 minutes to 1 hour. Wood ash is prepared using a burning procedure in a laboratory furnace under continuous air supply. The combustion temperature is set at 600 °C as suggested by ASTM EN 870-82. Xue [12] stated that generally, 6-10% of the burned wood produces ash. The obtained ash, according to Xue, then pounded until creating a homogeneous size that can pass the No. 200 filter (0.075 mm). Wood ash is used because the results of the combustion consist of 85% of silica, which is a compound that can react to water and form new substances that are adhesive to rocks. In other words, wood ash properties, in this case, are similar to portland cement. The density of wood ash was 2,9 gr/cm^3, which is based on Cahya's research [13]. Otoko [14] mentions some of the chemical properties contained in wood ash that are shown in the following table 2.

Table 2. Chemical composition of wood ash [13].

Element	Content (%)
Fe2O2	1.70%
Al2O3	2.70%
SiO2	85%
MgO	0.25%
CaO	3.50%

3.3. Selection of styrofoam waste

The utilization of waste as an alternative material is very beneficial for the environment, especially for the waste that is difficult to decompose like styrofoam or polystyrene. Styrofoam has to be treated before mixed with other additives. The initial treatment of styrofoam waste was carrying it out through heating and pressing it by using the hydraulic press at a temperature of 170°C and a pressure of 170 kg/cm^2 for 13 minutes. The purpose of this treatment is to remove the pores of styrofoam. After the pressing process, the process was continued with enumeration or cutting the styrofoam into a maximum of 2 x 2 mm^2 to facilitate the dissolution process of the aggregate mixture. Then, the object was put into the hot asphalt with the appropriate temperature and then stirred at high speed. Then, the hot aggregate is added and stirred again until the mix becomes homogeneous. When all plastic waste is dissolved into the aggregates and the hot asphalt, subsequently, the aggregate mixture can be used for concrete asphalt mixture.

3.4. Determination of asphalt content variations

Asphalt content was determined based on the initial estimation of asphalt content, called the medium asphalt content. The medium asphalt content is calculated using the following formula:

$$Pb = 0.035\ (\%\ CA) + 0.045\ (\%\ FA) + 0.18\ (\%\ filler) + K \quad (1)$$

In this study, based on designed aggregate gradation, CA = 36.50%, FA = 58.5%, filler = 6.2% and the selected constant value is 0.5. The medium asphalt content resulted from the formula is 5.5%. Then, the samples of asphalt concrete wearing course (AC-WC) mixture have to contain the asphalt content of 4.5%, 5%, 5.5%, 6%, or 6.5%.

3.5. The design of concrete asphalt mixture with the selected aggregate gradation
The gradation used in this study is continuous gradation based on the middle value of the technical specifications [5].

3.6. Producing and testing the test items
The test items of AC-WC mixture in this study consisted of the following three groups:
1. Test items contain the optimum asphalt content (OAC);
2. Test items contain the filler composed of 25% wood ash and 75% Portland cement and the styrofoam which is 7%, 9% and 11% of the wet weight of OAC;
3. Test items contain the best combination of wood ash and styrofoam based on the previous tests and tested for durability at 30 minutes and 24 hours immersion.

After the producing process of the test items, subsequently, the procedure was continued with Marshall testing. The total of the test items in this study is presented in table 3.

Table 3. Recapitulation of total test items.

No.	Description	Total
1	Test items to obtain OAC	15
2	Test items with the combination of wood ash and styrofoam	27
3	Durability test at 30 minutes and 24 hours immersion	6
	Total	**48**

4. Results and discussion

4.1. Test results
The results of properties examinations of the aggregates obtained from Leupung Baleu, Kuta Cot, Aceh Besar, Aceh, are presented in table 4.

Table 4. Physical properties of coarse aggregates.

Physical properties examined	Standard	Results	Unit	Specification of Bina Marga, 2018
Time-worn	SNI 2417-2008	9.63	%	Max. 40%
Tin Index	ASTM D-4791	27.35	%	Max. 10%
Oval Index	ASTM D-4791	20.96	%	Max. 10%
Specific Gravity	SNI 1969-2008	2.8	-	Min. 2,5
Absorption	SNI 1969-2008	0.263	%	Max. 3%
Weight of contents	AASHTO T-19-74	1.625	Kg/dm	Min. 1%
Impact	SNI 03-4426-1997	7.71	%	Max. 30%

Based on the results, the aggregate physical properties have met the requirements, except for the elongation index value that is above 10%. However, based on Bina Marga specifications [15], there is the requirement stated that if there is nonconformity for the elongation index, the value can be tolerated if the aggregates fulfill all other requirements, primarily the result of abrasion testing using Los Angeles machines. Besides these tests, the results of impact testing have met the requirements.

Table 5. Physical properties of fine aggregates.

Testing	Standard	Results	Unit	Specification of Bina Marga 2018
Specific Gravity	SNI 03-1970-1990	2.8	gr/cm^3	Min. 2.5
Absorption	SNI 03-1970-1990	0.263	%	Max. 3%

Table 6. Asphalt penetration 60/70 physical properties.

No.	Physical Properties of Asphalt	Standard	Unit	Specification of Bina Marga 2018
1	Specific Gravity (25 ºC)	SNI 2441-2011	1.029	≥ 1,0
2	Penetration (25 ºC; 5 s; 0,1 mm; 100 g)	SNI 06-2456-1991	65	Min. 60-70
3	Ductility (25 ºC; 5 cm/sec)	SNI 2432-2011	120	≥ 100
4	The soft spot; ºC	SNI 2432-2011	49	≥ 49

Table 7. Physical properties of asphalt penetration 60/70 containing styrofoam.

No.	Physical Properties of Asphalt	Unit	Styrofoam Content			Specification of Bina Marga, 2018
			7%	9%	11%	
1.	Specific Gravity	-	1.035	1.04	1.05	≥ 1,0
2.	Penetration	(0.1 mm)	58.6	57.8	55.8	Min. 60-70
3.	The Soft Spot	°C	48.5	49	49.5	≥ 49
4.	Ductility	cm	120	120	120	≥ 100

The values of specific gravity, penetration, softening point and ductility have met the requirements of General Specifications [5].

4.2. Marshall test results

Based on the Marshall test results, which is shown in table 8, the optimum asphalt content (OAC) is 6.05%. However, the asphalt content used to examine the properties of asphalt concrete containing wood ash and styrofoam was obtained by adding and reducing the OAC value (6.05%) with two somewhat equal values. As a result, the asphalt contents used in the test of asphalt concrete containing wood ash and styrofoam were 5.60%, 6.05%, and 6.50%.

Table 8. Recapitulation of Marshall test results with variations in bitumen content.

No	Mix Characteristics	Asphalt Content (%)					Specification of Bina Marga, 2018
		4.50	5.00	5.50	6.00	6.50	
1.	VIM (%)	6.88	6.82	4.66	4.30	4.07	3 – 5
2.	VMA (%)	17.55	18.58	17.79	18.56	19.44	Min. 15
3.	VFA (%)	60.94	63.41	74.39	76.95	79.16	Min. 65
4.	Stability (Kg)	1,321.72	1,071.68	1,108.75	929.05	1,002.61	Min. 800
5.	Flow (mm)	3.93	3.90	4.00	3.83	3.73	2 – 4

4.3. Results of examination of wood ash

Table 9. Marshall test results for asphalt concrete containing filler of 25% wood ash and 75% portland cement.

No	Mix Characteristics	Asphalt content (%)			Specification of Bina Marga, 2018
		5.60	6.05	6.50	
1	VIM (%)	4.87	4.39	3.86	3 - 5
2	VMA (%)	18.17	18.72	19.24	Min. 15
3	VFA (%)	73.21	76.57	79.99	Min. 65
4	Stability (Kg)	1,047.95	1,129.51	1,119.14	Min. 1,000
5	Flow (mm)	3.33	3.87	3.93	2 - 4

4.4. Styrofoam examination results

Table 10. Marshall test results for asphalt concrete containing filler of 25% wood ash and 75% portland cement and containing 7% styrofoam in the asphalt.

No	Mix Characteristics	Asphalt Content (%)			Specification of Bina Marga, 2018
		5.60	6.05	6.50	
1	VIM (%)	6.28	5.29	4.57	3 - 5
2	VMA (%)	19.33	19.43	19.78	Min. 15
3	VFA (%)	67.51	72.79	76.98	Min. 65
4	Stability (Kg)	1,173.32	1,052.41	1,075.54	Min. 1,000
5	Flow (mm)	3.23	4.43	4.30	2 - 4

Table 11. Marshall test results for asphalt concrete containing filler of 25% wood ash and 75% portland cement and containing 9% styrofoam in the asphalt.

No	Mix Characteristics	Asphalt Content (%)			Specification of Bina Marga, 2018
		5.60	6.05	6.50	
1	VIM (%)	5.66	5.93	4.35	3 - 5
2	VMA (%)	18.71	19.89	19.49	Min. 15
3	VFA (%)	69.73	70.18	77.75	Min. 65
4	Stability (Kg)	1,103.02	1,175.76	1,249.01	Min. 1,000
5	Kelelehan (mm)	3.50	3.20	3.13	2 - 4

Table 12. Marshall test results for asphalt concrete containing filler of 25% wood ash and 75% portland cement and containing 11% styrofoam in the asphalt.

No	Mix Characteristics	Asphalt Content (%)			Specification of Bina Marga, 2018
		5.60	6.05	6.50	
1	VIM (%)	5.73	5.47	6.15	3 – 5
2	VMA (%)	18.64	19.36	20.87	Min. 15
3	VFA (%)	69.26	71.73	70.51	Min. 65
4	Stability (Kg)	1,079.39	1,144.93	1,101.14	Min. 1,000
5	Flow (mm)	3.63	3.83	3.70	2 - 4

Based on the results, the best mix was asphalt concrete containing 6.5% asphalt content, 9% of which was styrofoam, and filler composed of 25% of wood ash and 75% of portland cement. The stability value of this asphalt concrete was 1,249.01 Kg, which meets the requirement of the General Specifications [5].

4.5. Results of durability value

The Marshall immersion test for 30 minutes and 24 hours at 60 °C was carried out to obtain the durability value of the asphalt mixture.

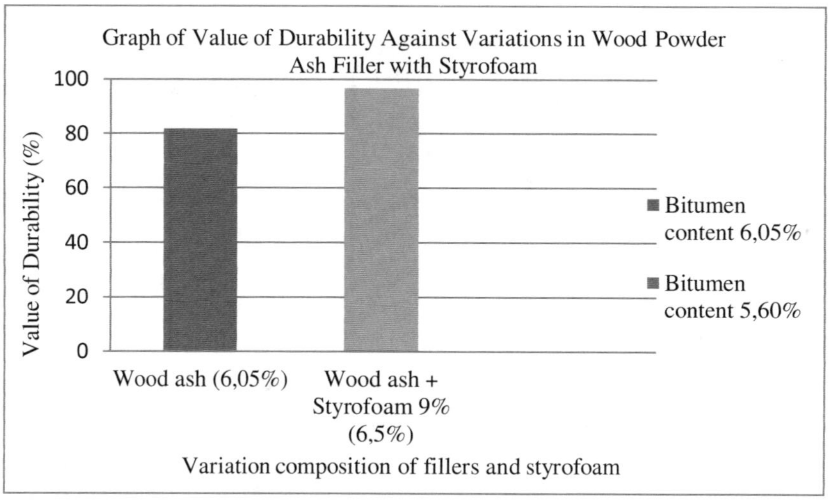

Figure 3. Durability testing charts.

Based on the result, the durability of asphalt concrete containing filler composed of 25% wood ash and 75% portland cement and containing 6.05% asphalt did not meet the requirement. The obtained value is 81.76%, while the required value is 90% or higher. It happens because wood ash fillers cannot fill the pores of the mixture. This fact can be observed from the asphalt concrete VIM which tends to be high. On the other hand, the asphalt concrete containing asphalt composed of 9% styrofoam and containing filler composed of 25% wood ash and 75% portland cement has the durability value of 96.51%, which meets the specification.

4.6. Discussion

The use of wood ash as a substitute for filler and styrofoam as a substitute for asphalt, at the best composition, can affect the value of Marshall properties, which are VIM, VMA, VFA, stability, and flow. VIM values tend to increase with the increasing content of styrofoam in asphalt. This fact happens because the asphalt gets thicker and harder. The value of VMA can also indicate the thickness of the asphalt blanket on the aggregate. The VMA value is influenced by the asphalt content that envelops the aggregate; a high asphalt content will form a thick blanket of aggregate resulting in larger pores between aggregates. The value of VFA increases with the increase of asphalt content and styrofoam content. The increase of VFA is because of the increasing amount of filled pores by the addition of asphalt content and styrofoam content. Adding styrofoam into the mixture can increase the stability of the AC-WC mixture; as a result, the shear resistance at high temperatures and damage prevention improves. High stability increases the ability of AC-WC mixtures to carry vehicle loads, while the increase of flow values increases the flexibility of the AC-WC mixture. It induces the favorable durability and flexibility of AC-WC. However, substituting wood ash to asphalt concrete's filler without adding the styrofoam into the asphalt does not improve the asphalt concrete durability and stability; conversely, substituting wood ash to asphalt concrete's filler and styrofoam to asphalt does improve the asphalt concrete durability and stability.

5. Conclusion and suggestion

5.1. Conclusion

1. Optimum asphalt content (OAC) is varied into three: 5.60%, 6.05%, and 6.50%. Based on the Marshall testing evaluation, the final OAC was 6.05%.
2. The best percentage of substituted materials for asphalt concrete's filler is 25% of wood ash and 75% of portland cement, performed at 6.05% of asphalt content.
3. The asphalt concrete stability increases by substituting the wood ash to filler and styrofoam to asphalt. The highest stability, obtained at 6.50% of asphalt content and 9% of styrofoam, is 1,249.01 kg, while the lowest stability, obtained at 6.05% of asphalt content and 7% of styrofoam, is 1,052.41 kg.
4. The durability of asphalt concrete substituted by wood ash is 81.76%. The value does not meet the required value, which is at least 90%, as stated in the 2018 fourth revision General Specifications of Bina Marga. Whereas, the durability of asphalt concrete substituted by wood ash and styrofoam is 96.51% and fulfill the requirement.
5. Wood ash as filler substitution and styrofoam as asphalt substitution in asphalt mixture affect the increase of stability and flow; styrofoam substitution in durability testing affects durability and stability values.

5.2. Suggestion

1. For future studies, other types of plastic with a different variation percentage should be used;
2. The asphalt mixture durability within 30 minutes and 24 hours immersion has a high VIM value; thus, in the future, porous asphalt should be used because styrofoam is appropriate for porous asphalt.

References

[1] Putri E E and Syamsuwirman 2016 *Tinjauan Subtitusi Styrofoam Pada Aspal Pen. 60/70 Terhadap Kinerja Campuran Asphalt Concrete-Wearing Course (Ac-Wc)* vol 6 (Syiah Kuala University Banda Aceh)

[2] Listiani A, Sembiring E and Rahman H 2015 *Evaluation Of Expanded Polystyrene (Eps) Plastic Waste Utilization As An Asphalt Substitution Material In Asphalt Concrete Wearing Course* (Bandung: Department of Environmental Engineering, Institut Teknologi Bandung)

[3] Fang C, Jiao L, Hu J, Yu Q, Guo D, Zhuo X and Yu R 2014 Viscoelasticity of asphalt modified with packaging waste expanded polystyrene *J. Mater. Sci. Technol.* **30** 939-43

[4] Hossain M F, Shuvo S N and Islam M A 2014 Effect of types of wood on the thermal conductivities of wood sawdust particle reinforced composites *Procedia Eng.* **90** 46-51

[5] Direktorat Jendral Bina Marga 2018 *Spesifikasi Umum Direktorat Jenderal Bina Marga Edisi 2010 Revisi 4 Divisi 6* (Jakarta: Kementerian Pekerjaan Umum)

[6] Fahmi R, Saleh S M and Isya M 2017 *Pengaruh Lama Rendaman Air Laut Terhadap Durabilitas Campuran Aspal Beton Menggunakan Aspal Pen.60/70 Yang Disubstitusi Limbah Ethylene Vinyl Acetate (Eva)* (Banda Aceh: Faculty of Engineering Syiah Kuala University)

[7] Sulaiman Y H 2010 Pengaruh penggunaan abu serbuk kayu terhadap kuat tekan beton *J. Portal* **2**

[8] Departemen Pekerjaan Umum 2009 *Modul Bahan Untuk Campuran Beraspal Panas* (Jakarta: Kementerian Pekerjaan Umum)

[9] Sukirman S 2003 *Beton Aspal Campuran Panas* (Bandung: Publisher Granit)

IOP Conf. Series: Materials Science and Engineering **933** (2020) 012003 doi:10.1088/1757-899X/933/1/012003

[10] Yuniarti R 2015 Modifikasi aspal dengan getah pinus dan fly ash untuk menghasilkan bio-aspal *J. Sains Teknologi dan Lingkungan* vol 1 (Mataram: Faculty of Engineering Mataram University)
[11] Dharma I B, Giri, Sudarsana IK and Tutarani N M 2014 Kuat tekan modulus elastisitas beton dengan penambahan styrofoam (Styrocon) *J Ilmiah Civ. Eng.* **12**
[12] Xue Y, Wu S, Cai J, Zhou M and Zha J 2014 Effects of two biomass ashes on asphalt binder: dynamic shear rheological characteristic analysis *Constr. and Build. Materials* **56** 7-15
[13] Cahya C Y 2018 *Pengaruh Penggunaan Abu Serbuk Kayu Sebagai Filler Pada Campuran Laston Lapis Aus (AC-WC)* (Banda Aceh: Syiah Kuala University)
[14] Otoko G R and Honest B K 2014 Stabilization of Nigerian delcatic laterites with saw dust ash *Int. J. of Scientific Research and Manag. (IJSRM)*
[15] Direktorat Jendral Bina Marga 2017 *Rancangan Spesifikasi Umum Bidang Jalan dan Jembatan Divisi VI Perkerasan Beraspal* (Jakarta)

IOP Conf. Series: Materials Science and Engineering **933** (2020) 012004 doi:10.1088/1757-899X/933/1/012004

The use of Sabang mountain rocks and substitution of LDPE in the AC-WC mixture with dry method

A Siswadi[1], S M Saleh[2] and Y Darma[2]

[1] Civil Engineering Postgraduate Program, Department of Civil Engineering, Universitas Syiah Kuala, Banda Aceh, Indonesia
[2] Department of Civil Engineering, Universitas Syiah Kuala, Banda Aceh, Indonesia

Corresponding e-mail: abdi.siswadi@gmail.com

Abstract. The aggregate used in the flexible pavement in the city of Sabang must be brought from Aceh Besar Regency. This can affect the high operational cost and the insufficient use of local materials. In anticipation of this, the local mountain rocks used are variated with coconut shell ash as a filler and low-density polyethylene (LDPE) plastic use as asphalt substitution. The method used in this study refers to the 2018 specification of Bina Marga (fourth revision of 2010 version). The research results showed that asphalt concrete wearing course (AC-WC), which filler varied with coconut shells (0% and 25% of the weight), meet all predefined parameters. In terms of AC-WC varied with coconut shell and LDPE, the best combination found in this result was 25% of coconut shell ash and 3% of LDPE. AC-WC varied with this best combination has a stability of 1,780.99 kg, the flow of 2.97 mm, VIM of 4.79%, VMA of 16.51%, and VFA of 70.97%.

1. Introduction

The sustainable road infrastructure in the city of Sabang, the most western outer island of Indonesia, should be developed to support tourism, the aspect which according to the Government of Sabang, can accelerate the economy of this city people. However, currently, the materials for asphalt concrete in the city of Sabang are brought from outside the city, which is Aceh Besar Regency. Based on the testing of the mountain reef material located in Ujong Kareung City Sabang, the material's value of abrasion and the impact meet the specification of Bina Marga Revision 3, 2014, while the specific gravity and the absorption value do not meet the requirements. Because of the low specific gravity, it is indicated that mountain reef material located in Ujong Kareung City Sabang has many large pores, so they will absorb more asphalt and make the asphalt mantle thinner. Mixing this aggregate with 60/70 penetration asphalt results in the stability value of 1,922.13 kg and the durability value of 90.60%, both at the optimum asphalt content of 5.75% [1].

On the other hand, the development of copra coconut business in Sabang has an impact on rising coconut shell waste. If this industry not utilized optimally, it would cause an environmental problem. The content of coconut shell charcoal such as mectosil, lignin, cellulose, and other minerals can increase the value of stability [2]. In addition, the plastic waste can be utilized as an additive in asphalt mixture called. The plastic used as an additive in asphalt mixture is called polymer modified asphalt. Asphalt modification is a mixture of asphalt which is added the substitute material from waste or natural things. One of the polymer materials used to raise the asphalt softening point is the low-density polyethylene

(LDPE) plastic, which, after substituted to asphalt, improved the dynamic stability by more than 250% and the deformation speed by lower than 24% [3].

Asphalt concrete is a mixture consisting of asphalt as a binder and coarse aggregate, fine aggregate, and filler, as the aggregate materials. The typical procedure of mixing is the hot mixing and compacting method. Based on its type, the concrete asphalt layers are divided into the asphalt concrete wearing course (AC-WC), the asphalt concrete binder course (AC-BC), and the asphalt concrete Base (AC-Base) [4]. The function of hot mixed concrete asphalt is as follows: as a surface coat that is resistant to weather, shear force, and wheel pressure, as a waterproof layer that can protect the layer underneath from water seepage, as the upper foundation layer, and as a layer of foundation in improvement or maintenance activities [5].

The aggregate, which is the main ingredient of the road structure, is a collection of grains of stone and sand or other minerals, both from natural and artificial results. The asphalt concrete contains 90-95% aggregates based on the percent of weight or 75-85% aggregated based on the percent of volume. The aggregate used must be in a clean state of dirt, organic materials, or other materials that are not desirable because they will reduce the performance of the asphalt mixture [4].

The Rock mountain of Weh Island, located in the mountain or hills, has a massive form so that it needs to be separated using human power and the stone-breaking machine. The aggregates from the mountains, hills, and rivers need to go through the process until they have broken fields, rough texture, and the desired size [6].

The filler is a component of asphalt concrete. It is a non-plastic material and must pass the No. 200 filter (0.075 mm), at least 75% of its total weight. Fillers are mixed materials that fill the space between the coarse and fine aggregates, thereby reducing pore volume and increasing density, as well as lowering the permeability value of asphalt mixture [7]. Coconut shell consists of cellulose, mectosil, lignin, and other minerals, while the content of these ingredients varies according to the coconut type. The weight of the coconut shell ranges from 12% of the total weight of the whole coconut. The effect of the hardness of the shell structure is the richness of the content of silica (SiO2) [8]. The chemical composition of the coconut shell is shown in table 1.

Table 1. Chemical composition of coconut shell ash [9].

Test Result	Unit	Testing parameters
Carbon	%	91.38
Ash content	%	4.79
Water content	%	3.83
Weight type	gr/cm^3	0.5722

The asphalt is a material of black or dark brown and has an adhesive substance with the main element of bitumen obtained from the natural or residue from the refining of petroleum. The example of the natural asphalt is the asphalt of Buton Island (asbuton) and the asphalt that exists in the Lake of Trinidad (Trinidad Lake asphalt). Residue asphalt, called oil asphalt, is the result of petroleum distillation [10]. The addition of low-quality plastic of LDPE can increase the asphalt softening point and reduce asphalt penetration. The resilient modulus of the asphalt mixture added with LDPE is better than the conventional asphalt in terms of density, stability, MQ, VFB [3].

The aggregate gradation is used to set the pore volume in the mixed aggregate. The aggregate gradation is produced by screening analysis, which can be done in dry or wet conditions. The aggregate gradation is expressed against the percentage of escaped aggregate to the total aggregate weight [10]. The aggregate gradation for the AC-WC mixture corresponds to the specifications shown in table 2.

Table 2. Aggregate gradation specifications for the asphalt concrete wearing course (AC-WC) [11].

Size of sieve		% Passing weight
ASTM	(mm)	AC-WC
3/4"	19	100
1/2"	12.5	90 – 100
3/8"	9.5	77 – 90
No. 4	4.75	53 – 69
No.8	2.36	33 – 53
No. 16	1.18	21 – 40
No. 30	0.6	14 – 30
No. 50	0.3	9 – 22
No. 100	0.15	6 – 15
No. 200	0.075	4 – 9

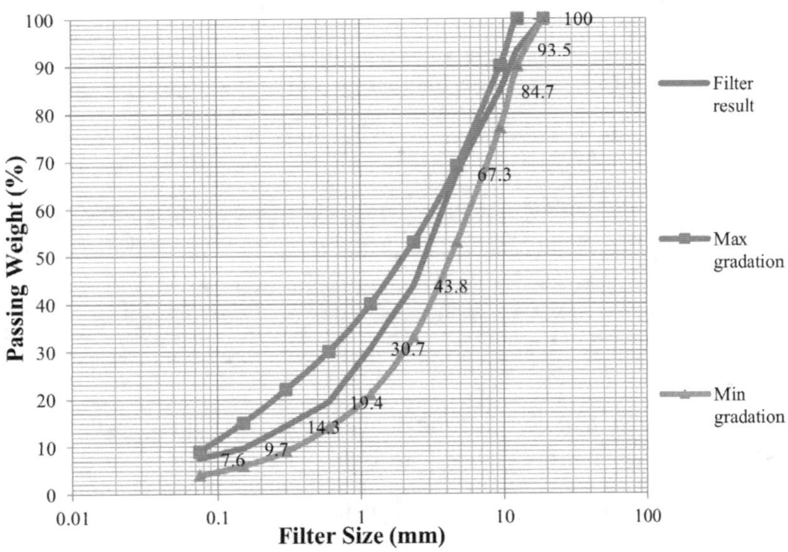

Figure 1. Aggregate gradation chart for test objects.

2. Methodology

The research methods used are the standard of Bina Marga, American Association of State Highway and Transportation Officials, and from other acknowledged institutions. This research will be conducted in the Road Laboratory of The Faculty of Engineering, Syiah Kuala University, Banda Aceh. The testing stages are as follows:

1. The testing of the physical properties of the aggregate, which are the test of impact, abrasion, flakiness index, elongated index, specific gravity, and absorption.
2. The testing of the physical properties of the asphalt, which are specific gravity, softening point, penetration, and ductility.
3. Calculation of the initial estimation of the middle asphalt rate (Pb) using equation (1) below.

$$Pb = 0.035(\%CA) + 0.045\ (\%FA) + 0.18(\%Filler) + K \quad (1)$$

Description:

P_b	= Middle/ ideal asphalt rate;
CA	= The aggregate held in No. 4 filter;
FA	= The aggregate passing No. 4 sieve and held No. 200 sieve;
Filler	= The aggregate, 75% of which passes No. 200 filter;
K	= Constants for asphalt concrete, which is 0.5 – 1.

The Pb value obtained in this study is 6%. Five types of AC-WC asphalt content (5%, 5.5%, 6%, 6.5%, 7%) were determined based on the Pb value which is added with 0%, 0.5%, 1%, -0.5%, -1%, as stated in the standard. Three test objects represented every asphalt content. The asphalt concrete wearing course (AC-WC) test objects in the research consist of 4 groups:

a. Test objects to determine the optimum asphalt content;
b. Test objects combined with coconut ash shell filler and portland cement, both for filler;
c. Test objects combined with the coconut ash and Portland cement, in the best combination, and LDPE substituted in Pen. 60/70 asphalt, using dry mixing method;
d. Test objects for durability tests.

There are 15 pieces of asphalt objects used to determine the optimum asphalt content. The number of test objects can be seen in table 3.

Table 3. Test items to determine optimum asphalt content.

Asphalt Content	Test Item code	Amount
5.0%	A_{11}, A_{12}, A_{13}	3 pieces
5.5%	A_{21}, A_{22}, A_{23}	3 pieces
6.0%	A_{31}, A_{32}, A_{33}	3 pieces
6.5%	A_{41}, A_{42}, A_{43}	3 pieces
7.0%	A_{51}, A_{52}, A_{53}	3 pieces
Amount		15 pieces

The optimum asphalt content was determined using the overlapping method. After that, 45 pieces of test objects were tested, as described in table 4.

Table 4. Test items with coconut shell ash filler and portland cement.

Filler combination	Asphalt content	Test item code	Amount
	$KAO_{(Bw)}$	B_{11}, B_{12}, B_{13}	
0% CSA : 100% PC	KAO	B_{21}, B_{22}, B_{23}	9 pieces
	$KAO_{(At)}$	B_{31}, B_{32}, B_{33}	
	$KAO_{(Bw)}$	C_{11}, C_{12}, C_{13}	
25% CSA: 75% PC	KAO	C_{21}, C_{22}, C_{23}	9 pieces
	$KAO_{(At)}$	C_{31}, C_{32}, C_{33}	
	$KAO_{(Bw)}$	D_{11}, D_{12}, D_{13}	
50% CSA: 50% PC	KAO	D_{21}, D_{22}, D_{23}	9 pieces
	$KAO_{(At)}$	D_{31}, D_{32}, D_{33}	
	$KAO_{(Bw)}$	E_{11}, E_{12}, E_{13}	
75% CSA: 25% PC	KAO	E_{21}, E_{22}, E_{23}	9 pieces
	$KAO_{(At)}$	E_{31}, E_{32}, E_{33}	
100% CSA: 0% PC	$KAO_{(Bw)}$	F_{11}, F_{12}, F_{13}	9 pieces
	KAO	F_{21}, F_{22}, F_{23}	

$KAO_{(At)}$	F_{31}, F_{32}, F_{33}	
Amount		45 pieces

Note:CSA: coconut shell additive
PC: portland cement
KAO: optimum asphalt content

After obtaining the best CSA-PC filler combinations, the test items were made with asphalt, 5.92% (optimum content) of AC-WC weight, containing LDPE, and with the best CSA-PC filler combinations. The description of the test objects is in table 5.

Table 5. Test items with LDPE waste substitutions.

LDPE Content	Asphalt Content	Test Item code	Amount
1%	KAO	G_{11}, G_{12}, G_{13}	3 pieces
2%	KAO	G_{21}, G_{22}, G_{23}	3 pieces
3%	KAO	G_{31}, G_{32}, G_{33}	3 pieces
4%	KAO	G_{41}, G_{42}, G_{43}	3 pieces
5%	KAO	G_{51}, G_{52}, G_{13}	3 pieces
	Amount		15 pieces

After obtaining the best combination of AC-WC with LDPE and CSA-PC, the test objects with the best combination of LDPE and CSA-PC were tested for durability, at a 30 minute and 24-hour immersion. All durability tests, including the durability tests for the former test objects, were shown in table 6.

Table 6. The number of test objects for durability testing with 30 minute and 24-hour immersion.

Test items types	30 minutes	24 hours	Amount
Test objects for optimum asphalt content determination	3	3	6 pieces
Test objects with the combination of CSA and PC as filler	3	3	6 pieces
Test objects added with LDPE with dry method	3	3	6 pieces
Amount			18 pieces

3. Results and discussion

3.1. Results of the examination of physical properties

The result of the inspection of the physical properties of mountain rock aggregate located in Ujong Kareung Sukakarya subdistrict, Sabang City, is presented in table 7. The results of the inspection of the physical properties of Pen. 60/70 asphalt is presented in table 8.

Table 7. Results of inspection of the physical properties of mountain rock aggregate.

Properties	Unit	Results	Specifications of Bina Marga (2018)
Specific gravity	-	2.42	Min. 2.5
Absorption	%	2.71	Maks. 3
Weight of contents	kg/dm^3	1.20	Min. 1

Flakiness index	%	15.00	Maks. 10
Elongated index	%	14.18	Maks. 10
Impact	%	13.51	Maks. 30
Abrasion	%	25.93	Maks. 40

Table 8. The results of the inspection of the physical properties of pen. 60/70 asphalt.

Properties	Unit	Results	Specifications of Bina Marga (2018)
Specific gravity	-	1.03	Min. 1
Penetration	0.1 mm	65	60 – 70
Softening point	^{o}C	48.5	Min. 48
Ductility, 25° C	cm	120	Min. 100

3.2. Marshall test results

Based on Marshall test results, which are stability, flow, VIM, VMA, and VFA, the optimum asphalt content of AC-WC composed of only aggregate, filler of portland cement, and asphalt, is 5.92%. Recapitulation of Marshall test results for determination of optimum asphalt content of the AC-WC mixture is presented in table 9.

Table 9. Recapitulation of Marshall test results to determine optimum asphalt content.

Mixed characteristics	Asphalt Content (%)					Specifications of Bina Marga (2018)
	5.00	5.50	6.00	6.50	7.00	
Stability (Kg)	1,908.21	1,888.93	1,850.38	1,869.66	1,792.56	Min. 800
Flow (mm)	3.07	3.30	3.63	3.80	4.33	2 – 4
VIM (%)	5.36	4.70	3.77	3.17	2.20	3 – 5
VMA (%)	15.97	16.37	16.55	17.01	17.15	Min. 15
VFA (%)	66.43	71.36	77.21	81.35	87.18	Min. 65

The Marshall characteristics obtained from the Marshall test were plotted on the several cross axis with coordinates of the asphalt content as the x-axis and the Marshall parameter as the y-axis, to facilitate the calculation of the regression analysis. Optimum asphalt content obtained is 5.92%.

In table 10-12, it indicates that the higher percentage of CSA, the lower the stability value. The stability value for all combinations of CSA filler variations still meets the requirements of min. 800 kg. Flow value inversely proportional to stability. The value of flow is influenced by the comparison of the asphalt mixture. In the table above, it shows that the flow value increases when the CSA level increases. This phenomenon occurs because of the influence of CSA substitution in asphalt which causes the asphalt to become mushier. The decrease in the value of VIM is caused by the increased percentage of the CSA. The VMA value tends to decrease as CSA fillers increase in asphalt mixtures. The VFA value is getting larger with the increased variance rate of CSA fillers in asphalt mixtures; this can occur due to the increased volume of CSA.

Table 10. Recapitulation of the Marshall test result of ACWC, containing 5.42% asphalt, substituted with coconut shell ash filler.

Mixed characteristics	Variety of filler CSA-PC					Specifications of Bina Marga (2018)
	0%	25%	50%	75%	100%	
Stability (kg)	1,676.91	1,638.36	1,607.52	1,665.35	1,229.73	Min. 800
Flow (mm)	2.27	2.70	3.07	3.30	3.53	2 - 4
VIM (%)	4.74	4.73	2.43	2.04	1.53	3 - 5
VMA (%)	16.25	15.60	12.97	12.08	11.14	Min. 15
VFA (%)	70.84	69.68	81.25	83.26	86.25	Min. 65

Note: CSA: coconut shell ash
PC: portland cement

Table 11. Recapitulation of the marshall test result of AC-WC, containing 5.92% asphalt, substituted with coconut shell ash filler.

Mixed characteristics	Variety of filler CSA-PC					Specifications of Bina Marga (2018)
	0%	25%	50%	75%	100%	
Stability (kg)	1,780.99	1,727.03	1,634.51	1,649.93	1,345.38	Min. 800
Flow (mm)	2.50	2.77	2.50	3.13	4.07	2 - 4
VIM (%)	4.23	4.06	2.06	1.51	1.23	3 - 5
VMA (%)	16.78	15.94	13.55	12.49	11.71	Min. 15
VFA (%)	74.87	74.61	84.81	87.95	89.55	Min. 65

Table 12. Recapitulation of the marshall test result of AC-WC, containing 6.42% asphalt, substituted with coconut shell ash filler.

Mixed characteristics	Variety of filler CSA-PC					Specifications of Bina Marga (2018)
	0%	25%	50%	75%	100%	
Stability (kg)	1,811.83	1,750.15	1,684.62	1,727.03	1,445.61	Min. 800
Flow (mm)	2.63	2.68	2.73	3.37	3.47	2 - 4
VIM (%)	3.62	3.33	1.66	1.19	1.15	3 - 5
VMA (%)	17.24	16.24	14.11	13.07	12.48	Min. 15
VFA (%)	79.04	79.48	88.29	90.92	90.81	Min. 65

Table 13 shows the value of stability increases because of the increase in the percentage of LDPE, which can provide a strong bonding between particles in the asphalt mixture. Contrastly, the flow value of the AC-WC increases as the percentage of LDPE plastic decreases. The table also shows the increase in the value of VIM caused by the increase in the amount of LDPE into the asphalt mixture, but this is only at the variation of 1% - 3%. It can be seen also that the percentage of LDPE in the asphalt mixture can increase VMA value and decrease VFA value.

Table 13. Recapitulation of the Marshall test result of AC-WC substituted with LDPE.

Mixed characteristics	Variations of LDPE substitutions					Specifications of Bina Marga (2018)
	1%	2%	3%	4%	5%	
Stability (kg)	1,730.88	1,765.57	1,780.99	1,773.28	1,804.12	Min. 1000
Flow (mm)	3.48	3.23	2.97	3.00	2.57	2 - 4
VIM (%)	4.12	4.42	4.79	5.22	5.57	3 - 5
VMA (%)	15.94	16.20	16.51	16.87	17.17	Min. 15
VFA (%)	74.23	72.76	70.97	69.09	67.59	Min. 65

Table 14. Result of durability testing.

Types of asphalt mixture	Soaking stability, 30 minutes	Soaking stability, 24 hours	Durability (%)
Test object with optimum asphalt content (5.92%)	1,719.32	1,567.43	91.17
Test object containing coconut shell ash (CSA) filler substitution with optimum asphalt content (5.92%)	1,754.01	1,480.31	84.40
Test object containing LDPE substitution with dry mixing method and with optimum asphalt content (5.92 %)	1,769.43	1,530.42	86.49

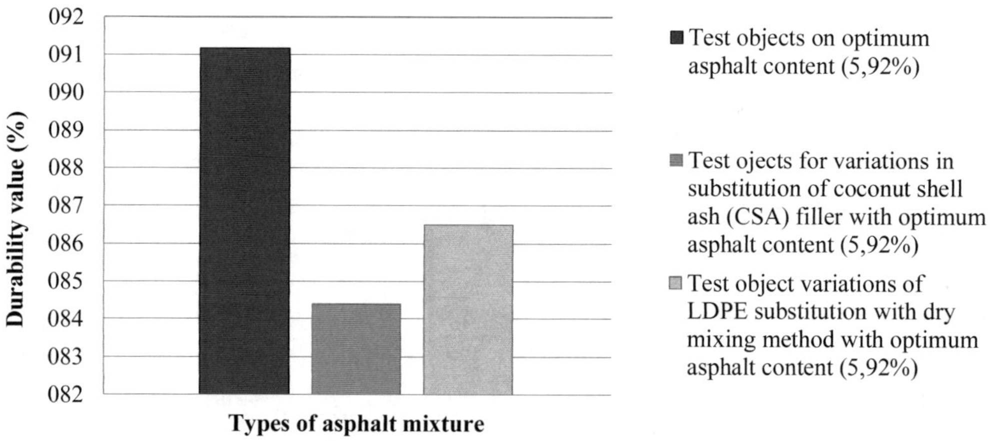

Figure 2. The durability of asphalt mixture.

The durability value of the asphalt mixture (see figure 2) with standard material meets the requirements, while the durability value of the asphalt mixture substituted with coconut shell ash and LDPE does not meet the requirements. The low durability of the asphalt mixture is caused by the coconut shell ash which properties as organic material are susceptible to water influence.

4. Conclusion

The Sabang Mount rock, used as the material of AC-WC substituted with coconut shell ash filler and LDPE penetration-60/70-asphalt substitution in dry method, meets the requirement of 2010 Bina Marga (fourth revision, 2018). Based on the Marshall test results, the best combination of the substitution in the AC-WC is 25% of CSA and 75% of PC for filler, 3% LDPE of asphalt content, and 5.92% of asphalt content; AC-WC in this combination satisfies all Marshall parameters. AC-WC with standard material obtains the best durability value, which meets the required value of 91.17%, while the durability value of AC-WC substituted with CSA and LDPE is less than 90%.

The results of this research can be used as a reference in the optimization of the usage of the Sabang Mountain rock as the primary material in the asphalt concrete. The recommended further research is a combination of coconut shell ash (CSA) filler and LDPE substitution on asphalt pen. 60/70 in AC-WC using a wet mixing method.

Acknowledgments
The authors would like to express his sincere thanks to the laboratory staff and students at the Highway Engineering Laboratory at the Syiah Kuala University for assistance in conducting research.

References

[1] Saleh S M, Anggraini R, Hermansyah H and Salmannur A 2018 Durability mix asphalt concrete using mountain coral aggregate from Sabang with asphalt binder Pen 60/70 and Retona Blend 55 *J. of Transp. Unpar* **18** 2
[2] Krisna A N 2012 *Influence of Powder Addition of Coconut Shell Charcoal to the Stability of Concrete Asphalt Mixture* (Malang: Building Engineering Education of the State University of Malang)
[3] Suroso T W 2008 *Effect of Addition of LDPE Plastic (Low-Density Polyethylene) Wet and Dry Way to Mix Performance Paved Road and Bridge Research* (Bandung)
[4] Hardiyatmo H C 2007 *Highway Maintenance 1st edition* (Yogyakarta: Gadjah Mada University Press)
[5] Sukirman S 1999 *Highway Bending* (Bandung: Nova Publisher)
[6] Hermansyah H 2018 *The Utilization of Coral Stone Mountain Pulau Weh for Asphalt Mixture of AC-WC with Asphalt Variation Retona Blend 55 and Asphalt Pen. 60/70* (The Civil engineering thesis of Syiah Kuala University)
[7] Salim A 2010 *The Use of Natural Aggregates of Simeulue Island and Portland Cement as Filler for Road Surface Layer* (The Civil engineering thesis of Syiah Kuala University)
[8] Isnanda, Saleh S M and Isya M 2018 Influence of substitution polystyrene (PS) and ash charcoal coconut shell as filler against the characteristics of mixed AC-WC *Civ. Eng. J. Syiah Kuala University* **1** 3
[9] Mashuri 2008 Effect of use of coconut shell charcoal powder and variation in the number of collisions against the characteristics of asphalt concrete mixtures *J. of Civ. Eng. Tadulako University* (Palu)
[10] Sukirman S 2003 *Hot mixed asphalt concrete* (Bandung: Granite Publisher)
[11] Directorate General of Bina Marga 2018 *General Specification of Directorate General of Bina Marga Edition 2010 Revision 4 Division 6* (Jakarta: Ministry of Public Works and Housing)

IOP Conf. Series: Materials Science and Engineering **933** (2020) 012005 doi:10.1088/1757-899X/933/1/012005

The influence of occupational safety and health program on the manpower performance of construction company in Banda Aceh city

M Jamil[1], F Maulina[1], Fachrurrazi[1], M Hasan[1] and Muslimsyah[2]

[1] Civil Engineering Postgraduate Program, Department of Civil Engineering, Universitas Syiah Kuala, Banda Aceh, Indonesia
[2] Department of Architecture, Syiah Kuala University, Banda Aceh, Indonesia

Corresponding e-mail: muttaqin@unsyiah.ac.id

Abstract. The implementation of occupational safety and health program in Banda Aceh city is still not optimal, as indicated by a high rate of work accidents. The purpose of this research is to study the effect of the occupational safety and health program on the manpower performance of construction companies in Banda Aceh city. Three construction companies were studied and compared to each other. Data collection was carried out by observation, using questionnaires, and interviewing the respondents of the research, in this case, the foreman, chief workman, workman, and worker. Data processing is conducted by using descriptive statistical analysis, validity test, reliability test, and regression analysis. The results of the analysis show that the application of occupational safety and health program has a significant influence on manpower performance.

1. Introduction

Occupational safety and health (OSH) issues in Banda Aceh are still often overlooked; this is indicated by the incessant infrastructure development carried out by the government which is directly proportional to the risk of labor accidents. The Indonesian Ministry of Manpower said the highest number of work accidents occurred in the construction sector [1]. In order to reduce the number of occupational accidents, the Ministry of Labor will increase supervision and law enforcement in the field of K3 [2,3]. Labor is an important factor for company activities because the company cannot be separated from the name of the workforce. Based on data from the Social Security Administering Agency (BPJS), there were 80,392 cases of work accidents in 2017 [4]. One of the causes of these accidents is the lack of optimal supervision and implementation of OHS, as well as OHS behavior in the workplace. Therefore, the International Labor Organization (ILO) estimates that losses from accidents reach 2-4% of a country's gross national product. This loss that happened due to accidents and other events is a risk that must be faced by every organization or company. Human factors are the main cause of workplace accidents (85% of total workplace accidents cause, according to ILO), therefore, workplace accidents are mostly caused by human factors.

In a construction project, OSH is a very important thing, one of which is the physical and mental of the workforce. The health of workers who are less than optimal can affect the high absenteeism and poor performance of the workforce, as well as the safety of the workforce. By meeting safety standards in

IOP Conf. Series: Materials Science and Engineering **933** (2020) 012005 doi:10.1088/1757-899X/933/1/012005

work will make the workforce safer and achieve the success of a project. If the workplace does not meet safety standards and there are many hazards, it will cause an accident while working.

The implementation of the OSH program on construction projects is expected to achieve a better quality of the workforce. One of the factors that influence the performance of a workforce is OSH; therefore, it must be considered more. The existence of the OSH program applied to a construction project can improve the performance of a better workforce, guaranteed in terms of safety and health in working in a healthy environment so that the overall workforce can work more optimally and the complete the construction project on time.

Based on the above background, the problem in this study is whether there is an effect of the implementation of the OSH program on the workforce performance of construction companies in Banda Aceh city. Three construction companies were studied. The purpose of this study is to see the effect of the OSH program on the performance of the three construction companies in Banda Aceh city.

2. Methodology

The research objects are three construction companies located in Banda Aceh city, which are PT. Raya Cipta Mulia, PT. Tanjung Harapan, and PT. Mirtada. The research variables were occupational safety and health (OSH) (X variable) and the manpower performance (Y variable). The indicators for variable OSH were workplace environment, machinery and equipment, guaranteed safety, condition of sense organ, stamina, emotions, and health insurance program. The indicators for variable manpower performance were quantity, timeliness, effectiveness, independence, and work commitment.

The research was conducted by observing the foreman, chief workman, workman, and worker during the construction project work. For PT. Raya Cipta Mulia, the observation was performed during the construction of Project Revitalisasi RKB SMAN 1 (Block A) Banda Aceh. For PT. Tanjung Harapan, the observation was performed during the construction of Project Revitalisasi RKB SMAN 1 (Block C) Banda Aceh). While for PT. Mirtada Sejahtera, the observation was conducted during the construction of Project Konstruksi Fisik Work Shop (Laboratorium Mini plant Turunan CPO dan Laboratorium Bengkel – Mekatronika) Sekolah Menengah Teknologi Industri (SMTI) Banda Aceh. All the project was constructed in 2018. In addition, the interviews were also conducted with the foreman, chief workman, workman, and worker regarding research variables.

The questionnaire was conducted by distributing a set of questionnaires that contained questions that were arranged according to the research variables to the workforce. The method used was a closed questionnaire. Questionnaire instruments were validated and reliability was measured so that the research produces valid and reliable data. The validation of the instrument was done by factor analysis, where each value in each question is correlated with the total value of all questions for one variable using the Pearson correlation formula t-test and as follows [5-7]:

$$r_{cal} = \frac{n\Sigma XY-(\Sigma X)(\Sigma Y)}{\sqrt{(n\Sigma X^2-(\Sigma X)^2)(n\Sigma Y^2-(\Sigma Y)^2))}} \quad (1)$$

$$t_{cal} = \frac{r\sqrt{n-2}}{\sqrt{1-r^2}} \quad (2)$$

where r_{cal} = correlation coefficient, t_{cal} = the calculation t, n = the number of respondents ΣX = total score of question items, and ΣY = the total score for a variable. The decision is made by comparing the of t_{cal} with t_{table}, if $t_{cal} > t_{table}$ means the instruments are valid and if $t_{cal} < t_{tabel}$ means the instruments are invalid. The reliability of the instruments was measured by the Cronbach Alpha method with a measuring scale from 0 to 1. The instruments are reliable if the Cronbach Alpha is greater or equal to 0.6 [8].

In the questionnaire, the research instruments for each indicator stated above are as follows:

a. For an indicator of workplace environment:
   - The company provides training and education for every workforce to maintain a safe and clean environment (X1)
   - Circulation, temperature, lighting in the workplace meet the standard operating procedures (X2)

IOP Conf. Series: Materials Science and Engineering **933** (2020) 012005 doi:10.1088/1757-899X/933/1/012005

b. For an indicator of machinery and equipment:
   - The company always provide protective work such as helmets, boots, gloves, masks (personal protective equipment, PPE) (X3)
   - All work equipment is in good condition and is suitable for use, and dangerous equipment is given a sign (X4)
   - Safeguarding work equipment is in accordance with the standard operating procedure of safety (X5)
   - Technology and how to use the machines are in accordance with the standard operating procedure and project requirements (X6)
c. For an indicator of guaranteed safety:
   - The company provides the instructions on the importance of using PPE when working and the instructions regarding the appropriate stages in completing work (X7)
   - The company provides a place to store the goods to maintain the quality of the goods (X8)
d. For an indicator of the condition of sense organ:
   - The influence of work, in the long run, has an impact on the deterioration of sense organ and lung function (X9)
e. For an indicator of stamina:
   - The company always guarantees the maintenance and improvement of the health and nutrition of the workforce (X10)
f. For an indicator of emotions:
   - Emotions are unstable due to the pressure of work that must be completed on time (X11)
g. For indicator health insurance program:
   - The company provides health insurance to every workforce (X12)
h. For an indicator of quantity:
   - The level of achievement of the work volume produced has been in accordance with company expectations (Y1)
   - Often completing work outside of working hours to achieve the targets charged by the company (Y2)
i. For an indicator of timelines:
   - Looking for other ways for the work to be completed on time when there is a deadlock in the process of completing work (Y3)
   - Doing a job deftly and not procrastinate (Y4)
   - Trying to prioritize work rather than personal interests (Y5)
j. For an indicator of effectiveness:
   - Can complete the work carefully as expected (Y6)
   - Do work accurately and rarely make mistakes (Y7)
k. For an indicator of independence:
   - Have initiatives that are useful to help resolve better jobs (Y8)
   - Able to provide creative ideas for the progress of the company (Y9)
l. For an indicator of work commitment:
   - Has responsibility for the work charged (Y10)
   - Responsible for the results of work achieved (Y11)

To determine the number of samples, the Slovin's formula is used as follows:

$$n = 1 / (1 + N e^2)$$

where n = the number of samples, N = the number of populations, and e = error. In this research, the value of $e = 0.1$ which means that the level of confidence of 90 % was used. According to the above formula, the minimum number of samples obtained is 46. The correction of the number of samples was done by adding several subjects so that the samples are fulfilled. A correction was done using the large drop-out correction formula as follows

$$n = n / (1 - f) = 46 / (1 - 0.1) = 51 \quad (3)$$

where f = estimated proportion of drop out = 0.1

To see the correspondence characteristics and perception of the implementation of the OSH program, the descriptive analysis, which uses frequency analysis and the obtained average values, was used. To see the relationship between variable OHS and variable manpower performance, the correlation analysis, and regression analysis were used. In order to see the difference and similarity between the three companies studied in this research, a Duncan test was used.

3. Results and discussion

The characteristics of the respondents based on gender, level of education, length of work experience, and position of work are shown in Table 1-4.

Table 1. Characteristics of the respondents based on gender.

Gender	Company name						Total	
	PT. Raya Cipta Mulia		PT.Mirtada Sejahtera		PT Tanjung Harapan			
	N	%	n	%	n	%	n	%
Man	18	100	17	100	16	100	51	100
Woman	0	0	0	0	0	0	0	0
Total	18	100	17	100	16	100	51	100

Table 2. Characteristics of the respondents based on the level of education.

Pendidikan	Frequency	Percentage (%)
SD	8	15.7
SMP	21	41.2
SMA	19	37.3
Diploma	3	5.9
Total	51	100

Table 3. Characteristics of the respondents based on the length of work experience.

Length of work experience	Company name						Total	
	PT. Raya Cipta Mulia		PT.Mirtada Sejahtera		PT Tanjung Harapan			
	N	%	n	%	n	%	N	%
0-3 years	0	0	5	29.4	5	31.3	10	19.6
>3-5 years	4	22.2	3	17.6	4	25.0	11	21.6
>5-7 years	8	44.4	3	17.6	3	18.8	14	27.5
>7 years	6	33.3	6	35.3	4	25	16	31.4
Total	18	100	17	100	16	100	51	100

Table 4. Characteristics of the respondents based on the position of work.

Work Position	Company name						Total	
	PT. Raya Cipta Mulia		PT.Mirtada Sejahtera		PT Tanjung Harapan			
	n	%	N	%	n	%	n	%
Foreman	1	5.6	1	5.9	1	6.3	3	5.9
Chief workman	2	11.1	2	11.8	2	12.5	6	11.8
Workman	7	38.9	6	35.3	7	43.8	20	39.2
Worker	8	44.4	8	47.1	6	37.5	22	43.1
Total	18	100	17	100	16	100	51	100

The results of the validity test for the X variable and Y variable are shown in Tables 5 and 6. From the test results, it is obtained that $r_{cal} > r_{table}$ which means that the instruments used in this study are valid.

Table 5. Result of validity test for X variable.

Variables	r_{cal}	r_{table}	Remarks
X1	0.729	0.444	Valid
X2	0.627	0.444	Valid
X3	0.620	0.444	Valid
X4	0.634	0.444	Valid
X5	0.610	0.444	Valid
X6	0.599	0.444	Valid
X7	0.613	0.444	Valid
X8	0.638	0.444	Valid
X9	0.603	0.444	Valid
X10	0.606	0.444	Valid
X11	0.628	0.444	Valid
X12	0.601	0.444	Valid

Table 6. Result of validity test for Y variable.

Variables	r_{cal}	r_{table}	Remarks
Y1	0.550	0.444	Valid
Y2	0.703	0.444	Valid
Y3	0.610	0.444	Valid
Y4	0.584	0.444	Valid
Y5	0.568	0.444	Valid
Y6	0.637	0.444	Valid
Y7	0.736	0.444	Valid
Y8	0.573	0.444	Valid
Y9	0.794	0.444	Valid
Y10	0.631	0.444	Valid
Y11	0.584	0.444	Valid

The results of the reliability test for the instruments used in this research are shown in Table 7. The C_{alpha} obtained from the analysis is greater than 0.6. This means that the instruments used in this research are reliable.

Table 7. The result of the reliability test.

Variable	n	C_{alpha}	Remarks
OSH Program	12	0,852	Reliable
Performance of manpower	11	0,853	Reliable

The relationship between the application of the OSH program and the performance of manpower is shown in Figure 1. The correlation test result on the influence of the application OSH program on the performance of manpower is shown in Table 8. From Table 8, it is known that the coefficient of the relationship between the application OSH program and the performance of the workforce is 0.541 which is positive relationships. From the significance test using the t-test, it is obtained that tcal is 4.503 which is greater than t table (at the level of 5%) which is 2.010. This means that the relationship between the two variables is significant. The regression model for the relationship between two variables is proposed as follows:

$$Y = 1.864 + 0,545X \tag{4}$$

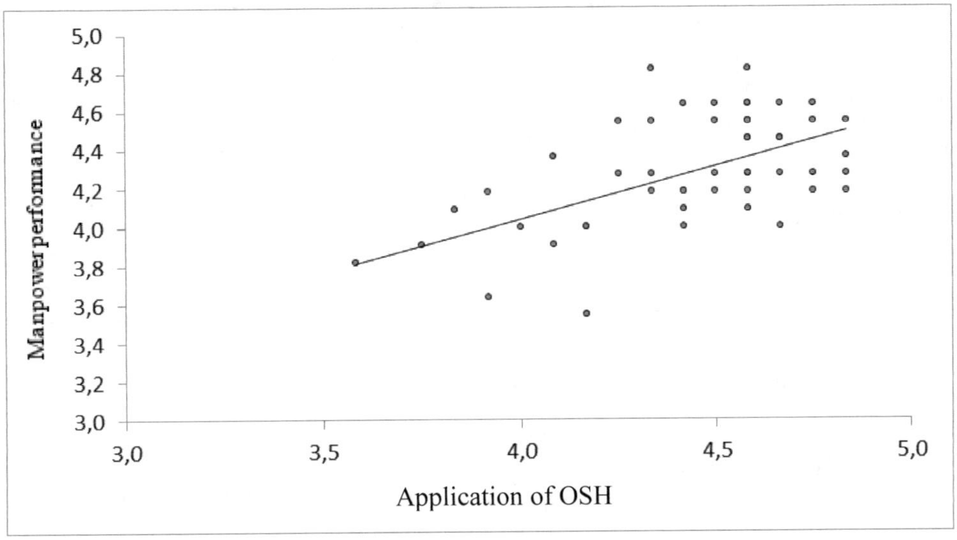

Figure 1. Relationship between the application of the OSH program and the performance of manpower.

Table 8. Result of correlation test on the influence of OSH program application on the performance of manpower.

The relation between OSH program application and manpower performance	Regression		R	R^2	t_{cal}	t_{table}
	B	Std. Error				
Constant	1.864	0.121	0.541	0.293	4.503	2.010
Coefficient	0.545					

Based on the hypothesis, the application of the OSH program has a significant influence on the workforce performance of the three companies. This is evidenced from a significant level of 0,000 smaller than a = 0.05. This means that there is an influence of the application of the OSH program on the workforce performance of the three companies.

4. Conclusions

Based on this research, the following conclusions can be drawn:

1. The results of the validity test show that all elements of the statement for data on the application of the OSH program and the performance of the workforce at the three companies having a value of r_{cal} greater than the value of r_{table}; this means that all instruments are valid.
2. The reliability test results show that all elements of the statement for each data of each variable are reliable based on the value of Cronbach's Alpha that is greater than 0.60.
3. The results of the hypothesis test are known that the application of the OSH program has a significant influence on the performance of the workforce of all three construction companies. This is evidenced from a significant level of 0,000 smaller than a = 0.05.
4. The simple linear regression has the result: Y = 1,864 + 0,545X (Y = performance of manpower, X = OSH program). This shows that there is a linear relationship between the implementation of the OSH program on the performance of the manpower of all three companies.

Acknowledgment

The authors wish to thank Mr. Fachri Jasri for his assistance in collecting data.

References

[1] Validnews 2016 *www.validnews.id*
[2] Anonim 2003 *Undang-undang No. 13 Tahun 2003 Tentang Ketenagakerjaan. Sekretariat Negara* (Jakarta)
[3] Anonim 2012 *PP No. 50 Tahun 2012 Tentang Penerapan Sistem Manajemen Keselamatan dan Kesehatan Kerja Sekretariat Negara* (Jakarta)
[4] Serambinews 2017 *www.aceh.tribunnews.com*
[5] Subagio P 2003 *Statistik Deskriptif* (Yogyakarta: BPFE UGM)
[6] Sugiyono 2003 *Metode Penelitian Bisnis* (Bandung: Alfabeta)
[7] Sugiyono 2013 *Metode Penelitian Kuantitatif, Kualitatif dan R&D* (Bandung: Alfabeta)
[8] Arikunto 2002 *Prosedur Penelitian Suatu Pendekatan Praktek* (Jakarta: Asdi Mahasatya)

Level of importance and housing satisfaction of Jeulingke residence housing on-road aspect

M Iqbal[1]

[1] Civil Engineering postgraduate program, Department of Civil Engineering, Universitas Syiah Kuala, Banda Aceh, Indonesia

Corresponding e-mail: iqbal1991@mhs.unsyiah.ac.id

Abstract. Jeulingke Residence is one of the subsidized housing located in Jeulingke Village, Syiah Kuala District, Banda Aceh City. Occupants after occupying, faced with problems regarding the performance of developers on aspects of roads. The road aspect has a bad condition because the road surface $\pm$ 246 m towards the housing complex is currently not pavement. This study aims to identify the level of importance and satisfaction of residents of Jeulingke Residence Housing on-road aspects. This study uses qualitative methods through observation and interviews and quantitative methods through questionnaires. The total population of houses that have been occupied in Jeulingke Residence is 102 units, and no sampling is taken. Data analysis used Importance Performance Analysis (IPA) through Minitab software 16. The results of the study showed that the level of importance and satisfaction that requires priority in improving performance in the aspect of the road is the condition of the entrance pavement.

1. Introduction

House is a building that functions as a human settlement, while housing is a collection of several housing units in an environment. Jeulingke Residence is one of the subsidized housing located in Jeulingke Village, Syiah Kuala District, Banda Aceh City. Based on [1], Jeulingke Residence Housing was built in the cultivation area, housing zone, sub-zone of medium density house (R-2). Jeulingke Residence Housing was built by PT. Matahari Cipta as the developer, which is divided into two stages of development. Phase I was built in 2017 in a total of 78 housing units, while phase II was built in 2018 as many as 205 housing units. The total housing units in the residence are 283 units, which can be seen in Appendix A Figure A.3.7 page 71. The building area of this house is 36 m^2, while the land area is 100 m^2. The house building specifications consist of 2 bedrooms, 1 living room, 1 bathroom, kitchen, terrace, and carport.

After occupying Jeulingke Residence Housing, residents are faced with problems of developer performance in several aspects. In the aspect of the road from the preliminary observations, the conditions were bad, because the road surface + 246 m towards the housing complex is currently not pavement. The pavement is only found in housing complexes made of rigid pavement. This is when rain occurs on roads that do not have pavement which can cause the surface of the road to become wet and muddy.

In planning the housing environment, aspects of roads have been well regulated by several regulations. The regulations are contained in [2]. These regulations should be a foundation for developers to plan housing complexes that meet standards for the residents. If the road aspects are below expectations then the residents feel dissatisfied with the service provided by the developer. Conversely, if they meet their expectations, the residents are satisfied with the service provided by the developer. In connection with these problems, this study aims to identify the level of importance and satisfaction of residents of Jeulingke Residence Housing on aspects of the road.

2. Methodology

2.1. Determining the population

The population is a generalization area consisting of objects/subjects that have certain qualities and characteristics set by researchers to be studied and then conclusions drawn. The sample is part of the number and characteristics possessed by the population [3]. The population in this study was intended for residents of the Jeulingke Residence Housing complex. Based on data from PT. Matahari Cipta, the number of houses in Jeulingke Residence Housing in phase I is 78 units, while phase II is 205 units, with a total of 283 units. Based on observations for phase I, the total number of houses has been occupied are 78 units, while for phase II the number of houses has been occupied are 24 units, with a total of 102 units. Given the minimal population of residents, no more sampling is taken, so that this study is determined as population research.

2.2. Descriptive analysis

Descriptive analysis is an analysis that describes data that will be made both alone and in groups [4]. Descriptive analysis in this study was used to describe respondents' perceptions of the level of importance and satisfaction of residents of Jeulingke Residence Housing on-road aspects. The steps of this descriptive analysis can be described as follows:

1. The questions in the questionnaire were analyzed descriptively through the help of SPSS version 22 software.
2. The output of the software is the frequency of answers to the importance and the level of satisfaction in the aspects of the road.
3. Calculating the mean indicator of the importance and the mean indicator of satisfaction. The mean obtained from the score is divided by the number of respondents on each indicator. Scores are obtained from the sum of all forms of answers that have been multiplied by the weight of the answers on each indicator.

Calculating the mean aspects of the level of importance and the mean aspects of the level of satisfaction. The mean aspect is obtained from the mean of each indicator added up then divided by the number of indicators.

2.3. Importance of performance analysis (IPA)

The IPA combines the measurement of importance level factors and satisfaction levels in a two-dimensional graph that facilitates the explanation of the data and gets practical proposals [5]. The IPA in this study was used to identify the level of importance and satisfaction of residents of Jeulingke Residence Housing on aspects of roads. The steps of this IPA can be described as follows:

1. Performing vertical axis intersection (Y-axis) through the mean aspect level of interest with the horizontal axis (X-axis) through the mean aspects of satisfaction level, so that IPA graphics can be made through the help of Minitab 16 software.
2. The output generated from the software is a 4-quadrant IPA-based graph containing indicators of road aspects. The position of the aspect indicator in a quadrant is determined by the meeting of the mean level interest with the mean level of satisfaction on the Y and X axes.
3. Interpreting the indicators of road aspects included in quadrants I, II, III, and IV. The interpretation of the IPA quadrant can be seen in table 1.

Table 1. Interpretation of the IPA quadrant.

No.	Quadrant	Interpretation
1	Quadrant I	Very important and very satisfied (maintain performance)
2	Quadrant II	Not important and very satisfied (tends to be excessive)
3	Quadrant III	Not important and not satisfied (low priority)
4	Quadrant IV	Very important and not satisfied (high priority / improve performance)

The quadrants in table 1 can show the level of importance and satisfaction of residents of Jeulingke Residence Housing on aspects of roads.

3. Results and discussion

3.1. Respondent's perception

Residents after occupying Jeulingke Residence Housing, faced with problems regarding developer performance in several aspects. Developer performance is the work of the developer in building housing units, their facilities, and infrastructure in a housing complex. Developers who perform well are developers who can pay attention to the interests of residents while meeting satisfaction for the residents. Based on observations of existing conditions, developer performance which raises significant problems in housing complexes is the aspect of roads. Residents certainly have their evaluation criteria for the level of importance and level of satisfaction in each of these aspects. Perceptions of residents of one house can also be different from the other. This is because one's perceptions are influenced by ways of thinking and feeling that are subjective. The respondents' perceptions of the level of importance and satisfaction in each aspect can be described as below.

1. Perception of the level of importance
   Based on the answers of 102 residents, the following can be shown the frequency of answers of importance ranging from Not Important at All (NIA), Not Important (NI), Slightly Important (SI), Important (I), Very Important (VI), score, and mean. This can be seen in table 2.

Table 2. Residents' perceptions of the importance.

No.	Statements	Resident's Perception					Score	Mean
		NIA (1)	NI (2)	SI (3)	I (4)	VI (5)		
1	Main road pavement conditions	-	-	-	12	90	498	4,882
2	Pavement condition of the entrance	-	-	-	28	74	482	4,725
3	Condition of road pavement in housing	-	-	-	27	75	483	4,735
4	Width of the main road	-	-	-	13	89	497	4,873
5	Entrance width	-	-	4	39	59	463	4,539
6	Road width in housing	-	-	-	26	76	484	4,745
7	Benefit Areas	-	-	-	68	34	442	4,333
8	Regional Road Supervision	-	-	-	29	73	481	4,716
9	Building Line	-	-	-	39	63	471	4,618
10	The width of the road for fire trucks access	-	-	-	14	88	496	4,863
	Mean level of importance of aspects of the road							4,703

Table 2 shows that the mean level of interest in aspects of the road is obtained at 4.703. The mean shows that overall residents perceive that housing complexes on aspects of roads have a very important level of importance.

2. Perception of the level of satisfaction
   Based on the answers of 102 residents, the following can be shown the frequency of answers of satisfaction ranging from Not Important at All (NIA), Not Important (NI), Slightly Important (SI), Important (I), Very Important (VI), score, and mean. This can be seen in table 3.

Table 3. Residents' perceptions of satisfaction.

No.	Statements	Residents' Perception					Score	Mean
		NIA (1)	NI (2)	SI (3)	I (4)	VI (5)		
1	Main road pavement conditions	-	-	-	12	90	498	4,882
2	Pavement condition of the entrance	22	27	53	-	-	235	2,304
3	Condition of road pavement in housing	-	-	2	60	40	446	4,373
4	Width of the main road	-	-	-	13	89	497	4,873
5	Entrance width	-	-	-	79	23	431	4,225
6	Road width in housing	-	-	14	64	24	418	4,098
7	Benefit Areas	7	25	69	1	-	268	2,627
8	Regional Road Supervision	-	-	-	88	14	422	4,137
9	Building Line	-	17	47	38	-	327	3,206
10	The width of the road for fire trucks access	-	-	16	70	16	408	4,000
	Mean level of satisfaction of aspects of the road							3,873

Table 3 shows that the mean level of satisfaction in the aspect of the road is obtained at 3.873. The mean shows that overall residents perceive that the housing complex in the aspect of the road has a "satisfied" level of satisfaction.

3.2. Levels of interest and satisfaction of occupants on the road aspect

The road is an infrastructure that functions as a vehicle and people traffic lane. Based on the perceptions of occupants on aspects of the road, the mean level of importance is obtained at 4.703 as the y-axis and the mean level of satisfaction is 3.873 as the x-axis. The mean generally means that residents feel very important and satisfied with aspects of the road. The two means are perpendicular to the axis intersection so that the road aspect indicators are spotted in each quadrant. The results of the level of importance and satisfaction of residents in the aspects of the road that have been analyzed through the Minitab software can be seen in figure 1.

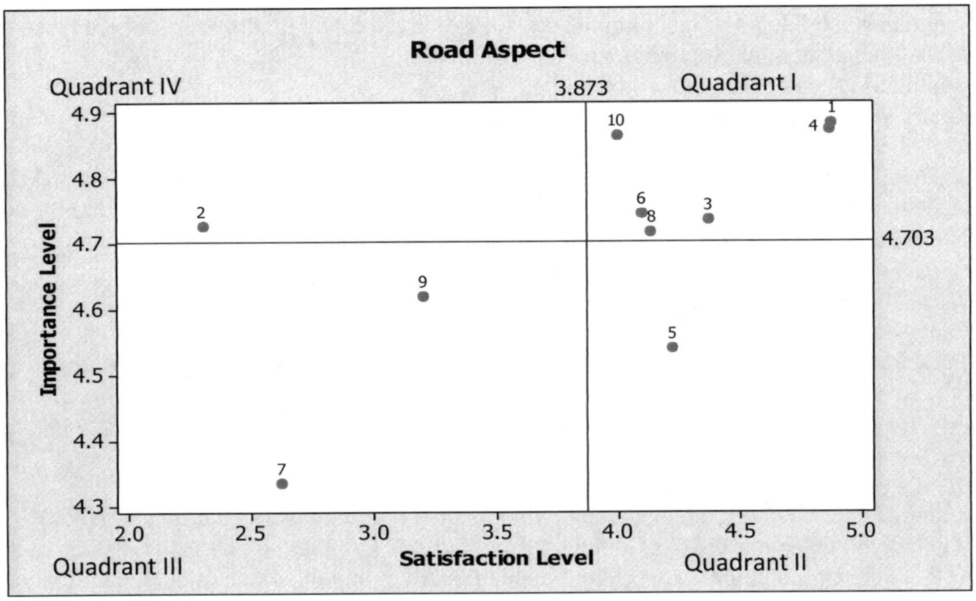

Figure 1. Level of importance and satisfaction on aspects of the road.

The distribution of indicators on aspects of the road from each quadrant can be presented in table 4.

Table 4. Distribution of indicators on-road aspect in each quadrant.

	Quadrant IV (High Priority)		**Quadrant I (Maintaining The Performance)**
2	Pavement condition of the entrance	1	Main road pavement conditions
		3	Condition of road pavement in housing
		4	Width of the main road
		6	Road width in housing
		8	Regional Road Supervision
		10	The width of the road can be accessed by fire trucks
	Quadrant III (Low Priority)		**Quadrant II (Overdoing)**
7	Benefit Areas	5	Entrance width
9	Building Line		

The quadrant interpretation in figure 1 and table 4 can be described as follows:

1. Quadrant I, there are 6 indicators in which performance needs to be maintained because residents assess the indicators as very important and very satisfied. Indicators that need to be maintained are as follows:
   a. Main road pavement conditions
   The main road of the Jeulingke Residence Housing complex is Teuku Nyak Arief Street which connects Kuta Alam District with Syiah Kuala District. According to the Spatial Detail Plan [1], this main road belongs to the classification of primary arterial roads. Primary arterial roads are roads that connect efficiently between national activity centers or between national activity centers and regional activity centers. Based on observation on the condition of the main road pavement is made of flexible pavement. Flexible pavement is a pavement that uses asphalt

material as a binder. The condition of the main road pavement can support occupant satisfaction so that the government can maintain its performance.

b. Condition of road pavement in housing
Roads in housing complexes are classified as secondary road classification II. Secondary local roads II are lanes that connect arterial / collector / local roads with the residential neighborhood, to access higher hierarchies. Based on observation, the pavement condition in a housing complex made of rigid pavement. Rigid Pavement is a pavement that uses cement as its binding material. The condition of pavement in this housing complex can support occupant satisfaction so that the developers can maintain their performance.

c. Width of the main road
According to Minister of Public Works Regulation Number 19/PRT/M/2011 concerning Road Technical Requirements and Technical Planning Criteria for Roads, the smallest primary arterial road (main road) is 21 m. Based on observations the main road has a width of 31 m. The width of the main road can support occupant satisfaction so that the government can maintain its performance.

d. Road width in housing
According to [6], the width of the secondary road II (the road in the housing complex) is 3-6 m. Based on observations the road in the housing complex has a width of 4 m. The width of the road in this housing complex can support occupant satisfaction so that the developers can maintain their performance.

e. Regional Road Supervision (RRS)
According to [6], the minimum width of RRS in a residential environment is 4 m. RRS is measured from the walls of the outermost building to the fence or the edge of the drainage outside the Area of The Road (AR). Based on the RRS observation on the housing complex, it has a minimum width of 4 m. RRS in this housing complex can support occupant satisfaction so that the developers can maintain their performance.

f. The width of the road can be accessed by fire trucks
According to [2], the inlet used for passing fire trucks should not be less than 4 m. Based on road observations the housing complex has a width of 4 m. The width of the road in this housing complex can support occupant satisfaction so that the developers can maintain their performance.

2. Quadrant II, there is one indicator in which performance tends to be excessive because residents assess the indicator as not important and very satisfied. Indicators that tend to label as an excessive performance are the width of the entrance. This entry is included in the collector's road classification. Collector road is a road that serves the transportation of collection/distribution with characteristics of medium distance travel, medium average speed, and a limited number of entrances. According to [6], the width of the collector road (entrance) is 7 m. Based on observation, the entrance has a width of 7.5 m. Therefore the government does not need to allocate resources in improving performance.
3. Quadrant III; there are 2 indicators whose performance has a low priority because residents assess that the indicator is not important and not satisfied. Indicators that need performance improvement are as follows:

a. Benefit Areas (BA)
According to [6], the width of housing' BA is 10-12 m. BA is measured from the edge of the left side of the drainage canal, the traffic lane, and the right side of the drainage canal. Based on BA observation, the housing complex only has a width of 4.6 m. Since the housing has been built from the planned site, the width of BA at the Jeulingke Residence Housing complex cannot be extended anymore. To improve their performance, developers should pay attention to the minimum width of BA for the next housing project.

b. Building Line (BL)
According to [6], the minimum width of BL is 10 m. BL is measured from the road axle to the outermost buildings on controlled land. Based on observations BL only has a width of 6 m. Considering that the housing has been built from the planned site, then the BL at the Jeulingke

Residence Housing complex cannot be expanded again. To improve their performance, developers should pay attention to the minimum width of BL for the next housing project.

4. Quadrant IV, there is one indicator of a performance that was categorized as a high priority, because the residents assess the indicator as very important and not satisfied. The indicator that needs to increase the performance in high priority is the condition of the entrance pavement. Based on the observation of the entrance + 246 m towards the housing complex there is currently no pavement. This is when rain occurs on roads that do not have pavement which can cause the surface of the road to become muddy. This situation can disturb the comfort of residents and road users. Therefore, the developer needs to allocate adequate resources to increase access to the road.

Research conducted by [7] showed that the transportation aspects that were taken into consideration in choosing the location of Villa Mutiara Cikarang Housing were smooth traffic, main road conditions, residential road conditions, location accessibility, public transportation availability, main road width, and road width in housing. Research conducted by [8] found that the factors that influence the preferences of residents of Pekanbaru in determining housing locations are accessibility factors, infrastructure, and facilities, environmental conditions, socio-economic conditions, the physical condition of the area, quality of housing, legality housing, and the findings of housing residents. Research conducted by [9] found that the factors that influence the value of residential buildings in the city of Denpasar are factors in soil characteristics, environment, location, and building characteristics.

4. Conclusion

Road aspect, the indicator that needs to be maintained is the condition of the main road pavement, the condition of the road pavement in the housing, the width of the main road, the width of the road in the housing, the Road Monitoring Area (RMA), and the width of the road accessible by fire trucks. Indicators that tend to be excessive are the width of the entrance. Indicators that have low priority are the Road Benefit Areas (RBA) and Building Border Line (BBL). The indicator that has a high priority is the condition of the entrance pavement.

It is recommended for the road aspect, the developer needs to increase the access road along + 246 m towards the housing complex.

References

[1] *Rencana Detail Tata Ruang (RDTR) Kecamatan Syiah Kuala* 2015

[2] Peraturan Menteri Negara Perumahan Rakyat 2008 *Nomor 22/PERMEN/M/2008 tentang Standar Pelayanan Minimal Bidang Peurmahan Rakyat Daerah Provinsi Dan Daerah Kabupaten/Kota*

[3] Sugiyono 2015 *Statistik Nonparametris untuk Penelitian* (Bandung :Alfabeta)

[4] Riduwan and Sunarto 2014 *Pengantar Statistika untuk Penelitian Pendidikan, Sosial, Ekonomi, Komunikasi, dan Bisnis* (Bandung: Alfabeta)

[5] Brandt D R 2000 *An "Outside-In" Approach to Determining Customer Driven Priorities for Improvement and Innovation* vol 2 (White Paper Series)

[6] SNI 2004 *Nomor 03-1733-2004 Tentang Tata Cara Perencanaan Perumahan di Perkotaan*

[7] Ruswandi Y 2013 Kajian Aspek Transportasi Terhadap Pemilihan Lokasi Perumahan Villa Mutiara Cikarang *J Bentang* **1** 2 (Bekasi: Universitas Islam)

[8] Asteriani F 2011 Preferensi Penghuni Perumahan di Kota Pekanbaru Dalam Menentukan Lokasi Perumahan *J Ekonomi Pembangunan* (Pekanbaru: Universitas Islam Riau)

[9] Rijasa M M 2014 Analisis Penilaian Bangunan Rumah Tinggal di Kota Denpasar *J Spektran* (Bali: Universitas Udayana)

IOP Conf. Series: Materials Science and Engineering **933** (2020) 012007 doi:10.1088/1757-899X/933/1/012007

Mix design and properties of reactive powder concrete with diatomaceous earth as cement replacement

T Saidi[1], M Hasan[1], A D D Riski[1], R R Ayunizar[1] and A Mubarak[1]

[1] Department of Civil Engineering, Universitas Syiah Kuala, Banda Aceh, Indonesia

Corresponding e-mail: muttaqin@unsyiah.ac.id

Abstract. Reactive powder concrete (RPC) consuming a large amount of Portland cement results in high cost. To reduce the cement used in RPC, the silica fume and ground granulated blast-furnace slag have been used as cement replacement. However, those materials are not available in Aceh Province – Indonesia. The objective of this research is to propose the mix design of RPC composed of locally available materials. In this paper, the mix design of RPC using calcined diatomaceous earth powder as partial cement replacement is presented. The iron ore powder was used as filler, and river sand with a diameter of less than 1 mm was used as aggregate. To maintain the workability of concrete, superplasticizer was added. The bulk density, specific gravity as well as particle size distribution of Portland cement, diatomaceous earth powder, iron ore powder and river size were tested. The modified Andreasen and Andersen model was utilized. The mix proportion of materials was determined using an optimization algorithm based on the least-squares method. Furthermore, all materials were mixed to produce RPC; and then the properties of fresh and hardened concrete i.e. workability, air content, compressive strength, splitting tensile strength, and flexural strength, were tested. The relative slump flow decreases with the increase of diatomaceous earth binder. The RPC with diatomaceous earth powder has a lower 7 days compressive strength but high flexural strength and splitting tensile strength.

1. Introduction

Many studies have been done on using the geopolymer as the cement replacement in concrete production [1-4]. Reactive powder concrete (RPC) was introduced by Richard and Cheyrezy in 1990 and published firstly in 1995 [5]. Since then, many studies have been done [e.g. 6-10]. RPC composed of fine materials such as micro sand as aggregate and a large amount of Portland cement results in high cost. For example, Rossi [11] in his experimental work to study the mechanical behavior of RPC, used 1050 kg/m^3 cement. Park et al. [12] used about 1000 kg/m^3 cement in studying the effect of hybrid fibers on the tensile strength of RPC. Considering the high cost of RPC, the silica fume and ground granulated blast-furnace slag (GGBS) have been used as partial cement replacement. For example, Hasan et al. [13] used about 650 kg/m^3 cement, 420 kg/m^3 GGBS and 120 kg/m^3 silica fume. El-Dieb [14] used about 900 kg/m^3 cement and 135 kg/m^3 silica fume. Tayeh et al. [15] used about 770 kg/m^3 cement and 200 kg/m^3 silica fume. Yu et al. [16] used about 875 kg/m^3 cement and 44 kg/m^3 micro-silica. However, silica fume and GGBS are not available in Aceh Province–Indonesia and it is very expensive for someone to order them from the other places. Based on this background, it is essential to find local material that can be used as the cement replacement. In this study, the calcined diatomaceous earth powder is used as partial cement replacement. The river sand with a diameter of less than 1 mm is used as aggregate. The iron ore powder,

IOP Conf. Series: Materials Science and Engineering **933** (2020) 012007 doi:10.1088/1757-899X/933/1/012007

which is the locally available material, is used as filler. Because of the different properties of the composed materials, it is important to determine the proportion of each material in RPC.

The purpose of this study is to propose the mix design of RPC composed of local materials. The mix proportion of materials was determined based on the modified Andreasen and Andersen model proposed by Funk and Dinger [17] to have the good packing density of the materials. The proportions of each material in the mixture are adjusted until an optimum fit between the composed mix, and the target curve is reached. The optimization algorithm based on the least-squares method was utilized.

2. Methodology

Several mix design tools have been proposed to design the mix proportion of concrete. De Larran and Sedran [18,19] postulated different approaches to design the concrete based on the properties of multimodal, discretely sized particles: the Linear Packing Density Model (LPDM), Solid Suspension Model (SSM) and Compressive Packing Model (CPM). Fennis et al. [20] later developed a concrete mix design based on the concepts of De Larrand and Sedran. However, all these design methods are based on the packing fraction of individual components and their combinations, and therefore it is complicated to include very fine particles in these mix design tools.

Based on the work of Fuller and Thompson [21], Andreasen and Andersen [22] developed a model in which a minimal porosity can be achieved by an optimal particle size distribution of all materials composing the concrete, as shown in the following equation:

$$P(D) = (D/D_{max})^q \qquad (1)$$

where P(D) = a fraction of the total materials with particle size is smaller than D, D = the particle size (μm), D_{max} = the maximum particle size (μm), and q = the distribution modulus.

Funk and Dinger [17] later modified the Andreasen and Andersen Model by taking into account the minimum particle size as follows:

$$P(D) = [D^q - (D_{min})^q] / [(D_{max})^q - (D_{min})^q] \qquad (2)$$

where D_{min} = the minimum particle size. The value of q determines the proportion between the fine and the coarse particles in the concrete mixture. Higher value (q > 0,5) leads to coarse mixture while lower value (q < 0,25) leads to fine mixture.

The so-called Modified Andreasen and Andersen Model in Eq. (2) has been successfully utilized in optimization algorithms for the mix design of normal concrete [23,24], lightweight concrete [25,26] and ultra-high-performance fiber reinforced concrete [16].

The materials used in this study were Portland cement, diatomaceous earth powder, iron ore powder, river sand, superplasticizer, and water. The Portland cement produced by PT Semen Andalas Indonesia of Lhoknga – Aceh Besar District was used. The specific gravity of the cement is 3.15. The particle size distribution of the cement was tested using the particle size analyzer (PSA).

The diatomaceous earth was collected from Beureunuet Village – Aceh Besar District in the form of the chunk. The chunk of diatomaceous earth was burned in the brick-burning furnace for 4 days, then ground using the Los Angeles machine to form a powder. The powder was then sieved to form the binder having the particle sizes less than 250 μm. The binder was calcined at the temperature of 600 °C for 5 hours. On the calcined diatomaceous earth powder, the bulk density, specific gravity, and particle size distribution tests were performed. The calcined diatomaceous powder is used as the pozzolanic material for cement replacement.

The iron ore was collected from Leupung – Aceh Besar District. Same as diatomaceous earth, the iron ore then ground using the Los Angeles machine and sieved to form the powder having the particle sizes less than 250 μm. The bulk density, specific gravity, and particle size distribution of the iron ore powder tests were then conducted. In this study, the iron ore powder is used as the filler.

The river sand was used as aggregate. The sand was sieved to have the particle sizes less than 1 mm. The bulk density, specific gravity, and particle size distribution tests were then performed. A polycarboxylic ether-based superplasticizer from PT Sika Indonesia was used to adjust the workability of RPC. The specific gravity of the superplasticizer is 1.06.

Based on the properties and particle sizes distribution of composed materials, the mix proportion of RPC was designed. The Modified Andreasen and Andersen Model was adopted. The Modified Andreasen and Andersen Model (Eq. 2) acts as a target function for the optimization of the proportion of the materials. The value of 0,31 for q was used. The proportion of the materials was determined by trial and error until the deviation between the target curve and composed particle size distribution of the mixture is minimum. A least square method expressed by minimizing the sum of the square of the residuals is adopted.

Three mix proportions of RPC, which are the different content of diatomaceous earth binder, were designed. The diatomaceous earth binder was used to replace 0%, 5%, and 10% of cement weight. The mix proportion of RPC with no diatomaceous binder (mix #1) was first designed. The water to cement ratio of 0,2 was used. After all the material proportion is determined, on the mix #2 and mix #3, the weight of cement was replaced by 5% and 10% of diatomaceous earth binder.

After the mix proportions were determined, the mixtures were mixed in a concrete mixer. First, cement, diatomaceous earth binder, iron ore binder, and sand were mixed with slow speed for 30 seconds. Then, 80% of mixing water was added, mixed with slow speed for 90 seconds, and stopped for 30 seconds. Later, the remaining water and superplasticizer were added and mixed with slow speed for 180 seconds following by high speed for 120 seconds. This mixing procedure was adopted from Yu et al. [16].

The workability of fresh RPC was determined by the flow table test. The concrete mixture was filled on the cone. Once the cone was rose up, the mixture was flown around. Two diameters perpendicular to each other (d_1 and d_2) were measured. The relative slump (ξ) is calculated as follows:

$$\xi = [(d_1 + d_2) / 2\, d_o]^2 - 1 \tag{3}$$

where d_o = the based inner diameter of the cone.

The air content of fresh RPC was determined as follows. The concrete mixture was filled in the cylindrical container with a known volume and vibrated for 30 seconds. The weight of the concrete mixture was measured. By known the composition ratio of composed materials in the mix, then the weight of each material can be calculated. The weight of each material then converted to its volume by dividing the weight with the density. The air content (φ) therefore, can be calculated as follows:

$$\varphi = [(V_c - V_{pc} - V_{de} - V_{io} - V_s - V_w - V_{sp})/V_c] \text{ x } 100\ \% \tag{4}$$

where V_c = volume of the container, V_{pc} = volume of Portland cement, V_{de} = volume of diatomaceous earth binder, V_{io} = volume of iron ore binder, V_s = volume of sand, V_w = volume of water, and V_{sp} = volume of superplasticizer.

Three cube specimens with 75 mm size, 3 beam specimens with 75 mm x 75 mm x 350 mm size, and 3 cylinder specimens with 50 mm diameter and 100 mm high were prepared. The specimens were then cured in water. The compressive strength, flexural strength, and splitting tensile strength tests were performed on the cube specimens, beam specimens, and cylinder specimens, respectively, at the age of 7 days.

3. Results and discussion

3.1. Properties of composed materials

The bulk density and the specific gravity of sand, diatomaceous earth binder, and iron ore binder used in this study are summarized in Table 1. The particle size distribution of solid materials in this study is shown in Figure 1. The particle size of the sand ranges from 0.575 µm to 766.2 µm; of the Portland cement ranges from 0.136 µm to 88.48 µm; of the diatomaceous earth binder ranges from 0.107 µm to 231 µm, and of the iron ore binder ranges from 0.107 µm to 231 µm. The specific gravity of the iron ore binder is almost the same as that of Portland cement. However, the specific gravity of diatomaceous earth is lower than that of Portland cement.

Table 1. Density and the specific gravity of sand, diatomaceous earth binder, and iron ore binder.

Materials	Bulk density (kg/m^3)	Specific gravity
Sand	1482	2.453
Diatomaceous earth binder	928	2.180
Iron ore binder	1211	3.571

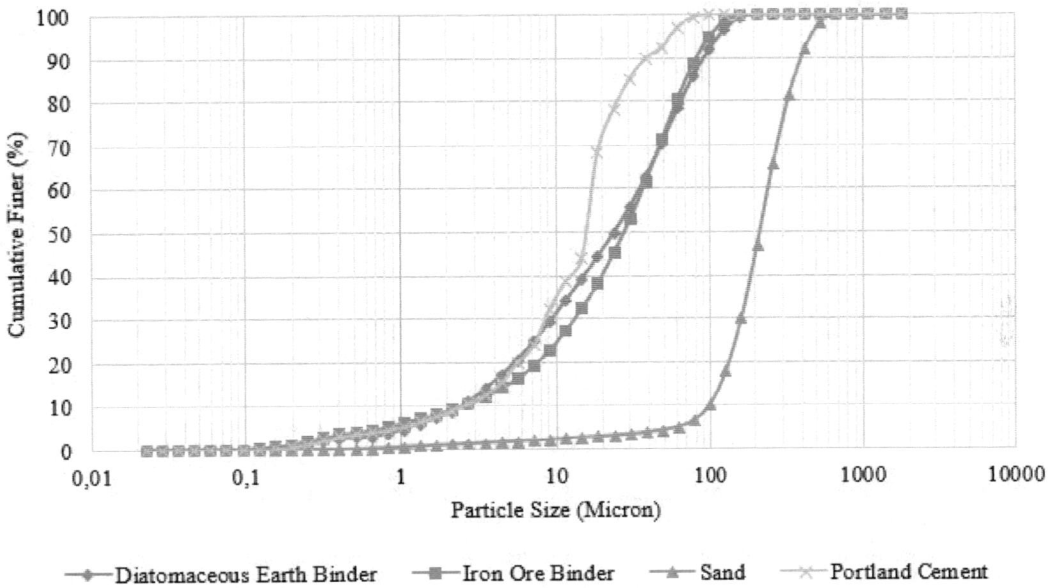

Figure 1. The particle size distribution of the materials.

3.2. Mix proportion of RPC

The mix proportion of RPC is shown in Table 2. In mixture #1, the weight of the iron ore binder was found of 10 % of the weight of cement. Since the weight of cement in mixture #2 and #3 was reduced and the weight of iron ore binder was maintained constant, the iron ore binder to cement ratio increased. The comparisons between mixture gradation and target curve based on Modified Andreasen and Andersen Model for mix #1, mix #2, and mix #3 are shown in Figure 2, Figure 3, and Figure 4, respectively. The gradings of all mixture closed to the target curve. Even though the weight of cement was replaced by 5% and 10% for mix #2 and mix #3, the good grading as mix #1 was found. For a particle size less than 10 μm, a little bit different which is about 3 % was found. This means that the mixture in this study has less particle finer than 10 μm. The use of finer particles in the RPC mixture with calcined diatomaceous earth powder will be studied in the future.

Table 2. The mix proportion of materials for 1 m^3 fresh RPC.

Materials	Mix #1 (kg)	Mix #2 (kg)	Mix #3 (kg)
Cement	872.00	828.40	784.80
Sand	1220.80	1220.8	1220.8
Diatomaceous earth binder	0.00	43.60	87.20

Iron ore binder	87.20	87.20	87.20
Water	174.40	174.40	174.40
Superplasticizer	13.08	13.08	13.08

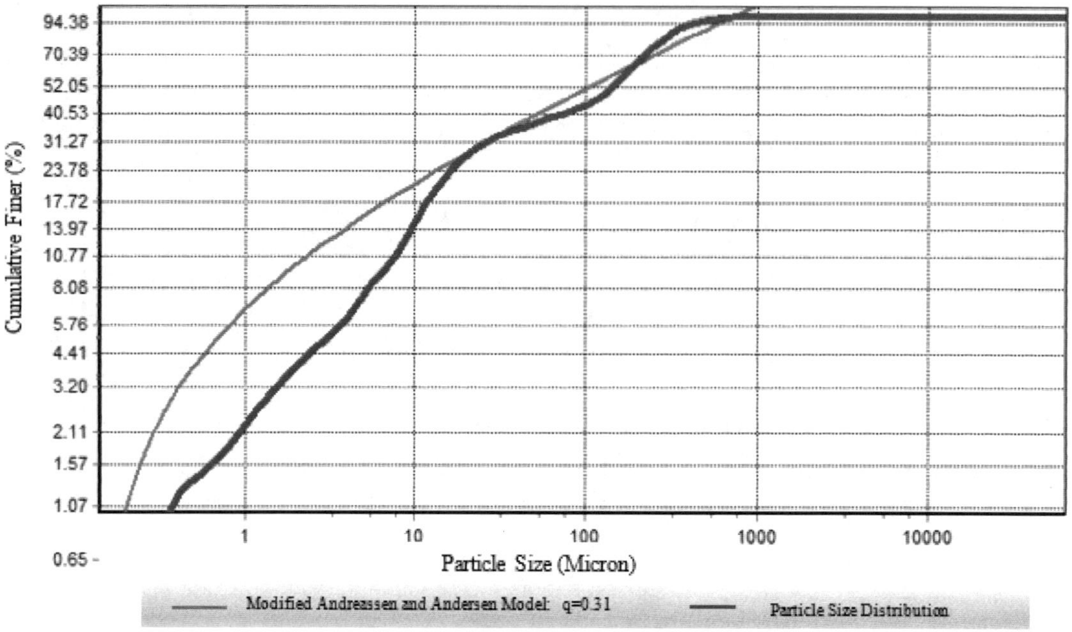

Figure 2. Comparison between the mix #1 grading and the target curve.

3.3. Properties of fresh RPC

The properties of fresh RPC are shown in Table 3. Replacing the partial cement content with diatomaceous binder causes the reduction of the relative slump. This is because of the higher absorption of the diatomaceous binder. The fresh RPC with diatomaceous binder has lower air content.

Table 3. Properties of fresh RPC.

Fresh RPC Properties	Mix #1	Mix #2	Mix #3
Relative slump flow	3.91	3.49	3.20
Air content (%)	0.81	0.74	0.76

3.4. Properties of hardened RPC

The average compressive strength, flexural strength, and splitting tensile strength of all mixtures at the age of 7 days are shown in Table 4. The RPC with diatomaceous earth powder has the lower 7 days compressive strength, but high flexural strength and splitting tensile strength. The higher flexural strength and splitting tensile strength in concrete make concrete structures experience cracking at a greater load [27]. The concrete strength at the age of 28 days and more will be studied in the future.

IOP Conf. Series: Materials Science and Engineering **933** (2020) 012007 doi:10.1088/1757-899X/933/1/012007

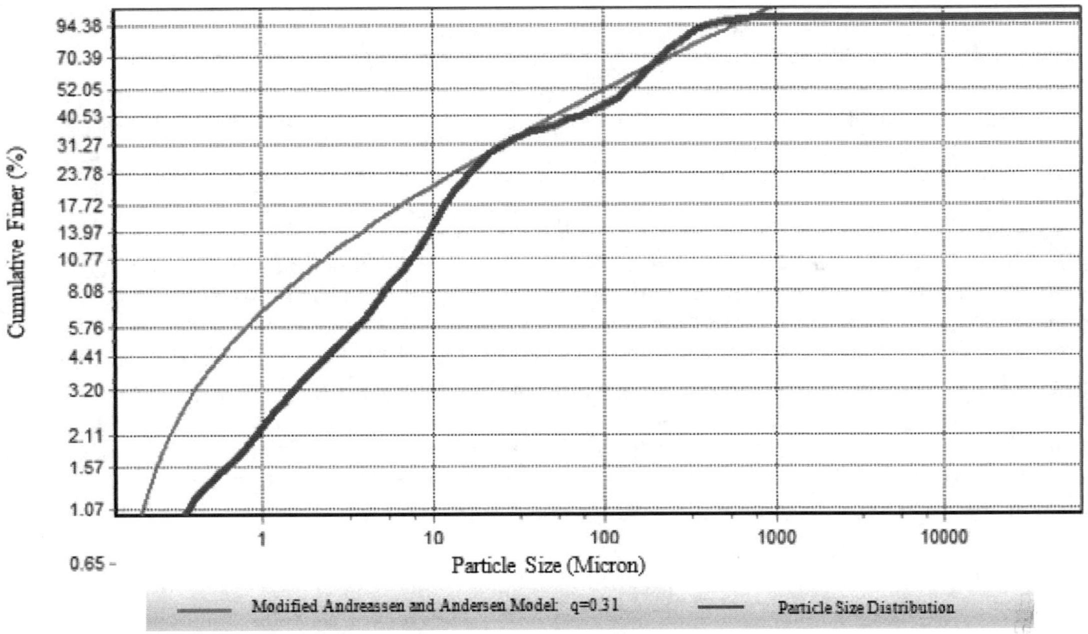

Figure 3. Comparison between the mix #2 grading and the target curve.

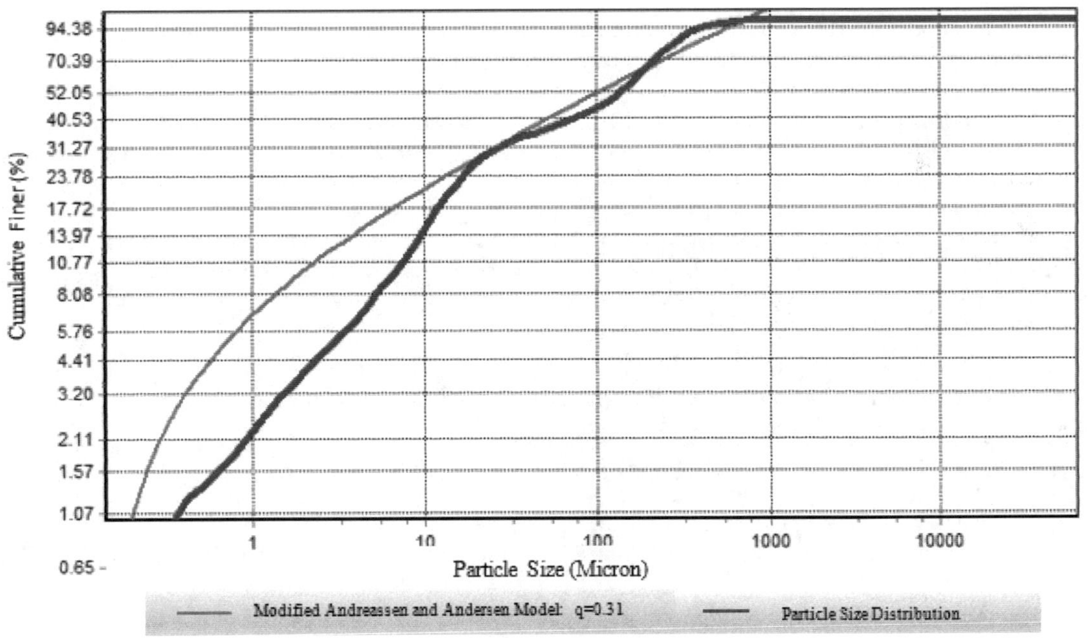

Figure 4. Comparison between the mix #3 grading and the target curve.

Table 4. Seven days average strength of RPC.

Hardened RPC Properties	Mix #1	Mix #2	Mix #3
Compressive strength (MPa)	80.21	65.41	72.73
Flexural strength (MPa)	7.63	8.15	7.78
Splitting tensile strength (MPa)	10.13	11.06	9.37

4. Conclusion

The mix design of reactive powder concrete (RPC) containing diatomaceous earth binder and iron ore binder is studied. The proportion of the materials was determined based on the Modified Andreasen and Andersen Model. Without using the diatomaceous earth binder, the cement content was 873 kg/m^3 concrete volume. The iron ore binder was 10% of the cement content. The cement was then replaced by diatomaceous earth by 5% and 10%. The good correlation between the mixture gradation and target curve was found except for the particles finer than 10 μm, which have a difference of about 3%. The adding of diatomaceous earth binder reduces the workability, air content, and 7 days compressive strength of RPC, but increases 7 days flexural strength and splitting tensile strength. The concrete strength at the age of 28 days and more will be studied in the future.

Acknowledgments
This work was supported by a grant provided by the Directorate of Research and Community Service, Ministry of Research, Technology, and Higher Education of the Republic of Indonesia under Contract No. 099/SP2H/LT/DPRM/2019 and LPPM Universitas Syiah Kuala under Contract No. 12/UN11.2/PP/SP3/2019.

References

[1] Chindaprasirt P, Homwuttiwong S and Jaturapitakkul C 2007 *Constr. Build. Mater.* **21** 1492-9
[2] Aulia T B, Muttaqin, Afifuddin M, Zaki M and Merriza S 2018 *IOP Conf. Series: Mater. Sci. and Eng.* **434** 012199
[3] Hasan M, Husin S and Nursaniah C 2016 *Key Engineering Materials* **711** 302-9
[4] Hamzani, Munirwansyah, Hasan M and Sugiarto S 2019 *IOP Conf. Series: Mater. Sci. And Eng.* **523** 012034
[5] Richard P and Cheyrezy M 1995 *Cem. Concr. Res.* **25** 1501-11
[6] Matte V and Moranville M 1999 *Cem. Concr. Compos.* **21** 1-9
[7] Yazici H, Yardimci, M Y, Yigiter H, Aydin S and Turkel S 2010 *Cem. Concr. Comp.* **32** 639-48
[8] Tuan N V, Ye G, Breugel K V, Fraaijm A L A and Dai B D 2011 *Constr. Build. Mater.* **25** 2030-5
[9] Wang C, Yang C, Liu F, Wan C and Pu X 2012 *Cem. Concr. Compos.* **34** 538-44
[10] Van V T A, Robler C, Bui D D and Ludwig H M 2014 *Cem. Concr. Compos.* **53** 270-8
[11] Rossi P 2013 *Cem. Concr. Compos.* **37** 246-8
[12] Park S H, Kim D J, Ryu G S and Koh K T 2012 *Cem. Concr. Compos.* **34** 172-184
[13] Hassan A M T, John S W and Mahmud G H 2012 *Constr. Build. Mater.* **37** 874-82
[14] Eldieb A S 2009 *Mater. Des.* 30 4286-92
[15] Tayeh B A, Abu Bakar B H, Megat Johari M A and Voo V I 2012 *Constr. Build. Mater.* **26** 538-48
[16] Yu R, Spiesz P and Brouwers H J H 2014 *Cem. Concr. Res.* **56** 29-39
[17] Fung J E and Dinger D R 1994 *Predictive Process Control of Crowded Particulate Suspensions, Applied to Ceramic Manufacturing*, Kluer Academic Publisher, Boston
[18] De Larran F and Sedran T 1994 *Cem. Concr. Res.* **24** 997-1009
[19] De Larran F and Sedran T 2002 *Cem. Concr. Res.* **32** 1688-1704
[20] Fennis S A A M, Walraven J C and Uijl D A 2009 Heron **54** 185-204

[21] Fuller W B and Thompson S E 1907 *Trans. Am. Soc. Civ. Eng.* **33** 222-98
[22] Andreasen A H M and Andersen J 1930 *Kolloid Z.* **50** 217-28
[23] Husken G 2010 *PhD Thesis* (Eindhoven: Eindhoven University of Technology)
[24] Hunger M 2010 *PhD Thesis* (Eindhoven: Eindhoven University of Technology)
[25] Yu Q L, Spiesz P and Brouwers H J H 2013 *Cem. Concr. Comput.* **44** 17-29
[26] Spiesz P Yu Q L and Brouwers H J H 2013 *Cem. Concr. Comput.* **44** 30-40
[27] Hasan M, Husaini and Abdullah N M 2019 *IOP Conf. Series: Mater. Sci. and Eng.* **523** 012037

Characteristic of calcined diatomaceous earth from Aceh Besar District – Indonesia as cementitious binder

M Hasan[1], T Saidi[1], A Muyasir[1], Y R Alkhaly[2] and M Muslimsyah[3]

[1] Department of Civil Engineering, Universitas Syiah Kuala, Banda Aceh, Indonesia
[2] Department of Civil Engineering, Universitas Malikussaleh, Lhokseumawe, Indonesia
[3] Department of Architecture, Universitas Syiah Kuala, Banda Aceh, Indonesia

Corresponding e-mail: muttaqin@unsyiah.ac.id

Abstract. Diatomaceous earth is a naturally occurring, soft, siliceous sedimentary rock, which is easily crumbled into a fine powder. In this paper, the characteristic of calcined diatomaceous earth from Aceh Besar District – Indonesia as a cementitious binder is studied. The chunk of diatomaceous earth was burned in the brick burning furnace for 4 days, then ground so that it becomes a powder. The powder was sieved to form the binder having the particle sizes less than 250 μm. The binder then calcined at the temperature of 600 °C for 5 hours in a laboratory furnace. The density, specific gravity, absorption, particle size analyzer, X-ray diffraction, X-ray fluorescence, and scanning electron microscopy tests were therefore performed. The test results show that the density, saturated surface dry specific gravity, and absorption of calcined diatomaceous earth binder are 767 kg/m^3, 2.000, and 6.535 %, respectively. The particle sizes range from 0,107 μm to 231 μm. The main minerals of the particle are graphite, quartz, and lead-silver thallium antimony sulphide. The main chemical compositions of calcined diatomaceous earth are SiO_2, 56.00 %; Fe_2O_3, 26.40 %; ZrO_2, 7.20 %; and Al_2O_3, 6.50 %.

1. Introduction

Many studies have lead-silver on the use of geopolymer such as fly ash, palm oil ash, rice husk ash, ground granulated blast furnace slag as cementitious materials for cement replacement in concrete production [1-7]. In recent years, the study on alkali-activated of such geopolymer materials is intensively conducted [8-12]. Diatomaceous earth, which is the silica-rich material, is one of the natural geopolymers and can be used as the supplementary cementing material. The properties of the blended cement containing diatomaceous earth have been studied [13-16]. The properties of concrete with diatomaceous earth as a partial substitution of cement have also been studied [17].

In Aceh Besar District – Indonesia, the deposit of the diatomaceous earth is quite high with an estimation of 40,353,700 tons weight. Since the main composition, the silica content and the structure of diatomaceous earth widely vary depend on its origin, the study of the characteristic of diatomaceous earth from Aceh Besar District – Indonesia as cementitious material is essential. This study aims to characterize the physical properties of calcined diatomaceous earth binder originated from Aceh Besar District – Indonesia such as density, specific gravity, absorption, particle size distribution as well as its mineralogical, crystal structure and chemical composition.

2. Methodology

The diatomaceous earth used in this study was collected from Beureunuet Village, Aceh Besar District, Indonesia, in the form of the chunk. The location where the material was collected is shown in figure 1.

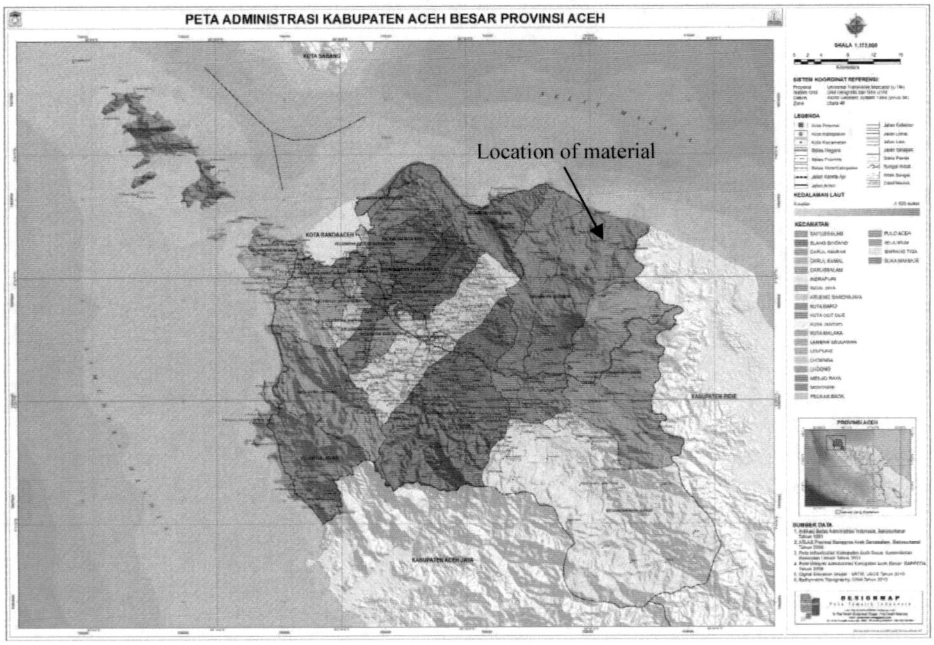

Figure 1. Map of the location where the material was collected.

2.1. Preparation of material

The chunk of diatomaceous earth was burned in the brick burning furnace for 4 days. When the temperature of diatomaceous earth chunk returned to room temperature, the material was ground in the Los Angeles test device. The material was then sieved to form a binder with the size of the particle less than 250 μm. The diatomaceous binder was calcined in the laboratory furnace at the temperature of 600 °C for 5 hours for used as a cementitious material.

2.2. Density test

The density of calcined diatomaceous binder was tested according to ASTM C188-17 [18]. Three samples were prepared. The sample was filled in a container that has a volume of 1.552 liters. The weight of the sample and container was measured as well as the weight of the container only. The density of the sample was then calculated as follows:

$$\text{Density} = (W_t - W_c)/V_c \tag{1}$$

where W_t = weight of sample and container, W_c = weight of the container, and V_c = volume of the container.

2.3. Specific gravity and absorption tests

The specific gravity and absorption of the calcined diatomaceous binder were tested on 3 samples using a pycnometer. The specific gravity was tested on the saturated surface dry (SSD) and oven-dry (OD) conditions.

2.4. Particle size analyzer (PSA) test

Particle size analyzer (PSA) test was performed to know the particle size distribution of calcined diatomaceous binder using MicroBrook 2000 L PSA test device manufactured by Brookhaven Instruments Corporation with laser diffraction method and has measured scale range from 0.02 μm to 2000 μm. The test was repeated 3 times on 3 different samples.

2.5. X-ray diffraction (XRD) test

X-ray diffraction (XRD) test was conducted to study the crystal structure and mineralogical characterization of calcined diatomaceous earth binder. In this study, a Maxima XRD-7000 device manufactured by SHIMADZU was used. XRD device uses X-ray radiation from the XRD tube with the length of the wave of 0.154184. The resulted X-ray diffraction will be in the form of X-ray diffraction intensity and the 2θ angles.

2.6. X-ray fluorescence (XRF) test

The chemical composition of calcined diatomaceous earth binder was characterized by X-ray fluorescence (XRF) test. An XRF Analyzer of TORONTOTECH TT-EDXPRT.XRF was used in this study for the XRF test.

2.7. Scanning electron microscopy (SEM) test

The surface texture (topography) and morphology (particle shape and size) of calcined diatomaceous earth binder were analyzed by using scanning electron microscopy (SEM) image. A JSM-6510 LA device manufactured by JEOL was used in this study.

3. Results and discussion

In this section, the characteristics of calcined diatomaceous earth obtained from the tests, as mentioned above, are described and discussed.

3.1. Density

The density of calcined diatomaceous earth binder originated from Aceh Besar District – Indonesia obtained from this study is summarized in table 1. Table 1 shows that the calcined diatomaceous earth has a low density compared to cement (density = 3150 kg/m^3). The low density of calcined diatomaceous earth has an implication on more high volume need when it used as cement replacement in concrete production. However, if the same volume of replaced cement is used, the concrete produced with the calcined diatomaceous earth has a low density compared with normal concrete.

Table 1. The density of calcined diatomaceous earth binder.

Sample number	The density of calcined diatomaceous earth (kg/m^3)
#1	760
#2	773
#3	767
Average	767

3.2. Specific gravity and absorption

The specific gravity (SSD and OD) and absorption of calcined diatomaceous earth binder obtained from this study are summarized in table 2. Table 2 shows that calcined diatomaceous earth binder has lower specific gravity compared with Portland cement and other cementitious materials. The absorption of calcined diatomaceous earth binder is 6.535 % which causes the more water need in mixing concrete with calcined diatomaceous earth compared to concrete with only original Portland cement.

3.3. Particle size distribution

The particle size distribution of calcined diatomaceous earth binder used in this study is shown in figure 2. The particle sizes range from 0.107 μm to 231 μm. More than 50 % of the sample has a particle diameter less than 33.90 μm. Based on the particle size distribution, the calcined diatomaceous earth binder in this study can be classified as microparticles that can be used for cement replacement in concrete production.

Table 2. Specific gravity and absorption of calcined diatomaceous earth binder.

Sample number	Specific gravity		Absorption (%)
	SSD	OD	
#1	2.000	1.880	6.383
#2	2.000	1.880	6.383
#3	2.000	1.872	6.838
Average	2.000	1.877	6.535

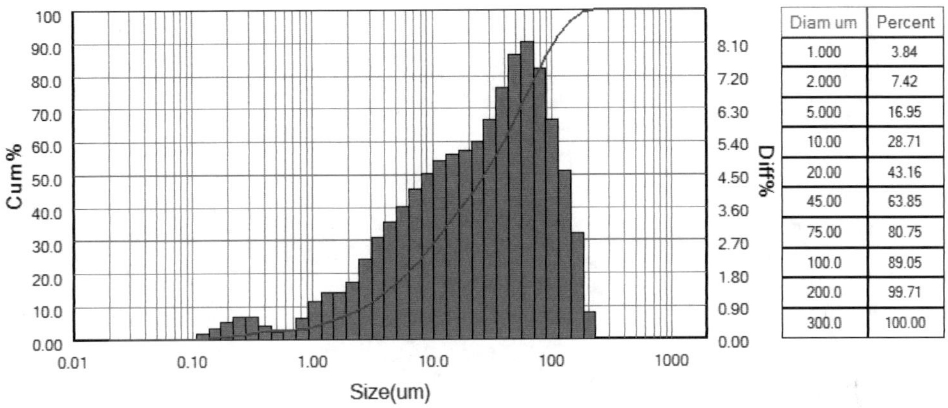

Figure 2. The particle size distribution of calcined diatomaceous earth binder.

3.4. Crystal structure and mineralogical characteristics

The XRD pattern of calcined diatomaceous earth binder is shown in figure 3. The strongest 3 peaks occur at 2θ = 27.755°, 26.583°, and 29.901°. The XRD pattern shows that graphite, quartz, and lead-silver thallium antimony sulphide are the main crystalline minerals of calcined diatomaceous earth binder. In addition, an amorphous silica phase (opal A) is also present. The other minerals present in the calcined diatomaceous binder are Plustite, lead, hydroxyl, calciobet, tousonit, algodonite, lueshite, macedonite, delafossite, lenoblite, krutaite (cobalti), eyl, frankdicksonite, khamrabaevite, dyscrasite, sorosite, zirconia, kurulite, tennantite, kerstenite, schreyerite, ikunolite, tetrahedrite, sein, berthie, toyohaite, sakharovai, bismutite, brunogeierite, bunsenite, schoenfliesite, wickmanite, cuprospinel, rhenium, Plustedtite, tellurium, tellurobismuthite, mushistonite, ferrucsite, arsen, ceryopiro, ferrowodginite, ferryhidrid, tsumoite, and zairite.

3.5. Chemical composition

The chemical composition of the calcined diatomaceous earth binder is shown in table 3. Based on the chemical composition, the binder can be classified as a cementitious binder with the content of SiO_2, Fe_2O_3, and Al_2O_3 is 88.90 %. According to ASTM C618-03, the binder is classified as Class N natural pozzolan [19].

IOP Conf. Series: Materials Science and Engineering **933** (2020) 012008 doi:10.1088/1757-899X/933/1/012008

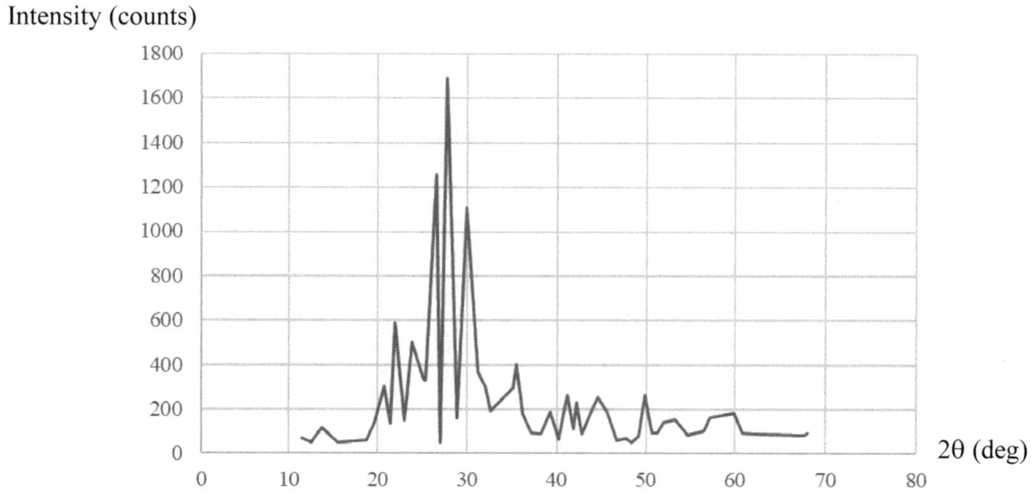

Figure 3. XRD pattern of calcined diatomaceous earth binder.

Table 3. Chemical composition of calcined diatomaceous earth binder (%).

SiO_2	56.00
Fe_2O_3	26.40
ZrO_2	7.20
Al_2O_3	6.50
TiO2	1.50
Sb_2O_3	1.30
MnO_2	0.29
ZnO	0.27
SnO_2	0.20
NiO	0.11
Cu_2O	0.05
PbO_2	0.02

3.6. Scanning electron microscopy (SEM) images

The SEM images of calcined diatomaceous earth binder are shown in figure 4. The shape of calcined diatomaceous earth particles is cylindrical with a circular cellular structure. The diameter of the cellular is around 2 μm. In addition, it also contains 15 μm size cube-shaped particles.

4. Conclusion

Based on the characteristic found in this study, the calcined diatomaceous earth binder from Aceh Besar District – Indonesia can be used as a cementitious material. The characteristics are as follows:

- The density of calcined diatomaceous earth is 767 kg/m^3.
- The saturated surface dry and oven-dry specific gravity is 2.000 and 1.877, respectively.
- The particle size has a diameter ranging from 0.107 μm to 231 μm. More than 50 % of the sample has a particle diameter less than 33.90 μm.

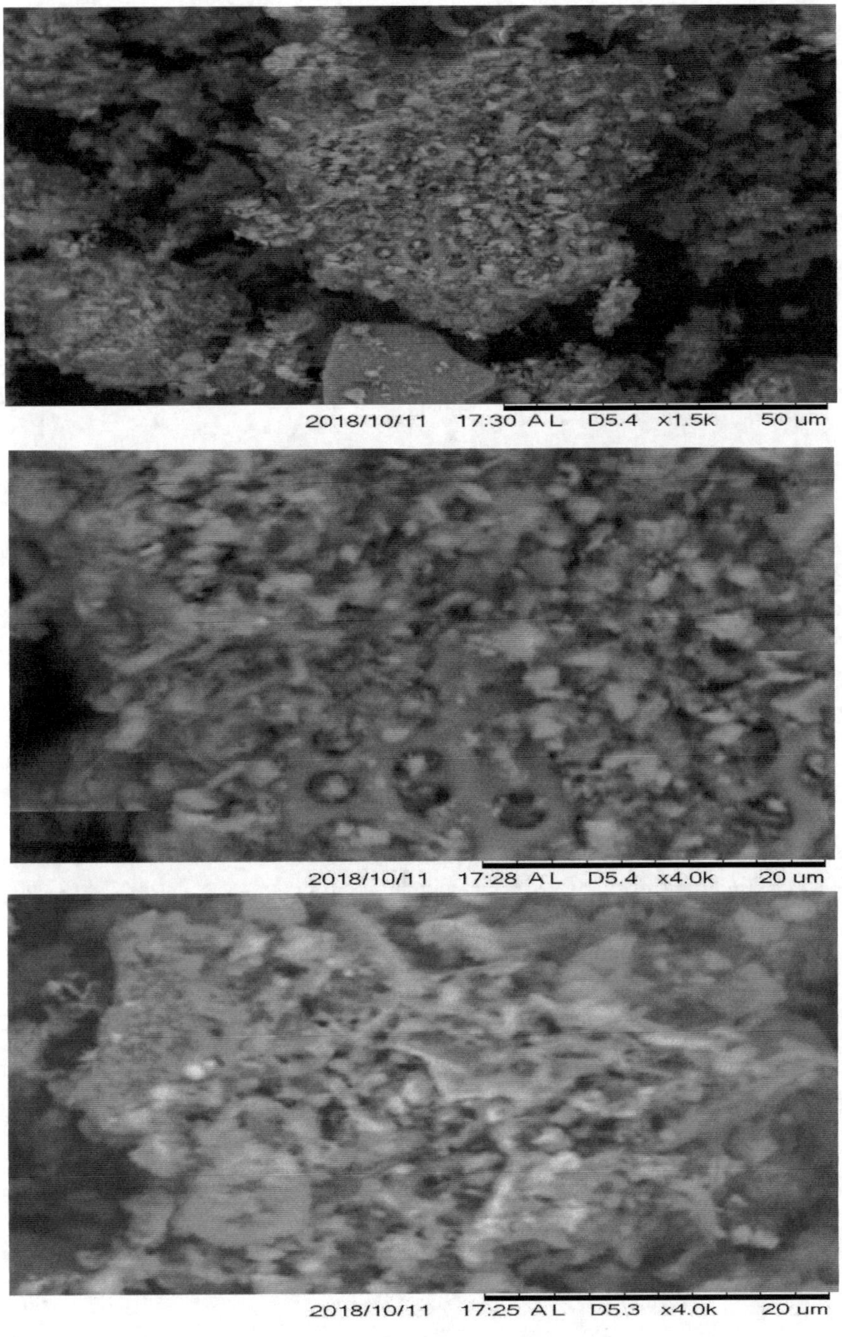

Figure 4. SEM images of calcined diatomaceous earth.

- The main minerals in calcined diatomaceous earth binder are graphite, quartz, and lead-silver thallium antimony sulphide. In addition, an amorphous silica phase (opal A) is also present.
- The main chemical compositions of calcined diatomaceous earth binder are SiO_2, 56.00 %; Fe_2O_3, 26.40 %; ZrO_2, 7.20 %; and Al_2O_3, 6.50 %.

- The shapes of calcined diatomaceous earth particles are cylindrical with 2 μm circular cellular and 15 μm size cube.

Acknowledgments
This work was supported by a grant provided by the Directorate of Research and Community Service, Ministry of Research, Technology, and Higher Education of the Republic of Indonesia under Contract No. 099/SP2H/LT/DPRM/2019 and LPPM Universitas Syiah Kuala under Contract No. 12/UN11.2/PP/SP3/2019. The authors thank Miss Nurul Husna Zubir, Miss Asmaul Husna Zubir, Miss Raudha Azzahra, Miss Putroe Nadilah Rahman, Miss Rifa Risky Ayunizar, Mr. Mahlil, Mr. M. Nasir and Mr. Razali for their assistance in experimental work.

References
[1] Sata V, Jaturapitakkul C and Kiattikomol K 2004 *J. Mater. Civ. Eng.* **16** 623-628
[2] Chindaprasirt P, Homwuttiwong S and Jaturapitakkul C 2007 *Constr. Build. Mater.* **21** 1492-1499
[3] Haque M N and Kayali O 1998 *Cem. Concr. Res.* **28** 1445-1452
[4] Aulia T B, Muttaqin, Afifuddin M, Zaki M, and Merriza S 2018 *IOP Conf. Series: Mater. Sci. and Eng.* **434** 012199
[5] Hamzani, Munirwansyah, Hasan M and Sugiarto S 2019 *IOP Conf. Series: Mater. Sci. and Eng.* **523** 012034
[6] Poon C S, Lam L and Wong Y L 2000 *Cem. Concr. Res.* **30** 447-455
[7] Hasan M, Husin S and Nursaniah C 2016 *Key Engineering Materials* **711** 302-309
[8] Luukkonen T, Abdollahnejad Z, Yliniemi J, Kinnunen P and Illikainen M 2018 *Cem. Concr. Res.* **103** 21-34
[9] Provis J L 2018 *Cem. Concr. Res.* **114** 40-48
[10] Myers R J, Bernal S A and Provis J L 2017 *Cem. Concr. Res.* **95** 30-38
[11] Ding Y, Dai J G and Shi C J 2016 *Const. Build. Mater.* **127** 68-79
[12] Kaur K, Singh J and Kaur M 2018 *Const. Build. Mater.* **169** 188-192
[13] Kastis D, Kakali G, Tsivilis S and Stamatakis M G 2006 *Cem. Concr. Res.* **36** 1821-1826
[14] Stamatakis M G, Fragoulis D, Csirik G, Bedelean I and Pedersen S 2003 *Cem. Concr. Compos.* **25** 177-184
[15] Fragoulis D, Stamatakis M G, Papageorgiou D and Chaniotakis E 2005 *Cem. Concr. Compos.* **27** 205-209
[16] Papadakis V G and Tsimas S 2002 *Cem. Concr. Res.* **32** 1525-1532
[17] Tagnit-Hamou A, Petrov N and Luke K 2003 *ACI Mater. J.* **100** 73-78
[18] ASTM C188-17 *Standard Test Method for Density of Hydraulic Cement*
[19] ASTM C618-03 *Standard Specification for Coal Fly Ash and Raw or Calcined Natural Pozzolan for Use in Concrete*

IOP Conf. Series: Materials Science and Engineering **933** (2020) 012009 doi:10.1088/1757-899X/933/1/012009

Analysis of unsignalized intersection upgrading at constrained area in the city of Banda Aceh

R Taufiqy[1], M Isya[2], Y Darma[2] and Thamalkhani[3]

[1] Civil Engineering Postgraduate Program, Department of Civil Engineering, Universitas Syiah Kuala, Banda Aceh, Indonesia
[2] Department of Civil Engineering, Universitas Syiah Kuala, Banda Aceh, Indonesia
[3] Department of Civil Engineering, Universitas Muhammaddiyah Aceh, Banda Aceh, Indonesia

Corressponding e-mail: rachmattaufiqy21@gmail.com

Abstract. The city of Banda Aceh is the capital city of Aceh province. The city's population is 259,913, and the city's annual population growth rate is 1.97% (BPS, 2018). Punge intersection is one of the intersections in Banda Aceh; this intersection is passed by the numerous and dense traffic. Based on the observation in certain hours at Punge intersection, it reveals several problems, namely: the intersection capacity and the performance level of the intersection. As a result, this study is aimed to evaluate the performance of Punge intersection. The intersection calculation was based on the Indonesia Highway Capacity Manual (MKJI, 1997). The benefit that can be obtained from this research is to eliminate traffic conflict. According to the survey, it was discovered that the dominant traffic (3,053 pcu/hours) was at the Sultan Iskandar Muda Street, D, from the central city to Ulee Lheu; on the contrary, only a few traffic (559 pcu/hours) passed on Cempaka Punge Jurong Street, A. By way of conclusion, the intersection capacity is inadequate because the degree of saturation has exceeded 0.85, viz 1.851. Therefore, Simpang Punge needs some improvements, such as repairing traffic lights and its geometric.

1. Introduction

The highway is one of the infrastructures for smoothing traffic either in a city or rural area. One part of the highway considered vitally important to be analyzed and evaluated is the intersection. The intersection is the meeting point of the highway network; this is because at the intersection often creates various traffic barriers.

The city of Banda Aceh is the capital city of Aceh province, with 259,913 citizens and an annual population growth rate reached 1.97% [1]. The origin-destination transportation occurred at Punge intersection was partly caused by the highway access permit of the city of Banda Aceh, access to offices, schools, historical sites, and tourist attractions. Such conditions cause a high concentration of movement to the Punge intersection area. The spatial pattern formed in that intersection consists of mixed areas filled with various types of activities such as trade in services, offices, public facilities, and settlements. At Punge intersection area, there is a node meeting for three secondary arterial roads, namely: Sultan Iskandar Muda Street, Punge Blang Cut Street, and Cempaka Punge Jurong Street.

From the mentioned problem, the purpose of the study is to evaluate the performance of the Banda Aceh Punge intersection to avoid traffic conflicts and discover intersection capacities, degree of saturation, delay and queue opportunities in 4-arm signal intersections, as well as solutions or alternative

IOP Conf. Series: Materials Science and Engineering **933** (2020) 012009 doi:10.1088/1757-899X/933/1/012009

solutions for traffic congestion problem at the intersection. The benefit of this research is to eliminate traffic conflicts disrupting traffic flow and driver safety. This is because there is no adequate infrastructure for road users at the intersection.

The scopes of the research are the capacity and performance level (degree of saturation and delay). Intersection calculations are based on the Indonesian Road Capacity Manual (MKJI 1997). Data retrieval is conducted for 3 days, namely: Monday, Thursday, and Sunday, in which the peak hours are in the morning office hours (07:00 – 09:00 WIB), noon (13:00 – 15:00 WIB), and afternoon (17:00 – 19:00 WIB).

2. Study literature

2.1. Intersection

The intersection is a point at which traffic flow conflict occurred. This is because the intersection commonly is a buildup of vehicles, especially during peak hours; this can cause congestion and accidents due to meeting one vehicle with another vehicle from the opposite direction [2]. The type of intersection based on the method of arrangement can be grouped into 2 (two) types, namely:

2.1.1. Unsignalized intersection. Intersections that do not use traffic signals. At this intersection road, users must decide whether they are safe enough to cross the intersection or must stop before passing the intersection [3].

2.1.2. Signalized intersection. At signalized intersections, road users can pass intersections according to the operation of traffic signals. So road users can pass it only when the traffic signal shows green on the arm of the junction.

Based on the type, the intersection, in general, consists of a plot of an intersection, intersection with splitting road lanes without a ramp, and interchanges. A plot of intersection is the intersection where two or more highways joined in which each highway leads out from a junction and forms part of it. These roads are called intersections [4].

2.2. The intersection point of conflict

In the intersection area, vehicles and pedestrians will intersect at a point of conflict; this conflict will reduce the movement and a potential location for accidents.

Traffic flow affected by the conflict at an intersection has a complex behavior: any movement turn left, turn right, or straight, each facing a different conflict and behavior associated with the movement [5]. The point and the kind of maneuver can be seen in figure 1 below.

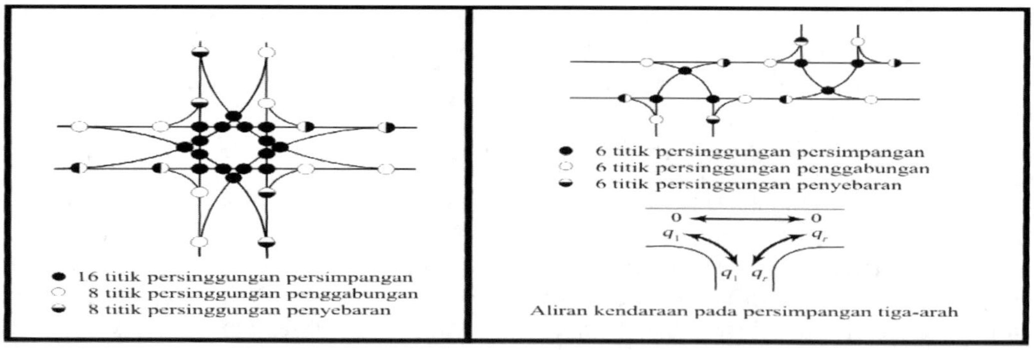

Figure 1. Potential points on the intersection [2].

2.3. Intersection setting objectives

The main purpose of setting the intersection is to maintain the security of the current flowing, to provide clear instructions, so it does not cause any doubt. Traffic control at intersections can be achieved using traffic lights, markers and signs, directions, and traffic islands [6].

From the selection of the intersection, it can be found the objectives that are as follows:

1. Both reducing and avoiding the possibility of accidents from various conflict point conditions.
2. Maintaining the capacity of the intersection so that the intersection can be achieved in its operation as planned.
3. In its operation of a deviation arrangement, it must provide clear and definite simple instructions that direct traffic in a suitable and safe place.

2.4. Level of service (LoS)

The level of service is a measure of road work performance or intersection that is calculated based on the level of road use, speed, density, and obstacle occurred [7]. LoS can be seen in the following table:

Table 1. Criteria level of services at unsignalized intersection [2].

Level Of Service (*Los*)	Capacity reserve (*Cr*)
A	> 400
B	300 – 399
C	200 – 299
D	100 – 199
E	0 – 99
F	-

2.5. Unsignalized-intersection capacity

The capacity is the maximum traffic flow that can be maintained on a section of road under specific conditions expressed in vehicles/hour or pcu/hour [2].

The total capacity of an intersection can be expressed by multiplying the basic capacity (Co) and adjustment factors (F). The formulation according to MKJI intersection capacity (1997) is written as follows:

$$C = C_o \times F_W \times F_M \times F_{CS} \times F_{RSU} \times F_{LT} \times F_{RT} \times F_{MI} \quad (1)$$

Information:

C = Actual capacity (according to existing conditions),
Co = The value of the base capacity,
FW = Approach width adjustment factor,
FM = The adjustment factor for the main street,
FCS = The adjustment size of the city factor,
FRSU = The adjustment types of side barrier road environment and unmotorized vehicles factor,
FLT = Turn Left adjustment factor,
FRT = Turn Right adjustment factor, and
FMI = Minor road flow factor.

2.6. Degree of saturation (DS)

The degree of saturation (DS) is the ratio of actual traffic flow (pcu/hour) to capacity (pcu/hour). The most appropriate way to assess the results of the calculations that we do is to look at the degree of

IOP Conf. Series: Materials Science and Engineering **933** (2020) 012009 doi:10.1088/1757-899X/933/1/012009

saturation (DS) for the observed condition. If the saturation obtained exceeds the received value (DS> 0.85), it is necessary to correct the geometric of an intersection, control the total drift current and adjust the current with the signs to maintain the desired degree of saturation (DS <0.85) [8]. The formulation according to MKJI degree of saturation (1997) is written as follows:

$$DS = \frac{Qtot}{C} \tag{2}$$

Information:
DS = Degree of saturation,
C = Capacity (pcu / hour), and
Qtot = real total current (pcu / hour).

2.7. Delay (D)
The delay at the intersection is the total average of time obstacles experienced by the vehicle when passing an intersection [6]. This obstacle arises if the vehicle stopped because the queue is deviated until the vehicle exit the intersection due to the influence of inadequate intersection capacity. The delay value affects the value of the travel time of the vehicle. The higher the value of delay, the higher the travel time.
Delays in traffic are divided into 5 types, namely:

1. Average traffic delay for all intersections (DTi)
   DTi traffic delay (seconds/pcu) is the average delay for all vehicles entering the intersection. Delay (DTi) is determined from the empirical relationship between delay (DTi) and degree of saturation.
   - For DS < 0.85 ;
   $$DTi = 2 + 8,2078 \text{ x } Ds - (1 - Ds) \text{ x } 2 \tag{3}$$
   - For DS > 0.85 ;
   $$DTi = 1,0504 / (0,2742 - 0,2042 \text{ x } Ds) - (1 - Ds) \text{ x } 2 \tag{4}$$
2. Average traffic delay for major roads (DT_{MA})
   Traffic delays on average for a major road traffic delay is the average for all vehicles coming in through the major intersection.
   - For DS < 0.85:
   $$DT_{MA} = 1.8 + 5.8234 \text{ x } Ds - (1 - Ds) \text{ x } 1.8 \tag{5}$$
   - For DS > 0.85:
   $$DT_{MA} = 1.05034 / (0.346 - 0.246 \text{ x } Ds) - (1 - Ds) \text{ x } 1.8 \tag{6}$$
3. Traffic delays average minor road (DT_{MI})
   The minor road traffic delay is determined based on the average traffic delay (DT_i) and the major road traffic delay (DT_{MA}).
   $$DT_{MI} = (Q_{TOT} \text{ x } D_{TI} - Q_{MA} \text{ x } DT_{MA}) / Q_{MI} \tag{7}$$
4. Geometric delay intersection (DG)
   Geometric delay intersection is the average geometric delay of all motorized vehicles entering the intersection. DG is calculated using the equation:
   - For DS $\leq$ 1.0 :
   $$DG = (1 - DS) \text{ x } (PT \text{ x } 6 + (1 - PT) \text{ x } 3) + DS \text{ x } 4 \tag{8}$$
   - For DS $\geq$ 1.0 :
   $$DG = 4 \text{ seconds/pcu} \tag{9}$$
5. Intersection delay (D)
   The intersection delay is calculated using the following equation:
   $$D = DG + DT \tag{10}$$

2.8. Opportunities queue (QP%)

The opportunities queue (QP %) is the possibility of a queue with two or more vehicles in any approaching area at an unsigned intersection. The limit of the opportunity queue value (%) is determined from the empirical relationship between the opportunity queue (%) and the degree of saturation (DS) [9]. Opportunities queue with an upper limit and a lower limit can be obtained using the following formula:

Upper limit:

$$QPa = (47.71 \text{ x DS}) - (24.68 \text{ x } DS^2) + (56.47 \text{ x } DS^2) \quad (11)$$

Lower limit:

$$QPb = (9.02 \text{ x DS}) + (20.66 \text{ x } DS^2) + (10.49 \text{ x } DS^2) \quad (12)$$

Information:

DS = Degree of saturation,
Qpa = Opportunity for upper queue (%), and
QPb = Opportunity for lower queue (%).

3. Research methods

3.1. Data processing and analysis methods

Data needed to analyze capacity and delays at the intersection of Punge are as follows:

1. Data on traffic volume for each group of crossings during rush hour
2. Geometric data of the road (width and number of lines)
3. Data about environmental conditions and land use in the intersection area.

Data collection was conducted at the Punge Banda Aceh intersection. At the intersection, there are 4 (four) arms. Each arm is placed in a post consisting of 3 (three) personnel. Each person must record the volume of the vehicle moving straight, the vehicle turning right, and the vehicle turning left, all on each crossing arm. For the calculation of non-signaled intersection, the MKJI 1997 form is used with the respective functions including:

1. Step A: input data.
   a. Geometric conditions.
   b. Traffic conditions.
   c. Environmental conditions.
2. Step B: capacity.
   a. The width of the approach and the type of the intersection.
   b. Basic capacity.
   c. The factor of median adjustment for the main road.
   d. The factor of city size adjustment.
   e. Adjustment factors for environmental types, side barriers, and non-motorized vehicles.
   f. The adjustment factor for turning left.
   g. The adjustment factor for turning the right.
   h. The minor road flows adjustment factor
3. Step C: level of performance.
   a. Degree of saturation.
   b. Delay

4. Results and discussion

4.1. Geometric intersection

This research was conducted by reviewing the volume of traffic flow vehicles on Sultan Iskandar Muda Street, Punge Blang Cut Street, Punge Jurong Street, and Punge intersection, in the city of Banda Aceh. The location is in the west of Banda Aceh which leads to tourist attractions, schools, and others. For more details, the location can be seen in figure 2. The intersection of geometric measurement results obtained directly on a research location by measuring the width of the effective arm closers on each sleeve using a tape meter.

The roads that are studied in this research have different types. The main road, which is Sultan Iskandar Muda Street (from Ulee Lheu to the city center) has the type of 6/2D road and secondary arterial road. In terms of Sultan Iskandar Muda Street (from the direction of the city to Ulee Lheu), the type is 2/1UD and secondary arterial roads. For Punge Blang Cut Street (minor), the type is 2/1UD roads and secondary arterial roads. For Punge Blang Jurong Street (minor), the type is 2/1UD and secondary arterial roads.

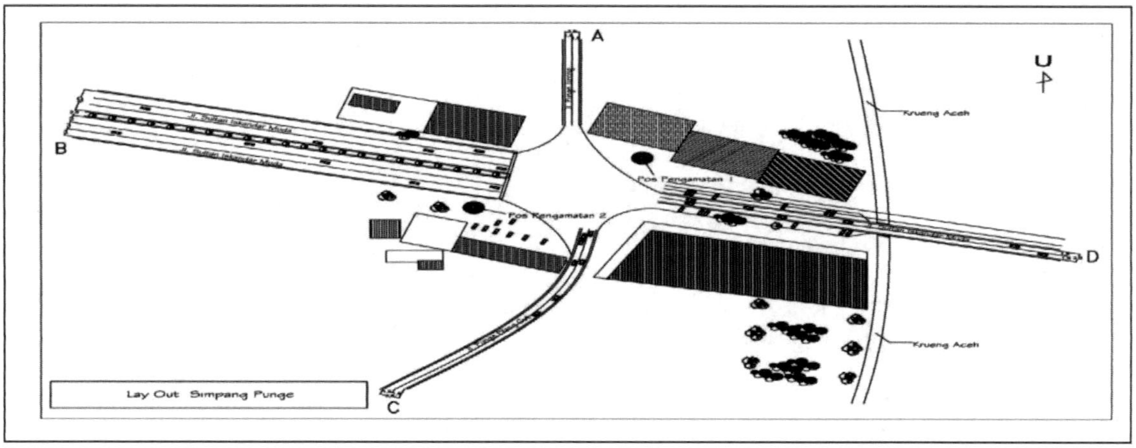

Figure 2. Geometric conditions of the Punge intersection.

4.2. Volume of traffic flow

Based on the results of a survey conducted for 3 (three) days, the peak traffic flow volume falls on Sundays, November 1, 2015, afternoon, at 17:00 – 19:00 WIB. The information is presented in table 2.

Table 2. Traffic flow volumes for each crossing arm.

Intersection arm	**The volume of traffic flow**							
	MC		**LV**		**HV**		**UM**	
	number of vehicles	**(%)**	**number of vehicles**	**(%)**	**number of vehicles**	**(%)**	**number of vehicles**	**(%)**
A	514	92	40	7	5	1	0	0
B	2,028	68	953	32	5	0	1	1
C	876	87	129	13	4	0	0	0
D	2,283	75	755	25	6	0	0	0

From table 2 above, at the intersection, the road most passed by vehicle is the Sultan Iskandar Muda Street (city center to Ulee Lheu), which coded as D. The most dominant vehicle passing the street is the motorized vehicle (MC) that number is 2,283 pcu/hour or 75% of total vehicles. The road, with the least amount of vehicles passing it, is Cempaka Punge Jurong Street, coded as A. The most dominant vehicle passing the street is the motorized vehicle (MC), which is 514 pcu/hour or 92%. The above table shows that the motorcycle (MC) still dominates on each arm closers while heavy vehicles (HV) are the smallest

IOP Conf. Series: Materials Science and Engineering **933** (2020) 012009 doi:10.1088/1757-899X/933/1/012009

percentage number of vehicles on each arm closers.

4.3. Calculation of survey data

The survey data of Sunday (afternoon), November 1, 2015 and the comparison are as follows:

1. The intersection capacity

a. Width of the approach

$W_I = (a/2 + b + c/2 + d/2) / 4$ (the median exists in the B arm)

$= (1.8/2 + 9 + 1.8/2 + 1.8/2) / 4$

$W_I = 2.93$.

a. Basic capacity (Co)

Because the intersection of Punge is not currently compatible with the type of intersection in MKJI (1997), the author considers that the type of intersection of Punge is suitable to the type of 422 which basic capacity value (Co) is 2,900.

b. Adjacent adjustment factor (F_W)

FW value is obtained from the calculation of the average width-types closers with 422.

$F_W = 0.70 + 0.0866 \times W_I$

$= 0.70 + 0.0866 \times 2.93$

$= 0.953$.

d. The main road median adjustment factor (F_M) is less than 3 m. As a result, the median type is narrow. F_M is 1.05.

e. The size of the city of Banda Aceh is 249,282 people; the city size class (F_{CS}) is 0.1 – 0.5 million people (a small category). The value (F_{CS}) is 0.88.

f. For adjustment factor type of commercial road environment, the obstacles $= \frac{UM}{mv} = 0{,}00$, which is a medium side. Unmotorized vehicles (P_{UM}) is 0.00, so F_{RSU} is 0.94.

g. Turning left adjustment ratio (F_{LT})

$$F_{LT} = 0.84 + 1.61 \times P_{LT} \quad (13)$$

Information ;

F_{LT} = Turning left adjustment ratio

P_{LT} = Turning left input variables (Form-I, using Line 20, Column 11).

Then:

$F_{LT} = 0.84 + 1.61 \times P_{LT}$

$F_{LT} = 0.84 + 1.61 \times 0.14$

$= 1.058$

h. The turning right adjustment ratio (P_{RT}) is 1.

i. The closers adjustment factor for the current minor road (F_{MI}).

This condition is related to the code of 422 intersections with PMI values of 0.1845.

$F_{MI} = 1.19 \times P_{MI}^2 - 1.19 \times P_{MI} + 1.19$

$= 1.19 \times 0.1845^2 - 1.19 \times 0.1845 + 1.19$

$= 1.011$.

j. Capacity (C)

$C = C_o \times F_W \times F_M \times F_{CS} \times F_{RSU} \times F_{LT} \times F_{RT} \times F_{MI}$

$= 2{,}900 \times 0.953 \times 1.05 \times 0.88 \times 0.94 \times 1.058 \times 1 \times 1.011$

$C = 2{,}569$ pcu/hour

The maximum capacity value at the intersection is 2,569 pcu/hour, which does not exceed the basic capacity of 2900 pcu/hour.

2. Level of performance

Level of performance includes:

a. The degree of saturation (DS)

Q_{tot} = 4,754 pcu/hour.

$DS = \frac{Qtot}{C} = \frac{4754}{2569} = 1.851$

From the calculation results, the DS value has exceeded the required DS value, namely DS > 0.85.

b. Delay (D)

• Delay traffic intersection

- For DS > 0.85

DTi = 1,0504 / (0.2742 – 0.2042 x Ds) – (1 – Ds) x 2

DTi = 1,0504 / (0.2742 – 0.2042 x 1.851) – (1 – 1.851) x 2

DTi = -8.43 det/smp.

The minus value (-) in calculating the traffic delay (DT_I) is obtained because the formula provided in MKJI is only valid for DS ≤ 1.2. Thus, the value of the actual delay is estimated to exceed 30 seconds/pcu.

• Delay the main street on traffic (DT_{MA})

- For DS > 0.85

DT_{MA} = 1.05034 / (0.346 – 0.246 x Ds) – (1 – Ds) x 1.8

DT_{MA} = 1.05034 / (0.346 – 0.246 x 1.851) – (1 – 1.851) x 1.8

DT_{MA} = -8.08 det/smp.

• Delay minor road on traffic (DT_{MI})

DT_{MI} = (Q_{TOT} x D_{TI} – Q_{MA} x DT_{MA}) / Q_{MI}

Information ;

Q_{MA} = taken from the cabbage USIG-I form.

Q_{MI} = taken from the cabbage USIG-I form.

DT_{MI} = (Q_{TOT} x D_{TI} – Q_{MA} x DT_{MA}) / Q_{MI}

DT_{MI} = (2,569 x -8.43 – 3,876.5 x -8.08) / 877

DT_{MI} = -9.96 det/smp.

value minus (-) in the calculation of the main road traffic delays (DT_{MA}) obtained for DT_I and DT_{MA}.

• Delay geometric of intersection (DG)

- For DS > 1.0 then:

DG = 4

• Intersection Delay (D)

D = DG + DT_I = 4 + (-8,43) = -4.43 second/pcu.

The minus value (-) in calculating the traffic delay (D) is obtained because the DT_I value is minus (-).

3. Opportunities queue (QP%)

Upper limit : QPa = (47.71 x DS) – (24.68 x DS^2) + (56.47 x DS^2)

= (47.71 x 1.851) – (24.68 x 1.851^2) + (56.47 x 1.851^2)

QPa = 326 %.

Lower limit : QPb = (9.02 x DS) + (20.66 x DS^2) + (10.49 x DS^2)

$= (9.02 \times 1.851) + (20.66 \times 1.851^2) + (10.49 \times 1.851^2)$

QPb $= 154$ %.

By the value of QPa and QPb above 100%, it can be concluded that the queues are inevitable at this intersection.

4. Level of services

The level of service at the intersection of Punge is categorized as B, which is the stable flow characteristics (speed slightly limited by traffic, the volume of services is used for the design of the outside lane of the city). Based on calculations, the intersection of Punge has a very high DS value which already exceeds 1.851 and DS that exceeds 0.85. If the width of the approach of Sultan Iskandar Muda Street (city center to Ule Lheu), coded as D, is changed or equal to the width of the approach to Sultan Iskandar Muda Street (Ule Lheu to city center), coded as B, the DS value still does not meet the requirements (DS> 0.85).

5. Conclusions and suggestions

5.1. Conclusions

From the analysis carried out, the following conclusions are obtained:

1. For traffic volume on Sunday (afternoon at 17:00 – 19:00 WIB), the highest number of vehicles is 3,053 pcu/hour, which passed on Sultan Iskandar Muda Street (city center to Ulee Lheu), and the road having the least traffic was the Cempaka Punge Jurong Street, with a total number of 559 pcu/hour.
2. On Sunday, the highest DS degree of saturation was occurred in the afternoon, at 17:00 – 19:00 WIB, which is equal to 1.851.
3. The maximum capacity value is 2,569 pcu/hour; it does not exceed the basic junction capacity that is equal to Co of 2,900 pcu/hour.
4. At the Banda Aceh Punge intersection, there is a chance of queuing at certain hours, especially in the afternoon.
5. Based on the results and observations obtained, the Punge Banda Aceh intersection is unable to accommodate traffic flows with conditions that do not have traffic control arrangements.

5.2. Suggestions

From the analysis carried out, the authors obtained some suggestions as follows:

1. A better traffic control system is needed at the intersection, such as repairment of the marker lines and installment of traffic signs.
2. It is necessary to reduce the conflict point at the intersection, by increasing the number of lanes at the intersection, the type of arrangement at the intersection, and the number of the direction of movement.
3. The personal awareness of drivers in driving vehicles and knowing traffic regulations, especially at the unsignalized intersections, is needed.
4. The junction capacity needs to be maintained.

References

[1] Badan Pusat Statistik Aceh 2018 *Pedoman Pendataan Survei Penduduk Antar Sensus 2018* (Banda Aceh)

[2] Direktorat Jendral Bina Marga 1997 *Manual Kapasitas Jalan Indonesia (MKJI)* (Jakarta: Department of Public Works)
[3] Morlok E K 1984 *Introduction to Civil Engineering and Transportation Planning* (Jakarta: Erlangga Publisher)
[4] Khisty C J and Kent L B 2005 *Transportation Engineering, An Introduction* (Pearson Education) 3rd edition
[5] Haryanto et al. 2015 *Performance Analysis of the Unsignalized Intersection Jln. Inspeksi Sal. Kalimalang-Jln Cipinang Indah Raya II, Kalimalang, East Jakarta* (Jakarta: Civil Engineering Study Program Faculty of Engineering University of Borobudur)
[6] Hobbs F D 1995 *Traffic Planning and Engineering* (Yogyakarta: Gajah Mada University Press)
[7] Tamin O Z 2008 *Planning, Modeling, and Transportation Engineering* (Bandung: ITB)
[8] Sumarni H 2013 *Performance Analysis of 3-Arm unsignalized Intersection jl. Urip Sumoharjo-Jl. Gunung BawaKaraeng-Jl. Mesjid Raya* (Makassar: Faculty of Civil Engineering Hassanuddin University)
[9] Wahyudin 2017 *Analysis of the Performance of the 3-Arm Unsignalized Intersection (Case Study: T-junction Jl. Pakuningratan Yogyakarta)* (Yogyakarta: Faculty of Engineering Muhammadiyah University Yogyakarta)

IOP Conf. Series: Materials Science and Engineering **933** (2020) 012010 doi:10.1088/1757-899X/933/1/012010

Use of bamboo fiber on the relationship between compressive strength and split tensile strength of high strength concrete

A Yusra[1], T Triwulan[2], M Safriani[1] and M Ikhsan[1]

[1] Engineering Department, Universitas Teuku Umar, Meulaboh, Indonesia
[2] Engineering Department, Institute of Technology Sepuluh November, Surabaya, Indonesia

Corresponding e-mail: yusra.andi@yahoo.com

Abstract. This study was to determine the influence of fiber bamboo on the split tensile strength of concrete and to know the relationship between compressive and tensile strength. In the concrete mixture, fibre bamboo is added with a percentage of 0%, 0.5%, 1% and 1.5% of cement. Concrete strength of the design is 60 MPa with *w/c* value of 0.25. The coarse aggregate used had a maximum diameter of 12 mm, while the plasticizer used was 2% of cement. Testing was carried out at 28 days. For 0% fiber bamboo, the obtained strength was 62.47 MPa and a split tensile strength was 4.15 MPa. For 0.5% fiber, the obtained strength was 59.83 MPa and a split tensile strength was 5.12 MPa. For 1% fiber, the obtained strength value was 59.07 MPa and a split tensile strength was 5.91 MPa. For 1.5% fiber, the obtained strength value was 54.92 MPa and a split tensile strength was 6.48 MPa. Using fiber bamboo of the test shows a decrease in compressive strength, where the use of 0.5% fiber bamboo in concrete was the most optimal value. The use of the highest-fiber bamboo content of 1.5% achieved the optimum value of split tensile strength of 6.48 MPa.

1. Introduction

One method to improve concrete performance is to add fibers to it to increase the tensile strength of the concrete. Concrete fiber is made from a mixture of cement, fine aggregate, coarse aggregate, and small amounts of fiber. Various types of fiber materials that can be used to improve the mechanical properties of concrete are steel fibers, glass fibers, polypropylene fibers (a variety of high-quality plastic), carbon and natural fibers derived from natural materials (natural fiber), such as palm fiber, coconut fiber, hemp fiber, bamboo fiber, and others. This research used bamboo fiber. Bamboo fiber has an excellent tensile strength, which is expected to increase the elasticity of high-quality concrete and be more resistant to cracking and increases the tensile strength of concrete.

The addition of filler is expected to cover the cavity or pore between the aggregate particles. The filled cavity increases the concrete area so that the concrete ability to withstand the load can be perfect. One of the materials that are used for fillers is palm shell ash because of the high content of silica (SiO2) used as a material pozzolan in concrete mixtures. Palm shell ash must pass sieve no. 200 in order to be used as a filler for the porosity cavity. The results also showed that the optimum compressive strength using palm shell ash was at the content of the additive of 15% [1]. The concrete quality used is $f'c = 60$ MPa with cylindrical specimens diameter of 150 mm and height of 300 mm. The purpose of this study was to determine the extent of the effect of bamboo fiber on the split tensile strength of high-quality

IOP Conf. Series: Materials Science and Engineering **933** (2020) 012010 doi:10.1088/1757-899X/933/1/012010

concrete and to find out how the relationship between compressive strength and high-strength concrete tensile strength.

2. Methodology

2.1. Bamboo

Bamboo is a plant that grows fast and can be harvest at around three years old. During growth, bamboo can grow vertically 5 cm per hour or 120 cm per day. In bamboo, there is the fiber in bamboo culms, causing the bamboo to have the strength and the ability for building purposes, namely high tensile strength [2]. Some types of bamboo having high tensile strength can be seen in table 1.

Table 1. Tensile strength of bamboo [2].

Type of Bamboo	Tensile Strength (Mpa) Without Nodia	Tensile Strength (Mpa) With Nodia
Ori	291	128
Petung	190	116
Wulung	166	147
Tutul	216	74

Bamboo generally has low natural durability, although there are differences regarding the type. Bamboo can be attacked easily by destructive organisms such as dry wood powder, dry wood termites, and subterranean termites. Furthermore, it is stated that anatomically and chemically bamboo and wood have similarities; therefore, the factors that influence wood will also affect the properties of bamboo. These properties include water content and the specific gravity of water [3]. Bamboo skin used as fiber concrete material is based on the consideration that the tensile strength is quite high, making from raw fiber materials is quite easy, and the bamboo population is quite large and scattered so that it is easily obtained [4]. Bamboo has physical properties, as shown in table 2.

Table 2. Physical properties of bamboo [5].

Physical properties	Bamboo Tali	Bamboo Hitam	Bamboo Kuning	Bamboo Andong	Bamboo Betung	Bamboo Ampel
Results	0.54	0.69	0.72	0.53	0.51	0.75
Quality Classification	Medium	Low	Low	Medium	Medium	Low
Value	4	3	2	5	6	1

2.2. High strength concrete

Concrete is a function of its constituent material consisting of hydraulic cement (portland cement), coarse aggregates, fine aggregates, water, and added materials (admixture or additive). Concrete is a set of mechanical and chemical interactions of its constituent element [6]. At present, the concrete is said to be the high-quality concrete if the compressive strength is above 50 MPa [7]. Several properties of high-quality concrete strength are shown in table 3.

The high-quality concrete is the concrete that has a strength higher than 6000 Psi or 41.4 MPa and is used to reduce the size of columns and beams to be more beneficial in a longer span and can ease the structure. For the nature of concrete itself, it is said that high-quality concrete has a unit weight higher than low-quality concrete [8].

Table 3. Strength properties of various high strength concrete [8].

Type	w/c	Compressive Strength (28 days old)	Remark
Normaly Consitency	0.35 - 0.40	35 – 80 Mpa	Slump 50 – 100 mm
No – Slump	0.30 – 0.45	35 – 50 Mpa	Slump > 25 mm
w/c low	0.20 – 0.35	100 – 170 Mpa	Use Admixture
Compacted	0.05 – 0.30	70 – 240 Mpa	Strength > 70 Mpa

2.3. *Fibre concrete*

Fibre Concrete is the concrete made from a mixture of cement, fine aggregates, coarse aggregates, and fiber at a small amount. Fibre material that can be used to improve the concrete properties is the fiber from natural materials, hemp, or fiber from other plants [9]. Concrete fiber is one of the developments of concrete technology. This type of concrete is one of the solutions to improve the quality of concrete [9]. Whereas according to [10], fiber concrete can be interpreted as concrete consisting of constituent materials such as water, cement, aggregate, or other additives and added fiber in the mixture. Fibre divided into four types of fiber, namely:

a. Fibre steel, for example, iron fiber and stainless steel.
b. Polymeric fiber, for example, nylon and fiber polypropylene.
c. Fiber mineral, for example, fiberglass.
d. Natural fibers, for example, coconut fiber, sugar cane fiber, banana tree fiber, and bamboo fiber.

Concrete is obtained from mixing fine and coarse aggregate materials, namely sand, stone, broken stone, or other similar material, by adding enough cement, and water as supporting material for chemical reactions during the concrete hardening and maintenance process [11]. Based on [12], concrete is a mixture of cement or other hydraulic cement, fine aggregates, coarse aggregates, and water, with or without additives which form a solid period. Normal concrete has a weight content of 2200 - 2500 kg/m^3 using natural aggregates which are broken down or not broken down. The aggregate content in concrete is very high, especially coarse aggregate or gravel, in which almost 78% becomes the main filler of concrete mixtures [8].

The good nature of concrete is if the concrete has a high compressive strength between 20 - 50 MPa, at the age of 28 days. In other words, it can be assumed that the quality of concrete is reviewed only from its compressive strength [13]. The value of compressive and tensile strength of concrete material is not directly proportional; every effort to improve the quality of compressive strength is only accompanied by a small increase in the value of tensile strength. The tensile strength of normal concrete material only ranges from 9% - 15% from its compressive strength [11].

2.4. *Palm shell ash*

Palm shell ash, the waste from the burning of palm oil shells, also contains inorganic cations such as potassium and sodium [14]. In the process of burning shells and fiber fruit at temperatures of 700 - 800 °C in the boiler furnace kitchen. The burning of shells and fruit fiber produces hard white - grey crust due to high-temperature combustion with 61% silica content. This high silica content makes the boiler crust ash can be used as a material pozzolan in concrete mixtures [14]. According to the results of the study [15], fiber and palm kernel shell waste in the form of ash has a useful element to increase mortar strength.

2.5. *Concrete tensile strength*

The value of compressive and tensile strength of concrete material is not directly proportional; every effort to improve the quality of compressive strength is only accompanied by a small increase in the value of tensile strength. A rough estimate can be used that the tensile strength of normal concrete materials is only between 9% -15% of its compressive strength (11).

The tensile strength of concrete is relatively low, a good approach to calculating the tensile strength of concrete (fct) is by using the formula of 0.1 f'c <fct <0.2 f'c [6]. There are a number of methods available to test tensile strength, and the most commonly used is the cylinder division test or the Brazilian test. This experiment is carried out by giving a load perpendicular to the longitudinal axis with the cylinder placed horizontally on the plate of the experimental machine. The tensile strength of concrete can be calculated by equation 1 (12).

$$f_{ct} = 2P / \pi ld \tag{1}$$

Where:
f_{ct} = Split tensile strength (MPa);
P = Crushed load (N);
l = Length of the test object in the depressed part (mm); and
d = Diameter (mm).

2.6. Research method

2.6.1. Material preparation and procurement. Palm shell ash was obtained from one of the palm oil processing factories in Purwodadi Village, Kuala Pesisir District, Nagan Raya Regency, Aceh. Concrete construction starts from the preparation and procurement of materials, the examination of aggregate physical properties, concrete mixing planning, concrete stirring work, concrete mortar inspection, manufacture of specimens, maintenance of test specimens and test specimens carried out in the Construction and Building Materials Department of Unsyiah. Bamboo is obtained from one of the local plantations in the Meulaboh area, West Aceh Regency. The material that will be used to make high-quality concrete is cement material, fine aggregate (sand), coarse aggregate (gravel), and water with added bamboo fiber. Then, as a comparison concrete, high quality 60 MPa concrete is also made without the addition of bamboo fiber (0% fiber bamboo). In this study, the physical and chemical properties of bamboo were not carried out. The physical and chemical properties of bamboo are based on the results of the research [10]. The water that will be used for the concrete mixture and its treatment comes from clean water obtained from the Syiah Kuala University Construction and Building Materials Laboratory. The water in this laboratory has met the standards of clean water that are suitable for use in concrete mixtures.

2.7. Mix design
It is estimated that the compressive strength of the plan is 60 MPa for cylindrical specimens of 15/30 cm using a cement water factor of 0.25. The plasticizer with a percentage of 2% of the weight of cement was used. The crude aggregate used is the broken stone with a maximum aggregate diameter of 12 mm, palm shell ash, and fly ash of 15% of the weight of cement. The percentage of bamboo fibre is 0%, 0.5%, 1%, 1.5%, of the weight of cement.

2.7.1 Work on the concrete mixture. The construction of high-quality concrete begins with mixing concrete forming materials (sand, gravel, cement, and water) and superplasticizer. Then, the material is inserted into a concrete mixer. The next job is to insert the concrete mold for 24 hours. The work on making fiber bamboo concrete is as follows: cement was mixed with coarse sand, fine sand, water, and superplasticizer so that it becomes a mixture of concrete material. Testing the compressive strength and tensile strength of the cylindrical concrete is carried out after the concrete reaches the planned age, i.e. at the age of 28. Concrete is given a vertical direction load or is parallel to the cylinder slowly until the specimen is destroyed. The total number of specimens, which is 24 cylindrical pieces (Ø15 cm, T = 30 cm) with various variations in the aggregate percentage, can be seen in table 4.

Table 4. Number of test object designs.

Type of Object Designs	Dimension	Number of Object Designs Compressive Strength (28 days)	Number of Object Designs Split Tensile Strength (28 days)
Cylinder (0% bamboo fibre)	Ø 15 x 30 cm	3	3
Cylinder (0.5% bamboo fibre)	Ø 15 x 30 cm	3	3
Cylinder (1% bamboo fibre)	Ø 15 x 30 cm	3	3
Cylinder (1.5% bamboo fibre)	Ø 15 x 30 cm	3	3

3. Results and discussion

The results obtained from the study began from examining the aggregate physical properties, planning the proportion of concrete mixtures, testing the tensile strength of concrete, and comparing the results of the test results of concrete compressive strength associated with the results of testing of split tensile strength. The results of the analysis are presented in the form of tables and graphs so that they are easy to understand. table 5 shows the proportion of one cylindrical specimen.

Table 5. Concrete mixtures for 1 test object.

Materials	Weight Based on Percentage of Bamboo Fibre			
	0%	0.5%	1%	1.5%
Portland cement (kg)	3.18	3.18	3.18	3.18
Bamboo fibre (kg)	0	0.02	0.03	0.05
Palm shell ash 15% (kg)	0.48	0.48	0.48	0.48
Fine sand 0 – 2 mm (kg)	2.34	2.32	2.30	2.28
Coarse sand 2 – 5 mm (kg)	0.78	0.77	0.77	0.76
Coarse aggregate 5 – 8 mm (kg)	1.56	1.55	1.53	1.52
Coarse aggregate 8 – 12 mm (kg)	3.12	3.10	3.07	3.04
Water (kg)	0.80	0.80	0.80	0.80
Super plasticizer 2% (kg)	0.06	0.06	0.06	0.06

The testing of the concrete tensile strength aims to obtain load data that can be carried by concrete cylindrical specimens until destruction occurs. The size of the split tensile strength is obtained from a ratio of twice the maximum load with the multiplication area between the diameter and height of the cylinder. Based on the results of testing the concrete tensile strength at the age of 28 days, the split tensile strength values are shown in table 6.

Table 6. Results of 28 days concrete split tensile strength.

Percentage Bamboo Fibre	w/c	Weight Of Specimen (kg)	Πx L x D (mm²)	P (KN)	2P (N)	(f_{ct}) (N/mm²)
0% (1)	0.25	12.77	141300	240	480000	4.09
0% (2)	0.25	12.92	141300	220	440000	3.75
0% (3)	0.25	12.93	141300	270	540000	4.60
Average				**243.33**	**486667**	**4.15**
0.5% (1)	0.25	12.80	141300	220	440000	3.75

0.5% (2)	0.25	12.84	141300	280	560000	4.77
0.5% (3)	0.25	12.99	141300	400	800000	6.82
Average				**300**	**600000**	**5.12**
1% (1)	0.25	12.90	141300	350	700000	5.97
1% (2)	0.25	13.05	141300	340	680000	5.80
1% (3)	0.25	12.97	141300	350	700000	5.97
Average				**346.67**	**693333**	**5.91**
1.5% (1)	0.25	12.93	141300	430	860000	7.33
1.5% (2)	0.25	13.06	141300	370	740000	6.31
1.5% (3)	0.25	12.86	141300	340	680000	5.80
Average				**380**	**760000**	**6.48**

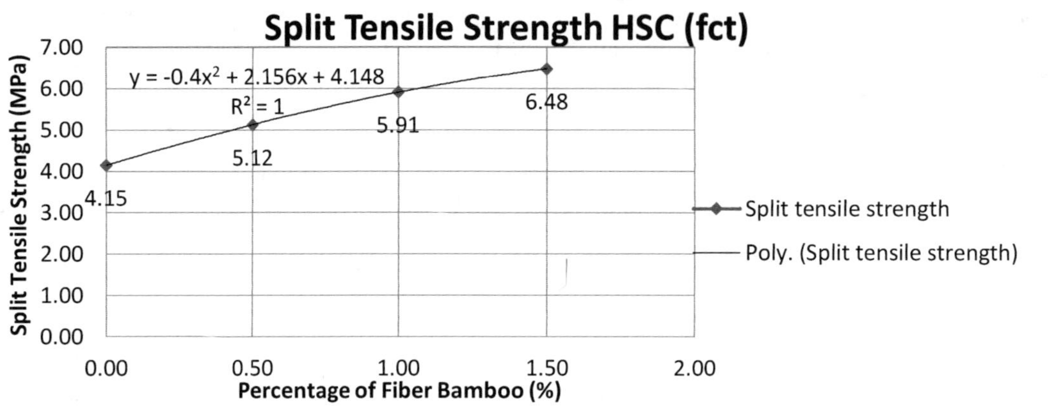

Figure 1. The graph of the concrete tensile strength at the age of 28 days.

Based on figure 1 of the graph of the concrete tensile strength at the age of 28 days, the average concrete tensile strength of concrete with the use of bamboo fiber of 0%, 0.5%, 1%, and 1.5% is 4.15 MPa, 5 MPa, 12 MPa, 5.91 MPa, and 6.48 MPa, respectively. The testing of concrete compressive strength aims to obtain load data that can be carried by concrete cylindrical specimens until destruction occurs. The compressive strength is obtained from the ratio of the maximum load to the cylinder cross-sectional area. Based on the results of testing the concrete compressive strength at the age of 28 days, the compressive strength values are shown in table 7.

Table 7. Results of 28 days concrete compressive strength.

Percentage Bamboo Fibre	w/c	Weight Content of Speciment (kg/m^3)	Cross Section Area (mm^2)	P (KN)	P (N)	(f'_c) (N/mm^2)
0% (1)	0.25	2477.36	17662.50	1080	1080000	61.15
0% (2)	0.25	2545.28	17662.50	1120	1120000	63.41
0% (3)	0.25	2512.09	17662.50	1110	1110000	62.85
Average				**1103.33**	**1103333**	**62.47**
0.5% (1)	0.25	2513.21	17662.50	1000	1000000	56.62
0.5% (2)	0.25	2515.09	17662.50	1050	1050000	59.45
0.5% (3)	0.25	2456.60	17662.50	1120	1120000	63.41

Average				**1056.67**	**1056667**	**59.83**
1% (1)	0.25	2535.85	17662.50	1120	1120000	63.41
1% (2)	0.25	2549.06	17662.50	1000	1000000	56.62
1% (3)	0.25	2509.43	17662.50	1010	1010000	57.18
Average				**1043.33**	**1043000**	**59.07**
1.5% (1)	0.25	2496.23	17662.50	1010	1010000	57.18
1.5% (2)	0.25	2520.75	17662.50	930	930000	52.65
1.5% (3)	0.25	2507.55	17662.50	970	970000	54.92
Average				**970.00**	**970000**	**54.92**

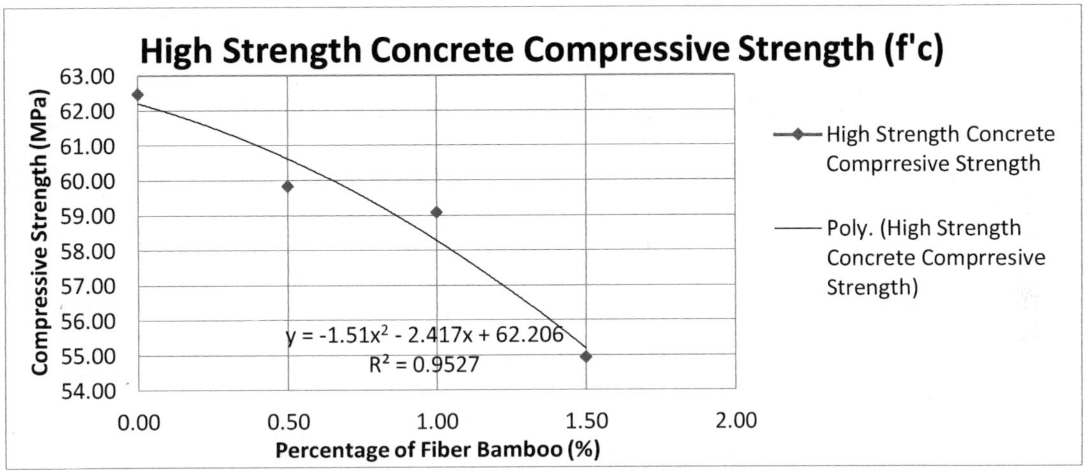

Figure 2. The graph of the compressive strength at the age of 28 days.

The value of compressive and tensile strength of concrete material is not directly proportional; every effort to improve the quality of compressive strength is only accompanied by a small increase in the value of tensile strength. A rough estimate can be used that the tensile strength of normal concrete materials only ranges between 9% - 15% of the compressive strength. Therefore, the writer wants to compare how the influence of the team's concrete compressive strength with the same proportion of mixture on the concrete tensile strength in theory. Figure 6 shows the graph of compressive strength and 28-day-old concrete tensile strength with the same mixed proposition and show a graph of the concrete tensile strength based on theoretical compressive strength of 0.1 f'c <fct <0.2 f'c [3].

Based on Graph 3, it can be seen that in testing the average compressive strength of 0% cylindrical concrete bamboo fiber is 62.47 MPa. Whereas concrete with bamboo fibre of 0.5%, 1%, 75% and 1.5% is 59.83 MPa, 59.07 MPa, and 54.92 MPa, respectively. The results of the study show a decrease in high-quality concrete compressive strength with the addition of bamboo fiber. The optimum concrete compressive strength, which is obtained with the addition of bamboo fiber at 0.5% and 1%, is 59.83 MPa and 59.07 MPa, respectively.

Based on the results of the tests, the value of the compressive strength for the split tensile strength equal to the theory cannot be obtained. Only the concrete with 1.5% bamboo fiber is equal to the theory. Based on [3], a good approach to calculating concrete tensile strength (fct) is the formula of 0.1 f'c <fct <0.2 f'c. Table 8 shows the relationship of compressive strength to the tensile strength of concrete.

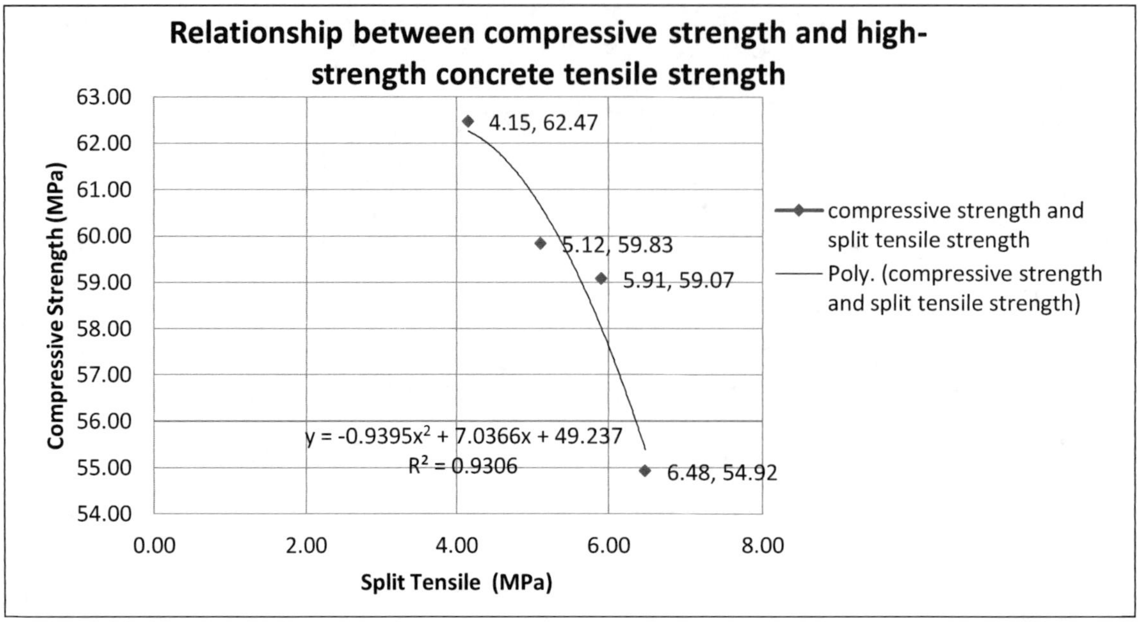

Figure 3. the graph of the relationship between compressive strength and split tensile strength.

Table 8. Relationship of compressive strength to concrete tensile strength.

Percentage of Bamboo Fibre	w/c	Compressive Strength (f'_c) (N/mm^2)	Split Tensile Strength (f_{ct}) (N/mm^2)	Tensile Strength (0.1 x f_{ct}) (Nawi, 1998) (N/mm^2)	Tensile Strength (0.2 x f_{ct}) (Nawi, 1998) (N/mm^2)
0% (1)	0.25	61.15	4.09	6.12	12.23
0% (2)	0.25	63.41	3.75	6.34	12.68
0% (3)	0.25	62.85	4.60	6.29	12.57
Average		**62.47**	**4.15**	**6.25**	**12.49**
0.5% (1)	0.25	56.62	3.75	5.66	11.32
0.5% (2)	0.25	59.45	4.77	5.95	11.89
0.5% (3)	0.25	63.41	6.82	6.34	12.68
Average		**59.83**	**5.12**	**5.98**	**11.97**
1% (1)	0.25	63.41	5.97	6.34	12.68
1% (2)	0.25	56.62	5.80	5.66	11.32
1% (3)	0.25	57.18	5.97	5.72	11.44
Average		**59.07**	**5.91**	**5.91**	**11.81**
1.5% (1)	0.25	57.18	7.33	5.72	11.44
1.5% (2)	0.25	52.65	6.31	5.27	10.53
1.5% (3)	0.25	54.92	5.80	5.49	10.98
Average		**54.92**	**6.48**	**5.49**	**10.98**

Based on table 8, it can be seen that the value of the split tensile strength obtained from the calculation does not fit into the theoretical range. The tensile strength of normal concrete without fiber is 4.15 MPa which is under the range of 6.25 – 12.49 MPa. The split tensile strength of concrete with the addition of bamboo fiber at the percentage of 0.5% is 5.12 MPa which is below the range of 5.98 – 11.97 MPa. The tensile strength of concrete with the addition of bamboo fiber at a percentage of 1% is 5.91 MPa which fits the range of 5.91 – 11.81 MPa. Also, the tensile strength of concrete with the addition of bamboo fiber at a percentage of 1.5% is 6.48 MPa which fits the range of 5.49 – 10.98 MPa.

The addition of bamboo fiber into high-quality concrete caused a decrease in the value of concrete compressive strength, but not too significant. On the contrary, the value of high-quality concrete tensile strength causes an increase in split tensile strength, even though the results of each variation have entered into the theoretical concrete tensile for 1% and 1.5% additional bamboo fiber.

4. Conclusion

Based on the results of research and data processing that has been done, several conclusions can be taken as the final results of this study:

a. From the results of the study, the use of bamboo fiber in the concrete mixture caused a decrease in the compressive strength of high-strength concrete, on the contrary, there was an increase in the quality of the concrete tensile strength due to the addition of bamboo fiber.
b. Overall, based on the testing of the concrete tensile strength at 28 days, the optimum concrete split tensile strength is recycled by substitution at 0.5%, which is equal to 6.48 Mpa.
c. Concrete with variations in the addition of bamboo fiber is still classified as high-quality concrete because it has a compressive strength greater than 50 Mpa [16].

Based on this research, the authors sincerely hope that further research on high strength concrete can be developed by adding other fiber naturals that are rattan or fiber abaca banana which are known to have high tensile strength. Then, this study should also be continued with the investigation of beam flexural capacity so that we can get how much the tensile strength of bamboo fiber contributes to the tensile strength of high-strength concrete beams.

Acknowledgments

The author would like to thank colleagues and family for the help and encouragement, Technical Staff from the Syiah Kuala University Civil Engineering Laboratory, and Teuku Umar University for financial support.

References

[1] Yusra A 2014 *Pengaruh Variasi Zat Tambah Terhadap Sifat Mekanis Beton Mutu Tinggi* (Banda Aceh: Program Pasca Sarjana Universitas Syiah Kuala)
[2] Morisco 1999 *Rekayasa Bambu* (Yogyakarta: Nafiri Offset)
[3] Liese W 1980 *Preservation of Bamboo. In Lessard, G & Chouinard, A (eds). Bamboo Research in Asia* (Canada: IDRC)
[4] Hidayat M E, Ismeddiyanto and Alex K 2016 *Pengaruh Penambahan Serat Kulit Bambu Terhadap Sifat Mekanik Beton. Jurnal* (Pekanbaru: Jurusan Teknik Sipil Fakultas Teknik Universitas Riau) **3** 1 p 2
[5] Fatriasari and Hermiati 2008 *Analisis Morfologi Serat Dan Sifat Fisis-Kimia Pada Enam Jenis Bambu Sebagai Bahan Baku Pulp dan Kertas. Jurnal Ilmu dn Teknologi Hasil Hutan* **1** 2 67-72
[6] Nawy G E 1985 *Beton Bertulang: Suatu Pendekatan Dasar* (Bandung: Eresco)
[7] Supartono 1998 *Beton Berkinerja Tinggi, Keunggulan dan Permasalahannya* (Jakarta: Seminar HAKI)
[8] Antoni and Nugraha 2007 *Teknologi Beton* (Yogyakarta: CV Andi Offset)
[9] ACI Committee544.1R-96 1996 *State Of The Art Report On Fiber Reinforced Concrete* (Farmington Hills: American Concrete Institute)
[10] Balaguru P N and Shah S P 1992 *Fiber Reinforced Composite Cement* International Ed. (Singapore: McGraw-Hill)
[11] Dipohusodo L 1994 *Struktur Beton Bertulang Berdasarkan SK SNI T-15-1991-03* (Jakarta: PT Gramedia Pusaka Utama)

[12] Badan Standarisasi Nasional 2013 SNI-2847-2013 *Tata Cara Perencanaan Struktur Beton untuk Bangunan Gedung* (Jakarta: Badan Standarisasi Nasional)
[13] Tjokrodimuljo K 2007 *Teknologi Beton* (Yogyakarta: Biro Penerbit Jurusan Teknik)
[14] Graille J, Lozano P, Pioch D and Geneste P 1985 *Essais d'alcoose d'huiles vegetales avec des catalyseurs Naturels Pour la Production de Carburants Diesel* (Oleagineux) **40** 5 217-76
[15] Muhardi, Sitompul I R and Rinaldi 2004 *Pengaruh Penambahan Abu Sawit terhadap Kuat Tekan Mortar* (Riau: Seminar Hasil Penelitian Dosen Program Studi S1 Teknik Sipil, Fakultas Teknik, Universitas Riau)
[16] Marrero R E, Soto H L, Benitez F R, Medina C and Suarez O M 2017 *Study of Hight-Strength Concrete Reinforced with Bamboo Fibers* ISBN 978-0-9975117-9-6

IOP Conf. Series: Materials Science and Engineering **933** (2020) 012011 doi:10.1088/1757-899X/933/1/012011

Risk analysis study on building projects in Pidie District

F Fahmi[1], T B Aulia[2] and M Isya[2]

[1] Master Program of Civil Engineering, Civil Engineering Department of Engineering Faculty, Universitas Syiah Kuala, Banda Aceh, Indonesia
[2] Civil Engineering Department of Engineering Faculty, Universitas Syiah Kuala, Banda Aceh, Indonesia

Corresponding E-mail: fauzan.fah@mhs.unsyiah.ac.id

Abstract. This study aims to determine the dominant risk factors occurring during the implementation of building construction and to know how to control and handle the risk factors that occur. The methodology used is a quantitative method with a questionnaire as a research instrument. The number of samples used was 55 respondents from a population of 121 companies. Based on the analysis of frequency index (FI), there are dominant risk factors that occur in the implementation of building construction in Pidie District. The dominant risk factors are skills and expertise, weather, access to the project location, unequipped workers with protective equipment, wrong implementation method, the expensive material, incomplete design data, low quality of material, damage to the tool, and a lack of workers. Furthermore, the ways to control and handle risks that occur during the construction project are increasing the trust of the owner, having knowledgable company management and competent workforce, reducing security disturbances, increasing collaboration with local village officials, minimizing equipment damage, managing material purchases before the prices rise, and mastering the implementation method.

1. Introduction

The development of construction projects in Pidie District is quite significant from year to year. This condition can be seen from the growing number of companies that are constantly increasing. The competition that occurs among contractors in obtaining work will be increasingly stringent. Therefore, companies that have good performance will continue to survive, while companies that have poor performance will continue to lag. In terms of improving performance, contractors as service providers are fully responsible for the success of a project and ensure that the development process goes according to plan. Uncertain events or conditions that occur in the field are a project risk that must be minimized. If the risk occurs, then it will have an impact both positively and negatively on the project objectives so that the conditions that occur will directly affect the performance of the company.

In 2016, besides local contractors, there were also contractors from outside the district who carried out the project in Pidie District. This condition is a form of decline in the performance of local companies that are unable to compete with other companies. Such conditions indicate that there is no increase in the performance of any local company in Pidie District. If this fact continues, then the existing local company would be inactive because one of the conditions in participating in the tender is the company's involvement in the construction project for the last three years. Therefore, a study and analysis are needed to find out the risk factors in the building construction project in Pidie District.

Based on the background, the formulation of the problems in this study are:

IOP Conf. Series: Materials Science and Engineering 933 (2020) 012011 doi:10.1088/1757-899X/933/1/012011

1. What are the dominant risk factors that occurred during the construction of building construction projects in Pidie District?
2. How to control and handle risk factors that occur during construction?

The objectives of this study are as follows:

1. To find out the dominant risk factors that occur during the construction of a building construction project in Pidie District.
2. To find out the control and the way to handle the risk factors that occur during construction

2. Literature review

2.1. Risk management

Risk management is a process of identifying, measuring, and controlling the financial risk that threatens the assets and income of a company or project; this financial risk can cause damage or loss to the company [1]. Risk management is a process of both identifying and measuring risk and forming strategies to prevent the occurrence of the risk. Risk management actions are taken by the company to respond to various risks. Risk management involves preventing and improving. The act of preventing being used to reduce, avoid, or transfer risk in the initial stages of a construction project [2].

2.2. Risk identification and classification

Risk identification and classification is a process of assessing risk and uncertainty that carried out systematically and continuously. The first step is to identify the types of risks: business risks and pure risk. Project risk is classified as pure risk. This project risk is identified based on potential sources of risk or the impact on the project objectives. The approach used in identifying this risk is the diagrams of cause and effects, namely by analyzing what will happen and the potential consequences that will be caused [3]. The most common risks found in practice are as follows [4].

1. Physical risk
   The risk relates to physical conditions, for example, the land, physical conditions that cause damage, damaged or destructive material and working result, costs of testing and sampling, unfriendly weather conditions, improper location provision, the absences of workers and staff, plant, time, and funding;
2. Risk of delay and dispute
   This risk includes the location ownership, land acquisition, late supply of information, implementation of ineffective work, delays outside the control of the parts, the disputes;
3. Risk of management and supervision
   The presence of incompetent supervisors, errors in documentation, unclear and unspecified requirements, inaccurate choice of consultants, etc;
4. Risk of loss to people and property
   Negligence and non-fulfillment of guarantees, limited time and events covered by insurance, and others;
5. Risk due to external factors
   Changes in government policies on taxes, labor, security laws and disturbances, social unrest, labor strikes, etc.;
6. Payment risk
   This risk includes devaluation, high inflation, currency fluctuations, the bankruptcy of one of the parts involved, costs for plant and management replacements, etc,; and
7. Legal risk and dispute resolution
   This risk involves the length of time for resolving disputes, injustices, uncertainty over court decisions, changes in applicable legislation, fees for judicial or arbitration proceedings, etc.

2.3. Identification of risk factors

Identifying risk factors based on the results of research conducted by previous researchers can be seen in table 1 as follows.

Table 1. Classification of risk factors in the implementation of construction.

No	Category	Risk factor	Source
1	Physical	Natural disasters	(Suharti 2009), (Sari 2011)
		Weather	(Suharti 2009), (Sari 2011)
2	Labor and Experts	Skills and expertise	(Kangari 1995), (Sari 2011)
		Discipline	(Suhani, 2009), (Sari 2011)
		Productivity	(Ahmed 1999), (Kangari 1995), (Santoso 2004), (E1 - Sayegh 2008), (Sari 2011), (Nizam 2012)
		The lack of workers	(Laia 2010), (Sari 2011)
3	Contractor	Contractor Ability	(Ahmed 1999), (Kangari 1995), (Kartam 2001), (Santoso 2004), (E1-Sayegh 2008), (Nizam 2012)
		Financial constraints Contractor	(Ahmed 1999), (Kangari 1995), (Kartam 2001), (Santoso 2004), (Nizam 2012)
4	Consultant	Design error	(Ahmed 1999), (Santoso 2004), (E1-Sayegh 2008), (Sari 2011)
		Design data is not complete	(Laia 2010), (Ahmed 1999), (Kangari 1995), (Kartam 2001), (Santoso 2004), (E1- Sayegh 2008), (Sari 2011)
		Delays in information from the planner	(Suharti 2009), (Sari 2011)
5	Owner	Owner financial failure	(Ahmed 1999), (Ahmed 1999), (Kangari 1995), (Kartam 2001), (Santoso 2004), (Sari 2011)
		Change order	(Suharti 2009), (Ahmed 1999), (Kangari 1995), (Kartam 2001), (Santoso 2004), (Sari 2011)
6	Material	Delays in delivery	(Suharti 2009), (Sari 2011), (Nizam 2012)
		Damage during shipment	(Suharti 2009), (Sari 2011), (Nizam 2012)
		Low material quality	(Suharti 2009), (Ahmed 1999), (Kangari 1995), (Kartam 2001), (Santoso 2004), (E1- Sayegh, 2008), (Sari 2011), (Nizam 2012)
7	Equipment	Number of equipment	(Suharti 2009), (Sari 2011)
		Condition of equipment that is not suitable for use	(Suharti, 2009), (Sari 2011), (Nizam 2012)
		Tool damage	(Suharti 2009), (Sari 2011), (Nizam 2012)
8	Construction Process	Quality (quality) of work is not good	(Ahmed 1999), (Kangari 1995), (Kartam 2001), (Santoso 2004), (El - Sayegh 2008), (Sari 2011)
		Communication and coordination problems	(Ahmed 1999), (Kangari 1995), (Kartam 2001), (Santoso 2004), (El - Sayegh 2008), (Sari 2011)
		Wrong implementation method	(Laia 2010), (Sari 2011), (Nizam 2012)

No	Category	Risk factor	Source
9	Project Neighborhood	Access to project location	(Suharti, 2009), (Sari, 2011), (Nizam. 2012)
		Security disorders	(Suharti 2009), (Sari 2011), (Nizam 2012)
10	Safety and Accidents	The machine is not checked before operating	(Laia 2010), (Sari 2011), (Nizam 2012)
		Workers are not equipped with protective equipment	(Laia 2010), (Sari 2011), (Nizam 2012)
		Safety regulations are violated	(Suharti 2009), (Sari 2011), (Nizam 2012)
11	Finance	The price of material is more expensive	(Suharti 2009), (Sari 2011)
		Additional equipment rental fees	(Ghosh 2004), (Sari 2011)
		Wages for workers are more expensive	(Suharti 2009), (Sari 2011)

2.4. Population and sampling techniques

Population can simply be said as all objects of research. Population value is all the values of both the results of calculations and the measurements [5]. Based on the number of members of the population, the population can be divided into limited population (finite) and unlimited population (infinite). Based on its nature, the population can be homogeneous and heterogeneous [6]. Sampling techniques are a technique for getting samples in a study so that the sample is representative of the population that represents it [7].

2.5. Frequency index (FI)

Frequency Index is calculated using equation (1) [8].

$$Frequency\ Index\ (FI) = \frac{\sum_{i=1}^{5} a_i n_i}{5N} \tag{1}$$

Information:

i : response category index (1, 2, 3, 4, and 5);
ai : weights associated with the first response value (1, 2, 3, 4 and 5 respectively);
n_i : frequency from the second response as a percentage of the total respondents for each factor; and
N : total number of respondents.

2.6. SWOT analysis

The tool used to develop strategic factors is the SWOT matrix. This matrix can clearly illustrate the interaction between the internal strategic analysis factors and external factor analysis strategies [9]. This matrix can produce four sets of possible strategic alternatives, as shown in table 2 as follows.

Table 2 Interactions between SWOT matrix factors.

	Strengths (S) Determine internal strengths factors	***Weakness (W)*** Determine internal weakness factors

Opportunities (O) Determine external factors	***Strategi SO*** Create a strategy that uses power to take advantage of opportunities	***Strategi WO*** Create strategies that minimize weaknesses to take advantage of opportunities
Threats (T) Determine external threat factors	***Strategi (T)*** Create a strategy that uses strength to overcome threats	**Strategi WT** Create strategies that minimize weaknesses and avoid threats

3. Research methodology

3.1. Research objects and locations
The research location is Pidie District, Aceh Province, which area is 3,087 km².

3.2. Primary and secondary data collection

3.2.1. Preliminary. The pre-survey was conducted to obtain secondary data, which would later become the research sample. The technique used is in the form of random sampling, in which the selection of prospective respondents is randomly selected. While secondary data is collected in the form of maps, literature studies, the company data is obtained from the institution in charge of construction service development—called LPJK. The company data is also obtained from the previous studies, which will later be used as guidelines in this study.

3.2.2. Fill out the questionnaire. The filling out of the questionnaire was carried out by the company director or project manager who worked on the building construction project in Pidie District. Questionnaires given to prospective respondents are closed; each question has been provided with an alternative answer.

4. Results and discussion

4.1. Characteristics of respondents
The characteristics of respondents, based on the position, were 30 directors and 25 project managers. Based on the respondents' highest education, senior high school alumni are 12 people, diploma alumni are 4 people, and the bachelors are 39 people. For the work experience of the respondents, 11 people have 3-5 years of experience, 21 people have 5-8 years of experience, and 23 people have more than 8 years of experience. Based on the company experience, 2 companies have 0-3 years of experience, 10 companies have 3-5 years of experience, 28 companies have 5-7 years of experience, and 15 companies have more than 7 years of experience. In terms of the project value, 11 companies have managed the project with a value of IDR 500 million - 1 billion, and 44 companies have managed the project with a value of more than IDR 1 billion.

4.2. Validity & reliability analysis
In determining the accuracy of a measuring instrument, the questionnaire was tested using validity analysis. In order to test the questionnaire feasibility, the C-alpha method was used. From the calculation results, it is obtained that the coefficient alpha value for the risk factor in the contractor is 1.01, which is higher than the requirement of 0.6. This means that the questionnaire is feasible and can be used as a tool for research.

4.3. Risk analysis

The results of the analysis of frequency index (FI) on 11 variables can be seen in table 3.

Table 3. Analysis of frequency index (FI).

No.	Variable	Number of answers					FI	Interpretation of scores
		VNI	NE	LI	TE	VI		
		1	2	3	4	5		
I	Physical							
P1	Natural disasters	18	6	13	6	12	0.56	Less influential
P2	Weather	0	0	17	2	36	0.87	Very influential
		Average FI					0.71	
II	**Labor and Experts**							
P3	Skills and expertise	0	0	3	20	32	0.91	Very influential
P4	Discipline	18	6	13	6	12	0.56	Less influential
P5	Productivity	3	4	31	8	9	0.66	Take effect
P6	The lack of workers	0	4	29	16	6	0.69	Take effect
		Average FI					0.70	
III	**Contractor**							
P7	Contractor Ability	5	11	26	3	10	0.61	Take effect
P8	Financial constraints contractor	26	19	10	0	0	0.34	No effect
		Average FI					0.47	
IV	**Consultant**							
P9	Design error	22	12	20	1	0	0.40	No effect
P10	Design data is not complete	0	0	21	11	23	0.81	Very influential
P11	Delays in information from the planner	11	20	24	0	0	0.45	Less influential
		Average FI					0.55	
V	**Owner**							
P12	Owner financial failure	22	2	3	21	7	0.56	Less influential
P13	Change order	12	6	11	16	10	0.62	Take effect
		Average FI					0.59	
VI	**Material**							
P14	Delays in delivery	28	5	2	7	13	0.50	Less influential
P15	Damage during shipment	23	10	13	3	6	0.45	Less influential
P16	Low material quality	0	0	23	27	5	0.73	Take effect
		Average FI					0.56	
VII	**Equipment**							
P17	Number of equipment	30	10	2	6	7	0.42	Less influential
P18	Condition of equipment that is not suitable for use	26	5	7	11	6	0.48	Less influential
P19	Tool damage	0	2	19	32	2	0.72	Take effect
		Average FI					0.54	
VIII	**Construction Process**							
P20	Quality of work is not good	25	8	3	7	12	0.50	Less influential

No.	Variable	Number of answers					FI	Interpretation of scores
		VNI	NE	LI	TE	VI		
		1	2	3	4	5		
P21	Communication and coordination problems	23	9	3	8	12	0.52	Less influential
P22	Wrong implementation method	0	0	2	41	12	0.84	Very influential
		Average FI					0.62	
IX	**Project Neighborhood**							
P23	Access to project location	0	0	10	17	28	0.87	Very influential
P24	Security disorders	18	6	13	6	12	0.56	Less influential
		Average FI					0.71	
X	**Safety and Accidents**							
P25	The machine is not checked before operating	18	6	13	6	12	0.56	Less influential
P26	Workers are not equipped with protective equipment	0	0	9	24	22	0.85	Very influential
P27	Safety regulations are violated	21	21	4	9	0	0.40	No effect
		Average FI					0.60	
XI	**Finance**							
P28	The price of material is more expensive	0	1	12	21	21	0.83	Very influential
P29	Additional equipment rental fees	11	19	23	0	2	0.47	Less influential
P30	Wages for workers are more expensive	22	2	3	21	7	0.64	Take effect
		Average FI					0.64	

4.4. Analysis of risk control and handling

Risk control and handling strategies are analyzed using SWOT analysis. Analysis of strategic factors is the beginning of the process of formulating strategies in controlling the risks that occur. This analysis aims to find compatibility between external opportunities and internal provisions while also pay attention to external threats and internal weaknesses. The SWOT analysis matrix controls and handles risks can be seen in table 4.

Table 4. Matriks SWOT.

Internal factors / **External factors**	***Strength (S)***	***Weakness (W)***
	1. As a local company 2. Access to the project location is not far away 3. Supervision of project implementation is easier because it is close to the company office 4. Reducing the possibility of security disturbances	1. Lack of workers 2. Damage to the tool 3. Low material quality 4. Incomplete design data 5. The price of material is more expensive 6. Wrong implementation method 7. Workers are not equipped with protective equipment 8. Access to the project location 9. Weather 10. Skills and expertise
Opportunity (O)	**Strategy (SO)**	**Strategy WO**

IOP Conf. Series: Materials Science and Engineering **933** (2020) 012011 doi:10.1088/1757-899X/933/1/012011

1. Efforts to maximize the construction process 2. Trust owner 3. Government support	1. Increase owner trust 2. Selection of company management in accordance with their knowledge in the field of construction. 3. Choosing a competent workforce 4. Reducing the occurrence of security disorders 5. Increasing cooperation with the local village apparatus	1. Minimizing tool damage 2. Manage material purchases before prices rise 3. Mastering the implementation method 4. Equip protective equipment for workers 5. Predict the weather in planning project implementation. 6. Improve skills and expertise
Threat (T)	**Strategy (ST)**	**Strategy (WT)**
1. Political situation 2. The entry of an outside company 3. Environmental issues	1. Improve company performance 2. Recruit supervisors who are competent with their work.	1. There needs to be an increase in HR for company managers to minimize the occurrence of risks during project implementation. 2. Selected competent workers

Based on the SWOT Matrix, the risk control strategies that can be carried out are as shown in table 5.

Table 5. Risk control strategies.

1. Strategy S-O	:	1. Increase owner trust 2. Selection of company management in accordance with their knowledge in the field of construction 3. Choosing a competent workforce 4. Reducing the occurrence of security disturbances 5. Increasing cooperation with the local village apparatus
2. Strategy W-O	:	1. Minimizing tool damage 2. Manage material purchases before prices rise 3. Mastering the implementation method 4. Equip protective equipment for workers 5. Compile a work schedule so that at any time it does not stop when it rains 6. Improve skills and expertise
3. Strategy S-T	:	1. Improve company performance 2. Recruit supervisors who are competent with their work.
4. Strategy W-T	:	1. There needs to be an increase in HR for company managers to minimize the occurrence of risks during project implementation. 2. Choosing competent workers

4.5. Discussion

The degree of skills and expertise of contractors is essential to improve the performance. Having strong soft skills will have a direct impact on the implementation of construction projects. The essential soft skills for contractors are problem-solving (the ability of managers to solve problems effectively and efficiently), budgeting and cost skills (the ability to budget project costs, make investment feasibility so that project finance can run optimally), scheduling and time management skills (the ability to arrange project scheduling), technical skills (the technical ability to cover the knowledge and experience in implementing the project itself), leadership skills (having leadership in influencing others in acting and reacting to project issues), resource management and human relationship skills (the use of resources and their placement), communication skills (having good and effective communication in the team), negotiating skills (reliable negotiation skills), marketing, contracting, customer relationship skills (the ability to market the results of their projects).

Predicting the weather in planning project implementation is very important to do especially related to earthworks such as basements, substructure, and roads. Many project conditions are misplaced due to not predicting the weather that will occur. So that it will have an impact on time and costs due to low productivity and manpower due to falling rain. Rain will also affect access to the project location; if the project location does not have road pavement, the access will be damaged and will have an impact on the inhibition of activities on site.

Protective equipment is very important when in the project location, the use of PPE (personal protective equipment) will reduce the risk of workplace accidents. Every worker is required to use PPE as stipulated in the Minister of Manpower and Transmigration Regulation of the Republic of Indonesia of Number Per.08/Men/VII/2010 concerning personal protective equipment. Workers need to anticipate any potential hazards that may arise during work. One of the anticipatory actions is to wear personal protective equipment while in the work environment. Syatauw [10] in his research also found that the risk impacts that occur in the implementation of construction projects in Jakarta and Depok are work accidents and non-operational procedures.

The dominant factor that occurs during the implementation of construction project work is that the method of implementing the project is wrongly implemented so that it impacts on the project being carried out. High material price increases also have an impact on the emergence of cost overrun. Furthermore, other risk factors can occur due to uncomplete the design data, low quality of material, damage to the tool, and a lack of workers. Therefore, there is a need for prevention that will affect the performance of contractors. So, the project being worked on can be completed properly.

The control and handle of the risks can be done by increasing the trust of the owner, selecting company managements in accordance with their knowledge in the field of construction, selecting competent workforce, reducing security disturbances, increasing collaboration with local village officials, minimizing equipment damage, managing material purchases before prices rise, and mastering the implementation method

5 Conclusions and recommendations

5.1. Conclusions

The conclusions that can be drawn from the results of research on risk factors in the implementation of building construction are as follows:

1. The dominant risk factors that occur in the implementation of building construction in Pidie Regency are obtained from the results of the analysis of frequency index (FI), namely:
    a. Skills and expertise;
    b. Weather;
    c. Access to the project location;
    d. Workers are not equipped with protective equipment;
    e. Wrong implementation method;
    f. The expensive material;
    g. Incomplete design data;
    h. Low quality of material;
    i. Damage to the tool; and
    j. Lack of workers.
2. Control and handling of risks that occur during the construction can be done as:
    a. Increasing owner trust;
    b. Selecting of company management in accordance with their knowledge in the field of construction;
    c. Selecting the competent workforce;
    d. Reducing security problems;

e. Increasing collaboration with local village officials;
f. Minimizing tool damage;
g. Manage material purchases before prices rise;
h. Mastering the implementation method.

5.2. Recommendations

1. This research has been carried out with a predetermined scope or limitation of the study.
2. In terms of reducing the risk of occurring in the field, the contractor or project manager must pay attention to the risks that occur in the field so that the planned project can run well.

References

[1] Smith C W Jr 1990 Corporate Risk Management: Theory and Practice *Journal De-rivatieves* **2** 4 21-30
[2] Anisa W G 2012 Analisis Faktor yang Mempengaruhi Pengungkapan Manajemen Risiko (Studi empiris perusahaan non keuangan yang terdaftar di BEI tahun 2010) *Diponegoro J of Account.* **2** 3 1-30
[3] PT PP 2003 *Buku Referensi Untuk Kontraktor Bangunan Gedung dan Sipil* (Jakarta: Gramedia Pustaka Utama)
[4] Kerzner H 2009 *Project Management* 10th Edition (New York: John Willey and Sons)
[5] Hasan M I 2004 *Analisa Data Penelitian Dengan Statistik* (Jogjakarta: Kanisius)
[6] Usman H dan Akbar S 1996 *Metodologi Penelitian Sosial* (Jakarta: PT. Bumi Aksara)
[7] Sugiyono 2005 *Metode Penelitian Bisnis* (Bandung: Alfabeta)
[8] Le-Hoai L, Lee Y D and Lee J Y 2008 Delay and cost overruns in Vietnam large construction projects: A comparison with other selected countries, *KSCE J. of Civ. Eng.* **12** 6, 367-77
[9] Rangkuti F 2005 *Analisa SWOT Teknik Membedah Kasus Bisnis* (Jakarta: PT. Gramedia Pustaka Utama)
[10] Syatauw C 2017 Analisa Pengelolaan Risko Kualitas pada Tahap Pelaksanaan Konstruksi Gedung Tinggi (Studi Kasus: Apartemen di Jakarta dan Depok) *J. Ilmiah Desain dan Konstruksi* **16** 1 10-20

Travel cost budget distributions among household life stages considering bus users in Banda Aceh

A Putri[1], S Sugiarto[2], S M Saleh[2] and D Basrin[3]

[1] Civil Engineering Postgraduate Program, Department of Civil Engineering, Universitas Syiah Kuala, Banda Aceh, Indonesia
[2] Department of Civil Engineering, Universitas Syiah Kuala, Banda Aceh, Indonesia
[3] Department of Civil Engineering, Universitas Samudra, Langsa, Indonesia

Corresponding e-mail: aqlima@mhs.unsyiah.ac.id

Abstract. This study aims to investigate the average travel cost budget (TCB) distributions among household life stages (HLS) groups considering the bus users of Trans-Koetaradja in Banda Aceh, capital of Aceh province. This research was conducted within three corridors of Trans Koetaradja. The onboard survey is performed to collect bus users related information such as their monthly travel budget, socio-economic attributes, and travel/mobility characteristics. In this study, we only consider to analysis the distributions and characteristics obtained data sets by deeply investigating socio-demographics and travel attributes of bus users. The analysis result reveals that the group 3 (families with pre-school children) spent the highest average of TCB (0.442 million IDR/month) and group 1 (single person) spent the lowest TCB (0.280 million IDR/month). While group 2 (childless couple) has portion the lowest average of TCB about 10.2% from their monthly income, and group 3 (families with pre-school children) has shared the highest TCB accounting up to 14.4% from their monthly income compared to the remaining groups. This result depicts that an average TCB across HLS groups is considerably high, especially early household life stages development. It is may partially due to a limitation on their monthly income and lack of affordable transport mode for them.

1. Introduction

To days, promising economic development, together with rapid urbanization, leads to increase people travel to satisfy their daily routines. In order to do that, people need to allocate more time and money to their transportations. Consequently, the more trips they are made, the greater the time and money allocated for traveling. The allocation of money for travel refers to the so-called travel cost budget (TCB). The TCB defines as the maximum amount of money that households/individuals willing to allocate for their transportation within a month [1, 2]. Furthermore, an investigation by Sugiarto et al. [1] and Saleh et al. [2] approved that the variation of TCB is significantly influenced by the household's monthly income as well as life stages development of the households. Sugiarto et al. [1] further revealed that changes in technology, incomes, attitudes, social demographics and travel behaviors could have a significant influence on travel cost expenditure.

In Banda Aceh, highly automobile ownership (i.e., car and motorcycle), which leads to private mode dependency, creates several transport-related problems such as increased travel time, additional fuel consumption and environmental deterioration [3-5]. This problem could be mitigated by reducing people's dependence on private mode by shifting their mode to public transport mode, such as the urban bus system. Public transportation is believed able to reduce people's dependence on private vehicles in Banda Aceh [6-8]. Furthermore, the government aims to implement bus reform so-called "Trans Koetaradja" with a full charge by the years of 2020. The Trans Koetaradja now is under trial runs, a free of charge is applied within running corridors [6]. It is argued that the willingness and ability to pay are significantly influencing by the TCB in order to determine the appropriate tariff of the bus.

The policy, in applying the tariff, will cause problems, especially for low income and captive commuters. Affordable public transport is necessary to be provided by the government, especially for those who highly dependent on public mode. Moreover, the TCB highly depends on household life stages (HLS). The understanding of the distribution of TCB across HLS could be used as a reference in the establishment of this bus tariff efficiently. Therefore, this study was conducted to determine the average TCB across HLS as well as their socio-economic and mobility attributes prior to the implementation of the tariff of the bus. Furthermore, the TCB could be used to determine the ability to pay of people in Banda Aceh prior to the tariff establishment. It should be noted that the HLS is classified based on the previous work done by [1].

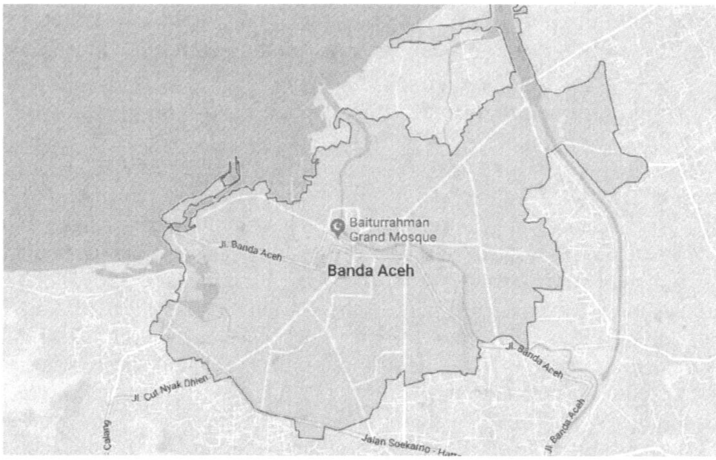

Figure 1. Banda Aceh city (map Data © 2019 Google).

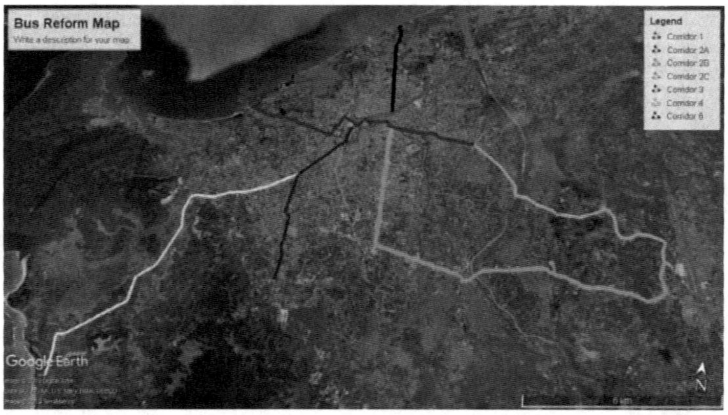

Figure 2. Study area within Trans Koetaradja corridors 1, 2B and 3 [8].

According to Sugiarto et al. [1], the HLS is classified into six groups: (1) group 1 (single-person household); (2) group 2 (childless-couple household); (3) group 3 (families with pre-school children);

(4) group 4 (families with young school children); (5) group 5 (families with college/university children); and (6) group 6 (families only with adults).

The onboard survey was conducted to collect respondent's socio-economic and travel attributes. This study was performed within Trans Koetardja corridors, namely corridor 1 (the City Center – Darussalam), corridor 2B (City Center - Ulee Lheue), and corridor 3 (City Center - Mata Ie), as illustrated in figure 2. Using onboard collected data, we investigate the average TCB distributions across HLS as well as their attributes of socio-economic and travel characteristics. The rest of this paper is systematized into several sections that describe the survey and data profiles, distribution data set, and discussion and conclusions of the study at the end of the paper.

2. Methodology

The survey was conducted in Banda Aceh (see figure 1) within three corridors of Trans Koetaradja. The target respondents were the householders who used the bus for their commuting. The survey was conducted by direct-interview to the householders. A total of 450 questionnaires with valid answers were collected and analyzed in this study. The summary of the questionnaire used can be seen in table 1 which consists of survey time, target area, distribution method, valid number of samples, and outline of the questionnaire. The questionnaire was designed to be able to gather information related to respondents' socio-economic attributes and travel attributes, which can be seen in table 1.

Table 1. Summary of SP questionnaire.

Description	Details
Time of survey	April 2019
Target location	Banda Aceh, Aceh Province
Distribution methods	A paper-pencil based direct interviews and collected by the enumerator based on the onboard survey
Number of valid questionnaires	450 samples
Questionnaire outline:	
Social-economic attributes	Gender, age, education, occupation, income monthly, monthly travel expenditure, housing owned status, household member, motorcycle ownership, car ownership.
Travel attributes	Mode use, public transportation use, frequency of public usage mode in daily life, willingness to use public transportation, travel destination.

This questionnaire was designed to obtain information related to respondents. As mentioned in table 1, information pertaining to socio-demographic characteristics and travel behavior is obtained by direct interviews adopted from previous works done by [8-11]. In this study, surveyors used 450 samples of data that had been collected. When the interview data has been collected, the data is separated according to the respective HLS group. Then, the data were analyzed for revealing the distribution of the data, especially for TCB across HLS groups. Prior to the distribution analysis of the respondent's attributes, we aggregated our datasets into several HLS.

3. Results and discussion

The aggregation result across HLS is described in figure 3. HLS group 4 (family with young school children) has larger portion of the sample (N=170) compared to the remaining HLS.

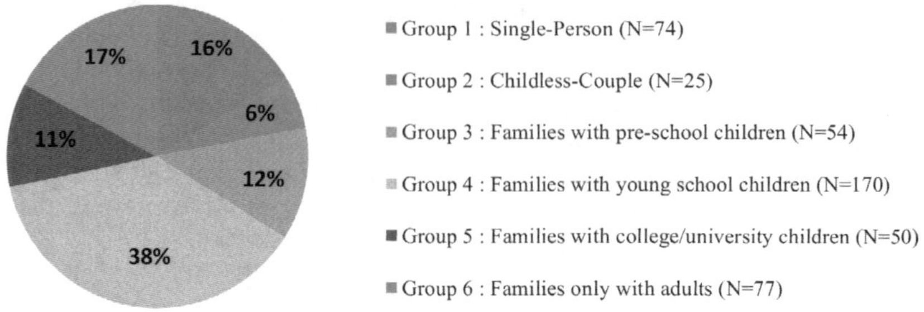

Figure 3. Aggregation of HLS group.

Table 2. Distributions of socio-demographic attributes.

Item	Category	Number of Samples	Share (%)
Gender	Male	138	30.7%
	Female	312	69.3%
Age	30-49 years	291	64.7%
	50-59 years	95	21.1%
	60 years or more	64	14.2%
Education	Senior High School	249	55.3%
	College	74	16.4%
	University / Bachelor	127	28.3%
Occupation	Government Employee	52	11.6%
	Private Employee	132	29.3%
	Retired	27	6.0%
	Trader	66	14.7%
	Housewife	173	38.4%
Housing Own Status	Owned	280	62.2%
	Family's Housing	66	14.7%
	Renting	101	22.4%
	Others	3	0.7%
Household Members	1-2 person	122	27.1%
	3-5 person	295	65.6%
	6-10 person	31	6.9%
	10 person or more	2	0.4%
Car Ownership	None	394	87.6%
	1-2 unit	56	12.4%
Motorcycle Ownership	None	208	46.2%
	1-2 unit	231	51.3%
	3 unit or more	11	2.4%

Table 2 contains data on the socio-demographic characteristics and travel attributes of targeted respondents. As for gender, female respondents (69.3%) have dominated our data sets, accounting for up to 69.3% compared to male respondents. Young respondents have larger proportion compared to older people in the data set, and they are between 30-49 years old or about 64.7%. Looking to education attribute, having high school education level and bachelor's degree are predominant up to 83.6% of respondents.

Concerning occupation attributes, respondents with private employees (29.3%) and housewives (38.4%) have a more substantial share in data sets. Interestingly, most of the respondents have not owned cars. Therefore they are more likely to use the bus as their daily mode, as shown in table 2. In addition, the car ownership level is nearly 12.5% of respondents. Furthermore, our datasets reveal that respondents who have their own homes are accounting more than 60%, and the remaining households are renting and stay with their parents. Moreover, the household member attribute shows close to 65% of households having 3 to 5 members, only 7.3% of the households have household member larger than five members.

Table 3. Distributions of travel data attributes.

Item	Category	Number of Samples	Share (%)
Representative Mode Use	Motorcycle	66	14.7%
	Car	15	3.3%
	Public Transportation	369	82.0%
Public Transportation Use	Trans Koetaradja (Bus)	355	96.2%
	Online Public Transport	3	0.8%
	Labi-Labi	11	3.0%
Frequency using public mode	1-2 days a week	27	7.3%
	3-4 days a week	129	35.0%
	5 days a week or more	213	57.7%
If they used a car for representative mode used, if they are asked to use a bus how frequently they willing to use a bus in a week	1 day a week	23	28.4%
	2 days a week	28	34.6%
	3 days a week	30	37.0%
Travel Destination	Work	166	36.9%
	Shopping	196	43.6%
	Pick up & drop to school	5	1.1%
	Social	47	10.4%
	Entertainment	36	8.0%

Looking at the respondent's travel attributes, table 3 discloses that most of the respondents use public transport, mainly bus for their daily routines accounting close to 82%. Amongst them are frequently using the bus (more than three times a week) for more than 90% of respondents. As for the trip, going to work and shopping has dominated activities dataset (80.5%).

Figure 4 illustrates the distribution of monthly income across households. figure 5 describes the monthly TCB corresponding HLS as well as their share of TCB. As for all data sets, we found that an average monthly income is 3.2 million (IDR), while the average TCB is 0.324 million (IDR). Looking at the highest average monthly income, the HLS of group 2 (childless-couple) is considered as the highest one, which is 3.7 million (IDR) per month, and the lowest average monthly income belongs to

HLS of group 1 (single-person) of about 2.7 million (IDR). For TCB, figure 5 reveals that the highest average TCB has corresponded to the HLS of group 3 (families with pre-school children) accounting up to 0.442 million (IDR) per month or 14.4% shared from their monthly income, and the lowest average TCB is the HLS of group 1 (single-person) for about 0.280 million (IDR) per month or 12.8% from their monthly income.

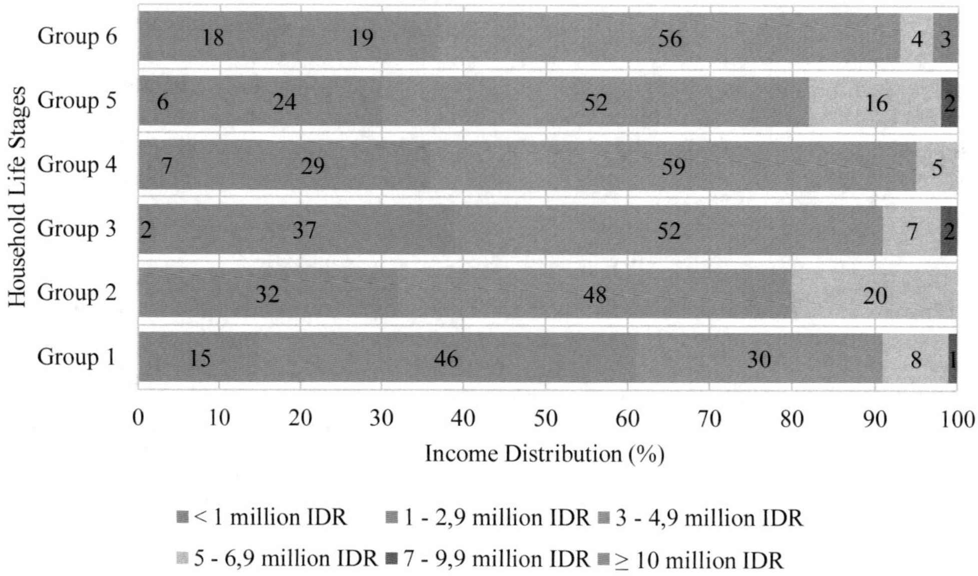

Figure 4. Distribution of monthly income.

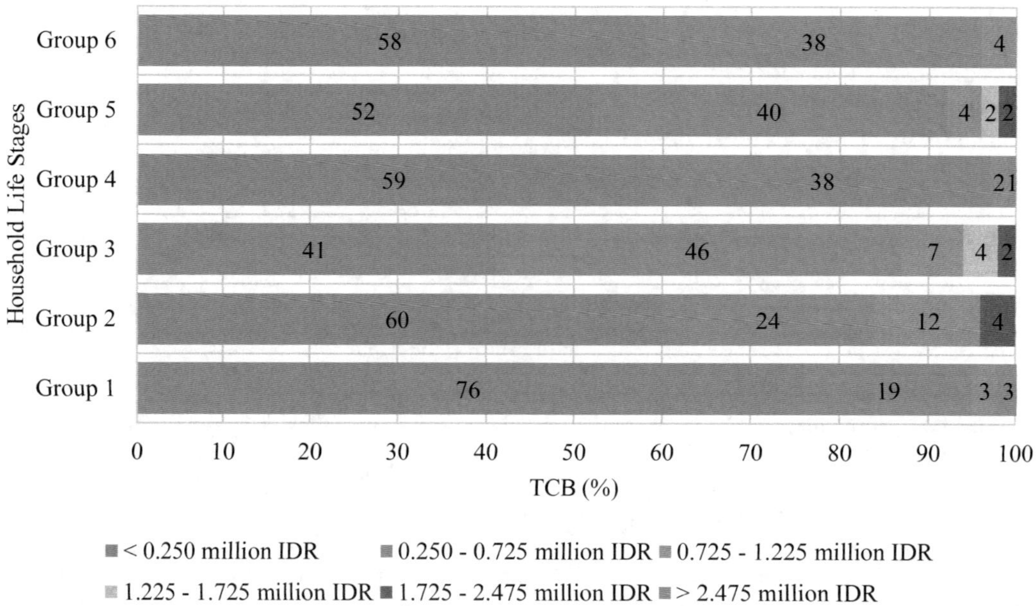

Figure 5. Distributions of TCB.

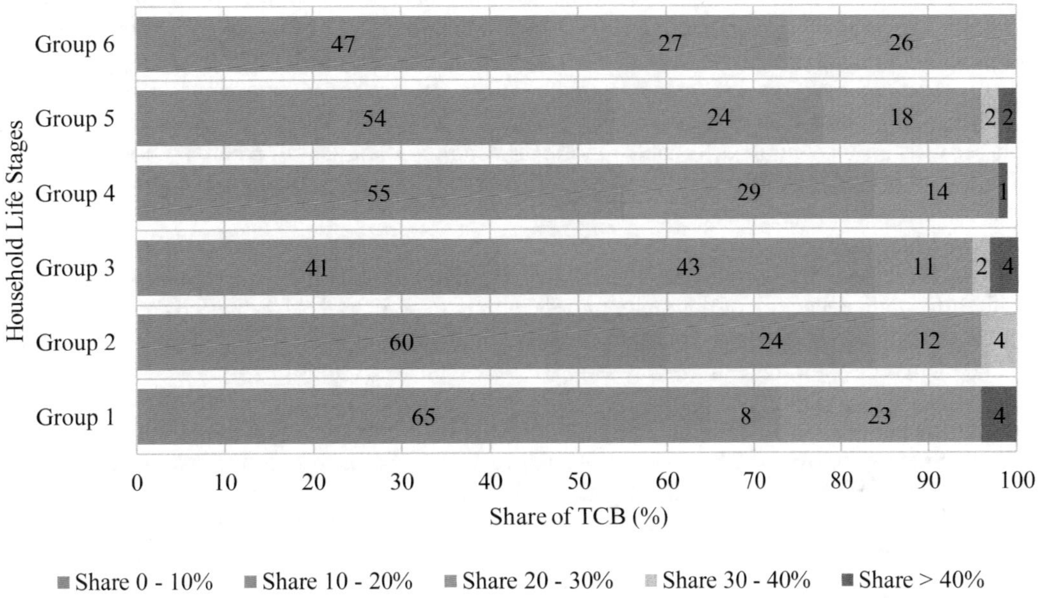

Figure 6. Distributions of share TCB.

Figure 4 indicates that 46% of group 1 (single-person) has the lowest income, with 1-2.9 million IDR, and the other life stages have 48% to 59% with 3-4.9 million IDR. Figure 5 indicates that 46% of group 3 (families with pre-school children) have the highest TCB with 0.250-725 million IDR and the other life stages have less 0.250 million IDR (52% to 76%). Figure 6 indicates that 43% of group 3 has the highest share of TCB with a share of 10-20%, and the other life stages have 47% to 65% with a percentage of 0-10%. It is presumed that several variables are affecting the variation of the TCB across HLS. For instance, table 3 depicts the number of family members and owned car characteristics corresponding to more disbursement for money expenditure on transport. The purpose of the trip for mandatory activities also results in more higher frequency in travel; as a consequence, it would like to add more money for traveling.

4. Conclusion

This study reveals that an average monthly income is 3.2 million (IDR), while the average TCB is 0.324 million (IDR). The highest average monthly income belongs to group 2 HLS (childless-couple) and the lowest average monthly income signs to group 1 (single-person) for 3.7 million (IDR) per month and 2.7 million (IDR) per month, respectively. This monthly income seems to be associated with the TCB, specifically for group 1 HLS. The distribution of the dataset further discloses that the highest average TCB has corresponded to the HLS of group 3 (families with pre-school children) and the lowest average TCB is the HLS of group 2 (childless-couple) accounting for 14.4% shared and 10.2% from their monthly incomes, respectively. This variation of share TCB may be affected by the limitation of their monthly income and their travel attributes such as owning a car, trip activity, and frequency. We hypotheses that the number of family members and owned car characteristic corresponds to more disbursement for money expenditure on transport; and, the purpose trip could result in higher frequency for traveling and consequently would like to add more money for traveling. However, further exploration is needed to confirm factors that effecting TCB across HLS, for example using the time-space prism diagram as it is applied by [1, 2] to explore money expenditure constraints in greater Jakarta.

Furthermore, the result shows that low-income HLS has a substantially higher share of TCB. It means that low-income HLS may have limited income or has limitations for spending the money on travel.

This evidence provides insight for the Government-related policymakers to consider monetary constraints when the government decides how much of the fare of the bus system must be applied. The government must consider to which income groups such policy will be imposed monetary, particularly for those who have the low-income bracket. The government policies must reflect the needs to be addressed for certain income groups in determining a more impartiality a full-charge implementation, thereby leading to more affordable public transport for citizens. A better understanding of travel behaviors is vital for successful transport planning to gain more efficient policies [10-12].

Acknowledgments
The authors would like to express their honest gratitude to Universitas Syiah Kuala for financially supported data collection of this study under Contract No. 522/UN11/SPK/PNBP/2019. Furthers thanks go to the staff and students of Transportation System Laboratory, Civil Engineering Department of Syiah Kuala University for their substantial help with the data collection.

References

[1] Sugiarto S, Miwa T, Sato H and Morikawa T 2014 Transportation expenditure frontier models in Jakarta Metropolitan area *Procedia-Social and Behavioral Sci.* **138** 148-58
[2] Saleh S M and Sugiarto S 2016 Stochastic production frontier models to explore constraints on household travel expenditures considering household income classes *Aceh Int. J of Sci. and Tech.* **5** 1 9-17
[3] Saleh S M, Sugiarto S, Hilal A and Ariansyah D 2017 *AIP Conference Proc.* 1903 1 06005.
[4] Anggraini R, Sugiarto S, Pramanda H 2017 *AIP Conf. Proc.* 1903 1 060011
[5] Sugiarto S, Miwa T, Sato H, Morikawa T 2014 *Proc. of 21st World Congress on Intelligent Transp. Systems* 12914
[6] Saleh S M, Sugiarto S and Anggraini R 2019 Analysis on the public's response toward bus reform policy in Indonesia considering latent variables *The Open Transp. J.* **13** 17-24
[7] Sugiarto S, Saleh S M, Anggraini R and Merfazi M 2019 *IOP Conf. Series: Materials Sci. and Eng* 523 012036
[8] Saleh S M, Sugiarto S and Salmannur A 2019 Attitudinal dataset for mediating the effects of public acceptance scheme on bus reform in a developing country context *Data in Brief.* **25** (104035) 1-10
[9] Sugiarto S, Miwa T and Morikawa T 2017 Inclusion of latent constructs in utilitarian resource allocation model for analyzing the revenue spending policy options in the congestion charging *Transp. Res. part A* **103** 36-53
[10] Sugiarto S, Miwa T, Sato H and Morikawa T 2017 Explaining differences in determinants acceptance towards the congestion charging policies in Indonesia and Japan, *J of Urban Plan. and Dev.* **143** 2 1-12
[11] Sugiarto S, Miwa T, Sato H and Morikawa T 2018 The tendency of public's attitudes to evaluate urban congestion charging policy in Asian megacity perspective: Case a study in Jakarta, Indonesia *Case Studies on Transp. Policy.* **online first**
[12] Sugiarto S, Lulusi, Isya M, Taufik and Haiqal M 2019 *IOP Conf. Series: Materials Sci. and Eng.* **523** 012032

IOP Conf. Series: Materials Science and Engineering **933** (2020) 012013 doi:10.1088/1757-899X/933/1/012013

The effect of public space availability on children's creativity (case study: Taman Sari kota Banda Aceh)

I Fitri[1], T B Aulia[2] and M Fuady[2]

[1] Civil Engineering Postgraduate Program, Department of Civil Engineering, Universitas Syiah Kuala, Banda Aceh, Indonesia
[2] Department of Civil Engineering, Universitas Syiah Kuala, Banda Aceh, Indonesia

Corresponding e-mail: bppd.idafitri@gmail.com

Abstract. The purpose of this study was to determine the effect of Taman Sari as a public space on children's creativity, what factors were the most influential of public space facilities on children's creativity, and how much the Taman Sari facility was structured for children. The research location was Taman Sari, a City Park, in Baiturrahman Sub-District, City of Banda Aceh. Primary data collection was done by dividing the questionnaire. This research was conducted by multiple linear regression analysis. The results showed that the existence of Taman Sari with its variables (accessibility, facilities, environment, and comfort), together, had a significant effect on children's creativity. The factors that have the highest impact on children's creativity were facilities and comfort. Taman Sari at present, 2019, has undergone improvements and handlings. There are many improvements in both quantity and quality facilities. The government has been performing marked improvement to make Taman Sari as a public space.

1. Introduction

A city park is one form of public spaces that are widely used by various communities to carry out various types of activities. Based on the Minister of Public Works Regulation of No. 05/PRT/M/2008, the city park is an open space that some of the purposes are for social and aesthetic functions such as recreational- and educational activities. City park also has a purpose as an urban green open space. A good city facilitates activities of the communities, including children [1].

In general, public space should have the ability to support human needs for gathering and interacting with each other. The good-handled public space can produce activities between people who do not know each other beforehand. Public space is expected to inspire the public to be smarter, innovative, and creative. Public space should be accessible for- and free of charge by the city people from various walks of life. Community and children need public space. Children who perform beneficial- and cultural activities in their leisure time would have a bright future; Contrastly, children, who spent time for negative activities, could disturb their achievement and the society's comfort and safety.

Taman Sari is one of the most popular public spaces in the city of Banda Aceh. Taman Sari is now a choice of family recreation destinations. Taman Sari has several advantages, including mileage that does not require a long time which is located in the middle of the city. Also, Taman Sari is equipped with public facilities that support the creativity of the community, especially children. In accordance with its function as a public space, Taman Sari, which is located in the city center and among the office areas, should be a comfort-, safe-, secure-, healthy place for children to develop their creativity.

For this reason, this study is intended to see the effect of Taman Sari as a public space on children's creativity. This study could be an input to the relevant agencies in implementing several strategies, namely, socializing and improving the functions of Taman Sari as the public and the open space city.

2. Methodology

Yudistira (2015) did research studying Taman Sari's existing conditions. Based on the research, Taman Sari has a score of 1-2 (bad-moderate), which means that many existing facilities in Taman Sari are damaged or not available. For this reason, further handling by the city government is needed to pay more attention to the maintenance of existing facilities in Taman Sari so that residents of the City of Banda Aceh can increase their visit to Taman Sari. As a result, in this case, the writer wants to examine the existence of the Taman Sari again nowadays. The author wants to connect several aspects that are related to creativity.

2.1. Definition of public space

Public space is an abstract space for the public to express opinions or oppose other opinions based on the principle of freedom of responsibility [2]. Public space is a place for communal activities where community interaction takes place in a variety of activities.

Public space is a space in which people's conversations, ideas, and thoughts meet [3]. Pro contra is the main element of the public sphere to find solutions for various social problems.

2.2. Criteria and indicators of public space

Criteria and indicators for designing public spaces can be seen in table 1 as follows.

Table 1. Criteria and indicators for designing public spaces.

No.	Criteria	Indicator
1.	Accessibility	Location is easily accessible
		The garden is accessible for children
		There is public transportation integrated into the park
2.	Facilities	Playing facilities
		Learning facilities
		Talent development facility
		Recreational facilities
3.	Environment	Availability of vegetation
		There is space to interact
4.	Convenience	There is a pedestrian path
		Cleaning facilities
		Adequate parking area

2.3. Children's creativity

Creativity is the ability of a person to produce to something new, in the form of ideas and the real works and in the form of new works and combinations of things that already exist, all of which are relatively different from what has been there before [4]

Aspects and indicators in assessing a child's creativity can be seen as follows:

1. Flexibility among them, generating varied ideas, answers or questions, ability to see problems from different perspectives and look for many different alternatives or directions.
2. Originality, which means that the person able to produce new and unique expressions, think of

unusual ways to express himself or herself, and have a strong will to complete the task.

3. Elaboration, which includes responding to questions passionately, completing tasks actively and eagerly, daring to accept or carry out heavy tasks, being happy to find practical methods or methods of learning, and asking questions aggressively.
4. Fluency, which includes sparking lots of ideas, answering, solving, or questioning problems, and being independent in learning.

2.4. Research method

The research method used is a combination of qualitative and quantitative approaches.

2.5. Research location

The research location was Taman Sari, a city park located in Baiturrahman Sub-District, the City of Banda Aceh. Taman Sari is one of the public spaces in the City of Banda Aceh.

2.6. Population and samples

The population in this study was 125 visitors per day. The sample calculation was done using the Slovin formula [5]. The sample in this study was 95 visitors.

2.7. Data processing

Data processing was done by multiple linear regression analysis using SPSS software.

3. Results and discussion

From the results, the linear regression calculations can be written in the following equation.

$$Y = 19{,}628 + 0{,}08X1 + 0{,}510X2 + 0{,}023X3 + 0{,}523X4$$

3.1. Correlation test and determination coefficient (R2)

To test and assess the significance of the estimates obtained from multiple linear regression equations, several correlation tests were conducted, and the coefficient of determination about the relationship and influence of accessibility variables (X1), facilities (X2), environment (X3) and comfort (X4) on children's creativity (Y) was discussed. Coefficient of determination and correlation model summary is shown in table 2 as follows.

Table 2. Coefficient of determination and correlation model summary.

Model	R	R Square	Adjusted R Square	Std. Error of the Estimate
1	0.710(a)	0.504	0.482	1.93165

a Predictors: (Constant), Comfort, Environment, Facilities, Accessibility

The correlation coefficient (R) was = 0.710, which indicates that the degree of relationship between the independent variable and the dependent variable was 71.0%. This means that based on the table of interpretation of the correlation coefficient, the value of 0.710 lies in the coefficient interval 0.60 - 0.799, then the level of the relationship was said to be strong between accessibility factors (X1), facilities (X2), environment (X3) and comfort (X4) with children's creativity.

Children's creativity in using Taman Sari as a public space is measured while the child is playing. Children's playing facilities in Taman Sari should be improved such as playground infrastructure for children, visitor accessibility, and an environment that makes visitors feel comfortable. Modern playgrounds usually have recreational tools such as seesaw, turntable, swing, slide, slide, sandbox, seesaw, game house, and labyrinth. By playing, children have the value of the opportunity to express something that he feels and thinks. Also, by playing, children are practicing skills and developing themselves. In playing, children can develop coarse and fine muscles, increase reasoning, understand the sustainability of their environment, form the power of imagination fantasy power, and creativity.

The determination coefficient (R2) was 0.504. This means that 50.4% of changes in creativity can be explained by changes in accessibility factors (X1), facilities (X2), environment (X3), and convenience (X4). While the rest, which was equal to 49.6%, was explained by other variable factors outside of this study.

3.2. F test

In order to test the hypothesis regarding the determination of accessibility significance (X1), facilities (X2), environment (X3), and convenience (X4) on creativity, F tests were carried out as well. Test results can be seen in the following table.

Table 3. Analysis of variance ANOVA[b].

Model		Sum of Squares	df	Mean Square	F	Sig.
1	Regression	341.490	4	85.372	22.880	0.000(a)
	Residual	335.816	90	3.731		
	Total	677.305	94			

a. Predictors: (Constant), Comfort, Environment, Facilities, Accessibility
b. Dependent Variable: Creativity

The ANOVA or F test results (F_{count} value) in the table above is 22.880, while F_{table} at the significance level α of 5% was 2.473 with a significance level of 0.000, which is smaller than 5%. As a result, $F_{count} > F_{table}$, H_a was accepted, and H_0 was rejected. This means that accessibility, facilities, environment, and comfort significantly influence children's creativity.

3.3. t test

The t test was performed in this research.

a. Accessibility variables (X1)
For accessibility variables (X1), t_{count} = 0.338, while t_{table} = 1.662. The results of this calculation indicate that $t_{count} < t_{table}$ with a significance of 0.000 or a probability above α = 5%. From these results, it can be concluded that partially the accessibility variable did not significantly influence children's creativity.

b. Facility variable (X2)
In terms of facility variables (X2), t_{count} = 3.249, while t_{table} = 1.662. The results of this calculation indicate that $t_{count} > t_{table}$ with a significance of 0.000 or a probability above α = 5%. From these results, it can be concluded that partially the facility variables significantly influence children's creativity.

c. Environment variable (X3)
In terms of environmental variables (X3), t_{count} = 0.117, while t_{table} = 1.662. The results of this calculation indicate that $t_{count} < t_{table}$ with a significance of 0.000 or a probability above α = 5%. From these results, it can be concluded that partially the environmental variables did not significantly influence children's creativity.

d. Variable convenience (X4)
For comfort variables (X4), t_{count} = 3.112, while t_{table} = 1.662. The results of this calculation indicate that $t_{count} > t_{table}$ with a significance of 0.000 or a probability above α = 5%. From these results, it can be concluded that partially the comfort variable has a significant effect on children's creativity.

IOP Conf. Series: Materials Science and Engineering **933** (2020) 012013 doi:10.1088/1757-899X/933/1/012013

Based on the results of the tests described above, the results that facilities and comfort in this study have significantly affected the creativity of children. This can be seen from the $t_{count} > t_{table}$ with a significance level below $\alpha = 5\%$.

4. Conclusion

Based on the results of the research, data processing, and discussions that have been carried out, several conclusions and suggestions can be taken, including:

1. The existence of Taman Sari, with its variables of accessibility, facilities, environment, and comfort, has a significant effect on children's creativity. F_{count} was 22.880, while F_{table}, at a significance level of $\alpha = 5\%$, is 2.473 with a significance level of 0.000, which is smaller than 5%. As a result, $F_{count} > F_{table}$, H_a was accepted, and H_0 was rejected.
2. The main factors that influence children's creativity are facilities and comfort. This can be seen from the fact that $t_{count} > t_{table}$ with a significance level below $\alpha = 5\%$. For facility variable (X2), $t_{count} = 3.249$, while $t_{table} = 1.662$ with a significance of 0.000. For comfort variable (X4), $t_{count} = 3.112$, while $t_{table} = 1.662$ with a significance of 0.000.
3. The facilities at Taman Sari has been improved. There are many improvements in facilities both in quantity and quality. The role of the government, in this case, the relevant agencies, is very intense in striving for Taman Sari to be a public space that can be utilized maximally by visitors.

References

[1] Rezasoltani M and Said I 2012 Methods for evaluating responses of children with outdoor environments *Procedia Social and Behavioral Sci.* **49** 39-46.
[2] Kaiser E J, Godschalk D R and Chapin F S 1995 *Urban Land Use Planning* 4th Edition (Urbana and Chicago: University of Illinois Press)
[3] McKee A 2005 The *Public Sphere: An Introduction* (Cambride: Cambride University)
[4] Munandar S C and Utami 1999 *Kreativitas dan Keberbakatan* (Jakarta: Gramedia Pustaka Utama)
[5] Sugiyono 2012 *Metode Penelitian Kuantitatif, Kualitatif dan R&D* (Bandung: Alfabeta)

Performance of reinforced concrete columns strengthened with carbon fiber reinforced polymer (CFRP) - case study of an abandoned school project

N M Amin[1] **and N A Ahmad**[1]

[1] Faculty of Civil Engineering, Universiti Teknologi MARA, 40450, Shah Alam, Selangor, Malaysia

Corresponding E-mail: norli830@uitm.edu.my

Abstract. This research presented a case study of an abandoned three-storey school building focusing on strengthening of Reinforced Concrete (RC) column with Carbon Fiber Reinforced Polymer (CFRP) jacketing. The constructed structural element of the building was assessed and evaluated by visual observation, destructive tests and non-destructive tests include carbonation tests, ferro scan tests, steel reinforcement verification, Ultrasonic Pulse Velocity (UPV), coring tests and rebound hammer tests. The modeling of the whole building was carried out using Orion 18, a structural analysis program based on the site test results in order to assess the structural integrity of the building. Based on the structural analysis conducted in Orion 18, a column at ground floor level was chosen to be strengthened with multilayer CFRP jacketing. This column has cross-section of 300mm x 300mm with 3600mm in height. Results were obtained through non-linear Finite Element simulation. The behavior of the column with and without CFRP jacketing was studied to observe its axial stress and strain behavior in ABAQUS Finite Element software. The result shows that increasing in number of CFRP layers has a significant effect on the behavior of such columns which was expected due to confinement effect.

1. Introduction

An abandoned school project consists of multiple building of classroom and administration building was chosen for the case study. The construction of this building has been started seen 2012 and due to delay in construction, the first appointed contractor was terminated in 2014 with almost all the RC structural members had been constructed. In 2016 a new contractor was appointed to continue the remaining work. Hence, there is a need for structural assessment and evaluation to be conducted in order to assess the structural integrity of the building. This research will only focus on Block B which has three-storey classroom building.

Block B which is a three-storey building was model using a computer program Orion 18 Software based on the result obtained from the destructive and non-destructive testing results. There are four carbonation tests, three ferro scan tests verified with one steel reinforcement verification test were conducted to identify the reinforcement arrangement, concrete cover, and concrete behavior when reacting with phenolphthalein. In order to identify the concrete strength of the structural members, there are 15 coring tests, 18 UPV tests and 32 rebound hammer tests were conducted. Reinforced concrete structural members and loads for the building analysis were assigned according to BS8110.

This research also analyzed and observed the behavior of the column strengthening technique using the new technologies that are less time consuming and have an easy installation method which is CFRP jacketing. A column with cross-sectional area of 300mm x 300mm at the ground floor level was randomly chosen to be modeled with and without CFRP sheets jacketing using nonlinear analysis in ABAQUS Finite Element Software.

1.1. Carbon fiber reinforced polymer (FRP) jacketing

Carbon Fiber Reinforced Polymer (CFRP) jacketing is widely used in construction industry in recent years. FRP is a composite material made of a combination of reinforced fibers and polymer matrix. The function of reinforced fiber is to provide strength and stiffness. There are three types of fibers alignment which are aligned fibers; the fibers are woven parallel to each other, random fibers; the fibers are woven randomly according to a specific density and woven fibers; the fibers are woven in the shape of a net and this is the strongest among the three. CFRP fiber fabrics are available in different fabric weights. Thousands of individual fiber filaments approximately 6 micro diameters are bonded as tows and formed into fabrics by gluing or weaving onto a plastic supporting mesh which does not contribute to the strengthening properties of the fabric.

It is necessary to consider the effect of the confinement provided by the CFRP in designing the strengthened column. As shown in figure 1 concrete in square column confined by hoop FRP displays an approximately bilinear stress-strain response. The stress-strain curve shows that the initial behavior of the unconfined column and confined column with CFRP is almost the same since the FRP wrap exerts a limited confining pressure on the concrete. The lateral deformation continues to increase as the axial stress increase which reduces the concrete stiffness. The FRP wrap is fully activated in transition zone once unconfined concrete reached its peak strength. At this stage, the stress-strain response becomes approximately linear with a slope dependent upon the stiffness of the hoop FRP.

The first slope of the curve is known as the initial elastic zone with the initial rigidity equal to EI and the second slope as the final plastic zone with a rigidity E_f. The slope in the plastic zone increase in accordance with the strengthening stiffness and the slope is closely connected with the strengthening stiffness E_f and is not influenced by the kind of FRP sheet [1] Hence that the strengthened column confinement governed by the failure of the FRP rather than the concrete itself. Based on [2] there are two factors contributing to the confined behavior which are confinement stiffness and confinement rupture strain capacity.

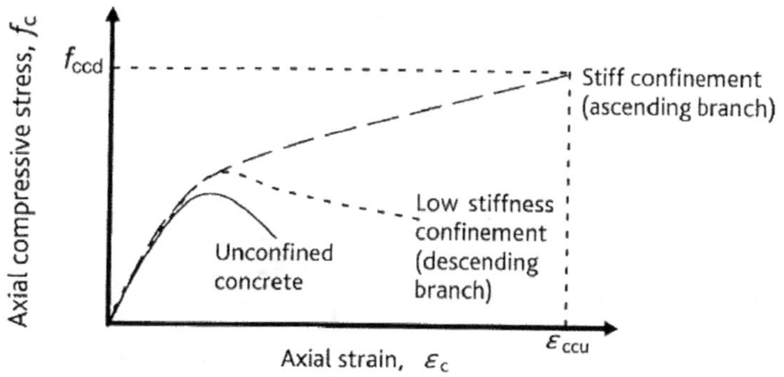

Figure 1. Idealized stress-strain curves for FRP-confined concrete [3].

1.2. Confined concrete for square column

The confinement action exerted by the FRP on the concrete core is a passive type and will be activated by lateral expansion of concrete under axial load. Lateral strain increase as the axial stress increases and the confining device develops a tensile hoop stress balanced by a uniform radial pressure which reacts again concrete lateral expansion. [4] When column strengthened with FRP subjected to axial compression, the concrete will expand laterally, and expansion is restrained by the FRP.

Based on [5] research, FRP confinement for the square and rectangular column not as effective in increasing the axial strength compared to the circular column and the effectiveness of FRP confinement in increasing the axial strength decrease as the unconfined concrete compressive strength increases. This is because the confining action for square and rectangular columns is mostly concentrated at the corners rather than around the entire parameter and thus level of confinement varies throughout the cross-section. Research conducted by [6] and [7] shown the confined square columns with reasonable levels of rounding at corners increase 50% maximum in compressive stress while for circular column increase 200% in compressive stress. The efficiency decreases further with columns of rectangular cross-section. Figure 2 shows the effective confinement areas in circular, square, and rectangular columns.

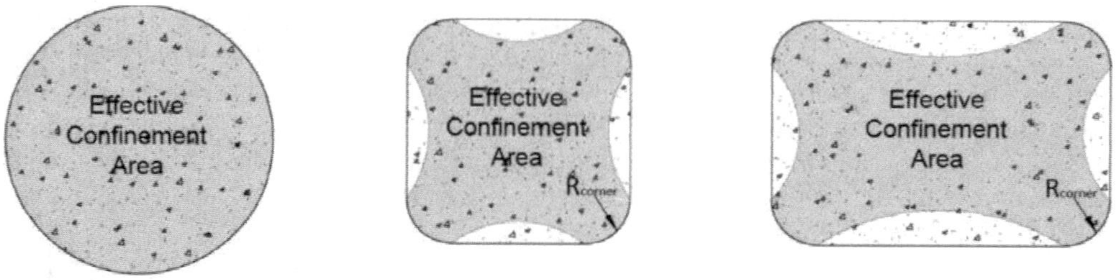

Figure 2 Effective confinement areas in circular, square, and rectangular columns [8].

2. Methodology

2.1. Structural assessment and evaluation

Based on visual observation, the overall condition for the beam and column are still in good contact despite being abandoned for about two years. Both column and beam do show a little significant quality of deterioration. There is honeycomb problem at both beam and column but it is not very serious and can still be repaired. However, there is present of crack on the slab and it can result in structural failure.

There are four carbonation tests conducted for the Block B which is at the first floor level for beam GH/1, column J/1 and D3 and at the corridor Column K/0. The carbonation test result was tabulated in table 1 shows that there is inconsistency in concrete structural members when reacting to phenolphthalein as a pH indicator. There is some part of the concrete structures turns to purple and there is some part of it does not change to purple. When concrete turns to purple shows that the concrete not carbonated with the carbon dioxide in the air and this concrete can still protect the reinforcement from corrosion and vice versa. This shows that there is problem with the preparation of the concrete mixture.

Table 1 Carbonation test results for block B.

Point	Structure	Location	Carbonation Depth	Note
1	Beam GH/1	First floor	< 5mm	Concrete not carbonated
2	Column J/1	First floor	< 2mm	Concrete not carbonated
3	Column K/0	Corridor	> 20mm	Carbonated Concrete
4	Column D3	Ground floor	< 5mm	Concrete not carbonated

Ferro scan testing was conducted at three different locations at the Block B which are one at ground floor level for staircase column and another two at the first-floor slab. Result shows that the steel reinforcement was installed according to the construction drawing. However, verification of steel reinforcement was conducted to double confirm the Ferro Scan result and identified the type and size of steel reinforcement.

The steel reinforcement verification was conducted at first-floor slab F1-G/2-3. Based on the steel reinforcement verification result it is found out that the type and spacing of the reinforcement used do not comply with the construction drawing. The type of the reinforcement used is BRC mesh reinforcement instead of high tensile strength reinforcement. Although the size of the reinforcement is the same but the spacing provided differ from the construction drawing. The comparison of steel verification with construction drawing is tabulated in table 2.

Table 2 Comparison between steel verification result with construction drawing.

	Main Reinforcement (mm)	Link (mm)	Concrete Cover (mm)	Result
Construction Drawing	T10-200	T10-200	25	Not comply the construction drawing
Steel reinforcement verification	BRC-10 (200 x 220)	BRC-10 (200 x 220)	20 (Bottom Cover)	

MS 1242:1991 (Clause 6.5.3) specified that the compression strength of in-situ concrete when tested should be 80% of the design strength. In order to determine the reinforced concrete strength, there are fifteen (15) coring test, eighteen (18) Ultrasonic Pulse Velocity (UPV) test and thirty-two (32) rebound hammer test is conducted. From the results obtained it is shown that there is concrete structure member does not achieve 80% of the design concrete (compression strength, f_{cu}=35N/mm^2). The reinforced concrete strength for slab range between 17.6N/mm^2 to 47N/mm^2 with almost half of the test specimens does not achieve the design compressive strength. For column structural there are about 27 testings conducted and about 20 percent of the specimen failed to achieve the minimum design compressive strength with range of 10.68N/mm^2 to 54.09N/mm^2. Even though the compressive strength for the beam achieved the minimum design compressive strength but it has raised question since there are only two tests conducted for the beam.

2.2. Structural analysis using Orion Software

Three-dimensional linear elastic behavior models of the reinforced concrete building were analyzed using Orion 18. The height of the building is 10.8m with length 50.3m and width of 11.9 m. Typical floor height is 3.6 m. In this research, the RC building will be simplified in the Orion 18 where only bare frame with slabs is constructed. Some assumptions and approximation modeling were made when creating a reinforced concrete in Orion 18.

The structural properties of the bare frame building were assigned in accordance with BS8110 design code consideration. The beams, columns, and slab thickness cross-section are tabulated in table 3 as described in as-built drawing.

Table 3 Material properties details for the structure model.

No.	Design Parameters	Values of Parameters
1	Compressive strength concrete, f_{cu} for Beam and Column	Minimum 20N/mm^2
2	Characteristics strength of high yield strength, f_y for longitudinal reinforcement	460 N/mm^2
3	Characteristics strength of high yield strength, f_y for shear link reinforcement	250 N/mm^2
4	Nominal concrete cover ground slab, ground beam, stump and column	40mm
5	Nominal concrete cover upper slab and upper beam	25mm
6	Density of concrete, ρ_c	24kNm^3

2.3. Finite element modelling for column strengthening

The nonlinear analysis is developed by means of ABAQUS Finite Element Software to simulate the nonlinear behavior of confined column. A three-dimensional deformable solid analysis is conducted to have realistic model. The model general configuration is a vertical column fixed to bottom which is restrained in all directions. For this study, a column at the Ground Floor area in grid line 2/D was chosen for the column strengthening with CFRP jacketing. This column has cross-sectional dimension of 300mm x 300mm with 3600mm in height with main reinforcement and shear reinforcement as shown in figure 3.

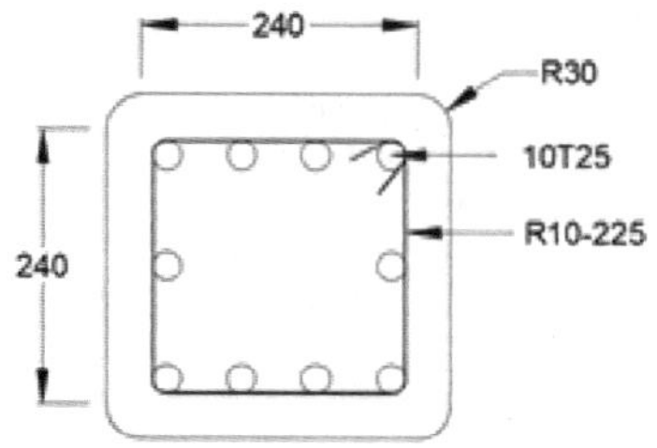

Figure 3 Cross-section of the strengthened column.

Each corner of the column was rounded with 30mm radius to avoid premature failure of CFRP due to shearing at sharp corner and to improve the RC column behavior. There are four (4) models of column with different number of CFRP layers was modeled and each modeled was named as tabulated in table 4. In all cases, the principal fibers were oriented perpendicular to the column axis, in a so-called 0° orientation. Based on [9], the epoxy-based bonding agent can be ignored due to its lower stiffness and for the simplicity of the model.

Table 4 Summary of column strengthening with CFRP jacketing modelled.

Column	Longitudinal steel	Lateral steel	No. of CFRP layers
RColumn0W	10T25	R10 – 225mm	-
RColumn1W	10T25	R10 – 225mm	1 layer
RColumn3W	10T25	R10 – 225mm	3 layers
RColumn5W	10T25	R10 – 225mm	5 layers

Concrete grade for the column was set to compressive strength 20N/mm^2 since the column compressive strength result tested with the destructive and non-destructive test varied from 54.09N/mm^2 to 10.68N/mm^2 and that is the lowest compressive strength can be set in Orion 18 Software. Concrete column was modeled using a solid eight-node element (C3D8R) due to the concrete core deformation characteristics when subjected to axial compression. The material properties for concrete column was set as elastic behavior and by taking into consideration of concrete damaged plasticity as suggested in ABAQUS software manuals. Young's Modulus of concrete, E_c, and stress-strain curves for compressive behavior was calculated by using equation (1), equations (2), and equation (3) respectively. Poisson's ratio, v, and concrete damage plasticity are inputted as shown in table 5 based on simplified concrete damage plasticity (SCDP) by [10]

$$E_c = 4700\sqrt{f_c'} \tag{1}$$

where f_c is compressive strength of concrete at 28 days in MPa.

$$\varepsilon_0 = \frac{2f_c}{E_c} \tag{2}$$

$$f = \frac{E_c\,\varepsilon}{1+\left(\frac{\varepsilon}{\varepsilon_0}\right)^2} \tag{3}$$

where ε_o is the strain at ultimate compressive strength, ε is the strain at stress f_c' and f is the stress for a value in the stress-strain relationship of concrete.

Table 5 Material properties for concrete with concrete damage properties [10].

Concrete Elasticity	Youngs Modulus, Ec	21019 MPa
	Poisson Ratio, v	0.2
Concrete Plasticity	Dilation angle	31
	Eccentricity	0.1
	f_{bo}/f_{co}	1.16
	K	0.67
	Viscosity parameter	0

	Stress (MPa)	Strain
Concrete Compressive Behavior	6.00	0.00000
	11.25	0.00030
	15.32	0.00060
	17.99	0.00090
	19.45	0.00120
	20.03	0.00150
	19.99	0.00180
	19.59	0.00210
	18.97	0.00240
	18.23	0.00270
	17.46	0.00299
	16.68	0.00329
	15.92	0.00359
	15.19	0.00389
	14.51	0.00419
Concrete Tensile Behavior	2	0
	0.02	0.000943396

The steel reinforcement was modeled using three-dimensional truss element (T3D2). The material properties for steel reinforcement was set as elastic perfectly plastic behavior with Young's Modulus and Poisson's ratio of 200GPa and 0.3 respectively. Steel reinforcement was embedded in the concrete column.

CFRP sheet is modeled with deformable three-dimensional shells obtained by extrusion elements (S4R). The CFRP was modeled with linear elastic behavior up to rupture as it has elastic brittle behavior. In this study, CFRP is considered as orthotropic element. Hence, it is necessary to consider the properties of the element in each direction separately. The details properties of the CFRP sheet are taken from [11] studies as modeled in validation modeling. The Young's Modulus in the main direction is set to 230GPa and the thickness of CFRP is 0.129mm. While for the orthotropic material model, in engineering constant in the ABAQUS software are as shown in table 6. Interaction of CFRP sheets and RC column was model with surface-to-surface contact elements. The surface of CFRP sheets was defined as a master surface whereas the concrete surface was defined as slave surfaces.

Table 6 Orthotropic material properties in each direction [11].

Young's Modulus		**Shear Modulus**		**Poisson ratio**	
E_{11}	230GPa	G_{12}	7GPa	v_{12}	0.3
E_{22}	12GPa	G_{13}	7GPa	v_{13}	0.3
E_{33}	12GPa	G_{23}	7GPa	v_{23}	0.45

3. Result and discussion

3.1. Structural assessment and evaluation results

The summary of testing results is tabulated in table 7. From the structural assessment and evaluation, it is shown that there is inconsistency in the RC structure construction. Therefore, structural strengthening is required in order to increase structural strength and preserve structural stability. Before structural strengthening is conducted, an analysis of the whole building is required in order to check the structural integrity of the building. In this case, the concrete strength of 20N/mm^2 was adopted in structural analysis. All structures members of the building were set to the worst case of 20N/mm^2 since the column compressive strength result tested with the destructive and non-destructive test was varies from 54.09N/mm^2 to 10.68N/mm^2 and 20N/mm^2 is the lowest compressive strength that can be set in Orion 18 Software.

Table 7 Summary of the destructive and non-destructive testing results.

Method of testing	Number of test	Results
Visual Observation	-	- honeycomb problem at both beam and column - present of crack on the slab
Carbonation Test	4 tests	There is inconsistency in concrete structural member when react to phenolphthalein as a pH indicator
Ferro Scan	3 tests	Steel reinforcement was installed according to the construction drawing
Steel Reinforcement Verification	1 tests	Type and spacing of the reinforcement used do not comply with the construction drawing.
Coring Test	15 tests	36 testing for slab shows concrete strength from 17.6N/mm^2 to 47N/mm^2 with almost half of the test specimen are not achieve the design compressive strength. 27 testing for column shows concrete strength from 10.68N/mm^2 to 54.09N/mm^2 with 20 percent of the test specimen are not achieve the design compressive strength. Beam shows that it is achieved the minimum design compressive strength. Only two tests conducted for the beam.
Ultrasonic Pulse Velocity	18 tests	
Rebound Hammer Test	32 tests	

3.2. Structural integrity of the building

From the analysis conducted in Orion 18, it was found that there is one line of the continuous beam (beam 2/A-L1) and seven columns (column 2/D, 2/E, 2/F, 2/G, 3/D, 3/E, and 3/F) fail the building analysis. This shown that these columns and beam need to be strengthened in order to restore the structure integrity and to ensure the safety of the building. The results obtained from the analysis also explained the column 2/D selection for the FEA as it carries the highest loading compared to the other fail column. The fail structure element was shown in figure 4.

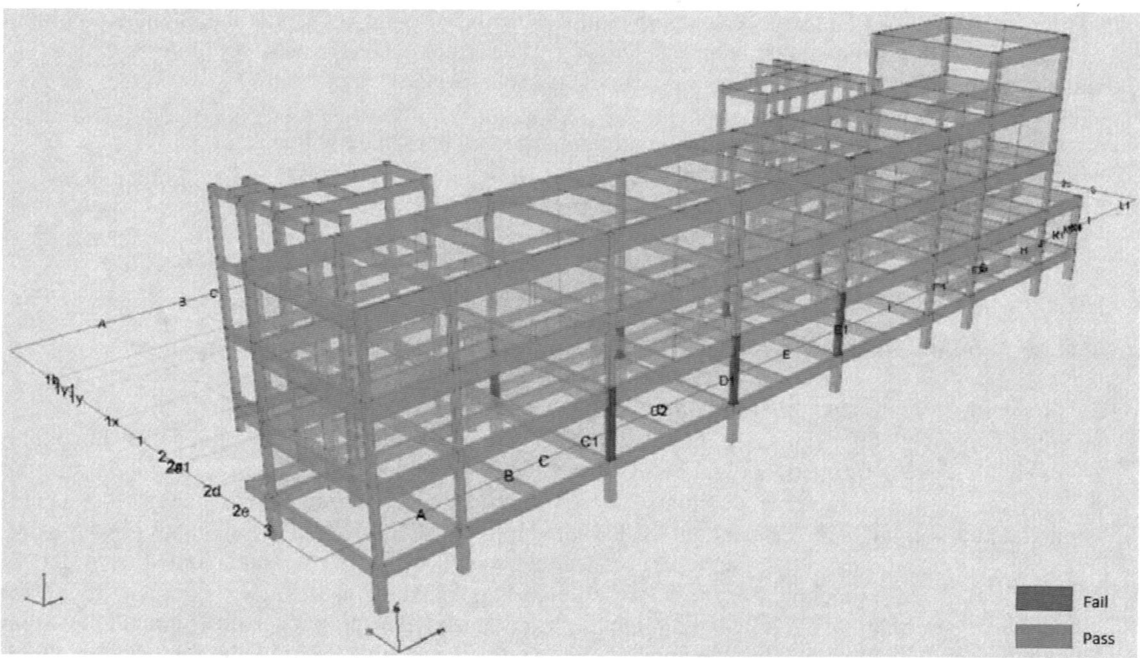

Figure 4 Result from the Orion 18 shows failure in columns and beam in block B.

3.3. Results for column 2/D in ABAQUS FEA

Nonlinear analysis in ABAQUS Finite Element software was used to study the behavior of the column with and without CFRP jacketing. Column RColumn0W was modeled without CFRP as the control model and the other three model RColumn1W, RColumn3W, and RColumn5W were modeled with one layer, three layers and five layers of CFRP respectively. Strain at the upper parts of the column were measured and plotted against stress for each model. The stress-strain curve was plotted and tabulated as in figure 5 and table 8.

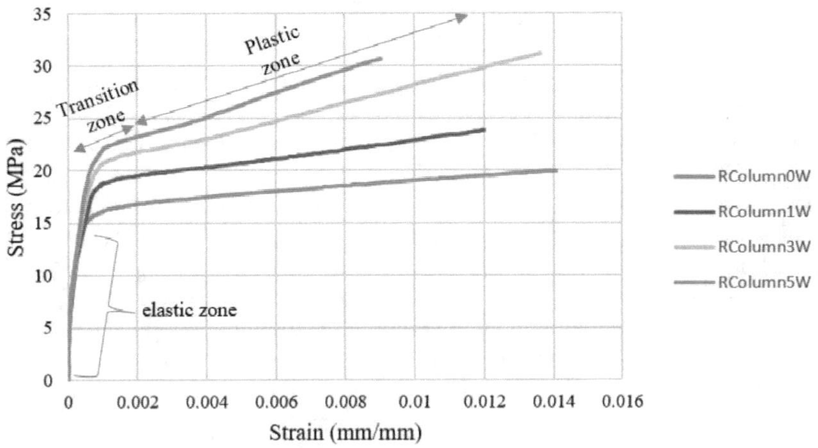

Figure 5 Comparison of stress-strain curve for column with and without CFRP Jacketing.

Table 8 Stress-strain value from ABAQUS FEM.

Column	Stress (MPa)	Strain (%)
RColumn0W	16.51	1.41
RColumn1W	23.82	1.20
RColumn3W	31.12	1.36
RColumn5W	30.69	0.90

4. Conclusion

Based on the results obtained it shows that it is important to conduct structural assessment and evaluation towards the abandoned structure before the new appointed contractor continues the remaining works. Due to the inconsistency of the structural performance, a structural strengthening needs to be done.

The analysis result of the CFRP jacketing shows good performance in the strengthened column. It is shown that the CFRP jacketing can improve the behavior of the RC column when subject to axial load. It can be concluded that the axial capacity of the column increases as the number of layers CFRP sheets increase with almost the same deformation level of the column.

Nowadays CFRP structural jacketing have high potential in construction industry. This study only focuses on column strengthening using full-length wrap with one direction CFRP orientation only. It is not necessary to wrap the whole length of the column with CFRP. Further study on partially wrap CFRP jacketing can be made to save the construction cost. On the other hand, the degree of fiber orientation of CFRP sheets also would give different results in structural strengthening. Another recommendation can be made is to combine the CFRP sheets and plate application in column strengthening technique. The combination of these two different materials might increase the column performance in axial stress and deformation. There are a few types of CFRP plate available in the market as compared to the CFRP sheet.

Acknowledgments

The authors gratefully acknowledge the financial support of the Faculty of Civil Engineering, Universiti Teknologi MARA, Selangor MALAYSIA.

References

[1] De Lorenzis L and Tepfers R 2004 Applicability of FRP confinement to strengthen concrete columns *Nordic Concrete Research Publication* (Oslo: The Nordic Concrete Federation 1/200) **31** 4 64-72A

[2] Teng J G, Jiang T, Lam L and Luo Y Z 2009 Refinement of a design-oriented stress-strain model for FRP-confined concrete *ASCE J. of Comp. for Constr.* **13** 4 269-78

[3] Technical Report No. 55 2012 *Design Guideline for Strengthening Concrete Structures Using Fibre Composite Materials* 3rd Edition

[4] Lorenzis L D and Tepfers R 2003 A comparative study of models on confinement of concrete cylinders with fiber-reinforced polymer composites *ASCE J. of Comp. for Constr.* **7** 3 219-37

[5] Harajli M 2006 Axial stress-strain relationship for FRP confined circular and rectangular concrete columns *Cement and Concr. Comp. 10.1016/j.cemconcomp.2006.07.005* **28** 938-48

[6] Darby A P, Coonan R M, Ibell T J and Evernden M 2011 FRP confined square and rectangular columns under concentric and eccentric loading *Proc. of Adv. Comp. in Constr. (ACIC)* (Warwick)

[7] Toutanji H, Han M, Gilbert J and Matthys S 2010 Behaviour of large scale rectangular columns confined with FRP composites *ASCE J of Comp. for Constr.* **14** 1 62–71

[8] Parvin A and Brighton D 2014 FRP Composites strengthening of concrete columns under various loading conditions *1040-1056. 10.3390/polym6041040* **6**

[9] Matias J V, Julio E and Silvestre N 2016 Numerical modelling of circular concrete columns strengthened with hybrid FRP jackets *Available from file:///C:/Users/user/Downloads /ExtendedAbstract_Jorge_Matias_final.pdf*

[10] Esfahani M R, Kianoush R and Tajari A R 2007 Flexural behavior of reinforced concrete beams strengthened by CFRP sheets *Eng. Str. 10.1016/j.engstruct.2006.12.008* **29**: 2428-44

[11] Ayyad J 2016 *Finite Element Modelling of Reinforced Concrete Columns Strengthened Externally With CFRP Sheets* (Palestine: The Islamic University of Gaza, Unpublished Master Thesis)

Topographic change and sediment transport on Maeshima Tombolo tidal flat

S Kato[1], I Kuritani[1], T Tabata[2], R Nakamura[3] and T Okabe[1]

[1] Department of Architecture and Civil Engineering, Toyohashi University of Technology, Toyohashi, Aichi 441-8580, Japan
[2] Nakanihon Engineering Consultants Co., Ltd., Nagoya, Aichi 460-0003, Japan
[3] Civil Engineering Program, Department of Engineering, Niigata University, Niigata, Niigata 950-2181, Japan

Corresponding E-mail: s-kato@ace.tut.ac.jp

Abstract. For the management of the tidal flat environment and the preservation of several important functions of the tidal flat, the accurate topographic monitoring is indispensable. However, the appropriate measurement of the tidal flat topography is very difficult because water level changes every moment due to tidal oscillation, and the accountable time for the measurement is limited within several hours under low tide condition. In this study, topographic measurement of tidal flat using Unmanned Aerial Vehicle (UAV) was conducted 5 times over a year, and Digital Elevation Model (DEM) obtained by UAV photogrammetry (UAV-DEM) of the tidal flat was constructed in each measurement. The area of 400m in the land-island (L-I) direction and 100m in the west-east (W-E) direction was set on the developed UAV-DEM to analyze the short-term topographical changes (2 weeks including typhoon events, half year and year). By the influence of typhoon passing, sand on the tidal flat was transported from east to west obviously within typhoon event. However, after the topographic change due to the typhoon event, the sand moved opposite direction. By the estimation of the sand volume on the tidal flat using UAV-DEM, it was found that the change of the volume was quite small in one year. These results indicate that the sand volume on this tidal flat is mostly preserved although active sand transport occurs on the tidal flat.

1. Introduction

Tidal flat has several important functions, such as formation of fine biological environment, the water quality purification, and economic value creation [1], [2]. To manage the tidal flat environment, it is indispensable to carry out the accurate topographic measurement [3], [4], [5]. However, the topographic measurement on the tidal flat is difficult to conduct appropriately because water level changes from hour to hour by tidal condition, and the accountable time for the measurement is limited within two to three hours under low tide condition. On the other hand, case studies of UAV monitoring on coastal topography have been reported recently. Mancini et al. [6], Kuroiwa et al. [7], and Matsubayashi et al., [8] conducted topographic measurements by UAV in the beach area. They examined their UAV measurement accuracy and compared their accuracy with those of conventional methods. However, there are few examples of UAV topographic measurements on the tidal flats. Long et al. [9] showed that the tie points between photographs are smaller in white and relatively flat areas like tidal areas, making it difficult to create a Digital Surface Model (DSM). To overcome this difficulty, Tabata et al. [10] utilized UAV to conduct short-time topographical measurements of a tidal flat. The UAV measurement result was compared with Real-Time Kinematic - Global Navigation Satellite System (RTK-GNSS) measurement result to verify the UAV measurement accuracy. The comparison demonstrated the possibility of topographical measurements during the low tide by UAV with a high accuracy within ± 5 cm of error. In this study, based on the outcomes by Tabata et al. [10], Digital Elevation Models by UAV (UAV-DEM) of the tidal flat were developed, and the short term topographic changes by about two weeks including typhoon event, half year and one year were investigated by UAV-DEMs.

2. Study area

The topographic measurement was conducted on a small tidal flat located at Nishio city, Aichi prefecture, Japan (figure 1). This tide flat appears between the offshore island (Maeshima) and the mainland at low tide. This tide flat forms a tombolo that connects the island to the mainland. Therefore, this tidal flat is called Maeshima Tombolo tidal flat. This tidal flat is famous as the place of shell gathering at low tide. Many people visit this tidal flat every year. However, local people, especially fishery persons, have been recognizing the changes in tidal flat topography and biological environment in recent years. It is hard to find the source of sediment supply to this tidal flat, such as big rivers, near this tidal flat. It means that it will be hard to restore the tidal flat environment once sediment on the tidal flat decreases and its topography changes, especially shrink, significantly.

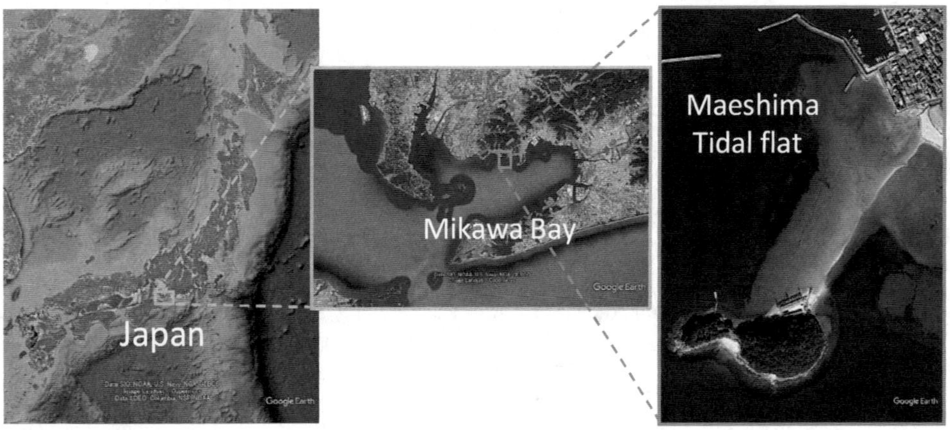

Figure 1. Study area (Left : Japan, Middle : Mikawa Bay, Right : Maeshima Tidal flat).

3. Summary of topographic measurement

The topographic measurement was conducted five times from September 2017 to August 2018 (table 1) at low tide. Since a wide area of the tidal flat dries out at the lowest tide, the aerial photographs of the tidal flat were taken about one hour before and after the lowest tide time to survey the topography. During our monitoring period, Typhoon 18 in 2017, which is cited as "TY1718" (Talim), from 17th to 18th September 2017, TY1812 (Jongdari) from 18th to 19th July 2018 and TY1820 (Cimaron) from 23rd to 24th August 2018 passed near the research fields (figure 2). These atmospheric impacts resulted in the apparent topographic changes on this tidal flat. Network type RTK-GNSS survey was carried out simultaneously with aerial photographing by UAV to obtain the cross-sectional (East-West and Land-Island) profiles of the tidal flat (figure 3) and positional information (X,Y,Z) of Ground Control Points (GCPs) placed on the tidal flat. The results of RTK-GNSS survey were used to evaluate the accuracy of UAV-DEM.

Table 1. List of UAV measurements.

Measurement	Date	Period
Meas-1	5th, September, 2017	
		Prd-1
Meas-2	20th, September, 2017	
		Prd-2
Meas-3	3rd, February, 2018	
		Prd-3
Meas-4	11th, August, 2018	
		Prd-4
Meas-5	28th, August, 2018	

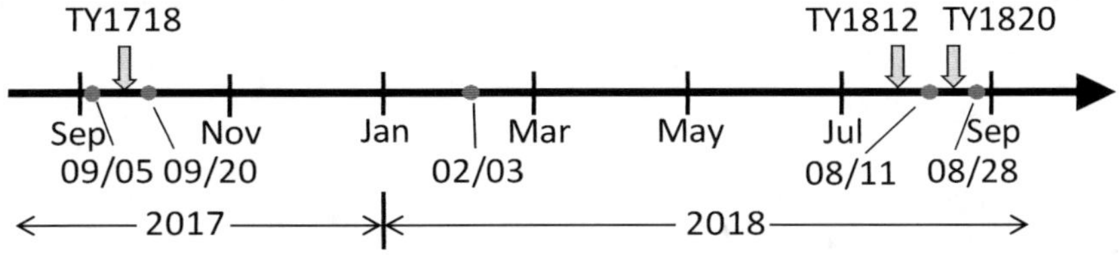

Figure 2. Time series of UAV topographic measurement and Typhoon event.

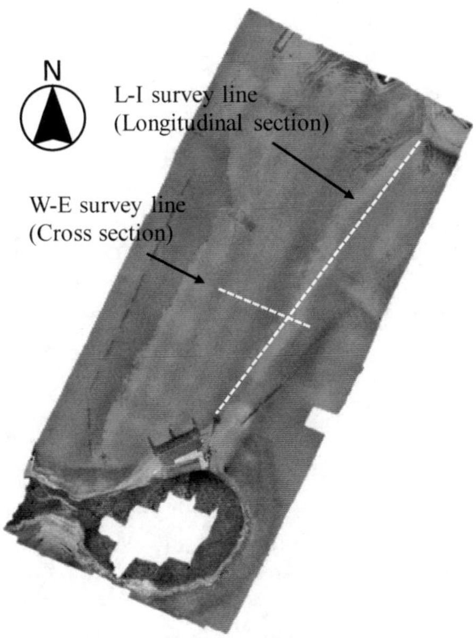

Figure 3. RTK-GNSS survey lines.

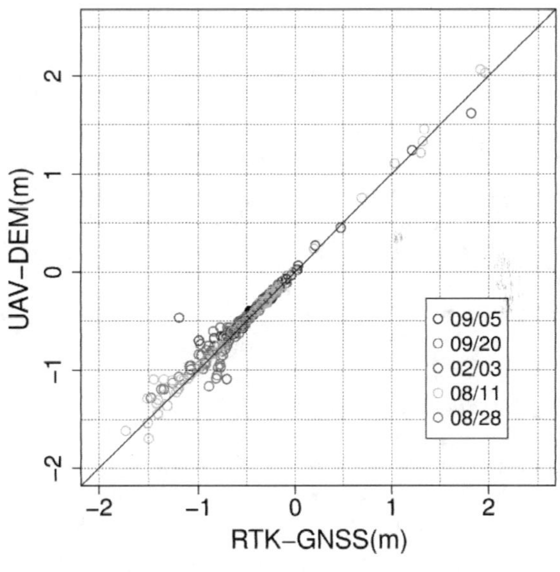

Figure 4. Comparison of elevation on the W-E and L-I survey lines by RTK-GNSS and UAV-DEM.

DJI Phantom 4 Pro was used to conduct the aerial photography in all measurements of this study. Because a manual operation of UAV photographing needs high manipulation skills and its operation has a high possibility to cause the unevenness of overlapping of UAV pictures [11], the automatic navigation system, DJI-GS-Pro, was used to perform aerial photographing appropriately in this study. An altitude of UAV flight was demonstrated at 50m over the tidal flat, in order to obtain a high spatial resolution in a picture. The 50m flight could take a picture of the tidal flat surface of about 90m by 60m with using 5472×3648 pixels, and realized a spatial resolution of about 1.7cm/pixel. In the standard aerial photography condition given in the manual for public survey using UAV by Geospatial Information Authority of Japan (GSI) [12], three or more outside GCPs, which are located outside the target area and are within 100 m of each other, are recommended. It is also recommended to place at least one GCP in the target area and keep the distance from the surrounding GCPs to 100m or less. However, it is difficult to set an outside GCP on the tidal flat because the area is surrounded by the sea. In addition, Rehak et al. [13] have found that the error increases when there is a bias in a GCP arrangement. In our measurement, only the inside GCPs are arranged at the interval of 30m to reduce the disadvantage of the lack of the outside GCPs. The ratios of over-lap (OL) and side-lap (SL) are set to 90% and 65% in the aerial photography although 80% and 60% are proposed in the manual by GSI as the general setting. The photos obtained by UAV are analyzed by Structure from Motion (SfM) processing using Photoscan Professional (Agisoft), and the Digital Elevation Model (DEM) of the tidal flat was developed. These conditions are same as Tabata et al. [10].

4. Results

4.1. Verification of accuracy

Figure 4 demonstrated the comparison of the results of UAV-DEM with RTK-GNSS data obtained on the survey lines (W-E line and L-I line) shown in figure 3. The UAV-DEM data on the RTK-GNSS survey lines are selected. On the whole, the UAV-DEM data indicate good agreement with the results of the RTK-GNSS survey. The root-mean-squared error (RMSE) was evaluated by about 5cm. However, some points indicate a large difference (error) between UAV-DEM and RTK-GNSS. Figure 5 shows the comparison of the elevation along the W-E survey line (cross section) and L-I survey line (longitudinal section) on February 3rd, 2018. The error, a value of UAV-DEM minus RTK-GNSS, is also indicated in this figure. It is identified that large errors occur at both ends of the W-E survey line. The height of the lowest water level on the survey date, February 3rd, 2018, was around 70cm below the mean water surface, which height was estimated using water elevation data at a tide station near this tidal flat because the water level was not measured on this tidal flat. Then, the areas indicating a large difference would be submersible areas when the photos were taken by UAV. On the other hand, a large difference occurred at the right-hand side, which is close to a mainland, in the figure 5(b) and (c). This area was the end of aerial shooting range (figure 3). Number of overlapped photo for SfM procedure in this area is less than in other areas. It caused low accuracy of UAV-DEM. These large errors correspond to the data greatly deviating from the 1:1 line in figure 4.

4.2. Morphological changes

To investigate the morphological changes during the UAV monitoring period, from September 2017 to August 2018, an analyzed area of 100m in the East-West (E-W) direction and 400m in the Land-Island (L-I) direction was set on UAV-DEM (figure 6). The mean elevation at 5m interval in the W-E direction and 10m interval in the L-I direction was obtained in this area (figure 6(a)) from UAV-DEM with higher resolution. In the mean elevation dataset, there were the data which seems to be too deep (much lower than the mean level). It is expected that these data are in submersible area mentioned in figure 5(a) and they have a slightly large error. Therefore, the elevation of -1.5m was set as the minimum value (lowest elevation) in this dataset. In addition, this mean elevation distribution was divided into small areas of 25m in the E-W direction and 50m in the L-I direction, so that the total small blocks are 32 (figure 6(b)). The volumetric changes were estimated in each small area, and the feature of the change of tidal flat morphology was investigated by the distribution of the sediment volume and its change.

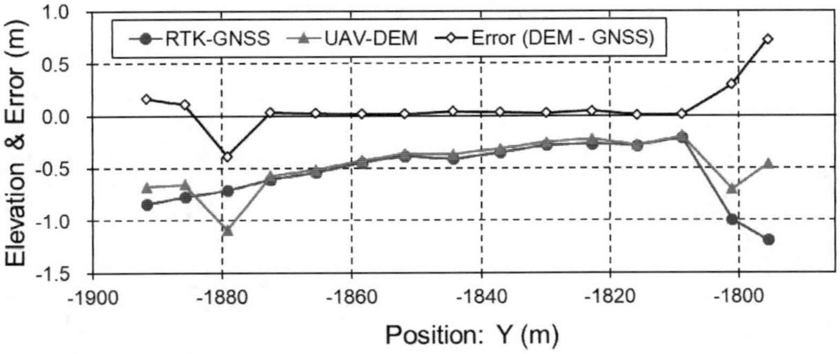

(a) Comparison and error on W-E (cross-sectional) survey line.

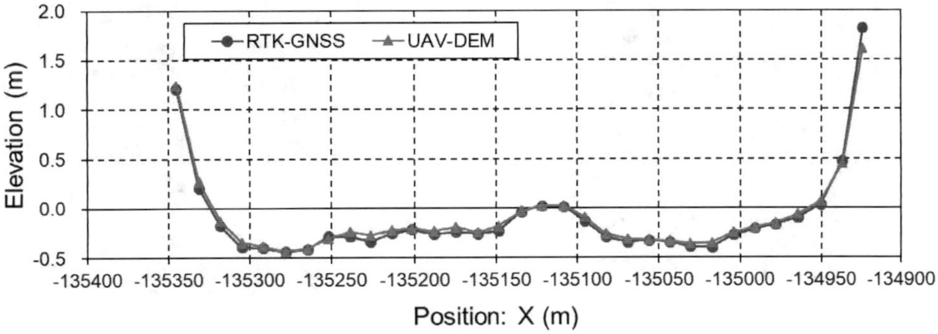

(b) Comparison of RTK-GNSS and UAV-DEM on L-I (longitudinal) survey line.

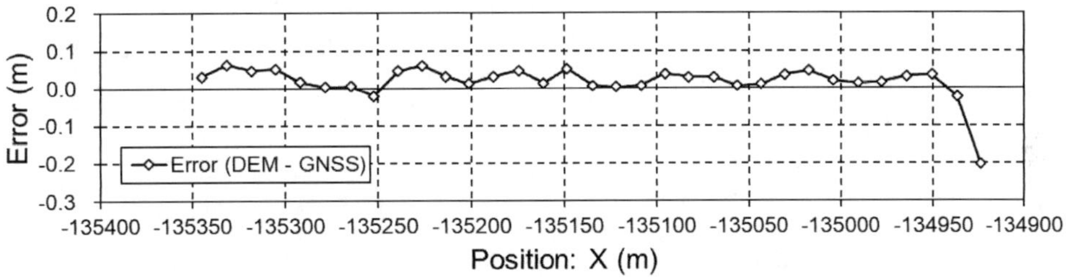

(c) Error between RTK-GNSS and UAV-DEM of (b).

Figure 5. Verification of UAV-DEM accuracy on 3rd, February, 2018 (Meas-3 in table 1).

4.2.1. Total volume of sediment on tidal flat. Figure 7 shows the time series of the total amount of sediment volume on the tidal flat. The sediment volume in each small block was estimated by multiplying a mean elevation by an area of a small block (50m by 100m). And the total volume of sediment within the analyzed area was calculated by adding the sediment volumes of all the small blocks. The red broken line in the figure represents the average of the estimated sediment volumes during the measurement period, and the solid blue lines indicate the sand volume of ± 5.0% from the average. From this figure, it can be confirmed that the total sand volume in the analyzed area has not changed substantially. In Tabata et al. [10], the occurrence of the topographic change of this tidal flat in about 6 months was confirmed. It is known that there is no source of sediment supply, such as a river, near the tidal flat. However, figure 7 indicated that no significant sediment volume changes occurred in one year.

These results would conclude that sediment transport occurs only in the tide flat (the analyzed area) and total sediment volume on the tidal flat is almost conserved.

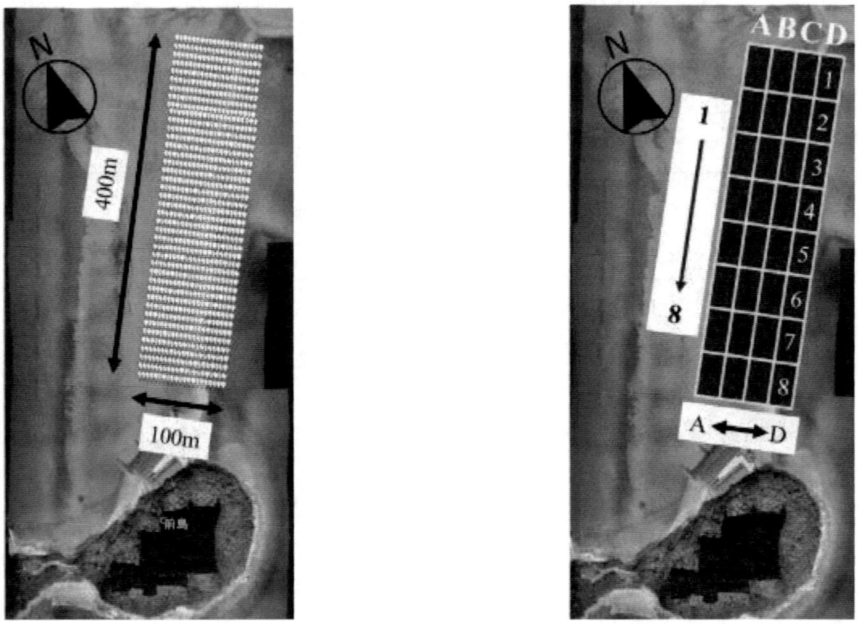

(a) Mean elevation point. (b) Small block for sediment volume estimation.

Figure 6. Analysis area of sediment volume in UAV-DEM.

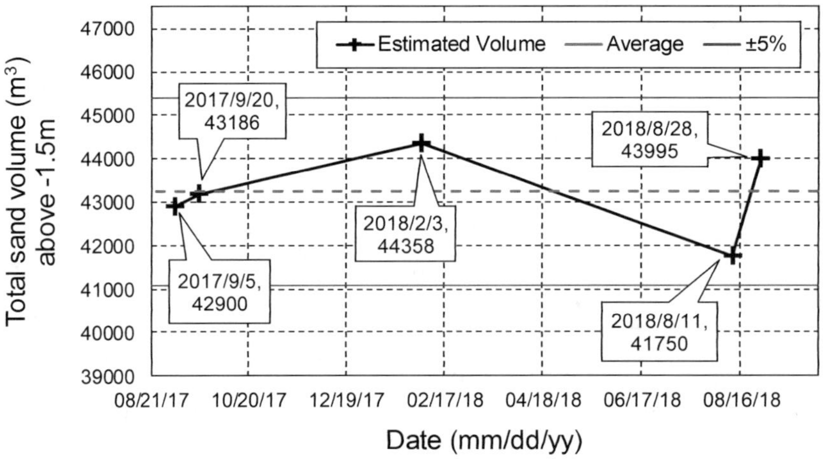

Figure 7. Change of sand volume in one year.

4.2.2. Topographic changes in short term due to typhoons. Tabata et al. [10] demonstrated that the deposition on the western area of the tidal flat and the erosion on the eastern area was caused by TY1718. The analyzed area (32 small blocks) estimating the sediment volume is divided into 4 longitudinal sections, (A) westernmost, (B) western, (C) eastern and (D) easternmost, as shown in figure 6(b), to investigate the spatial characteristics of the volume changes. Figure 8 shows the volume changes in Prd-

1 (2017/9/20-2017/9/5) and Prd-4 (2018/8/28-2018/8/11) to identify the influence of typhoon on the topographic changes. The volume changes of sediment due to a typhoon along the longitudinal sections, (A), (B), (C) and (D), are compared separately. The same tendency was displayed in (A), (B) and (C). However, the opposite one was shown in (D). The decrease of sediment volume (erosion) occurred in the island side (block of 4, 5, 6 and 7) although the increase of sand volume (deposition) occurred in the same area. The distinct difference between TY1718 and TY1820 was found in the trajectory of typhoon. The typical track of a typhoon around the main island of Japan was from west to east, and TY1718 followed that route. However, TY1812 approached the main island from east to west. That track was in the opposite direction. It was the extraordinary event in Japan and it was very difficult to estimate the influence of this typhoon on the topographic condition in the research area. It is inferred that the reverse trend on volume changes results from this rare phenomenon. It is also indicated the necessity of further monitoring of tidal flat topography to reveal the characteristics of topographic change and sediment dynamics due to a typhoon on this tidal flat.

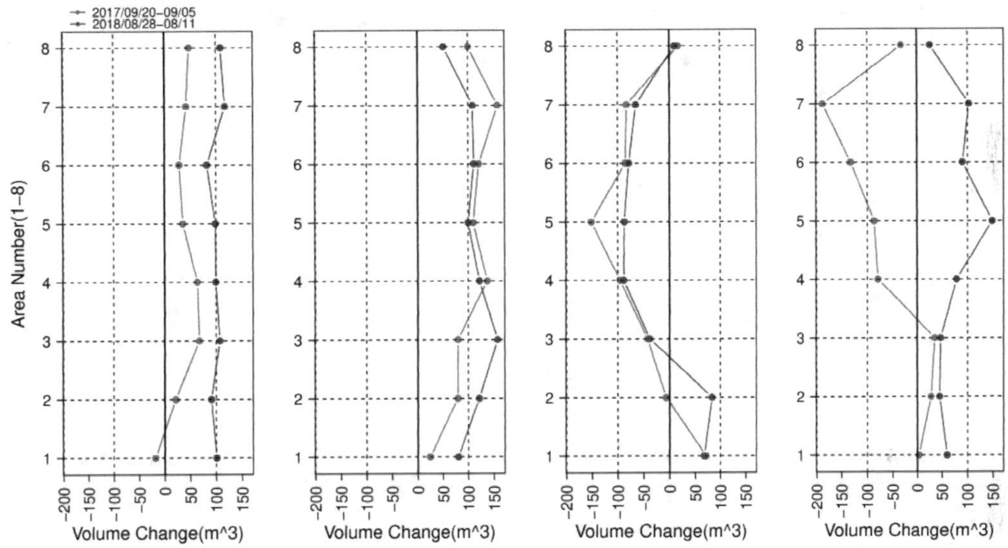

(a) Area A (Westernmost area) | (b) Area B (Western area) | (c) Area C (Eastern area) | (d) Area D (Easternmost area)

Figure 8. Distribution of volume change caused by typhoon passing along the L-I direction (Land: 1, Island: 8).

4.2.3. Topographic change of tidal flats in half a year. Figure 9 shows the volume changes of sediment in the period of Prd-1&2 (2018/2/3-2017/9/5), and Prd-3&4(2018/8/28-2018/2/3). Each period is almost half a year. Accumulation on the east-side area (C, D) and very small changes in the west-side area (A, B) in Prd-1&2 (red line) were confirmed. In the study by Tabata et al. [10], it was confirmed that the westward sediment transport occurred due to a typhoon, and the sediment transport in the opposite direction was generated after the typhoon. In this study, TY1718 passed close to the study area before Prd-2 (figure 2) and caused the topographic changes on the tidal flat. Then, it can be considered that the topographic change in Prd-1&2 included the recovery process from the changes caused the typhoon. This topographic recovery seems to continue further in Prd-3 after Prd-2. However, two typhoons passed near the research area and the opposite trend was demonstrated at the end of Prd-4. Significant erosion in the east-side area and slight deposition in the west-side area were found. As mentioned above, the track of TY1821 was opposite direction to the normal track. It is supposed that the eastward sediment transport was generated due to TY1812 before the end of Prd-3. About one month later, next typhoon TY1820 approached Japan and landed in Prd-4. The strength of TY1820, the lowest central pressure and the maximum wind speed, was greater than that of TY1812. But the track of TY1820 was far from this tidal flat compared to the track of TY1812. Therefore it is suggested that the influence of TY1812 which was the eastward sediment transport remained at the end of Prd-4. However, the same characteristics that the volume changes of sediment in both periods were larger on the island side than on the land side, especially in east-side area.

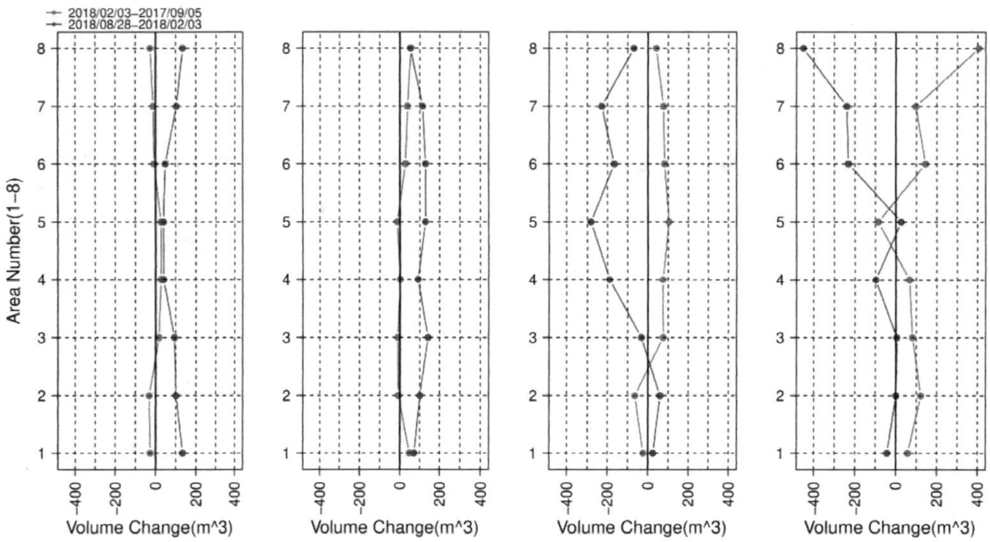

(a) Area A (Westernmost area) (b) Area B (Western area) (c) Area C (Eastern area) (d) Area D (Easternmost area)

Figure 9. Distribution of volume change in half a year along the L-I direction (Area 8 to 1).

4.2.4. Morphological change during the investigation period (1 year). Figure 10 shows the annual change of tidal flat morphology derived from Meas-1 (2017/9/5) to Meas-5 (2018/8/28). This Figure shows erosion in the east-side area and deposition on the west-side area. This feature is the same as the volume change after the typical typhoon passing. However, as mentioned in 4.2.1, the total amount of sediment on the tidal flat has hardly changed in one year. These results indicate that the existence of this tidal flat is maintained although active sand movement on the tidal flat causing the topographic changes are generated by natural phenomena.

About 1000m3 increase of total sediment volume in one year was estimated. The analyzed area is 40,000m2 (= 400m x 100m). And the mean increase of the tidal flat topography is calculated to be 0.025m. On the other hand, the accuracy of UAV-DEM is about 5cm (section 4.1). Then the mean topographic change is less than the accuracy (limitation) of UAV-DEM. Figure 11 shows the change of the cross-sectional profiles in the analyzed area on the tidal flat. These profiles changed more than 5cm apparently. In terms of the total sediment volume, no significant change can be confirmed in the sediment volume of this tidal flat, but the cross-sectional profile has clearly changed and it has been confirmed that active sediment movement has occurred on this tidal flat.

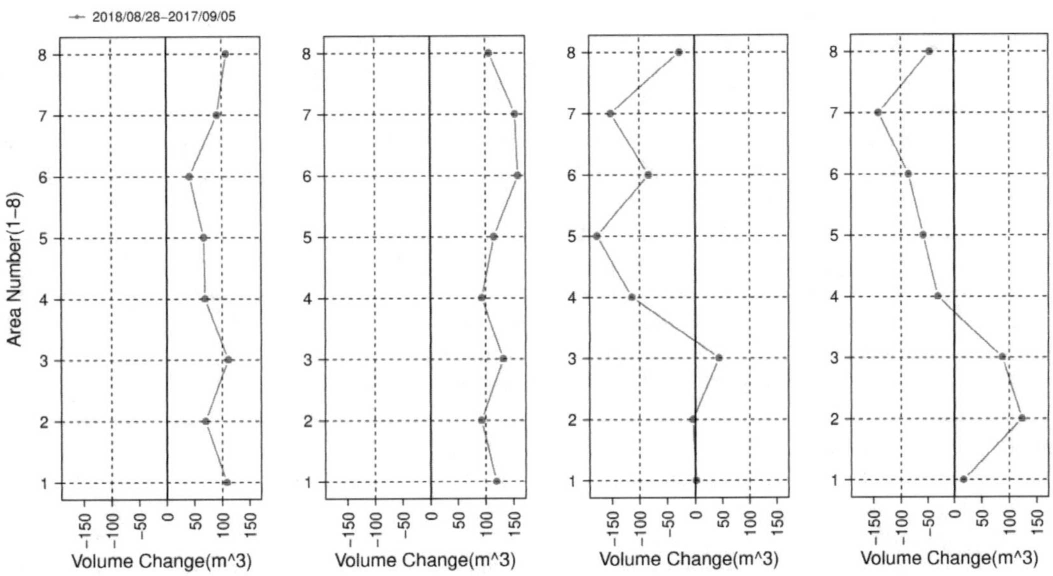

(a) Area A (Westernmost area) (b) Area B (Western area) (c) Area C (Eastern area) (d) Area D (Easternmost area)

Figure 10. Distribution of volume in one year along the L-I direction.

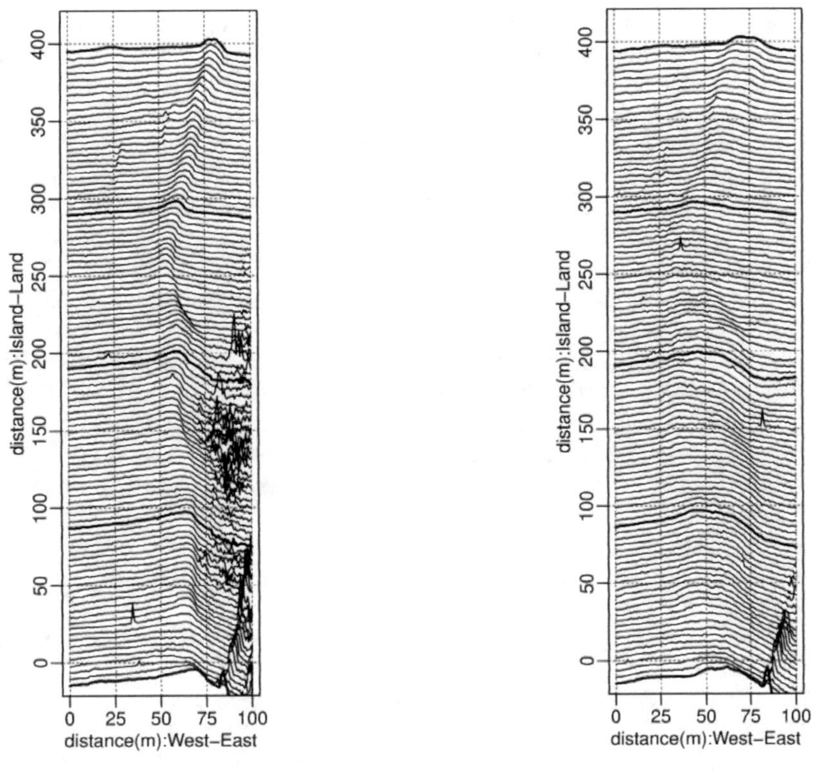

(a) 2017/09/05 (b) 2017/09/20

Figure 11. Change of cross-sectional profiles on the tidal flat.

5. Conclusions

In this research, the detailed topography of tidal flat was measured using UAV, and the measurement was conducted 5 times in one year. The topographic changes were investigated by analysis of UAV-DEM in some periods.

From the estimation of total sand volume on the tidal flat, it was demonstrated that sediment transport occurs only in the tideland and total sand on the tidal flat is almost conserved although topographic changes were caused by natural phenomena, such as typhoon.

By the influence of typhoon, sand on the tidal flat was transported from east to west in case of typical typhoon tracks. However, during the monitoring period of this research, the extraordinary event which typhoon moved in the opposite direction occurred. The different trend of topographic change and sand transport was resulted from two typhoon events in one year. Further monitoring of the tidal flat topography is necessary to grasp the characteristics of its change.

The morphological change of this tidal flat will occur continuously in the future. However, if the amount of sand does not change (does not decrease), the existence of this tidal flat will be maintained in the future. Therefore, in order to maintain and preserve the quality of the tidal flat, it is indispensable for investigating its topography in detail.

Acknowledgments

This research was conducted by the financial supports of JFE 21st Century Foundation and The Iwatani Naoji Foundation. Field survey on the tidal flat was supported by Higashi-Hazu fisheries cooperative association. Topographic survey using Network type RTK-GNSS was conducted by the support of JENOBA CO., LTD.

References

[1] Mitsch W J and Gosselink J G 2000 The value of wetlands: importance of scale and landscape setting *Gosselink : Ecological Economics* **35** 25-33
[2] Bridges T, Henn R, Komlos S, Scerno D, Wamsley T and White K 2013 Coastal risk reduction and resilience *US Army Corps of Engineers*
[3] Thom B G, Wright L D and Coleman J M 1975 Mangrove ecology and deltaic-estuarine geomorphology: Cambridge Gulf-ord river, western Australia *The J. of Ecology* **63** 1 203-32
[4] Wilson C A, Hughes Z J, FitzGerald, D M, Hopkinson C S, Valentine V and Kolker A S 2014 Saltmarsh pool and tidal creek morphodynamics: dynamic equilibrium of northern latitude saltmarshes? *Geomorphology* **213** 99-115
[5] Marsooli R, Orton P M, Georgas N and Blumberg A F 2016 Three-dimensional hydrodynamic modeling of coastal flood mitigation by wetlands *Coastal Engineering* **111** 83-94
[6] Mancini F, Dubbini M, Gattelli M, Stecchi F, Fabbri S and Gabbianelli G 2013 Using unmanned aerial vehicles (UAV) for high-resolution reconstruction of topography: The structure from motion approach on coastal environments *Remote Sensing* **5** 12 6880-98
[7] Kuroiwa M, Sueyoshi R, Ichimura Y and Fukuoka K 2016 Study on coastal topographic change analys utilizing UAV *J. of Japan Society of Civ. Eng. Ser. B3 (Ocean Engineering)* **72** 2 I_784-89 (in Japanese)
[8] Matsubayashi Y, Komatsu H and Ogasawara T 2017 Simplified method for coastal topography measurement with UAV *Journal of Japan Society of Civil Engineers Ser. B2 (Coastal Engineering)* **73** 2 I_1627-32 (in Japanese)
[9] Long N, Millescamps B, Guillot B, Pouget F and Bertin X 2016 Monitoring the topography of a dynamic tidal inlet using UAV imagery *Remote Sensing* **8** 5 387
[10] Tabata T, Kato S, Nakamura R, Oda T, Nishizono H and Okabe T 2018 Verification of measurement accuracy of tidal flat topography by UAV and applicability to continuous monitoring of tidal flat *Journal of Japan Society of Civil Engineers, Ser. B2 (Coastal Engineering)* **12** 2 I_961-6 (in Japanese)
[11] El Meouche R, Hijazi I, Poncet P A, Abunemeh M and Rezoug M 2016 UAV Photogrammetry

Implementation to Enhance Land Surveying, Comparisons and Possibilities *International Archives of the Photogrammetry, Remote Sensing & Spatial Information Sciences* 42

[12] Geospatial Information Authority of Japan (GSI) 2017 *Manual for public survey using UAV (draft)* revised in March, 2017 (in Japanese)

[13] Rehak M and Skaloud J 2015 Fixed-wing micro aerial vehicle for accurate corridor mapping *ISPRS Annals of Photogrammetry, Remote Sensing & Spatial Information Sciences* 10.5194/isprsannals-II-1-W1-23-2015

IOP Conf. Series: Materials Science and Engineering **933** (2020) 012016 doi:10.1088/1757-899X/933/1/012016

Slope stability analysis in The Babahrot-Blangkejeren road section sta. 13 + 885

Kurniawan[1], Y Yunus[2] and S M Saleh[2]

[1] Civil Engineering Postgraduate Program, Department of Civil Engineering, Universitas Syiah Kuala, Banda Aceh, Indonesia
[2] Department of Civil Engineering, Universitas Syiah Kuala, Banda Aceh, Indonesia

Corresponding e-mail: kurniawan.mts@gmail.com

Abstract. The Babahrot - Blangkejeren road section located in Aceh Barat Daya Regency is one of the roads in Aceh Province that often experience landslides because of its location, especially on the Singgah Mata Mountain track, which is in hilly areas. Landslides often occur at STA 13 + 885 at the foot of Singgah Mata Mount. Access to this lane is used for economic, educational, and social activities from Gayo Lues Regency to Blang Pidie, Aceh Barat Daya Regency. The impact of landslides is the disconnection of Gayo Lues community access to Southwest Aceh and vice versa. Therefore, it is necessary to do a slope reinforcement study at the top of the road construction. The analysis used to find the safety factor (FK) was Bishop Slicing Method. For comparison, the safety factor was also analyzed by using an application program (Plaxis). The results of the analysis of slope stability in the initial conditions using manual methods (Bishop Slicing Method) and using the application program (Plaxis) show that the average safety factor against landslides is smaller than 1. Thus, it can be said that, in the existing conditions, the slope is unstable. So, it is necessary to do slope reinforcement to prevent landslides. Based on the results obtained after the improvements to the slope angle and the use of reinforcement slopes with cantilever retaining walls, the value of the safety factor for landslides increased from 0.810 to 1.620.

1. Introduction

Indonesia is a country that experiences landslides many times. This is because most parts of Indonesia are in mountainous and hilly areas, so it is very vulnerable to landslides, specifically in the rainy season. As a result, it can cause enormous losses for the Indonesian people, especially the loss of life.

The Babahrot-Blangkejeren road section located in Aceh Barat Daya Regency is one of the provincial road segments in one of the Provinces in Indonesia, namely Aceh. This road section experiences landslides many times because of its location in hilly areas, especially on the Singgah Mata Mountain track. Landslides often occur, resulting in the obstruction of the middle lane of the Aceh Province road system and the obstruction of economic, educational, or social activities from the Gayo Lues Regency to Blang Pidie, Aceh Barat Daya Regency. The impact of landslides resulted in the disconnection of Gayo Lues community access to Southwest Aceh and vice versa. It is necessary to handle the slopes to prevent the occurrence of landslides on these roads.

The slope stability analysis on the existing condition of the lower slope in Babahrot-Blangkejeren road section using manual method (Fellenius method) and application (Geoslope) obtained that safety factors at all points of the review were unsafe. After strengthening with the counterfort type retaining wall and changing the slope angle < 20°, a safety factor becomes more than 1.5, which is safe [1].

The calculation of the initial slope conditions with the Fellenius method obtained a safety factor of 1.4701 and, with the application (Plaxis), obtained the value of the safety factor of 1.3476. The value of the safety factor still less than the requirement for slope stability, which is 1.5. Alternative handling is done in two ways: the first one uses a counterfort wall plus bored piles along 10 meters, and the second one uses bored piles along 11 meters. The results show that the use of 11 meter bored piles can increase the safety factor to 1.6383, which is more than 1.5 and safe [2].

To support the analysis, supporting data is needed in the form of soil parameter data. The soil parameter data used is the secondary data obtained from the Aceh Highways Agency.

2. Research method

The study was conducted at STA 13 + 885 Babahrot-Blangkejeren section in the Southwest Aceh District. This road is a provincial road link that connects Southwest Aceh District and Gayo Lues District. Two methods were used to analyze slope stability. The methods are a manual method using the Bishop slice method and the application using Plaxis. This is to obtain the safety factors of existing conditions and safety factors of the post-reinforced conditions.

2.1. Sampling and testing of soil samples

The study was conducted at STA 13 + 885 Babahrot-Blangkejeren section in the Southwest Aceh District. This road is a provincial road link that connects Southwest Aceh District and Gayo Lues District. Taking undisturbed samples is done when testing boring at a certain depth. The sampling method follows the applicable SNI and ASTM procedures. The number of samples at that point is 1 (one) tube that taken at a depth close to the end of drilling. The tube is then taken to the laboratory to be tested for soil physical and mechanical properties. The tube carried will be coded according to the drill point and its depth. In addition, soil sampling is disrupted using the test pit method. The test pit work is to find out the type and thickness of the layer below the topsoil. Soil sampling for the pit 1 test was taken at STA 13 + 885 on the upper slope of coordinates of 03°56'41.8" N and 096°45'53.3" E.

Laboratory tests carried out on all disturbed soil samples and undisturbed soil samples. The disturbed soil samples were obtained at the test wells in Test Pit (TP) testing, whereas the undisturbed soil samples were obtained at Bored Hole (BH) in boring tests. Laboratory testing consists of testing the physical and mechanical properties of the soil. The whole test is based on the American Society for Testing and Materials (ASTM) standard.

Examination of the physical properties or the properties of soil indices, called properties test index, includes:

- Unit weight,
- Natural moisture or water content,
- Specific gravity,
- Atterberg limit, and
- Grain size distribution.

Meanwhile, for testing the mechanical properties of the soil (mechanical properties test), the testing of shear strength with the direct shear test includes cohesion (c), shear angle (ϕ), angle dilatancy (Ψ), modulus young (E), and poison ratio (ν).

2.2. Test result data

Based on the soil parameter data obtained, it is assumed that the land in that location is homogeneous. The homogenous land assumption was formed by averaging existing soil data into three types of soil, as found in table 1. The soil parameter values obtained in table 1 are used as input to the Plaxis program. The soil permeability coefficient 'kx' and 'ky' are assumed to be the same and are obtained from the results of interpretation with the soil type.

IOP Conf. Series: Materials Science and Engineering **933** (2020) 012016 doi:10.1088/1757-899X/933/1/012016

Table 1. Data on soil parameters.

Soil Parameters	Soil Layers			Unit
	Layer 1 silt clay soil (Blue)	Layer 2 sedimentary rock with a silt (Red)	Layer 3 solid sedimenatay rocks (Yellow)	
Material model	MC	MC	MC	-
Type of bahaviour	Drained	Drained	Drained	-
Dry soil weight (γ_{unsat})	16.31	21.56	23.52	kN/m^3
Saturated soil weight (γ_{sat})	18.91	23.52	25.48	kN/m^3
Horizontal permeability (k_x)	0.000198	0.0831	0.0831	m/day
Vertical permeability (k_y)	0.000198	0.0831	0.0831	m/day
Young's modulus (E_{ref})	2942.1	1271618	12000000	kN/m^2
Poisson's ratio (v)	0.4	0.3658	0.3658	-
Cohession (c)	2.03	1915	3351	kN/m^2
Friction angle (φ)	24.32	22	25	o
Dilatancy angle (ψ)	0	2	4	o
Shear Modulus(G)	-	465520	4270934	kN/m^2

The distribution of soil types on the slopes can be seen in figure 1. The blue layer is clayey silt soil; the red layer is a layer of sedimentary rock with silt; the yellow layer is a layer of solid sedimentary rock.

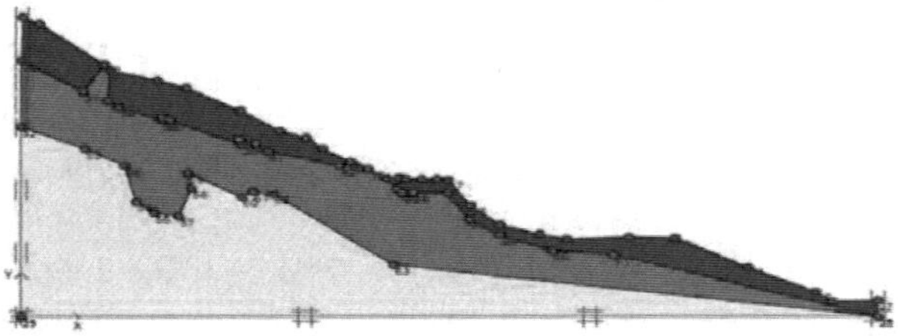

Figure 1. Distribution of types of soil on slopes.

2.3. Application program (Plaxis)

Plaxis is a program in the geotechnical field that uses the Finite Element Method for geotechnical applications; in this program, soil models are used to simulate the behavior of the soil. The material model described in the Plaxis program is a form of a mathematical equation that describes the relationship between stress and strain. This material model is often expressed in terms of added stress and added strain [3].

Plaxis can be used to conduct modeling and analysis of all geotechnical problems such as slope stability, seepage, and consolidation. In addition, Plaxis can also model and analyze geotechnical structures and soil interactions with structures such as shallow foundations, deep foundations, retaining walls, anchors, etc. It should be noted in the use of Plaxis that the user must master the theories and concepts concerning soil mechanics and foundation engineering. This is very important because Plaxis will still perform calculations and output even if the input data is incorrect.

The next discussion is the analysis of the initial slope conditions and slope conditions after the improvement of the slope angle and slope reinforcement using a retaining wall on the slope above the road. The initial conditions of the slope, for more details, can be seen in figure 2 below.

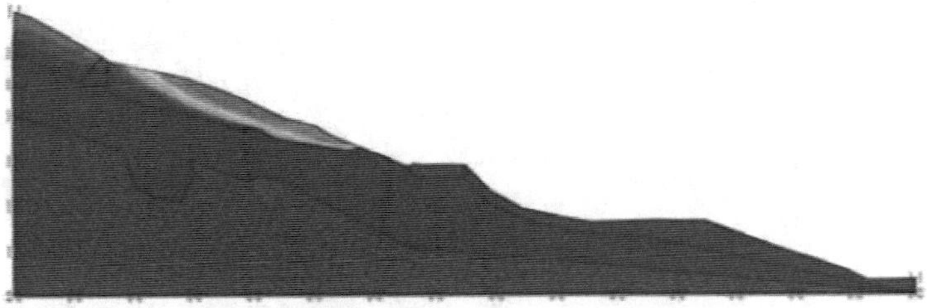

Figure 2. Initial slope conditions.

In figure 2, it shows that the landslide field in the initial condition of the slope. A landslide is found on the slope above the road body. The red layer shows a potential landslide layer of soil, while the blue layer shows a stable or safe soil or rock layer.

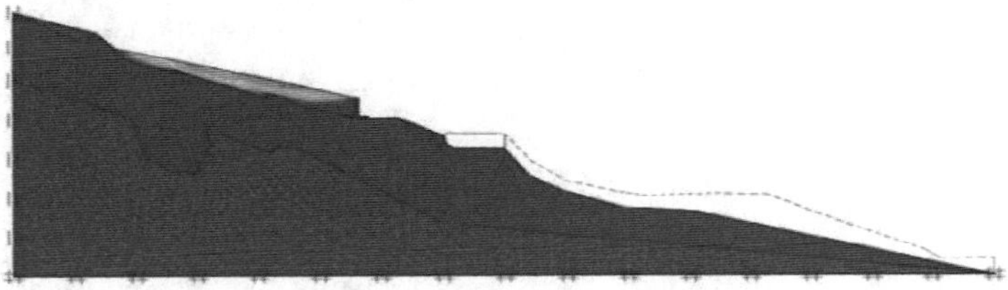

Figure 3. Slope condition after improvement of slope angle and addition of soil retaining walls.

Figure 3 shows the slope conditions after improvements to the slope angle and the addition of a retaining wall. It shows the condition of the slopes that have been stable against the threat of landslides.

2.4. Manual method (Bishop)

The Calculation of slope stability analysis on the body of this road uses a simplified Bishop slice method because the solution is more precise and will get a more appropriate safety factor. The Bishop method is the method introduced by A.W. Bishop, who uses a cut method in which the forces act on each piece, as shown in figure 4 and figure 5. The Bishop method is used to analyze the slip surface in a circular shape. In this method, it is assumed that the total normal forces work at the center of the cut and can be determined by describing the forces on the pieces vertically or normally. Balance requirements are used on the pieces that make up the slope. The Bishop method assumes that the forces acting on the slices have a resultant zero in the vertical direction. The cross-section of the landslide is divided into several sections or slices in the vertical direction, and the effect of forces on the edges of each slice is also taken into account [5].

Safety factor equation of the Bishop slice method used is as follows:

$$F = \frac{\sum_{i=1}^{i=n}[c'b_i + W_i(1 - r_u)\, tg\, \emptyset'] \left(\frac{1}{cos\, \theta_i (1 + tg\, \theta_i\, tg\, \emptyset'/F)}\right)}{\sum_{i=1}^{i=n} W_i\, sin\, \theta_i} \tag{1}$$

IOP Conf. Series: Materials Science and Engineering **933** (2020) 012016 doi:10.1088/1757-899X/933/1/012016

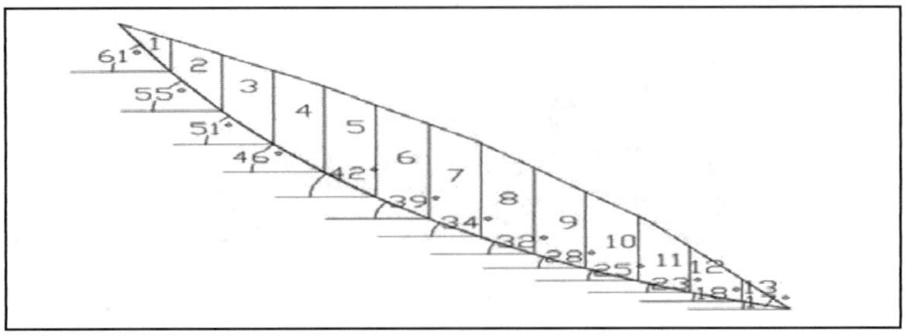

Figure 4. Slice the slope with a simplified Bishop slicing method.

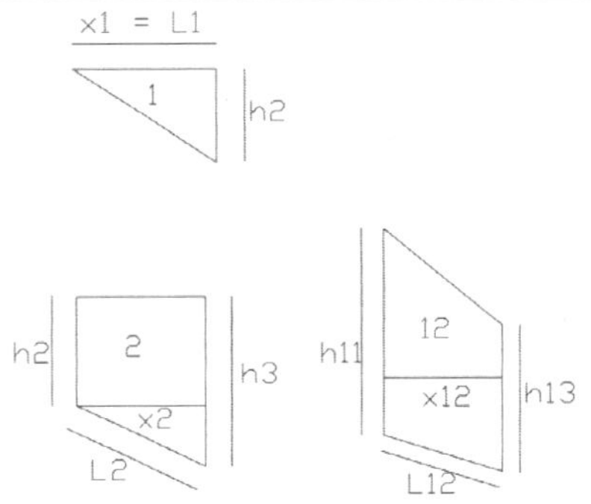

1, 2 and 12 = Incision number
x = Wide of the incision
L = Slip plane length
h = Side height of the incision

The area of the incision triangle 1
= (1/2.x1) (h2)

The Area of the incision trapezoidal 2 and 12
= (1/2.x2) (h2+h3)

Figure 5. The calculation of slope slices.

3. Result and discussion

The results of this method are the safety factors obtained using the Bishop method and using the Plaxis applications. The results can be seen in table 2 below.

Table 2. Safety factors for initial slope conditions.

Point (STA)	Safety Factor (Bishop)	Safety Factor (Plaxis)
13 + 885	0.822	0.810

From table 2, it can be observed that, in the initial condition of the slope, the safety factor obtained both by manual method and application is below 1 (SF <1). This means that all slopes are in critical condition. Therefore, efforts must be made to deal with it, so that there is no landslide.

The results of slope stability analysis in three conditions, which are initial condition, after reinforcement condition, and additional handling condition, using Plaxis applications, can be seen in table 3 below.

Table 3. Safety factors for initial slope conditions and after strengthening.

Cantilever Retaining Wall Stability	Safety Factor	Information
Initial Condition	0.810	Unsafe
After Strengthening Condition	1.620	Safe

From table 3, it can be observed that the initial conditions of the slope are not safe because, based on the results of calculation, the safety factor is smaller than 1 (SF <1). Then, after repairing the slope angle and using reinforcement slopes with retaining walls, the value of the safety factor for landslides increased from 0.810 to 1.620 or increased to double.

Based on the data displayed and the calculation results on potential landslides, it can be explained that there are three causes of potential landslides on the Babahrot-Blangkejeren road section in STA 13 + 885, namely:

- The condition of the slopes is steep about 35°
- There is a slip (rock layer) under the water-resistant soil surface
- There is enough water (rain) that enters the soil pores so that the soil pressure increases

These results are consistent with those stated by Brook [5] that landslides will occur if three conditions are met, namely: the slope is quite steep, there is a field of launch (rock) under the water-resistant soil surface, and there is enough water (rain) which enters the soil pores above the impermeable rock layer so that the soil pressure on the slope increases. This result is also supported by Bowles [6], stating that rainfall as one component of the climate will affect water content and saturation. Rain can increase the water content in the soil, and further will cause the physical condition of the slope's body to change. The increase in soil water content will weaken the physical-mechanical properties of the soil (affect the internal condition of the body of the slope) and reduce the slope safety factor. The analysis of the landslides in cililin also found that landslides occurrence are influenced by two factors, namely the control factor (the physical property of rock or soil making up slope, geological structure, slope, and hydrology) and the trigger factors (rainfall, vegetation cover, land use, and earthquake vibration) [7].

Based on the data used in the analysis of slope stability on each slope reviewed, in the initial conditions, the slopes at the Babahrot-Blangkejeren road section are in a critical condition. After repairing the slope angle and utilizing slope reinforcement with retaining walls, then the slopes are in a safe condition. This is in accordance with the function of the retaining wall, as stated by Bowles [8], that the retaining wall is one of the concepts of soil reinforcement that is widely used in civil engineering work to hold soil loads vertically or to certain slopes. Retaining walls are the constructions that are used to provide stability to the soil or other materials whose conditions do not have a natural slope. This result is also in line with the research conducted by Ciptaning [1] that the use of soil retaining walls by improving the slope angle can increase the safety factor of the slope to be more than 1.5, which is safe. Research conducted by Pratama [2] also found a similar thing, that the use of bored piles with a height of 11 meters can increase the safety factor to 1.6343, which is more than 1.5 (safe).

4. Conclusions

Based on the results and discussion of slope stability analysis on the Babahrot-Blangkejeren road section on STA 13 + 885, conclusions can be drawn as follow:

- Factors causing landslides on the Babahrot-Blangkejeren road include a reasonably steep slope, a sliding field which is a rock layer under the water-resistant soil surface, and enough water or rain that enters the soil pores and increases the pressure land;
- In the initial conditions, after being calculated using manual Bishop method and the application program (Plaxis), slope safety factors become less than 1 (SF <1); and
- Safety factor, calculated after slope reinforcement with soil retaining wall and after cutting or improving slope angle, increased from 0.810 to 1.620, which is more than 1.5 (safe).

References

[1] Ciptaning K, Yunus Y and Saleh S M 2018 Analisis stabilitas lereng dengan konstruksi dinding penahan tanah tipe counterfort *JARSP* **1** 2 58-68

[2] Pratama R B, Muhibbi I M, Dwi A I and Hardiyati S 2014 Analisis stabilitas lereng dan alternatif penanganannya: studi kasus longsoran jalan alternatif Tawangmangu Sta 3+150 – Sta 3+200, Karanganyar *J. Karya Teknik Sipil* **3** 3 573–585

[3] Brinkgreve R B J 2002 *PLAXIS: finite elemen code for soil and rock analyses: 2D-version 8: [user's guide]* (The Netherlands: Delf University of Technologi and PLAXIS b.v)
[4] Bishop A W 1955 The use of slip surface in the stability of analysis slopes *Geotechnique* (London) **5**
[5] Brook K N, Folliott P F, Gregersen H M and Thames J K 1991 *Hydrology and the Management of Watersheds* (Ames: Iowa State University Press)
[6] Bowles E J 1986 *Sifat-sifat Fisis dan Geoteknis Tanah (Mekanika Tanah)* (Jakarta: Erlangga)
[7] Tuhua, Changjang M L, Zhim Singapore ing L and Qiyun L B 2015 Rainfall intensity duration thresholds for initiation of landslides in Zhejiang Province, China, *J. of Geomorphology* **245** (Elsevier)
[8] Bowles J E 1982 *Foundation Analysis And Design* 3rd Edition (Singapore: McGraw-Hill Book Company)

The analysis of roadside obstacles to the performance of Syiah Kuala Street

R Musita[1], R Anggraini[2] and S Sugiarto[2]

[1] Civil Engineering Postgraduate Program, Department of Civil Engineering, Universitas Syiah Kuala, Banda Aceh, Indonesia
[2] Department of Civil Engineering, Universitas Syiah Kuala, Banda Aceh, Indonesia

Corresponding e-mail: renimusita@mhs.unsyiah.ac.id

Abstract. Urban roads are the segments that have permanent and continuous development, at least on one side of the road. The purpose of this study is to evaluate the performance of the road due to side constraints on Syiah Kuala Street, to find out how much the operational costs of the vehicle (BOK) based on the side obstacles on Syiah Kuala Street section, and to find out traffic management recommendations by considering subscriber scenarios. The research result indicates that the maximum volume of the vehicle at peak hour is 1987 pcu/hour (on Monday), and the corresponding roadside obstacle is 551.7, which is categorized in the high class. Based on this data analysis result, it is shown that the capacity on Monday is 1998 pcu/hour with the corresponding degree of saturation as much as 0.99. As a result, the level of service is E (unstable flow, obstructed, with intolerable delay). The average travel speed is around 25 km/hour; v/c ratio is 0.85 – 1.00.

1. Introduction

Drivers' behavior and road users become the main factor of congestion since the lack of discipline of road users. The flow of traffic at the road could be influenced by several factors; one of them was the roadside obstacle.

The roadside obstacle often significantly affects the capacity and performance of the road, especially in an urban area. These roadside obstacles generally are related to pedestrians, people who cross the roads, street vendors, running vehicles, vehicles that stop carelessly, on-street parking vehicles, and other vehicles that go in and out the road-surrounding area through the roadside.

2. Methodology

2.1. Traffic volume

Traffic volume is the number of vehicles that pass a road's point or cross-section at one time. For the use of design and evaluation, the traffic volume applied is the number of vehicles per hour [1]. The traffic volume defined as equation (1) below:

$$q = n/T \tag{1}$$

where:
q = traffic volume (vehicle/hour/lane);

n = the number of vehicle crossing a section of road at a line, in time interval of T, and
T = time interval observed.

2.2. Capacity
The capacity defined as the maximum flow of vehicles crossing a point in a road maintained hourly at a specific condition. For a two-lane and two-way road, the capacity determined by the two-way flow (two-way combination); However, for the multi-lane road, the flow is divided per way, and the capacity is determined by the number of the line [2]. The basic equation to determine the capacity is:

$$C = C_O \text{ x } FC_W \text{ x } FC_{SP} \text{ x } FC_{SF} \text{ x } FC_{Cs} \tag{2}$$

where:
C = Capacity (pcu/hour),
C_O = Basic capacity (pcu/hour),
FC_W = Road width adjustment factor,
FC_{SP} = Adjustment factor of separated way (only for undivided highway or single carriageway),
FC_{SF} = Adjustment factor of roadside obstacle or side friction, and
FC_{CS} = Adjustment factor of city size.

2.3. Roadside obstacles
The roadside obstacle is defined as the roadside activity that could evoke conflict and affect the movement of traffic flow and decrease road performance. The types of road obstacles are described as:

a. A total number of pedestrians passing or crossing throughout the road;
b. A total number of vehicles stopping and parking;
c. A total number of motor vehicles going in and out of the roadside and on the roadside; and
d. The flow of slow-paced vehicles, which is the total flow "vehicle/hour" of the bicycle, pedicab, 'delman', cart, tractor, and others.

The weight of the influence of roadside obstacles is designated in table 1.

Table 1. Roadside obstacle weighting factor [2].

Type of roadside obstacle	Weighting factor
Pedestrian	0.5
Stopping, parking vehicle	1.0
Slow-paced vehicle	0.4
Street vendor	1.0

To simplify the roadside obstacles for the benefit of calculating road performance analysis, the Indonesian Road Capacity Manual (MKJI) [2] group them in five different classes from the lowest to the highest. These classes are defined in such a way as shown in the following table 2 below:

Table 2. Roadside obstacle class [2].

Roadside Obstacle Class (SF_C)	Code	Total number of weighting factor per 200 m (both sides)	Specific condition
Very Low	VL	<100	Residential area; streets with a roadside

Low	L	100-299	Residential area; several public transportations
Moderate	M	300-499	Industrial area; several activities at the roadside
High	H	500-899	Commercial area; high activity at the roadside
Very High	VH	>900	Commercial area; with market activity at the roadside

By defining the class of roadside obstacles, the value of road width adjustment could also be obtained. The capacity adjustment factor of roadside obstacle (FCsf) with a defined road shoulder is shown in table 3 below:

Table 3. Capacity adjustment factor of roadside obstacle or side friction (FC_{SF}) [2].

Road type	Side Obstacle/ Friction Class (SFC)	Adjustment factor of roadside obstacle and shoulder-width			
		Average effective shoulder-width Ws (m)			
		≤ 0.5 m	1.0 m	1.5 m	≥2 m
Two undivided lane (2/2 UD) or one-way road	Very Low	0.93	0.95	0.97	0.99
	Low	0.90	0.92	0.95	0.97
	Moderate	0.86	0.88	0.91	0.94
	High	0.78	0.81	0.84	0.88
	Very High	0.68	0.72	0.77	0.82

2.4. Degree of saturation

Degree of Saturation (DS) is defined as the ratio of traffic flow to the capacity which is used as the main factor in determining the performance level of intersection and road segment. The DS value indicates whether the road segment has any capacity segment or not. To measure a degree of saturation of an urban road, an equation is used from MKJI [2] published by the Directorate General of Highway as follows:

$$DS = Q/C \tag{3}$$

where:
DS = Degree of saturation;
Q = Maximum flow (pcu/hour); and
C = Capacity (pcu/hour).

2.5. Level of service

LOS (level of service) is one of the methods used to measure road performance, which becomes the indicator of congestion. A road considered as having congestion if the calculation result of LOS value is close to 1. The correlation between the level of service, characteristic of traffic flow, and volume ratio to capacity (DS ratio = q/c) is shown in the following table 4.

Table 4. The relationship between the level of service, characteristic of traffic flow, and volume ratio to the capacity (ratio Q/V) [3].

Level of Service	Explanation
A	Free flow, Drivers can maintain the desired speed without delay;

	Average travel speed ≥ 80 km/hour; v/c ratio 0.0 – 0.20.
B	Free flow, The driver is free enough to choose the speed and lane that is desired; Average travel speed decreased to ≥ 40 km/hour; v/c ratio 0.21 – 0.44.
C	Free flow, The driver has limitations to choose speed, move lanes or overtake; Average travel speed decreased to ≥ 30 km/hour; v/c ratio 0.45 – 0.74.
D	Close to unstable flow, The driver has very limited freedom in running the vehicle; Average travel speed decreased to ≥ 25 km/hour; v/c ratio 0.74 – 0.84.
E	Unstable flow, obstructed, with intolerable delay ; Average travel speed around 25 km/hour; v/c ratio 0.85 – 1.00.
F	The flow is restrained, stuck; Average travel speed < 15 km/hour; V/C ratio requested exceeds the saturated intersection.

3. Results and discussion

3.1. Research location

The research was conducted on Syiah Kuala Street in Kuta Alam sub-district. The road is a collector road that has two-lane and two way (2/2 UD) and around 250 meters of road length. The road is located in an education area.

3.2. Data collection

At this research, data collection specifically conducted on Syiah Kuala Street by adopting the Indonesian Highway Capacity Manual, called MKJI. The supporting data needed in this research were grouped into two types: the secondary data and the primary data. The primary data is data acquired directly from relevant instances.

The survey was conducted for 4 days (Monday, Tuesday, Thursday, and Sunday), at peak hours in the morning, afternoon, and evening, from 07.00 – 09.00 WIB, 12.00 – 14.00 WIB, and 16.30 – 18.30 WIB.

3.3. Data processing

3.3.1. Traffic volume. Traffic volume was obtained by recording every type of vehicle passing the observation point during the time interval of 15 minutes. Then, the traffic volume was processed to be the hourly interval traffic volume. Then, the data was equated into the passenger car unit (PCU) by multiplying the total number of each type of vehicle with the equivalent number of each type of vehicle (passenger car equivalent—PCE).

3.3.2. Roadside obstacle. The type of event and weighting factor is classified as follows:

a. Total number of pedestrian passing or crossing the road segment (weight=0.5);
b. Total number of stopping or parking vehicle (weight=1);
c. Total number of the motor vehicle that was going in and out (weight=0.7);
d. Traffic flow of slow-paced vehicle (weight=0.4); and
e. Street vendor (weight=1).

IOP Conf. Series: Materials Science and Engineering **933** (2020) 012017 doi:10.1088/1757-899X/933/1/012017

3.4. Data analysis

The analysis conducted includes calculating several parameters affecting the road, such as traffic volume, capacity, roadside obstacle, and degree of saturation. The method implemented is MKJI [2] for the urban road. Result of calculation using this method further calibrated to gain a result that approaches the condition according to the real condition.

3.5. Geometric data of the road

Syiah Kuala Street is a collector road having a length of ± 250 m, two-lane and two way, the pavement width of 7 meters, each lane width of 3.5 meters, and the road shoulder of 1.5 meters each side.

3.5.1. Traffic volume. Retrieval of traffic volume data is carried out at intervals of 15 minutes, for 4 (four) days of observation, 6 hours each day—morning at 07.00 - 09.00 WIB, afternoon at 12.00-14.00 WIB, and evening at 16.00-18.00 WIB. Based on the survey result, table 5 lists the traffic composition obtained during four observation days.

Table 5. Traffic volume calculation at the peak hour on Syiah Kuala Street.

Day / Date	Observation time	Type of vehicle			Total Two-way
		LV	MC	HV	Q (pcu/hour)
Monday/ June 25, 2018	07:00 – 08:00	1318	657.75	11.7	1987
Tuesday / June 26, 2018	07:00 – 08:00	1221	554.75	7.8	1783
Thursday / June 28, 2018	07:00 – 08:00	1208	560	16.9	1785
Sunday / July 01, 2018	12:00 – 13:00	1177	413.5	9.1	1599

3.5.2. Roadside obstacle. The calculation of side barriers on Syiah Kuala Street was performed at peak hours based on the MKJI procedure. Each of the factors causing the roadside obstacle is multiplied by the side drag coefficient. The peak hour roadside obstacle occurring at three observation days on Syiah Kuala Street is presented in table 6 below.

Table 6. Roadside obstacle calculation at peak hour (two-way) on Syiah Kuala Street.

Day	**Time**	Event's frequency/hour					Weighting frequency					**Total**	**Class**
		PK	**KP**	**KM**	**KL**	**PKL**	**PK**	**KP**	**KM**	**KL**	**PKL**		
							0.5	**1.0**	**0.7**	**0.4**	**1.0**		
Monday	12.00 – 13.00	83	63	586	45	19	41.5	63	410.2	18	19	551.7	H
Tuesday	12.00 – 13.00	62	85	522	71	18	31	85	365.4	28.4	18	527.8	H
Thursday	12.00 – 13.00	64	111	518	47	14	32	111	362.6	18.8	14	538.4	H
Sunday	16.00 – 17.00	34	43	238	24	0	17	43	166.6	9.6	0	236.9	M

where:

PK : Pedestrians
KP : Parking vehicles
KM : Vehicle exit/ enter side road
PKL : Street vendors
KL : Slow vehicles
H : High
M : Medium

3.6. Capacity

The value of the road capacity is influenced by several factors such as Co (basic capacity), FCW (adjustment factor of the width of the traffic lane), FCSP (adjustment factor for direction separation),

IOP Conf. Series: Materials Science and Engineering **933** (2020) 012017 doi:10.1088/1757-899X/933/1/012017

FCSF (adjustment factor due to side barriers), and FCCS (adjustment factor for city size). The capacity value of the observed road is shown in table 7 below.

Table 7. Road capacity value on Syiah Kuala Street on Monday, Tuesday, Thursday.

Day	Co cpu/hour	FC_W	FC_{SP}	FC_{SF}	FC_{CS}	C cpu/hour
Monday	2900	1	1	0.84	0.9	2192
Tuesday	2900	1	1	0.84	0.9	2192
Thursday	2900	1	1	0.84	0.9	2192
Sunday	2900	1	1	0.91	0.9	2375

3.7. *Degree of saturation*

According to the calculation of capacity mentioned above, the degree of saturation obtained is listed in table 8 below.

Table 8. Degree of saturation values and road performance.

Day	Maximum flow (cpu/hour)	Road capacity (cpu/hour)	Degree of saturation	LOS
Monday	1987	2192	0.91	D
Tuesday	1783	2192	0.81	C
Thursday	1785	2192	0.81	C
Sunday	1599	2375	0.68	A

Based on the calculation obtained, the traffic volume (cpu/hour) at the peak hour (07.00 –08.00 WIB) on Monday, June 25, 2018, is 1987 cpu/hour. On Tuesday, June 26, 2018, the maximum flow recorded, at 07.00 – 08.00 WIB, is 1783 cpu/hour. On Thursday, June 28, 2018, the maximum flow, which was at 07.00 – 08.00 WIB, is 1785 cpu/hour. On Sunday, July 01, 2018, at 12.00 – 13.00 WIB, the maximum flow is 1599 cpu/hour.

The roadside obstacle on Monday was 551.7, which is categorized as a high obstacle; on Tuesday was 527.8, which is considered as a high obstacle; on Thursday was 538.4, which is considered as a high obstacle; on Sunday was 305.7, which is considered in the moderate class. The type of roadside obstacles that dominated this road are vehicles going in or out the roadside. This event causes decreasing vehicle speed, disrupting the traffic flow, and increasing vehicle operating cost (VOC).

According to the result of the data analysis, it is indicated that, on Monday, the capacity is 2192, the saturation degree is 0.91, the LOS is D; on Tuesday, the capacity is 2192, the degree of saturation value is 0.81, and LOS is C; on Thursday, the capacity is 2192, the degree of saturation value is 0.81, and LOS is C; on Sunday, the capacity is 2375, the degree of saturation value is 0.68, and the LOS is A.

4. Conclusion

It can be concluded that the excellent level of service on Syiah Kuala Street will be achieved if the side obstacles are moved to a particular place. Based on the result of research and data analysis conducted, several recommendations are suggested as follows:

1. The one-way street could be applied if the traffic flow is heavy. However, when the flow is normal, the one-way road could be returned into a two-way system (applied at the specific hours).
2. A rigid rule should be required for the street vendor who transacts at the roadside and does wild parking which often inhibits the speed of vehicles.
3. Mass transport should be encouraged, and school transportation should be organized. The bus will take the route scheduled based on the data of students' residential area and stop at some particular points.

IOP Conf. Series: Materials Science and Engineering **933** (2020) 012017 doi:10.1088/1757-899X/933/1/012017

4. The government should apply the strict rules regarding a rearrangement of parking by parallel parking only on one side of the road, so the road or line that will be passed will be wider and the road performance improves.

References

[1] Morlok E K 1991 *Pengantar Teknik dan Perencanaan Transportasi* (Jakarta: Penerbit Erlangga)
[2] Direktorat Jendral Bina Marga 1997 *Manual Kapasitas Jalan Raya Indonesia (MKJI)* (Jakarta: Bina Marga)
[3] Peraturan Menteri Perhubungan No.16 2008 *tentang Tingkat Pelayanan.*

IOP Conf. Series: Materials Science and Engineering **933** (2020) 012018 doi:10.1088/1757-899X/933/1/012018

The reliability of fire protection system case study of Aceh Besar district regency office

F Anwar[1]**, T B Aulia**[2] **and M Afifuddin**[2]

[1] Master of Civil Engineering, Faculty of Engineering, Syiah Kuala University, Banda Aceh
[2] Civil Engineering Department, Faculty of Engineering, Syiah Kuala University, Banda Aceh

Coressponding e-mail: fadlianwar.73br@yahoo.co.id

Abstract. The Regency Office is a building where the central government is administered in an administrative area at the district level. According to the Pd-T-11-2005-C regulation on building fire safety inspections, in order to ensure the safety and comfort of the building, the value of building safety system reliability (VBSSR) should not be less than 80%. This study aims to evaluate the value of the reliability of the office building safety system and propose recommendations on the reliability conditions of the Building of Aceh Besar District Regency Office. This study uses qualitative methods through observation. Evaluation of the reliability of building safety systems and recommendations on building reliability conditions is carried out based on technical guidelines of the Pd-T-11-2005-C regulations. The objects of the observations are the completeness of the site, rescue devices, active protection systems, and passive protection systems. For data analysis, the researcher used the scoring analysis (weighting). The results showed that the reliability value of the safety system of the Building of Aceh Besar District Regency Office was 74.97%, and the reliability was at a sufficient level.

1. Background of study

The Regency Office is a building that functions as the governmental center of an administrative area at the district level. In regulating the center of government, the Building of Aceh Besar District Regency Office is located on Prof. Dr. A. Ibrahim street, city of Jantho. This building is a new building having a 3-story. The Building of Aceh Besar District Regency Office must meet the building's reliability criteria for fire hazards to provide protection for building users so that the government can run as it should.

Fire, the unwanted event and sometimes cannot be controlled, is a result of burning material in the air and emitting heat and flame. Fire can cause death and can also cause a collapse of building structures. Fire control failures in buildings generally occur because they do not have a good value of building safety systems (VBSSR) and override applicable regulations. According to the Pd-T-11-2005-C [1] regulation on building fire safety inspections, VBSSR includes site completeness parameters, rescue facilities, active protection systems, and passive protection systems. To be able to guarantee the safety and comfort of buildings, the VBSS must not be less than 80%.

Based on preliminary observations, the Building of Aceh Besar District Regency Office does not have good reliability because it has not met the parameters in the VBSSR. First, the site has no yard hydrants. Second, the active protection system has no detection and alarm, siamese connection, building hydrants, overflow extinguishing systems, smoke controllers, smoke detection, smoke disposal, directions, and operating control rooms. Third, passive protection systems do not have good openings protection. Therefore, the Aceh Besar District Regency Office Building's reliability of the building safety system needs to be evaluated so that the percentage value of the VBSSR is identified. The percentage of VBSSR can recommend appropriate forms of handling so that buildings can provide protection for users. This study aims to evaluate the value of the reliability of the safety system of the Building of Aceh Besar District Regency Office.

2. Research methods

2.1. Fire protection system

Minister of Public Works Regulation of Number 26/PRT/M/2008 [2] concerns the technical requirements of fire protection systems in buildings and environments. According to the regulation, the fire protection systems in buildings and the environment are systems consisting of tools, equipment, and facilities, all of which are used to protect buildings and their environment against fire hazards through active protection systems, passive protection systems, and internal management methods. Fire protection systems are used to detect and extinguish fires as early as possible with manually or automatically driven equipment.

2.2. Value of building safety system reliability (VBSSR)

According to Pd-T-11-2005 C [1] regarding building fire safety inspection, the VBSSR is the result of system performance measurement based on applicable building safety standards or the knowledge of the inspection team. The steps to evaluate the NKSKB are as follows:

1. Criteria for assessment

The condition of each component or part of a building must be assessed or evaluated. The condition of the building's fire protection components is divided into three levels, which can be seen in table 1.

Table 1. Level of fire audit assessment [2].

No.	Score	Compatibility	Reliability
1	> 80 – 100	Qualified	Good (G)
2	60 – 80	Installed but there are a small number of installations that do not meet the requirements	Fair (F)
3	< 60	Not suitable at all	Poor (P)

2. Weighting

The weighting of each component must be done using the analytical hierarchy process (AHP) method. This method was chosen to reduce the element of subjectivity at weighting. AHP is a systematic method for comparing a list of observations or alternatives. Hierarchy is a special type of system based on the assumption that existing units, which have been identified, can be grouped into separate collections, in which units of a group influence the unit of another group that is also influenced by another group. Elements of each hierarchical group are assumed to be independent of each other. The weighting of the parameters of the building safety system component (BSSC) can be seen in table 2.

Table 2 The weighting of the parameters of building safety system components [2].

No.	BSSC parameters	Quality of BSSC and Sub-BSSC
1	Site completeness	25%
A	Water sources	27%
B	Road environment	25%
C	Distance between buildings	23%
D	Yard hydrant system	25%
2	Rescue facilities	25%
A	Way out	38%

B	Exit construction	35%
C	Helipad	27%
3	Active protection system	24%
A	Detection and alarm	8%
B	Siamese connection	8%
C	Light fire extinguisher	8%
D	Hydrant system	8%
E	Sprinkler	8%
F	Overflow extinguishing system	7%
G	Smoke controller	8%
H	Smoke detection	8%
I	Smoke disposal	7%
J	Fire elevator	7%
K	Emergency light	8%
L	Emergency electricity	8%
M	Operation control room	7%
4	Passive protection system	26%
A	Fire-resistance of building structures	36%
B	Compartmentation	32%
C	Openings protection	32%

3. Interpretation
The safety reliability of a building can be judged to be good if the reliability of a building component or the reliability of a building safety system or NKSKB value is not less than the lower limit of good categories, as listed in table 1. The required building safety and comfort rate is 80%.
4. Recommendations
Based on the results of the NKSKB examination that has been calculated, the recommendations can be submitted by the inspection team to improve the condition of less (K) or enough (C) to be good (B). Recommended steps include:

a. periodic checks,
b. periodic maintenance or maintenance,
c. periodic maintenance and repairs,
d. adjustment or repair of elements, and
e. completion of the components that are lacking.

3. Result and discussion

3.1. Completeness of site
Completeness of site is the completeness of the place where a building is established. The completeness of the site includes water sources, road environment, the distance between buildings, and yard hydrants. The results of observations and assessments of the completeness parameters of the site in the Building of Aceh Besar District Regency Office can be described as follows:
1. Water source
The building has been equipped with water sources having capacities that meet the minimum requirements for the function of the building. Based on observation, the source of water condition can be considered as good.
2. Road environment
The building has been equipped with an environmental road having a width of 7.5 m, which passes the required road width of 6 m. The structure of the environmental road surface is rigid pavement. The width of the entrance is 7.5 m, where the required width is at least 4 m. Based on the observation, the environmental road condition can be assessed as good.

3. Distance between buildings
The distance between buildings, which have a height of more than 14 m, must be more than 8 m. The observed building has a height of 22.98 m. The building distances is 45.41 m. The Building of Aceh Besar District Regency Office is adjacent to the Aceh Besar District Financial Management Agency Office Building. Based on the observation, the distance between buildings condition can be assessed as good.
4. Hydrant yard
The building was not equipped with a yard hydrant. The hydrant yard or hydrant pillar is a fire prevention system that requires a water supply and is installed outside the building. Based on the observation, the yard hydrant condition was assessed as poor.

3.2. Rescue facilities
Rescue facilities are facilities that need to be prepared to save building users and property in the event of a fire in a building. Rescue facilities include an exit, construction of an exit, and a helipad. The rescue facilities in the Regent Office Building can be described as follows:
1. Exit
Based on the observation on the building, each floor has two exits. The exits of the first and second floors have a height of 4.35 m, while the third-floor exit has a height of 5.67 m. The required height of the exit is at least 2.5 m. Every exit is protected from fire because every corridor space has concrete walls, and there are no flammable objects. The maximum distance from the exit is 57.55 m, where the maximum distance required is 20 m. The minimum width of the exit is 2.87 m, where the minimum width required is 2 m. The distance from an exit to another exit is 5 m. The building is not provided with a smoke-free lobby. Exit on each floor is not blocked by walls or objects. The exit goes to a safe open space outside the building. Based on observations, the way out condition was assessed as good.
2. The structure of the exit
The structure of the exit must resist fire hazards at least for 2 hours. The way out is free from something that can hinder building users. The minimum width of the exit is 2.87 m, which passes the minimum required value of 2 m. The canal (exit release) road is protected from fire hazards. The interior walls of the building are not flammable, and the ceiling, which is made of gypsum and equipped with fire propagation resistance is not less than 60 m. At a certain level, building elements can maintain structural stability if there is fire. Buildings cannot prevent the spread of smoke from fires because they are not provided with smoke control equipment. Building users have enough time to carry out the evacuation process. Based on observations, the construction of the exit is assessed as good.
3. Helipad
The building is not equipped with a helipad because the building only has a height of 22.98 m, where the maximum height required is 60 m. The helicopter platform is useful for carrying out the evacuation process if there is a fire in the building. Based on observations, the helipad cannot be assessed.

3.3. Active protection system
The active protection system is a protection system to detect and reduce fires both manually and automatically with complete equipment. Active protection systems include detection and alarm, siamese connection, light fire extinguishers, building hydrants, sprinklers, overflow extinguishing systems, smoke controllers, smoke detection, smoke disposal, fire lifts, emergency lights, emergency electricity, and operating control rooms. The results of the observation and evaluation of the active protection system in the Building of Aceh Besar District Regency Office can be described as follows:
1. Detection and alarm
Based on the observation, the building is not equipped with a detection device and alarm, as required. Detection and alarm are useful for finding the location of the presence of fire hazards and providing emergency signals to building users so that they can immediately carry out the evacuation process. Based on observation, the detection system and alarm cannot be assessed.

2. Siamese connection
As a result of the observation, the building was not equipped with convention siamese as required. Siamese connection is useful for supporting water supply in buildings if the water in the reservoir is no longer sufficient. The water carried by fire engines will be channeled through this component for the water needs of the sprinkler and hydrant components. Based on observations, the siamese connection can be assessed as less.
3. Light fire extinguisher
The results of the observation show that the building has been equipped with a light fire extinguisher (LFE), which type is a chemical powder (dry chemical powder). The Regent's Office Building has a length of 60 m, which every floor has 1 LFE. The number of APAR is not in accordance with the length of the building. Based on observations, LFE can be assessed as less.
4. Building hydrants
As a result of the observation, the building was not equipped with a building fire hydrant. Hydrant building is useful as a storage place for equipment to extinguish fires such as fire hose, hydrant nozzle, and valve. Based on observations, the building hydrants cannot be assessed.
5. Sprinkler
The observation results show that the building is not equipped with a sprinkler because the building only has an area of 7,144 m, where the required area must be more than 18,000 m^2. The sprinkler is useful for extinguishing fires automatically by emitting a water discharge when a fire is detected. Based on observation, the sprinkler cannot be assessed.
6. Overflow extinguishing system
The building is not equipped with an overflow extinguishing system. Overflow extinguishing systems are useful for reducing fire propagation using gas, foam, or dry powder. Based on observations, the overflow extinguishing system can be assessed as less.
7. Smoke controller
The building is not equipped with smoke control. Smoke control is useful for blocking the flow of smoke into the way out, shelter, and others. Based on observations, the smoke controllers have not been able to be assessed.
8. Smoke detection
The observation results show that the building is not equipped with smoke detection. Smoke detection is useful for giving warnings to building users when fires occur through alarms so that they can immediately carry out the evacuation process. Based on observations, smoke detection can be assessed as less.
9. Disposal of smoke
The results of the observation show that the building is not equipped with the exhaust disposal system. Based on observations, the discharge of smoke can be assessed as less.
10. Fire escape elevator
The building is not equipped with a fire escape elevator. Based on observations, the fire escape elevator cannot be assessed.
11. Emergency directional lighting
The building has been equipped with emergency lights on each stair. Moreover, the stair handrail is made of iron, so it is protected against fire. The emergency light is useful to direct the building users in the evacuation process. However, the building is not equipped with directions or exits. Based on observations, emergency directional lighting in this building can be assessed as less.
12. Emergency electricity
The results of observation show that the building has been equipped with emergency electricity, which power supplied from the national electricity company and internal generators. The generator has a capacity of 250 KVA and can operate automatically if the electricity goes out with a 10 minute time delay. Temporary fire-resistant cables are used in the emergency electrical system; this cable is resistant for 60 minutes. Emergency electricity is useful to provide a supply of backup electricity while the fire is happening. As a result, the equipment can be activated to control the building during the evacuation process. Based on observation, emergency electricity in this building can be assessed as good.

13. Operation control room

The building was not equipped with an operating control room. The operating control room is used to monitor potential fire hazards. The Regent Office Building is only equipped with simple equipment, namely closed-circuit television (CCTV) to monitor the fire hazards. CCTV equipment is ready in all rooms on each floor of the building. For outside the building, the CCTV equipment is only installed on the front and rear sides of the building. Based on observations, the emergency operation control room can be assessed as sufficient.

3.4. Passive protection system

The passive protection system is a protection system to reduce fires that are directly built on buildings without using the equipment. Passive protection systems include the resistance of building structures to fire, compartmentalization, and protection of the openings. The results of observations of passive protection system in the Building of the Office of the District Head of Aceh Besar are as follows:

1. The building structures' resistance to fire

Based on observations, the building has been equipped with the fire resistance of type A of building structures. According to SNI 03-1736-2000, the fire resistance of type A of building structures are those whose construction elements are fireproof, capable of structurally resisting building loads, and preventing the fire from spreading on room or building. Based on observation, the fire resistance of building structures can be assessed as good.

2. Compartmentalization

Based on observations, the building has been equipped with compartments, in the form of walls and room partitions, with aluminum iron frames and glass as a separator between workspaces. Compartemenisation limits the fires of a room and hampers the fire spreading to other rooms in a building. In addition, the width of the entrance of this building is 7,5 m and suitable for the requirement of 6 m. This entrance width can be passed by a fire engine. Based on observation, compartmentalization can be assessed as good.

3. Protection of openings

Based on the results of observations, the building has been equipped with the protection of the openings. The windows and doors are completed with the fire stoppers. The Regent's Office Building was completed with the fire suppressors using aluminum and glass iron material on windows and doors on each floor of the building. However, protection facilities for openings such as fire doors, fire windows, smoke barrier doors, and fire covers that comply with standards are not available. The door leaf can rotate on one side. The door has a 35 mm density. The wall on the exiting path is a concrete wall, so it is resistant to fire. The door on the exit can close automatically. Based on observations, the protection of openings can be considered as sufficient.

3.5. Assessment of VBSSR

VBSSR assessment includes site completeness parameters, rescue facilities, active protection systems, and passive protection systems. Each of these parameters has a number of individual conditions. VBSSR assessment is obtained by totalizing the total number of parameter values that exist. The VBSSR assessment of the Building of Aceh Besar District Regency Office is shown in table 3.

Table 3 Assessment of VBSSR.

No.	**BSSR**	**Rating Result**	**Standard Value**	**Weight**	**Value of Condition**
(1)	**(2)**	**(3)**	**(4)**	**(5)**	**(6) = (4) x (5) x 25%**
1	Water source	Good	100	27%	6,75%
2	Road environment	Good	100	25%	6,25%
3	Distance between buildings	Good	100	23%	5,75%
4	Yard hydrant	Poor	60	25%	3,75%
Total value of site completeness condition					22,50%

1	Escape route	Good	100	38%	9,50%
2	Escape route construction	Good	100	35%	8,75%
3	Helipad	-	-	27%	0,00%
Amount of condition of the rescue facility					18,25%
1	Detection and alarm	-	-	8%	0,00%
2	Siamese connection	Poor	60	8%	1,15%
3	Light fire extinguisher	Poor	60	8%	1,15%
4	Hydrant building	-	-	8%	0,00%
5	Sprinkler	-	-	8%	0,00%
6	Overflow extinguishing system	Poor	60	7%	1,01%
7	Smoke controller	-	-	8%	0,00%
8	Smoke detection	Poor	60	8%	1,15%
9	Disposal of smoke	Poor	60	7%	1,01%
10	Fire escape elevator	-	-	7%	0,00%
11	Emergency light and directions	Poor	60	8%	1,15%
12	Emergency electricity	Good	100	8%	1,92%
13	Operation control room	Fair	80	7%	1,34%
The total value of the active protection system condition					9,89%
1	Fire-resistance of building structure	Good	100	36%	9,36%
2	Compartmentation	Good	100	32%	8,32%
3	Opening protection	Fair	80	32%	6,66%
The number of passive protection system condition					24,34%

Table 3 shows that for the site completeness parameters, the condition value is 22.50% of the weight of 25%; for the rescue facility, the condition value is 18.25% of the weight of 25%; for the active protection system, the condition value is 9.89% of the weight of 24%; and for the passive protection system, the condition value is 24.34% of the weight of 26%. The total value of the VBSSR condition is 74.97%. Based on table 3, VBSSR of 74.97% indicates a sufficient level of reliability. The evaluation results indicate that the level of reliability of the safety system of Aceh Besar District Regency Office is sufficient. The Building of Aceh Besar District Regency Office has no guarantee of safety and comfort because the VBSSR is less than 80% (74.97%). This finding recommends that site completeness, rescue facilities, active protection systems, and passive protection systems of the Building of Aceh Besar District Regency Office to be completed.

Based on research conducted by Kurniawan [3], the Orthopedic Hospital Prof. Dr. R. Soeharso Surakarta has an average VBSSR of 92.77%. This means that the reliability of the building against fire hazards is good. In addition, the research conducted by Pratama [4] found that the VBSSR on the Eria Mother and Child Hospital of Pekanbaru City Pekanbaru is 57.24%. This means that the fire protection system in this building is classified as less reliable. Technical recommendations that should be performed are periodic maintenance and repair to each existing fire protection facility, the repair of damaged elements, and the addition of fire protection elements that are not yet available. Research conducted by Ruspianof [5] showed that the VBSSR at PT. PLN, at Riau Region and Riau Islands, was 86.47%. This means that the fire protection system in this building is classified as reliable. The research conducted by Trikomara [6]showed that the VBSSR was 79.07%. This finding showed the reliability of buildings in sufficient categories. This finding recommended that it is necessary to complete components that are not yet complete and to carry out inspections, repairs, and maintenance of components of the fire protection system periodically. The research conducted by Zulfiar and Gunawan [7]showed that the VBSSR on five floors UNY Hotel Buildings in Yogyakarta was 91.60%. This means that the condition of the building is included in the good category so that the building can be used optimally, and this building gets good protection from the dangers of building fires.

IOP Conf. Series: Materials Science and Engineering **933** (2020) 012018 doi:10.1088/1757-899X/933/1/012018

4. Conclusion and suggestion

4.1. Conclusion
The value of the reliability of the safety system of the Building of Aceh Besar District Regency Office is 74.97% with a sufficient level of reliability. The building has no guarantee of safety and comfort because the VBSSR is less than 80%. It is necessary that the lacking components in site completeness, rescue facilities, active protection systems, and passive protection systems of the Building of Aceh Besar District Regency Office to be completed.

4.2. Suggestion
It is recommended that the Aceh Besar District Government improve the reliability of the Building of Aceh Besar District Regency Office by completing components that are lacking in the completeness of the site, facilities for rescue, active protection systems, and passive protection systems. It is suggested to the next researcher to be able to evaluate the reliability of other buildings in the Aceh Province.

References
[1] Pd-T-11-2005 C Tentang Pemeriksaan Keselamatan Kebakaran Bangunan Gedung
[2] Peraturan Menteri Pekerjaan Umum 2008 Nomor 26/PRT/M/2008 Tentang Persyaratan Teknis Sistem Proteksi Kebakaran pada Bangunan Gedung dan Lingkungan
[3] Kurniawan P A 2014 Evaluasi penerapan sistem proteksi kebakaran pada bangunan rumah sakit (Studi Kasus RS. Ortopedi Prof. Dr. R. Soeharso Surakarta) *e-Jurnal Matriks Teknik Sipil* (Surakarta: Universitas Sebelas Maret)
[4] Pratama R 2017 Analisis keandalan sistem keselamatan bangunan terhadap bahaya kebakaran (studi kasus: gedung rumah sakit ibu dan anak Eria Bunda kota Pekanbaru) *Jom FTEKNIK* **4** 2 (Pekanbaru: Universitas Riau)
[5] Ruspianof 2017 Evaluasi keandalan sistem proteksi kebakaran pada bangunan gedung (studi kasus gedung PT. PLN wilayah Riau dan kepulauan Riau) *J. Saitis* **17** 2 (Pekanbaru: Universitas Islam Riau)
[6] Trikomara R 2012 Keandalan sistem proteksi kebakaran pada bangunan gedung (studi kasus gedung kantor bupati Indragiri Hilir) *Jurnal* (Pekanbaru: Universitas Riau)
[7] Zulfiar M H dan Gunawan A 2018 Evaluasi sistem proteksi kebakaran pada bangunan hotel UNY 5 lantai di Yogyakarta *J. Semesta Teknika* **21** 1 65-71

The effect of porous zigzag-type breakwaters on the coastline changes in Neuheun village, Aceh

P Pagihariadi[1], E Fatimah[2] and S Syamsidik[2]

[1] Civil Engineering Postgraduate Program, Department of Civil Engineering, Universitas Syiah Kuala, Banda Aceh, Indonesia
[2] Department of Civil Engineering, Universitas Syiah Kuala, Banda Aceh, Indonesia

Corresponding e-mail: putra.p@mhs.unsyiah.ac.id

Abstract. The process of changing the coastline has always been a challenge for coastal communities. Beach erosion has been a threat with the coastline withdrawing from its previous position. Coastal protection buildings are needed to overcome coastal erosion problems, and alternative construction also needs to be developed. The zig-zag type porous breakwater is quite effective in reducing erosion based on several small-scale physical models that have been developed. Therefore a prototype was built and tested to observe its usefulness, and how it was applied in the field. This study aims to assess the effect of installing zigzag type porous breakwaters on coastline changes. This research began with the installation of construction and continued with observing changes in the coastline. Observations are carried out periodically at any time, to see changes in the contour of the beach that occurs. Observations were made by leveling the beach surface against several transverse profiles. Changes in coastline due to construction will be analyzed by looking at the relation to the wind direction. The results obtained from several observations show that the construction of a zig-zag type porous breakwater has a positive effect in reducing erosion, as long as the breakwater distance with the coastline does not exceed the breakwater length.

1. Introduction

The coastline is a meeting point between land and sea, tides, wind, and waves attacking land and where the land responds to attacks with various acts of giving and taking that are effectively aimed at removing energy from the sea [1]. As a result of these conditions, coastline changes are a common thing. Coastline changes are generally caused by erosion and sedimentation. Erosion during certain wind seasons and sedimentation in other seasons is a common cycle because erosion and sedimentation are part of the dynamism that occurs on the coast.

To deal with the problem of erosion and sedimentation, a coastal security building is generally made. The coastal protective buildings show effective results in resisting coastal erosion. Based on these problems, a study was carried out in the form of prototyping of coastal security structures in the form of porous breakwaters installed in zigzag ways [2].

The selection of breakwaters as the object of research is because these buildings are already commonly used in protecting beaches, breakwaters built in parallel on the coast have proven to be effective ways to reduce coastal erosion [2]. The existence of pores or holes in the breakwater building functions to absorb some of the incoming wave energy. In a previous study carried out in a laboratory

IOP Conf. Series: Materials Science and Engineering **933** (2020) 012019 doi:10.1088/1757-899X/933/1/012019

concluded that an increase in breakwater porosity reduces the reflected wave height and the effect of porosity becomes significant in reducing wave energy when the breakwater is made higher [3]. Besides that, the porous breakwater structures can attract marine life and ultimately improve environmental quality [4].

The selection of zigzag type breakwaters is done because this type of building is considered to be more effective in resisting erosion rates, it is based on previous research conducted by conducting a study of physical models of breakwater not submerged porous zigzag types against wave diffraction patterns [5]. In that study the breakwater model with width (L) 28.3-56.6 cm was able to reduce wave energy by 52.5-65.18%, the value of the diffraction coefficient increased with reduced wave steepness, whereas the diffraction coefficient value decreased with increasing wave steepness. The relationship of the wave period with the diffraction coefficient is directly proportional.

The study of a porous zigzag type breakwater model to change the velocity distribution and wave-current pressure had been carried out [6]. The research was carried out by using a breakwater model with a porosity of 47%, 46% and 45% and the waves arranged with wave height (H) of 10 cm and wave period (T) 1,268 seconds. From the simulation results, it can be concluded that porosity in the breakwater building can affect the speed and pressure generated by the wave currents. The average current velocity before the construction of a porous zigzag type breakwater model is 0.2823 m/ s and the current velocity after construction is 0.2141 m/ s, this indicates that the construction of porous zigzag type breakwaters can reduce the velocity distribution of currents that will hit the coast.

2. Methodology

This research was conducted at ADB Beach, which is located in Gampong Neuheun of Masjid Raya District Aceh Besar Regency. This location is chosen because the beach has dynamic characteristics, so it is considered suitable as the location of the prototype of a built-in-pore breakwater building zigzag that aims to see how big the building influences Change of coastline. It is hoped that the building will be able to withstand the rate of erosion effectively to help the coastal protection process.

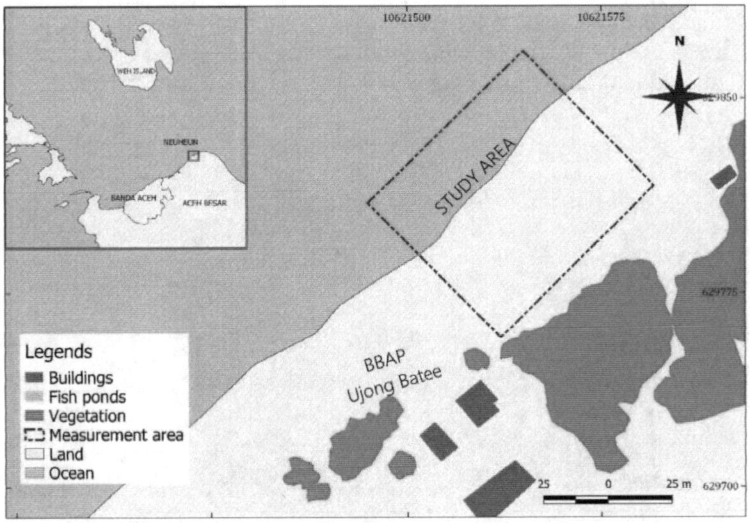

Figure 1. Study area, Neuheun Village, Aceh Besar district.

The offshore breakwater is one of the alternative constructions to protect the coastal. The breakwater to be installed has a porosity rate of 47% [7]. This study was done in three stages; installation of offshore porous zigzag breakwater prototype, observation (measurement) of coastal morphology change, and data analysis.

2.1. Zigzag and porous breakwater installation

The installation of the breakwater was done by using the excavator by first moving the breakwater units measuring 1.0x1.0x1.0 m from the temporary placement location (Fig 2) to the point Installation. The lifting of a breakwater using an excavator is done by tying a breakwater or clasp with a two tons strong strap belt which is then attributed to the excavator bucket. After all breakwater and hooks are located around the mounting point then the installation according to the research plan is done. This research, the breakwater is installed only one breakwater building, not a series consisting of several breakwaters. This is done due to the limited funds available.

Figure 2. Units breakwater at the temporary site.

Before the installation is done first to expand the geotextile in the installation location. Geotextile installation aims to allow if the built-in breakwater decreased due to unstable soil or sand then the decline becomes uniform so that the breakwater arrangement is expected to remain as early as Installation. The breakwater installation was done by compiling the breakwater in a zigzag line with ten lines in which each row was filled by three breakwater units. One unit of the breakwater with the other is associated with a hook that has been made to prevent it from being installed later. The retrofitted breakwater has a length of 17. 0 m, width of 2.8 m, and height of 1.0 m as can be seen in figure 3.

Figure 3. Breakwater when it is installed in the field.

2.2. Observation and data collection

The data collected includes secondary and primary data. Secondary data used include data on wind speed and direction sourced from Blang Bintang BMKG Station. Primary data collected includes topographic and bathymetry maps. Topographic surveys aim to make topographic maps that contain the latest information on the state of the surface of the mapped coast, the information presented includes the state of relief (high and low) on the surface of the land or area of the measurement area. Topographic measurements were carried out onshore and land at the research location. The area of measurement is 60 m x 80 m, which is 60 m along the coast and 80 m perpendicular to the coast. The area of measurement is divided into five cross profiles. The elevation measurements for each transverse profile are carried out starting from the mainland towards the sea using a water pass. The results of the water pass readings for each measurement of the transverse profile are recorded on the measurement form and then analyzed to obtain an elevation value for each transverse profile. The elevation value is then poured into the image to produce a map of the location of the research site. Topographic measurements are carried out in the initial conditions on 29 October 2017, then measurements are taken periodically at certain times. The breakwater layout after being installed from the measurement results can be seen in figure 4.

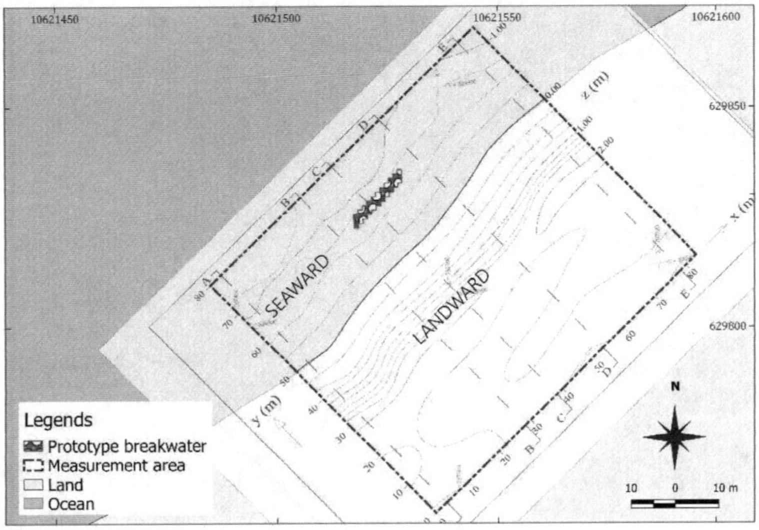

Figure 4. Breakwater installation layout.

2.3. Data analysis

Wind data were analyzed to determine the velocity and dominant wind direction during observation. The analyzed wind data were presented in a wind rose diagram. From the wind rose diagram you get information about the winter season that occurs in the measurement period, so it is useful to see how it affects the coastline change. The topography measurement is done several times in a certain period, it aims to see how much the coastline changes occur from one-time measurement to the next measurement. Based on the results of topographical measurements obtained the data in the form of reading threads on the topographical measuring device which is then analyzed resulting in the value of elevation. The value of elevation is then poured into the image so that it becomes the topographical map of research locations and images of transverse pieces of the beach. From the image of the topographical measurements can be calculated how much sedimentary movement occurs both erosion and sedimentation.

3. Results and discussion

3.1. Coastline changes during observation

The coastline was changed after breakwater installation compared to the initial condition on October 29th, 2017. The deterioration of the coastline from the initial conditions occurred from the measurements on December 10, 2017, until the most severe on May 12, 2018, which was an average of 40.9 m behind the coastline at the initial conditions. On the July 31st measurement of 2018, the coastline is still behind the initial condition of an average of 2.8 m, but the conditions have been progressing the coastline from the previous measurements on 12 May 2018, which is 38.1 m. At the measurement conducted on 8 October 2018, the coastline was progressing from the initial condition with an average distance of 6.1 m. The coastline returned to a decline of an average of 6.4 m on the measurement Conducted on 22 December 2018. Figure 5 shows the shape of coastline changes during observation. The distance of coastline changes during the observation can be seen in table 1.

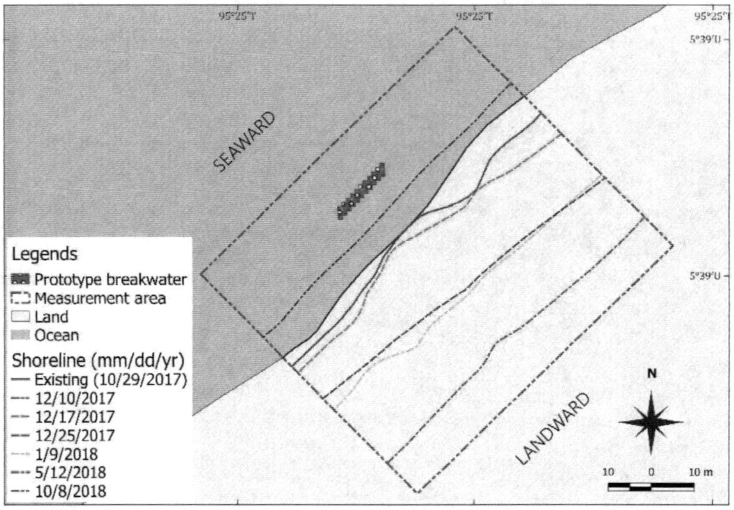

Figure 5. Coastline changes during observation.

Table 1. The distance of coastline changes during observation.

	Coastal morphology induced by a breakwater									
Measurement time	Left side			Backside			Right site			Average
	m		m	m		m	m		m	m
10 December 2017	2.0	-	3.5	0.0	-	2.0	7.0	-	8.0	3.8
17 December 2017	4.3	-	5.1	2.0	-	6.5	9.0	-	14.0	6.8
25 December 2017	15.0	-	19.0	20.0	-	24.0	26.0	-	30.0	22.3
09 January 2018	16.5	-	24.0	23.0	-	25.0	27.0	-	30.0	24.3
15 March 2018	35.0	-	38.0	36.0	-	38.0	40.0	-	42.0	38.2
12 May 2018	38.5	-	39.0	38.0	-	41.0	43.4	-	45.5	40.9
31 July 2018	2.8	-	3.2	1.7	-	5.5	1.5	-	2.3	2.8
08 October 2018*	**-6.6**	-	**-8.7**	**-6.2**	-	**-6.6**	**-3.3**	-	**-5.0**	**-6.1**
22 December 2018	0.2	-	4.6	5.6	-	6.4	9.0	-	12.8	6.4

*) The coastline is more advanced than the initial conditions

3.2. Coastline change due to construction impact

Based on observation can be seen the impact of the porous breakwater that is mounted zigzag against the change of coastline. This is evidenced by the measurement results carried out on December 10 and 17, 2017. From the results of the measurements made on December 10, 2017, the coastline behind the breakwater did not multiply the setbacks in the left and right positions. Similarly, when measurements were made on 17 December 2017. Here the breakwater was able to maintain the coastline position during the distance between the coastline to the breakwater does not exceed the breakwater length. Unlike the measurements on December 25, 2017, onwards, where the breakwater is no longer able to defend the coastline from setbacks due to the distance between the breakwater and the coastline has been very long breakwater. The effect of the breakwater on changing coastline can be seen in figure 6.

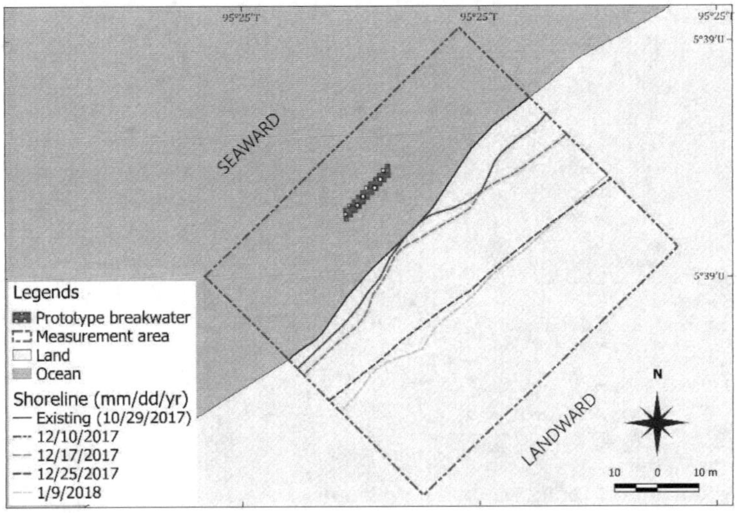

Figure 6. Breakwater impact on changing coastlines during observations.

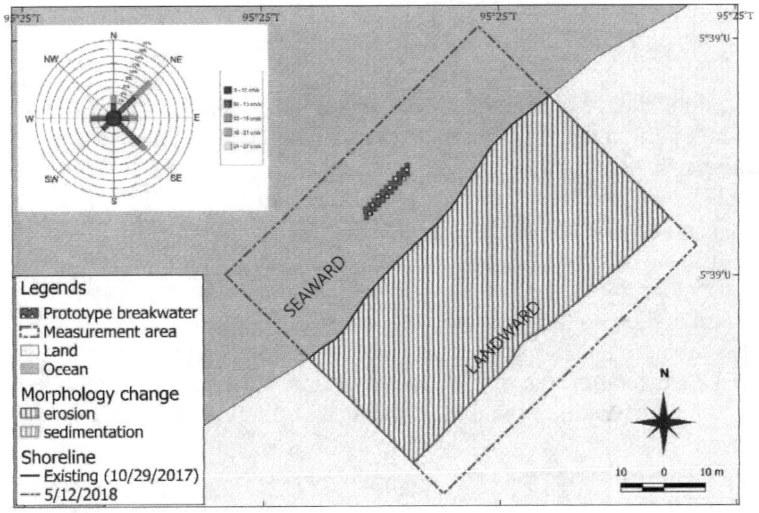

Figure 7. Coastline change from existing conditions 29 October 2017 until 12 May 2018 which is in the East wind season.

3.3. Changes in coastline due to the wind season

The deterioration of the coastline from the initial conditions occurred from the initial observation condition of 29 October 2017 until 12 May 2018 as seen in figure 7. In this period the winds are in the eastern season, generally blowing from east to west. From the observation done in the blend with the wind rose can be concluded that in the east season the coast is erosion evidenced by the pullback coastline with significant distances as seen in table 1.

Different from the coastline changes that occurred between the measurements made from 12 May 2018 to 8 October 2018 where the coastline goes forward as seen in Figure 8. During this period the wind was in the West season, generally blowing from west to east. From the observation done in the blend with the wind rose can be concluded that in the eastern season the coast was subjected to sedimentation evidenced by the coastline with a significant distance.

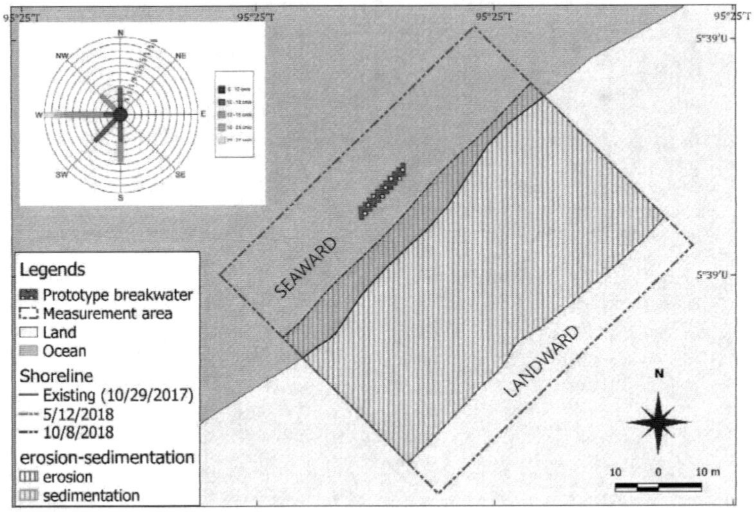

Figure 8. Coastline change from May 12, 2018, to 8 October 2018 which is in the West wind season.

4. Conclusion

Based on observations that have been done, the impact of the breakwater that was installed with a length of 17.0 m, the width of 2.8 m and a height of 1.0 m against the coastline change is only seen when the distance breakwater with the coastline does not exceed the length of breakwater. In this condition breakwater was able to withstand the decline of the coast amounted to 0.0 – 2.0 m compared to the beach without a breakwater that has decreased by 2.0 – 8.0 m. Breakwater becomes a no impact on the coastline change when the distance between Breakwater with the coastline exceeds the length of breakwater. In the observation site, the influence of the wind seasons is more dominant against the coastline change. During the east wind, there is a large erosion that causes the coastline to reach an average of 40.9 m. During the West wind, sedimentation was caused by the coastline up to an average of 47.0 m. The impact of the breakwater in accelerating the occurrence of sedimentation is not seen due to the formation of tombolo or salient on the coastline behind the breakwater, although in general sedimentation occurs in the west wind season.

In this study, the constructions were mounted only one row of the breakwater building so that the effect of the breakwater on changing coastline is not very noticeable. It is hoped that the future-related research can be made in the form of a series consisting of several breakwater buildings so expect the influence of breakwater to change coastline can be more noticeable. The high breakwater in this study is only 1.0 m so it is not very effective in holding the wave compared to the higher buildings, expected in future related research can be made higher so that it will be more to protect the coastline.

References

[1] CERC 1984 *Shore Protection Manual Volume I* (Washington DC: Department of The Army US Army Corps of Engineers)

[2] Alvarez F and Pan S 2016 Predicting coastal morphological changes with empirical orthogonal function method *Water Sci. and Eng.* **9** 1 14e20

[3] Ting C L et al. 2004 Porosity effects on non-breaking surface waves over permeable submerged breakwaters *Coastal Eng.* **50** 213–24

[4] Metallinos A S et al. 2019 Hydrodynamic conditions in a submerged porous breakwater *Ocean Eng.* **172** 712–25

[5] Nufus S H 2015 *Physical Model Study of Undrowned Breakwaters Type Zig-zag Porous to the Wave Diffraction Pattern*, (in Bahasa) (Banda Aceh: Syiah Kuala University)

[6] Fajri I Z 2015 *Study of Numerical Breakwater Type Waveform Solver for the Change of Speed Distribution and Wave-current*, (in Bahasa) (Banda Aceh: Syiah Kuala University)

[7] Fatimah E et al. 2015 The influence of single zigzag type porous groin in the change of beach profile, *Procedia Eng.* **125** 257–62

IOP Conf. Series: Materials Science and Engineering **933** (2020) 012020 doi:10.1088/1757-899X/933/1/012020

Analysis of flexural capacity in hybrid high strength reinforced concrete beams with additive, fine aggregate, and coarse aggregate substitutions

D Hastuti[1], T B Aulia[2] and M Muttaqin[2]

[1] Civil Engineering Postgraduate Program, Department of Civil Engineering, Universitas Syiah Kuala, Banda Aceh, Indonesia
[2] Department of Civil Engineering, Universitas Syiah Kuala, Banda Aceh, Indonesia

Corresponding e-mail: hastutideni.pupr@gmail.com

Abstract. High-strength concrete is formed with a low water-cement ratio value, so it requires large amounts of cement, aggregate, and additives (silica fume) and admixture superplasticizer. Therefore it is necessary to do further research on the use of additional substitute materials. This study aims to analyze the flexural capacity of high-strength hybrid concrete beams using materials from Charcoal combustion fly ash, palm, pozzolanic sand, and hematite nanomaterials as additives and aggregate substitutions. In this study, 7 beams measuring 15 x 30 x 220 cm were tested for each additive variation and aggregate substitution of one specimen and one normal beam specimen. The test object is designed to experience flexural failure. The results showed that all beams failed flexibly as planned. The highest flexural capacity in reinforced concrete beams with additive variation of Charcoal fly ash and substitution of palm shell fine aggregate (*BBMTH BB - KCS*) is 257.707 KN or 14.217% higher than reinforced concrete beams without additive and substitution variations (*BBMTH* Normal) with a value of flexural capacity of 225.628 KN. It can be concluded that additive variation and aggregate substitution can increase the value of flexural and deflection capacity but are unable to add ductility values to high-strength hybrid concrete.

1. Introduction

High-strength concrete can be interpreted as high strength concrete which considers the durability of the concrete as well as the ease of workability. Based on SNI Pd-T-04-2004-C, high-strength concrete is concrete with required compressive strength, f'c, of 40 MPa - 80 MPa. High-strength concrete is formed with a low water-cement ratio value, so it requires large amounts of cement, aggregate, and additives (silica fume) and admixture superplasticizer. Concrete strength improvement can be done by adding or substituting other materials (additives) into the concrete mixture.

In this study, researchers substituted cement, fine aggregates and coarse aggregates on high-strength concrete mixtures with charcoal fly ash, palm oil clinkers, pozzolanic sand, and added hematite nanomaterials as fillers. As previously investigated in the shear capacity experiment of high-strength reinforced concrete beams using geopolymers fly ash [1], this study will review the flexural capacity of high-strength hybrid reinforced concrete beams.

The purpose of this study was to determine the flexural capacity of high-strength reinforced concrete beams using alternative materials, namely charcoal fly ash, palm shell chunks and pozzolanic sand instead of cement, fine aggregates and coarse aggregates and the additive of hematite nanomaterial fillers. In addition, the ductility and crack beam destruction patterns of the high-strength reinforced concrete beams that use a combination of variations in cement substitution, fine aggregate, and coarse aggregate were compared to those of the high-strength reinforced concrete beams without variation (normal).

2. Methodology

2.1. High-strength concrete

High-strength concrete criteria change through times and progress in strength levels. In the 1950s, concrete was categorized as high-strength if the compressive strength was 30 MPa, but from 1960 to 1970, the criteria rose to 40 MPa. At present, the concrete is said to be of high-strength concrete if its strength above 55 MPa, to be of very high-strength if the strength is above 80 Mpa, and to be of ultra high-strength if the strength is above 120 MPa [2].

2.2. Additional materials on concrete

The use of additional materials is intended to improve and add the properties of the material in accordance with the desired concrete properties. Additional materials used in concrete can be divided into two, namely chemical additives or chemical admixture and mineral additives, known as additives. Additives are added to improve the performance of concrete strength, while admixture functions for the ease of the work (workability). Admixture is added during stirring and when the casting is carried out [2].

2.3. Flexural beam analysis

The flexural analysis of multiple reinforced concrete blocks involves determining the nominal moment strength of a cross-section (Mn) with values a, b, d, d ', As, As2, f'c and fy, which can be written with the following equation [3]:

$$M_{n1} = A_s . fy \left[d - \frac{a}{2} \right] \quad (1)$$

$$M_{n2} = A_s' . fy \, (d - d') \quad (2)$$

$$M_n = M_{n1} + M_{n2} \quad (3)$$

ergo:

$$a = \frac{A_s . f_y .}{0,85 . f'c . b} \quad (4)$$

$$C = \frac{a}{\beta} \quad (5)$$

Information:

M_n = Nominal strength of flexural moment (kg.cm),
a = High block compressive voltage (cm),
C = The distance of the outer fiber to the neutral line (cm),
d = Distance from the outermost fiber to the center of the tensile reinforcement (cm), and
d' = Distance from the outermost fiber to the center of the compressive reinforcement (cm).

2.4. Planning high-strength hybrid reinforced concrete beam planning

Based on the analysis carried out to obtain specimens that failed to resist flexural, the size and number of reinforcement were obtained, as shown in table 1. The planned specimens were hybrid high-strength reinforced concrete blocks with additional materials and aggregate substitution in the form of charcoal fly ash, palm oil clinkers, sand pozzolan, and hematite nanomaterial with 70 MPa plan strength. The beam used is 15 cm x 30 cm x 220 cm.

Table 1. Size and amount of reinforcement of beam test items.

Test object	Basic Reinforcement		Reinforcement Shear	Total
	Compressive	Tensile		
Hybrid high-strength concrete (*BBMTH*)	2 D 11.9	4 D 15.8	D 11.9 - 100	7

2.5. Concrete mixing planning

The composition plan of the concrete mix (concrete mix design) was based on the method of comparing the volume weight of the concrete forming material. For the design of this high-strength hybrid concrete mixture, it is planned to have 70 MPa concrete strength with 0.3 water-cement ratio. The composition of the hybrid high-strength concrete (*BBMTH*) plan can be seen in the following table 2.

Table 2. Composition of hybrid high-strength concrete.

Composition of Hybrid High-Strength Concrete							
Test Object (TO) 1 (Normal *BBMTH*)	**Cement (100 %)**		**Fine aggregate (100%)**		**Coarse aggregate (100%)**		**Nanomaterial hematite 6% as filler**
TO 2 *BBMTH BB-PP*	Charcoal Fly ash (15%)	Cement (85%)	Pozzolanic sand (10%)	Fine aggregate (90%)	Palm oil clinker coarse aggregate (40%)	Coarse aggregate 60%	Nanomaterial hematite 6%
TO 3 *BBMTH BB-KCS*	Charcoal Fly ash (15%)	Cement (85%)	Fine palm oil clinker aggregate (20%)	Fine aggregate (80%)	Palm oil clinker coarse aggregate (40%)	Coarse aggregate 60%	Nanomaterial hematite 6%
TO 4 *BBMTH AP-PP*	Pozzolanic fly ash (10%)	Cement (90%)	Pozzolanic sand (10%)	Fine aggregate (90%)	Palm oil clinker coarse aggregate (40%)	Coarse aggregate 60%	Nanomaterial hematite 6%
TO 5 *BBMTH AP-KCS*	Pozzolanic fly ash (10%)	Cement (90%)	Fine palm oil clinker aggregate (20%)	Fine aggregate (80%)	Palm oil clinker coarse aggregate (40%)	Coarse aggregate 60%	Nanomaterial hematite 6%
TO 6 *BBMTH CS-PP*	Palm oil clinker fly ash (15%)	Cement (85%)	Pozzolanic sand (10%)	Fine aggregate (90%)	Palm oil clinker coarse aggregate (40%)	Coarse aggregate 60%	Nanomaterial hematite 6%
TO 6 *BBMTH CS-KCS*	Palm oil clinker fly ash (15%)	Cement (85%)	Fine palm oil clinker aggregate (20%)	Fine aggregate (80%)	Palm oil clinker coarse aggregate (40%)	Coarse aggregate 60%	Nanomaterial hematite 6%

2.6. Testing of objects

This test aims to obtain data on the maximum load that can be held by beams, strain, and deflection that occur. Steel and concrete deflection and strain are monitored for every 100 kg load increase using the LVDT (transducer) and strain gauge. The test set for *BBMTH* beam flexural capacity can be seen in the following picture.

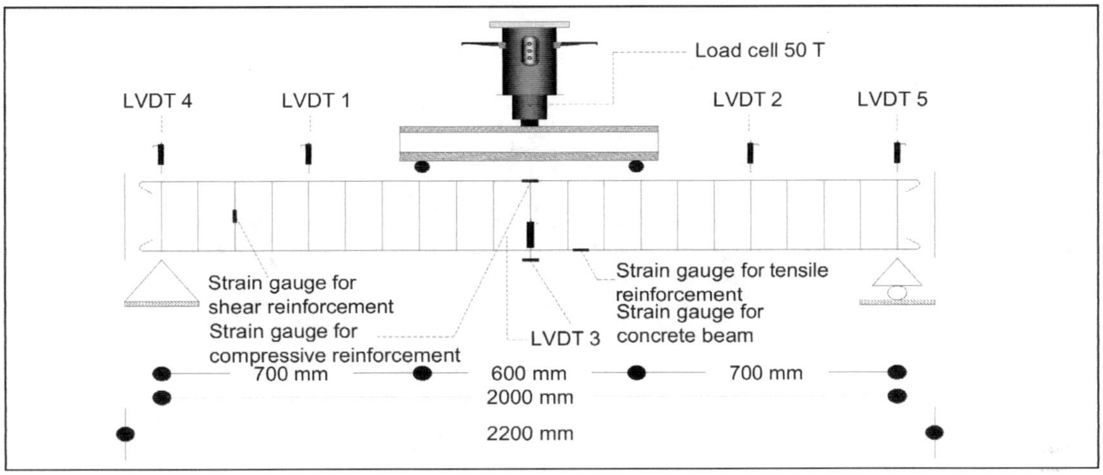

Figure 1. Strength testing of reinforced concrete.

3. Results and discussion

3.1. Compressive strength and flexural tensile strength

Based on figure 2, it can be seen that the value of the compressive strength of the BMTH control cylinder with additive variation and aggregate substitution having mixed results when compared with the value of compressive strength in Normal *BBMTH. BBMTH BB - PP*, *BBMTH*, *BB - KCS BBMTH AP - PP*, and *BBMTH AP - KCS* have higher compressive strength values than Normal *BBMTH* whereas the last two specimens, namely Normal *CS - PP* and *BBMTH CS - KCS* have lower compressive strength values. However, the compressive strength of *BBMTH CS - PP* and *BBMTH CS - KCS* can still be classified as high-strength concrete [3].

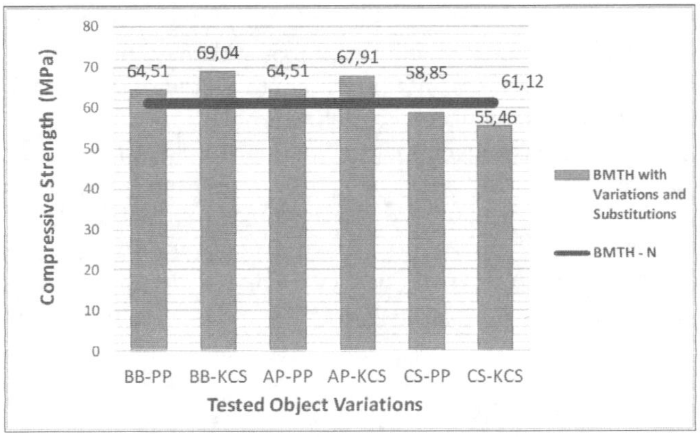

Figure 2. Compressive strength of concrete cylinder testing.

In contrast to compressive strength, flexural tensile strength *BBMTH* additive variation and aggregate substitution have uniform results. This can be seen in figure 3. The value of flexural tensile strength is higher than that of Normal *BBMTH*. The average maximum load produced in this flexural tensile strength test is used as a reference to determine the initial cracks in *BBMTH* beams and serve as a reference during testing.

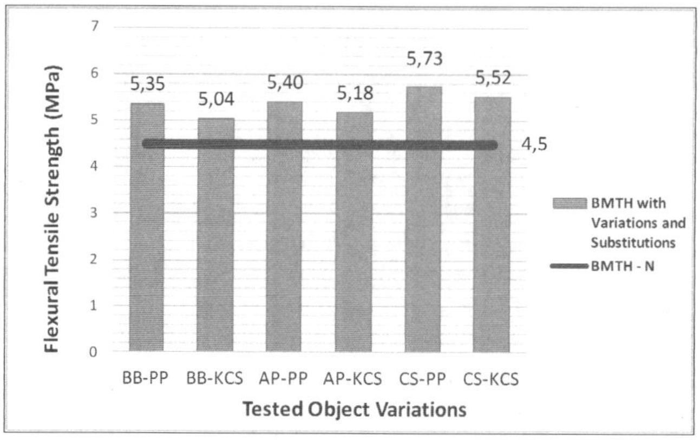

Figure 3. Strength testing of flexural concrete beams.

Based on the previous analysis on *BBMTH* with additive variation and aggregate substitution, the value of compressive strength is not always followed by the value of the *BBMTH* flexural tensile strength itself. In other words, in this case, the relationship of the value of compressive strength with flexural strength is not linear. The results obtained are in accordance with the research which states that the relationship between the compressive strength of the flexural strength of the concrete is always parabolic [4]. This means that at a certain point, the increase in the value of concrete compressive strength will be accompanied by an increase in the flexural strength of the beam, but at certain points, it shows the opposite result.

3.2. Load and deflection

The comparison of load and deflection of *BBMTH* with additive variation and aggregate substitution can be seen in figure 4. In figure 4, it can be seen that the graphs of *BBMTH AP - PP* and *BBMTH CS - KCS* are more sloping compared to the other five graphs. This shows that the high-strength hybrid reinforced concrete beams with a variety of additional materials have a smaller flexural capacity value compared to the other five beams. The results of the graph reading for the loading area at the initial crack, transitional load, and ultimate load, can be seen in table 3.

Comparison of *BBMTH* beam deflection with additive variation and aggregate substitution on *BBMTH* Normal beams can be seen in table 4. It can be seen from the deflection value in the table, *BBMTH* beams with additive variation and aggregate substitution can increase the value of high-strength concrete beam deflection. In general, the largest increase in deflection values occurs in *BBMTH* beams with additive variations of fly ash pozzolan (code: AP). The increase in the highest deflection value is in the *BBMTH AP - PP* beam, which is equal to 252.157% of the deflection of Normal *BBMTH*. This is because the fly ash pozzolan has a small grain diameter so that the spread in the mixed matrix is more evenly distributed. The increase in deflection that occurs in *BBMTH* with additive variation and aggregate substitution varies greatly between 104.575% - 252.157%.

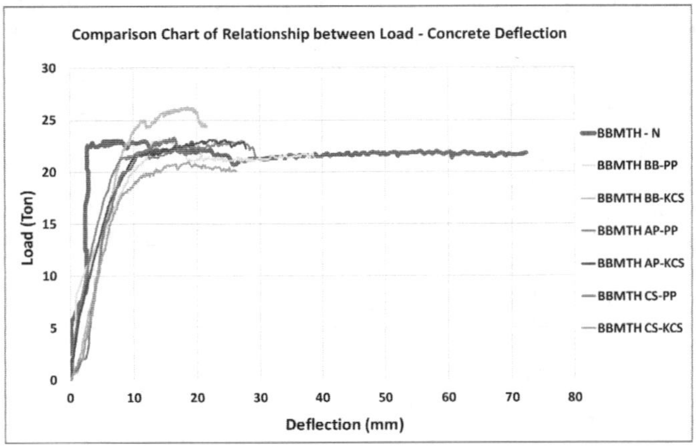

Figure 4. Comparison chart of the relationship between load - concrete deflection.

Table 3. Results of reading deflection charts.

Beam Test Objects	First crack		Transition		Maximum Load	
	P (ton)	Deflection (mm)	P (ton)	Deflection (mm)	P (ton)	Deflection (mm)
BBMTH-N	7.550	1.750	22.830	3.450	23.000	7.650
BBMTH BB-PP	4.550	0.900	19.580	9.250	23.070	15.650
BBMTH BB-KCS	5.100	2.500	22.650	9.250	26.270	18.500
BBMTH AP-PP	3.030	0.430	20.140	7.060	22.960	26.940
BBMTH AP-KCS	4.100	0.910	21.480	10.100	23.100	24.530
BBMTH CS-PP	3.380	2.840	19.860	8.300	23.380	16.440
BBMTH CS-KCS	6.590	3.460	19.690	12.170	20.960	18.710

Table 4. Comparison of *BBMTH N* beam deflection with *BBMTH* beams additive variations and aggregate substitution.

Beam Test Objects	Theoretical deflection (mm)	Deflection at the *P* Ultimate (mm)	Comparison of Deflection Against Normal Deflection *BBMTH*	
			Percentage (%)	Difference (%)
BBMTH-N	17.02	7.650	100.000	-
BBMTH BB-PP	17.74	15.650	204.575	104.575
BBMTH BB-KCS	18.68	18.500	241.830	141.830
BBMTH AP-PP	17.74	26.940	352.157	252.157
BBMTH AP-KCS	18.44	24.530	320.654	220.654
BBMTH CS-PP	16.53	16.440	214.902	114.902
BBMTH CS-KCS	15.79	18.710	244.575	144.575

3.3. Concrete load and strain

Based on figure 5, it can be seen that the Normal *BBMTH* Beams have a concrete strain of 197.650 με at a load of 23.00 tons. The *BBMTH BB - PP* beam has a concrete strain of 218.742 με at a load of 23.07 tons. The *BBMTH BB - KCS* beam has a concrete strain of 106.017 με at a load of 26.27 tons. *BBMTH AP - PP* beam has a concrete strain of 78.153 με at a load of 22.96 tons. *BBMTH AP - KCS* beam has a concrete strain of 387.550 με at a load of 23.10 tons. The beam of *BBMTH CS - PP* has a concrete strain of 33.196 με at a load of 23.38 tons. Whereas the *BBMTH CS - KCS* beam has a concrete strain of 64.656 με at a load of 20.96 tons.

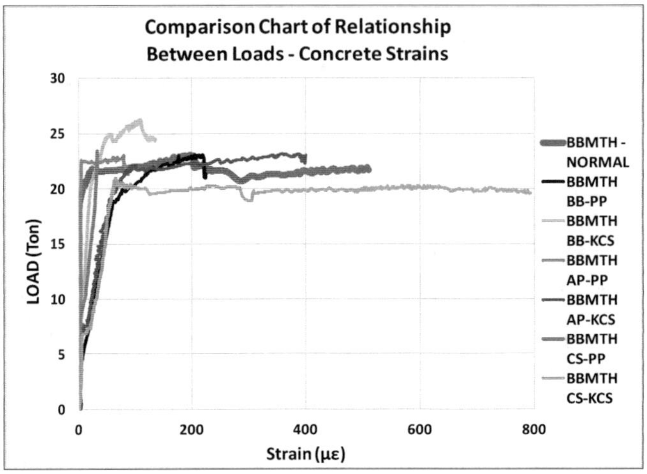

Figure 5. Comparison chart of the relationship between loads - concrete strains.

3.4. High-strength hybrid concrete reinforced capacity (BBMTH)

The *BBMTH* beam test results are shown in table 5 below.

Table 5. Results of testing of the flexural capacity of *BBMTH*.

Beam Test Objects	f'c (Mpa)	Comparison of Results					
		Maximum Load (KN)			Flexural Moment (KN.m)		
		Pu Teoritis	Plab	Plab/Pu	Mn Teoritis	Mlab	Mlab/Mn
BBMTH-N	61.115	206.697	225.628	1.092	72.344	86.539	1.196
BBMTH BB-PP	64.511	207.721	226.315	1.090	72.702	86.387	1.188
BBMTH BB-KCS	69.038	208.988	257.707	1.233	73.146	96.589	1.320
BBMTH AP-PP	64.511	207.721	225.236	1.084	72.702	86.345	1.188
BBMTH AP-KCS	67.906	208.681	226.609	1.086	73.038	86.462	1.184
BBMTH CS-PP	58.852	205.973	229.356	1.114	72.090	87.031	1.207
BBMTH CS-KCS	55.457	204.816	205.616	1.004	71.686	78.890	1.100

From table 5, the highest increase in flexural capacity in *BBMTH* with additive variation and aggregate substitution on BBMTH Normal, i.e., on the *BBMTH BB-KCS* specimen with a value of 257.707 KN, increased by 14.2% from Normal *BBMTH* at 225.688%. From table 5, it can be seen that the Plab ratio ranges from 1.084 to 1.233Pu. This explains that the laboratory results are in accordance with the theoretical results which are calculated by the difference in a difference of 8.4% - 23.3%. While from

Table 6, it can be seen that the theoretical comparison ratio between Shifts to Plots is more than 1 (Placement < Shift). This is in accordance with the initial planning regarding the behavior of the *BBMTH* beam with variations and planned substitutions which failed to flex.

Table 6. Comparison of flexural and shear loads of theoretical calculation results.

Beam Test Objects	*f'c (Mpa)*	Based on the Preliminary Design of Beam Cross Section Capacity				
		Flexural Strength		P Shear (kN)	Theoretical Deflection (mm)	Ratio
		Moment (kNm)	P Flexural (kN)			(P Shear / P Flexural)
BBMTH-N	61.115	72.344	206.697	557.293	17.018	2.6962
BBMTH BB-PP	64.511	72.702	207.721	559.794	17.737	2.6949
BBMTH BB-KCS	69.038	73.146	208.988	563.047	18.675	2.6942
BBMTH AP-PP	64.511	72.702	207.721	559.794	17.737	2.6949
BBMTH AP-KCS	67.906	73.038	208.681	562.244	18.443	2.6943
BBMTH CS-PP	58.852	72.090	205.973	555.564	16.531	2.6973
BBMTH CS-KCS	55.457	71.686	204.816	552.927	15.788	2.6996

From the results of the comparison of the beam capacity with additive variation and aggregate substitution shown in table 5 and table 6, it can be concluded that high-strength concrete with additive variation and aggregate substitution can increase the high-strength concrete flexural strength.

3.5. Ductility

Based on table 7, *BBMTH* with additive variation and aggregate substitution does not show an increase in ductility values in *BBMTH* beams. The ductility of *BBMTH* with additive variation and aggregate substitution is lower when compared to Normal *BBMTH*, which ranges from 74.089% - 87.582%. *BBMTH AP - PP* specimens, which have the highest ductility value, are higher when compared to Normal *BBMTH;* the value is 167.100% or 67.1% higher than those of Normal *BBMTH.*

Table 7. *BBMTH* beam ductility calculation.

No.	Test Object Variations	Tear condition		Ultimate condition		Ductility μ = Δu/Δy	Actual Ductility for *BBMTH* Normal (%)
		Load (P)	Deflection (Δy)	Load (P)	Deflection (Δu)		
		ton	mm	ton	mm		
1	*BBMTH-N*	22.830	3.350	23.000	7.650	2.284	-
2	*BBMTH BB-PP*	19.580	9.250	23.070	15.650	1.692	74.089
3	*BBMTH BB-KCS*	22.650	9.250	26.270	18.500	2.000	87.582
4	*BBMTH AP-PP*	20.140	7.060	22.960	26.940	3.816	167.100
5	*BBMTH AP-KCS*	21.480	10.310	23.100	24.530	2.379	104.189
6	*BBMTH CS-PP*	19.860	8.300	23.380	16.440	1.981	86.738
7	*BBMTH CS-KCS*	19.690	10.310	20.960	18.710	1.815	79.469

3.6. Crack pattern

The formation of cracks in general from each specimen varies, but the destruction that occurs is the same, namely flexural destruction or failure of flexural. This is indicated by the predominance of cracks in flexible areas.

Figure 6. The pattern of *BBMTH* beam destruction.

4. Conclusion

1. *BBMTH* beams with additive variation and aggregate substitution increase the value of flexural and deflection capacity but generally are not able to increase the ductility value of hybrid high-strength concrete beams. The highest flexural capacity on *BBMTH BB - KCS* beams is 257.707 KN or 14.217% higher than Normal *BBMTH* with a flexural capacity value of 225.628 KN.
2. The maximum deflection for additive variation is obtained from test specimens with additive variations of fly ash pozzolan (*BBMTH AP - PP*) with the percentage of deflection to Normal *BBMTH* amounting to 252.157%.
3. The crack pattern between the *BBMTH* beams of additive variation and substitution indicates that the beam has failed flexibly. It is proven by cracks that occur in flexible areas, and almost no cracks occur in the support area. This is also in accordance with the calculation of the shear ratio: placement which ranges from 2.964-2.700.
4. Ductility in *BBMTH* beams with additive variation and aggregate substitution, in general, does not increase. The ductility of *BBMTH* with additive variation and aggregate substitution is lower when compared to Normal *BBMTH*, which ranges from 74.089%-87.582%. *BBMTH AP-PP* specimens that have the highest ductility value are higher when compared to Normal *BBMTH*; the value is 167.100% or 67.1% higher than those of Normal *BBMTH*.

References

[1] Aulia T B et al. 2018 Shear capacity analysis of high strength reinforced concrete beams using geopolymer fly ash and palm oil blast furnace slag as additives and aggregate substitution *IOP Conf. Series: Materials Science and Eng* 434 012199
[2] Mulyono 2005 *Teknologi* Beton (Yogyakarta: Andi)
[3] Nawy E G 1985 *Reinforced Concrete a Fundamental Approach*, (Sidney: McGraw-Hill Book Company)
[4] Dady Y T 2015 *Pengaruh Kuat Tekan Terhadap Kuat Lentur Balok Bertulang* (Manado: Fakultas Teknik, Jurusan Sipil, Universitas Sam Ratulangi Manado)

The study of the operation of Trans Koetaradja public bus transportation of the corridor II A and II B in Banda Aceh city

F Fachrurrozy[1], E Wulandari[2] and Y Darma[3]

[1] Civil Engineering Postgraduate Program, Department of Civil Engineering, Universitas Syiah Kuala, Banda Aceh, Indonesia
[2] Lecturer Advisory Committee the Architectural Study Programs in the Field of Urban Area Development, Engineering Faculty, Universitas Syiah Kuala, Banda Aceh, Indonesia
[3] Lecturer Advisory Committee in the Field of Transportation the Civil Department, Civil Engineering Faculty, Universitas Syiah Kuala, Indonesia

Corresponding e-mail: rozy@mhs.unsyiah.ac.id

Abstract. Aceh Government is trying to plan the integrated transportation system. Aceh Government has planned the construction of 6 Corridors of Trans Koetaradja Bus, some of them are Corridors II A and II B. The previous public transport cannot accommodate the transportation service needs of the route of Sultan Iskandar Muda Airport - city center - Ulee Lheue Crossing Port. However, currently, the Banda Aceh Trans Koetaradja Public Bus Transportation is projected to be able to improve the services optimally. The purpose of this study is to assess the level of performance and to improve the quality of the service up to the customers' satisfaction expectations. The method of this research comprises the survey, which is sampling Kouto and distributing questionnaires and data analysis. The questionnaire distribution uses statistic test method and Importance Performance Analysis (IPA) through the SERVQUAL approach, namely: reliability, responsiveness, assurance, empathy, and physical evidence. The result of the study is that the current performance of the services of Trans Koetaradja public bus transportation is rated as good; the importance of the service regarding the customers' expectation is rated good; the quality of service provided to customers is rated 94.52%. The conclusion of this study is that there are several shortages of transportation performance, such as the timeliness of arrival, waiting time at bus stops, passenger assistance, and available seats.

1. Introduction

Aceh Government seeks to plan the integrated transportation system. Trans Koetaradja, the public bus transportation system in Banda Aceh City, is expected to provide optimal service to the community. Previous public transportation was not enough to accommodate the service needed in Corridor II A and II B.

Banda Aceh is a capital city, which has been being developed since the days of the Aceh kingdom. Banda Aceh is a center of activity for the surrounding suburbs, with radial-shaped city structures that reach out in all directions [1]. At present, the city has developed as a modern city, which has various forms of public transport services organized by both the government and the community. One of the public transportation services is the Trans Koetaradja Banda Aceh public transportation.

The Trans Koetaradja Bus begins in 2017. The plans will operate until 2021. The bus transportation system serves 6 corridors. Some of the corridors are the Corridor II A and Corridor II B that serve the routes connecting the Greater Aceh Region and Banda Aceh City, namely: Sultan Iskandar Muda Airport - Type A Batoh Terminal - APK Keudah Terminal - Baiturrahman Grand Mosque - Ulee Lheu Ferry Port.

According to Black [2], the public wants good transportation, so it is very important to involve the public in reviewing public transport services in order to obtain recommendations that need to be considered by transportation service providers. The aspects to be reviewed are security, accuracy, regularity, comfort, speed, pleasure, and user satisfaction. The performance of public transportation services should match users' expectations. The intended performance includes a sufficient level of service, travel time, waiting time, guaranteed safety, and comfort.

The purpose of this research is to conduct to identify customer perceptions of the performance of the Trans Koetaradja bus – corridor II A and II B and to review the bus system's level of service performance and level of service quality performance. The public perception was considered in the SERVQUAL approach: reliability, responsiveness, assurance, empathy, and tangible evidence.

2. Methodology

The study began with a preliminary survey, identification of problems, the study of literature, determination of data collection methods, and determination of population and samples using kouto sampling. According to Sugiyono [3], kouto sampling is a technique for determining samples from populations that have certain characteristics up to the number desired. According to Margono [4], kouto sampling is the technique of population numbers not taken into account but classified in several groups.

The samples were several groups in the research area. The samples were divided into two groups, namely: the respondents at the bus stop and the respondents inside the Trans Koetaradja Bus serving the corridors II A and II B. The questionnaires were based on Likert scale measurement. The questionnaire consisted of questions about respondent characteristics and public perceptions about the level of service performance and the level of importance of service quality using the SERVQUAL indicator according to Parasuraman [5], namely: reliability, responsiveness, assurance, empathy, and physical evidence.

2.1. Validity test

The collected data will be tested for validity using the formula below which is according to Sugiyono [3].

$$r_{xy} = \frac{N\sum XY - (\sum X)(\sum Y)}{\sqrt{\{N\sum X^2 - (\sum X)^2\}\{N\sum Y^2 - (\sum Y)^2\}}} \tag{1}$$

Information:

r_{xy} = Correlation coefficients between variables X and Y;
N = Number of respondents;
ΣX = Number of scores obtained from respondents tested; and
ΣY = Total total score of all items from all respondents tested.

Table 1. r value distribution.

df = (N-2)	**The level of significance for one-way tests**				
	0.05	0.025	0.01	0.005	0.0005
	The level of significance for the two-way test [3]				
	0.1	0.05	0.02	0.01	0.001
117	0.1515	0.1801	0.2131	0.2353	0.2979
118	0.1509	**0.1793**	0.2122	0.2343	0.2967
119	0.1502	0.1786	0.2113	0.2333	0.2955
120	0.1496	0.1779	0.2104	0.2324	0.2943
121	0.1490	0.1771	0.2096	0.2315	0.2931

2.2. Reliability test

The reliability test was calculated using the formula below, which is according to Sugiyono [3].

$$r_i = \frac{k}{(k-1)}\left[1 - \frac{\Sigma\sigma_b{}^2}{\sigma_t{}^2}\right] \tag{2}$$

Information:

r_i = Instrument reliability coefficient (alpha cronbach);
k = Number of questions;
$\Sigma\sigma_b^2$ = Amount of variance; and
σ_t^2 = Total variance.

Table 2. Interval coefficients [3].

Interval Coefficients	Relationship Level
0.00 – 0.199	Very low
0.20 – 0.399	Low
0.40 – 0.599	Enough
0.60 – 0.799	Strong
0.80 – 1.000	Very strong

2.3. Statistic test

The Importance Performance Analysis (IPA) method was calculated using statistical tests to find out the quadrants I, II, II, IV that interprets the characteristics of respondents and public perceptions about current service performance and service quality. According to Martilla and James [6], the IPA method was represented by variables X and Y, where:

X = Current performance services of Trans Koetaradja Bus; and
Y = The important aspect of the service according to the customers of the Trans Koetaradja Bus.

The formula used are:

$$T_{ki} = \frac{X_i}{Y_i} \times 100\% \tag{3}$$

Where:

T_{ki} = The level of suitability of respondents;
X_i = Score of the performance level of the Trans Koetaradja Bus; and
Y_i = Score of the value of customer interests.

The horizontal axis (X) will be filled with the score of the implementation of the performance level. The upright axis (Y) will be filled with the importance level score. The formula is as follow:

$$\bar{X} = \frac{\Sigma X_i}{n} \tag{4}$$

$$\bar{Y} = \frac{\Sigma Y_i}{n} \tag{5}$$

Where:

$\bar{X}$ = Average level of implementation/performance;
$\bar{Y}$ = Average level of importance score; and
n = Number of respondents.

IOP Conf. Series: Materials Science and Engineering **933** (2020) 012021 doi:10.1088/1757-899X/933/1/012021

The Cartesian diagram is divided into four sections that are bounded by two lines that intersect perpendicular to the average point X and Y, which is the average score of performance/importance of passenger satisfaction from all factors or attributes. The formulas are as follows:

$$\bar{\bar{X}} = \frac{\sum_{i=1}^{N} \bar{X}_i}{K} \tag{6}$$

$$\bar{\bar{Y}} = \frac{\sum_{i=1}^{N} \bar{Y}_i}{K} \tag{7}$$

Where:
K = The attributes or facts that can affect customer satisfaction.

The final stage is the collaboration of each attribute in the Cartesian diagram as shown in figure 1.

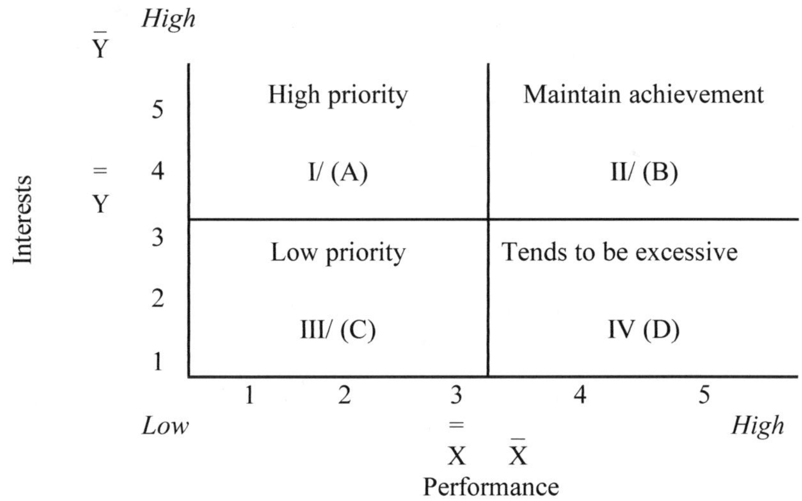

Figure 1. Cartesian diagram.

Table 3. Cartesian quadrant [6].

Quadrant	Interpretation
I	Very important and not satisfied (High priority/increase performance)
II	Very important and very satisfied (Maintain performance)
III	Not important and dissatisfied (Low priority)
IV	Not important and very satisfied (tends to be overdone)

3. Results and discussion

The survey was conducted in the operational area of Trans Koetaradja Bus of Corridor II A and II B, which routes can be seen in the below table.

Table 4. Routes.

Routes	Street Name				Bus Stop		Information
					Go	Return	
Airport – City center	1	Bandara Sultan Iskandar Muda St.	5	T. Mohammad Daud Beureueh St.	17	9	Permanent

IOP Conf. Series: Materials Science and Engineering **933** (2020) 012021 doi:10.1088/1757-899X/933/1/012021

	2	Soekarno Hattta St.	6	Jend. Ahmad Yani St.			
	3	Mr. Mohammad Hasan St.	7	Cut Meutia St.	5	8	Portable
	4	Hasan Dek St.	8	P. Diponegoro St.			
City center - Port	9	S. Alaidin Mahmudsyah St.	12	Sultan Iskandar Muda St.	8	7	Permanent
	10	Mohammad Jam St.	13	Pelabuhan Ulee Lheu St.			
	11	Tgk. Abu Lam U St.					

Source: Primary Data (2019)

3.1. Sample

The samples were collected using kouto sampling for three days, which can be seen in the below table.

Table 5. Sample.

Sub-District		Kouto Sampling in Bus Stop	in Bus	Sample
Blang Bintang		7	8	15
Ingin Jaya		7	8	15
Darul Imarah		7	8	15
Lueng Bata		7	8	15
Kuta Alam		7	8	15
Kuta Raja		7	8	15
Baiturrahman		7	8	15
Meuraxa		7	8	15
	Total	56	64	120

Source: Primary Data (2019)

3.2. Respondents questionnaire

120 respondents answered the questions from the questionnaires. The respondents' characteristics can be seen in tables 6, 7, 8 and 9.

Table 6. Gender.

Gender	Total	%
Male	48	40
Female	72	60
Total	120	100

Table 7. Age.

Age	Total	%
< 18 Years old	16	13
19 - 29 Years old	60	50
30 - 39 Years old	27	23
40 - 49 Years old	12	10
50 - 59 Years old	5	4
> 60 Years old	0	0
Total	120	100

Table 8. Work.

Work	Total	%
Employees (PNS)	18	15
Private employees	22	18
Lecturer/Teacher	16	14
Entrepreneur/Trader	10	8
Laborers/Farmers	12	10
Student	42	35
Total	120	100

Table 9. Income.

Income	Total	%
< 1.000.000	54	45
1.000.000 - 2.000.000	40	33
2.000.000 - 3.000.000	20	17
> 3.000.000	6	5
Total	120	100

Source: Primary Data (2019)*6,7,8 and 9

3.3. Validity and reliability tests

The form of measurement to determine the validity and reliability of data for 120 respondents with a significant amount of 5% in detail can be seen in table 1 of Rtable Distribution. The results of the detail can be seen in table 10, 11, 12 and 13 below.

Table 10. SERVQUAL questionnaire indicators.

No		Item Questions	R count of performance	R count of satisfaction	Rtable (df = N - 2)	Information performance	Information satisfaction
I	**Reliability**						
1	x	Timeliness departure of Trans Koetaradja Bus	0.713	0.116		Valid	Invalid
2	1	Timeliness arrival of Trans Koetaradja Bus	0.747	0.596	0.179	Valid	Valid
3	2	Waiting time for the Trans Koetaradja Bus at the bus stop	0.695	0.596		Valid	Valid
4	3	Travel time for Trans Koetaradja Bus	0.757	0.681		Valid	Valid
5	4	The accuracy of Trans Koetaradja Bus officers in providing information	0.757	0.255		Valid	Valid
6	5	Availability of Trans Koetaradja Bus schedule and route information	0.418	0.219		Valid	Valid
II	**Responsiveness**						

7	6	The skill / ability of the driver in operating the Trans Koetaradja Bus	0.713	0.402	0.179	Valid	Valid
8	7	Driver discipline (driver's obedience to traffic regulations)	0.401	0.402	0.179	Valid	Valid
9	8	The Trans Koetaradja Bus operator assistance to the passengers	0.713	0.596		Valid	Valid
10	x	The Trans Koetaradja Bus operator reminds the destination of the stop	0.031	0.377		Invalid	Valid
11	9	The operator regulates passenger density in the Trans Koetaradja Bus	0.713	0.291		Valid	Valid
12	x	Trans Koetaradja Bus Operators provide guidance to customers on the ease of switching routes to other corridors	0.058	0.130		Invalid	Invalid
III	**Assurance**						
13	10	Passenger safety/security while inside the Trans Koetaradja Bus	0.757	0.681	0.179	Valid	Valid
14	11	Safety/security for Trans Koetaraja Bus passengers when they stop at the bus stop	0.205	0.402		Valid	Valid
15	12	Security in Trans Koetaradja Bus and at the stops	0.401	0.681		Valid	Valid
16	13	Comfort in t Trans Koetaradja Bus or at stops	0.713	0.402		Valid	Valid
IV	**Emphaty**						
17	14	Good service from the driver (friendly, polite and caring) while te user inside the Trans Koetaradja Bus	0.695	0.413	0.179	Valid	Valid
V	**Tangible**						
		Feasibility of the physical condition of the Trans Koetaradja Bus					
18	15	Available seating is decent	0.757	0.413	0.179	Valid	Valid
19	16	Trans Koetaradja Bus Feasibility (color, glass and door)	0.401	0.681		Valid	Valid
20	x	The availability of security facilities such as Light Fire Extinguisher (APAR) and CCTV	0.334	0.095		Valid	Invalid
21	17	Additional convenience facilities are provided such as AC or WiFi	0.205	0.402		Valid	Valid
		Condition of Trans Koetaradja Bus Stop					
22	18	Bus stops that are currently available are easy to reach	0.195	0.681	0.179	Valid	Valid
23	19	Availability of a waiting room with a decent number of seats	0.205	0.314		Valid	Valid

24	x	Availability of standing space, fences, and stairs up and down at the bus stop	0.091	0.237		Invalid	Valid
25	x	Bus stop building that is friendly to disabled passengers	0.334	-0.017		Valid	Invalid
		Cleanliness and tidiness					
26	20	Trans Koetaradja Bus	0.757	0.338	0.179	Valid	Valid
27	x	Halte	0.151	0.318		Invalid	Valid
28	x	**The Trans Koetradja Bus trip fees/fees**	0.757	0.171		Valid	Invalid

Table 11. Results of SERVQUAL questionnaire indicators.

No	**Indikator SERVQUAL**	**Number of Questions**	**After Test** *Stage Valid I*	**After Test** *Stage Valid II*	**Item Elimination Questions**
1	Reliability	6	1	5	1
2	Responsiveness	6	2	4	10, 12
3	Assurance	4	0	4	0
4	Emphaty	1	0	1	0
5	Tangible	11	5	6	20, 24, 25, 27, 28
	Total	28	8	20	

Table 12. Results of SERVQUAL questionnaire indicators.

No	Item description	Interval Coefficients	Variable	Information
I	*Cronbach Alpha* > 0.6 Performance Level	0.753	Reliable	Strong
II	*Cronbach Alpha* > 0.6 Satisfaction Level	0.891	Reliable	Very strong

Source: Primary Data (2019)*[10, 11, 12 and 13]

3.4. Community perception of service level

The results of the research on community perceptions of the level of public transport services are carried out by the SURVQUAL approach, along with the results of the questionnaire distribution recapitulation, which can be seen in the below table.

Table 13. Perception of Trans Koetaradja's BRT public service level.

Item	Current Service performance (X)	Service Interest importance (Y)	Suitability Level (%)
Reliability	3.60	3.62	100.27
Responsiveness	3.32	3.70	89.79
Assurance	3.64	3.33	109.37
Empathy	3.28	4.38	74.90
Tangible	3.81	4.18	92.08
rata-rata	**3.58**	**3.84**	**94.52**

Source: Primary Data (2019)

IOP Conf. Series: Materials Science and Engineering **933** (2020) 012021 doi:10.1088/1757-899X/933/1/012021

Based on table 14, the average value of the level of suitability is in the range of values 90-100 which is 94.52%, so that it can be concluded as a whole that these attributes fall into the "satisfied" category.

3.5. Cartesian diagram

The diagram shows that the distribution of question items about the Trans Koetaradja Bus for Corridors II A and II B. The average value of services from community perceptions about the level of performance and the importance of customer expectations can be seen in the cartesian diagram in figure 2.

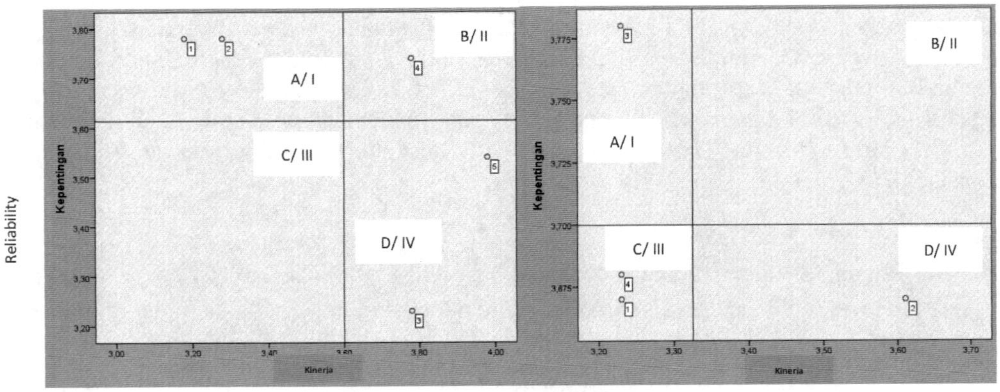

Figure 2. Results of the cartesian diagram.

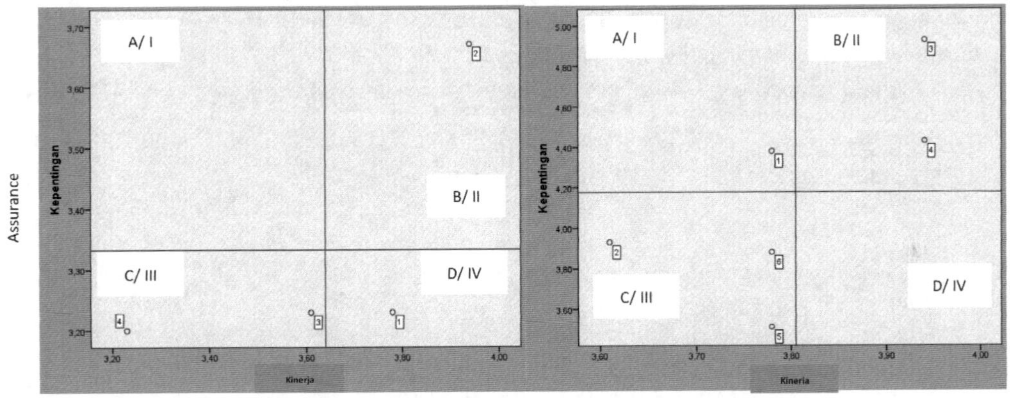

Figure 3. Results of the cartesian diagram.

(1) Quadrant I(B): The question items of number 1, 2, 8 and 15 for service elements that are considered very important, but the management has not implemented the interests of customer desires.
(2) Quadrant II(D): The question items of number 4, 12, 17 and 18 which constitute the main service elements that have been successfully carried out by companies.
(3) Quadrant III(C): The question items of number 6, 9, 12, 13, 16, 19 and 20 are not important for customers, but the implementation is classified as mediocre, considered to be less important and less satisfying
(4) Quadrant IV(D): The question items of number 3, 5, 7, and 10 that were factors that considered as less important, but the implementation tends to be excessive.

4. Conclusion

Based on the results of the study, it was found that the public assessed that the current performance service on Trans Koetaradja Bus public transportation was considered good at 3.54. The community

IOP Conf. Series: Materials Science and Engineering **933** (2020) 012021 doi:10.1088/1757-899X/933/1/012021

assesses that service interests for the expectation of Trans Koetaradja Bus Public Transport services to customers are valued hight rated at 3.84 with the quality of services provided to customers on Trans Koetaradja Bus public transportation being satisfied meeting expectations of 94.52%. But there are some things that there are still deficiencies that need to be considered to improve performance for the better, on the punctuality of arrival, waiting time at the bus stop, the operator helps passengers and available seats, becoming a very satisfying service of hope.

Acknowledgments
First of all, my gratitude for Allah SWT. Then, my thanks to the father/mother, Dr. Ir. Elysa Wulandari, MT as the First Advisory Committee and Dr. Yusria Darma, ST., M.Eng. Sc as the Second Advisory Committee and all teaching staff, and colleagues. My thanks to Civil Engineering Masters Program, Civil Engineering Study Program, Civil Associates, and Implementers of the Aceh International Symposium On Civil Engineering 2019 Syiah Kuala University Faculty of Engineering Department of Engineering Civil.

References
[1] Elysa W, Sugiono S, Joesron A, Syahbana and Asnawi M 2017 The ecology character of Banda Aceh City in the 17th century *J. of Architecture* **4** 3 93-101
[2] Black A 1995 *Urban Mass Transportation Planning* (USA: University of Kansas)
[3] Sugiyono 2010 *Educational Research Methods Quantitative, Qualitative and R & D Approaches* (Bandung: Alfabeta)
[4] Margono S 2004 *Educational Research Methodology* (Jakarta: PT. Rineka Cipta)
[5] Parasuraman A, Zeithaml V A and Berry L L 1998 SERVQUAL: A multiple-item scale for measuring consumer perception of service quality *J. Of Retailing* **64** 1 12-37
[6] Martilla J A and James J C 1977 Importance performance analysis *J. of Marketing* **41** 77-79

Numerical modeling of suspended sediment concentration (SSC) due to water surface fluctuations of the Krueng Langsa estuary, Aceh Province

F Isma[1], Y Ismida[1], M Purwandito[1] and H Fajri[1]

[1] Department of Civil Engineering, Universitas Samudra, Langsa, Indonesia

Corresponding e-mail: faizisma.ts@unsam.ac.id

Abstract. Krueng Langsa estuary is a meeting zone between the sea and the river that crosses the Langsa city. As a result, the Krueng Langsa estuary had a physical change due to tides and river discharge. The Indonesian Minister of Trade Regulation of Number 24 of 2019 has announced that the Krueng Langsa estuary will resume its activities as international shipping. This indeed becomes an important problem for the continuity of shipping activities. An initial study needs to be carried out related to changes in the physical waters of the estuary due to high and low tides. This study is focused on the estuary physical system—especially in water depth, tidal current, salinity, temperature, and SSC along the estuary due to tidal dynamics processes and river discharge runoff—which are modeled numerically. The influence of the Malacca Strait and the Langsa River flow on SSC at each measurement point have also been observed. In this study, the model results were evaluated using the in-situ measurements. The results show that numerical modeling is capable of providing useful information about the physical condition of the Krueng Langsa estuary and agree well with the measurement data.

1. Introduction

Estuaries are a place where two water masses, interrelated seas, and rivers, are mixed. The mixing of the seas and the rivers causes water to experience changes in the estuary physical system against time, especially in bathymetry (depth and width), tidal currents, temperature, salinity, and suspended sediment concentration (SSC). Due to tides that move towards the river upstream and the river that flows to the sea, estuary becomes a complex environment and could affect water quality including SSC content, which in this case needs to be monitored [1]. UNESCO, 1977 once compiled a list of rivers that flowed into the ocean. Although 260 major rivers flow into the ocean, only 25% have been regularly monitored for water quality. Complex problems from estuary make all fields of science participate in solving problems. In order to solve the problem, the method must be improved, and it takes years—when they are applied [2]. However, the direct measurement process is difficult to see as the estuary's physical state has a complex relationship, so that observations can be made by modeling the interconnected physical parameters. This model also able to provide information in decision making on the estuary physical system that prioritizes estuary quality. Saltwater intrusion in estuaries is influenced by seasons; as a result, [3] conducted a numerical modeling of the Indus estuary in Pakistan in seeing the tidal characteristics in the upstream estuary; in the dry season, the tidal reaches 65 km more than those in rainy season because of the resistance from the river to the sea as a barrier to saltwater intrusion in estuary.

In his study of the Belawan estuary physical system modeling, [4] stated that the main consideration in the estuary was a sedimentation increase, which led to the need for high costs of dredging to achieve the draft requirement of the ship. the increase in sedimentation is common, especially in estuaries, where the upper cross the city and causes an increase in the content of the SSC. [5] studied the flow characteristics of the effects of saltwater intrusion in areas where there are mangroves or no mangroves around the coast. It was found that the comparison of observations and numerical simulation results in Liao China estuary obtained good results. The continuity of activities in a port and increased resources around the estuary waters have reached alarming levels, especially the Krueng Langsa estuary, which will again become an international shipping channel. Krueng Langsa estuary is predicted to have experienced sedimentation, as evidenced by the large draft vessel unable to enter the port of Kuala Langsa port [6]. With the issuance [7], the waters in the port of Kuala Langsa must fulfill the Indonesian National Port Order. The service requirements for the depth of the primary international port shipping line are at least 12 meters Mean Sea Level (MSL) and for secondary and tertiary international ports, the minimum value is 9 meters MSL [8]. In line with the predictions made, [9] stated that the erosion of the Langsa river basin partially flown to the Krueng Langsa estuary because the estuary-sediment load elevated 10,512.25 m^3/year.

The main objective of this study was to determine the physical changes of estuary due to fluctuations in water level by forming a numerical model simulation compared to the measurement data. Thus, numerical modeling is needed to observe the physical characteristics of Krueng Langsa estuary. The model results are then compared with the measurement data based on numerical model validation that has been done by [10] and [11]. Then, the value of the Correlation Coefficient (CC), Mean Error (ME), and Root Mean Square Error (RMSE) were determined to obtain information about the relationship between the results of modeling and measurement data.

2. Methodology

The study area is the Krueng Langsa estuary, where the mouth part is connected to the Malacca Strait and the downstream part is in the Krueng Langsa River. The observation points, which start from the estuary mouth and go upstream, were checked every 1 km by observing the salinity parameters close to 0 mg/L. The estuary mouth was coded as the STA A (0 + 000), and the upstream of the estuary was coded as STA Q (16 + 000), as shown in figure 1.

Measurements were taken at each station point by taking the physical parameters of the estuary, namely, flow depth using Fish Finder GPS MAP 585. Handled GPS is utilized as a station measurement at every 1 km distance. The current meter is used to measure flow rate while salinity and temperature are measured with Handheld IP67. Furthermore, SSC is collected by Van Dorn Water Sample tools at 0.2 h, 0.6 h, and 0.8 h and tested using the evaporation method at the laboratory. Krueng Langsa estuary has no tidal recording data. The tides measurements are examined for 15 days using a gauge through the lowest ebb datum which aims to obtain the amplitude of the solar (A_{S2}) and amplitude of the moon (A_{M2}) through the admiralty method.

In this study, the model is simulated by computational programs using the physical computational estuary model given by [2], namely:

2.1 Depth and width of estuary

The first step in numerical modeling is determining the stream depth (D) and the width of the Krueng Langsa estuary in an exponential variation from the mouth towards the upstream. This step was determined by dividing each measuring distance to produce a bathymetric geometry form. The formula used in this step was as follows:

$$\text{Width: } W_x = W_o e^{-a\left(\frac{x}{L}\right)} \text{ and Depth: } D_x = D_o e^{-b\left(\frac{x}{L}\right)} \qquad (1)$$

With, W_x = the estuary width at the point of measurement, x (m); D_x = the estuary depth at the measurement point, x (m); W_0 = the estuary width at the mouth (m); D_0 = the estuary depth at the mouth

IOP Conf. Series: Materials Science and Engineering **933** (2020) 012022 doi:10.1088/1757-899X/933/1/012022

(m); x = the distance from the mouth of the estuary (m), L = the estuary length from the estuary mouth (m); a and b are the coefficient of estuary width and depth, respectively.

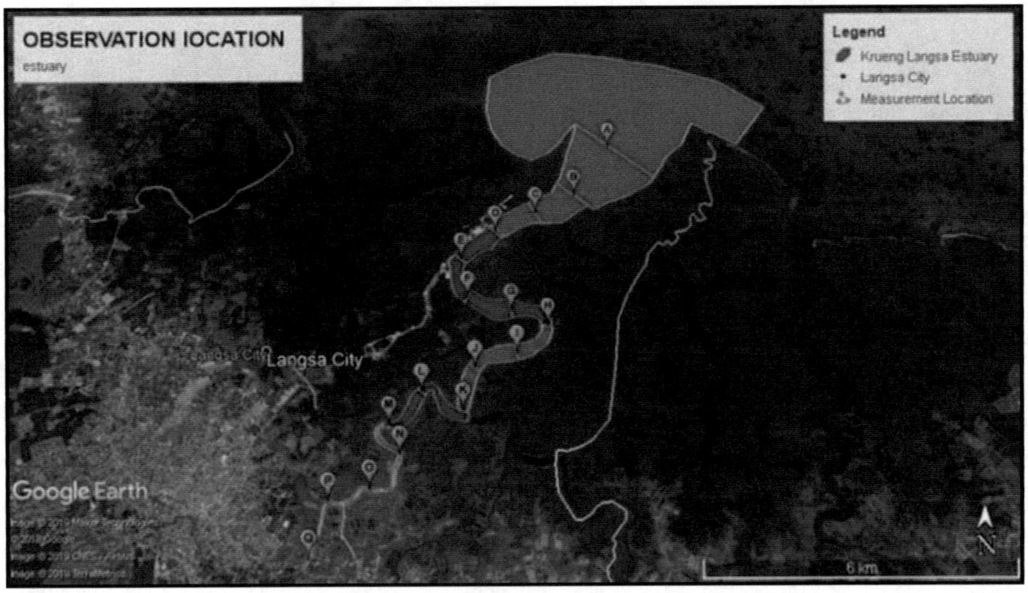

Figure 1. Observation site locations performed in this study.

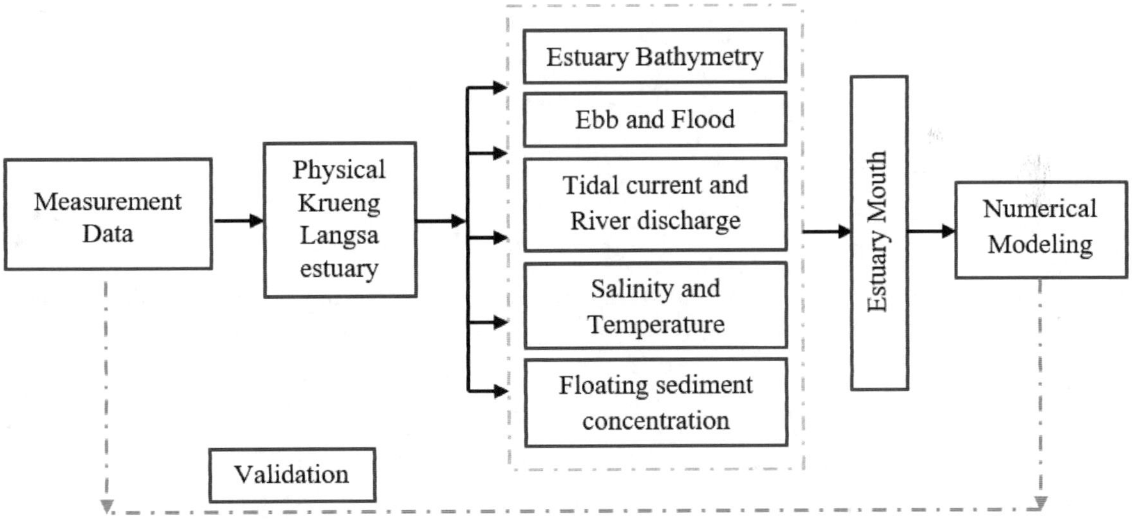

Figure 2. Flow chart of the numerical model of the physical estuary.

2.2 Tidal currents

The tidal type is determined with Formzal number according to tidal measurement data for 15 days. The changes in estuary depth $h(t)$ due to tidal harmonic is stated as follow:

$$h(t) = A_{S2}Sin(2\pi t/T_{S2}) + A_{M2}Sin\,(2\pi t/T_{M2}) + \;A_{M4}Sin\,(2\pi t/T_{M4}) + DT \qquad (2)$$

Tidal water when reaching shallow water will experience an increase (in-depth) affected by the A_{M4} component which stated by:

$$A_{M4} = \frac{3.xA_{M2}^2}{4\lambda Th} = \frac{3xA_{M2}^2}{4hT\sqrt{gh}} \tag{3}$$

With, $h(t)$ = total water level with time t hour (m); A_{S2} = tidal amplitude due to solar attraction (m); T_{S2} = tidal period due to the sun (12 hours); 2π = the rotation angle of the moon against earth and earth against the sun; A_{M2} = tidal amplitude due to lunar attraction (m); T_{M2}= tidal period due to the moon (12.42 hours); A_{M4} = amplitude of quarter-diurnal lunar tides (m); T_{M4} = period due to diurnal lunar quarter (6.21 hours); DT = water depth for each measurement point, D_x (m), and g = gravitational acceleration (9.81 m/s^2)

The combination of two tidal flows from the sea and the river results in current variations in each distance and time of U (x, t) to be:

$$U_{(x,t)} = \frac{\int_{x=X}^{x=L} W_x \Delta h_t dx}{W_x D_x} - \frac{Q}{W_x D_x} \tag{4}$$

With, U (x, t) = Estuary current for each distance x and time t (m/s); Δh = change in depth due to tides per hour per second (m/s); and Q = River discharge (m^3/s).

2.3 Temperature and salinity

The distribution of temperature and salinity from the estuary mouth to the upstream of the estuary varies numerically from the temperature distribution of each measurement point (T_x) and the salinity of each measurement point (S_x). T_x and S_x can be obtained through the following gauss distribution equation:

$$T(x) = (T_R - T_S)\, exp\left[-\frac{x^2}{2\sigma_x^2}\right] + T_S \tag{5}$$

and

$$S(x) = S\, exp\left[-\frac{x^2}{2\sigma_x^2}\right] \tag{6}$$

By, T_S = temperature at sea (°C); T_R = temperature on the river (°C); S = salinity in the mouth of the estuary; and σ_x= Gauss distribution variance.

2.4 Suspended sediment concentration (SSC)

Due to the exchange of tidal currents and river flow in the estuary, the content of the floating sediment concentration will change with time (t) as the final step of the numerical model predicted by the following equation:

$$C(t) = \left(\frac{C_{Max}-C_B}{2}\right)\left(1 + cos\left(\frac{2\pi t}{6.21}\right)\right) + C_B \text{ ; where: } C_{Max} = \frac{M}{S_p \omega_s}\left(\frac{u_m^2}{u_{cr}^2} - 1\right) + C_B \tag{7}$$

With, $C(t)$ sediment concentration at each tide time (mg/dm^3); C_B = background of floating sediment concentration (mg/L); C_{Max} = maximum concentration in estuary; S_p = suspension parameters that depend on the type of estuary, S_p = 2 (estuary perfectly mixed), S_p = 4 (partially mixed estuary), and S_p = 8 (estuary start); u_{cr} = critical threshold currents of 0.2 m/s (silt), 0.4 m/s (fine sand) and 0.6 m/s (moderate sand); and u_m = maximum current speed (m/s).

This study also continued from the studio [4] in Belawan estuary with the same numerical model and was not validated, so numerical models were carried out at different estuary locations and validated between the model and field measurements. In general, numerical models on simple estuaries are widely available, one of them is Box Model/ Finite Segment [12], but it must have complete data to see the intrusion conditions of each point to the upstream of the estuary. While the exponential model used in this model only requires physical data parameters in the estuary mouth, for other conditions, it can directly give an overview of each parameter to the upstream of the estuary.

3. Results and discussion

The beginning of the measurement was to collect salinity data from the estuary mouth to the upstream in order to find salinity with a value close to 0 mg/L. From the measurement results, salinity with 6.85 mg/L is then used as the final measurement station. The observation points were examined every 1 km, starting from estuary mouth (STA A) toward the upstream (STA Q). The result has shown the length of the estuary (L), which ranges at 16 km. Then, the numerical model is formed on six estuary physical parameters, namely depth, width, tide, current, temperature, salinity, and elevated sediment concentration. The model is carried out from bathymetry (depth and width) using equation (1). Based on measurements in the estuary mouth with width (W_0 = 1,869 m) and depth (D_0 = 8,263 m), the relationship of the numerical model and in situ measurement is shown in figure 3.

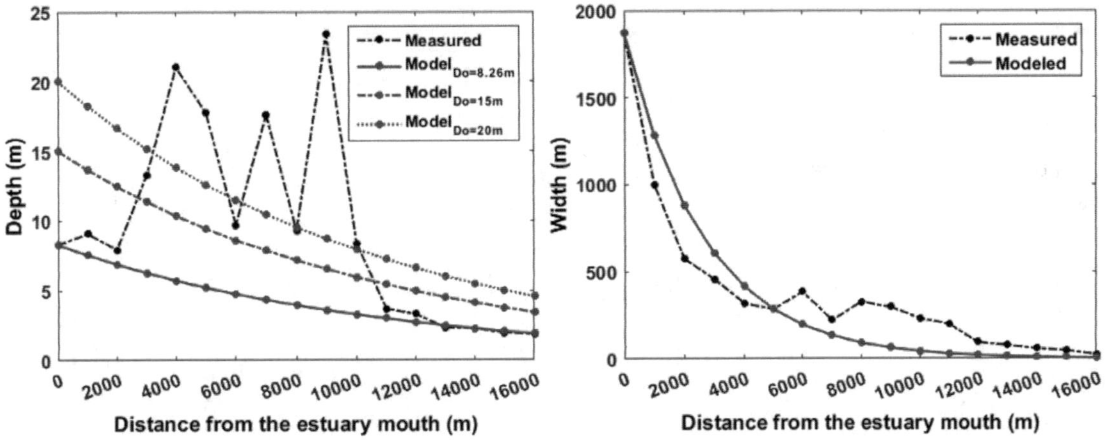

Figure 3. Comparison between the numerical model and in situ measurement for depth (left) and width (right) of Krueng Langsa estuary.

The numerical model results are compared with the measurement data by looking at the error value density of Mean Error (ME) and Root Mean Square Error (RMSE) with good Correlation Coefficient (CC) values as shown in table 1.

Table 1. MEs (E[Model – Measurement]), RMSEs, and CCs of the physical parameters of the Krueng Langsa estuary.

Parameter	Coefficient	ME	RMSE	CC
Width	6.03	-29.79	159.34	0.959
Depth	1.48	-5.11	8.02	0.421
Temperature	10.62	0.09	0.27	0.962
Salinity	11.19	-2.07	3.98	0.941
Tide	-	-6.06	25.12	0.830

Table 1 shows the result of the minimum error value of the width, depth, temperature, salinity, and tide from error density data compared between the numerical model and in situ measurement. The result of the width coefficient value was 6.03. The value of Coefficient Correlation (CC) equals 0.959 which indicated a good relationship between the model and measurement, as shown in figure 3 (right section). Meanwhile, the depth coefficient value was 1.48 with a low CC of 0.421. This low correlation indicated irregular variation between model and measurement as shown in figure 3 (left section).

The results of irregular depth at each station from the estuary mouth to the upstream estuary are shown in table 1. Variation of depth showed that the estuary mouth has an average depth of 8.26 m MSL, which did not satisfy the requirements of the national port standard [4] with a minimum depth of 9 m MSL. Nevertheless, STA D, STA E, and STA F—which located at the port of Kuala Langsa—have significant depth of the shipping channel of 13,287 m-, 21,033 m- and 17,763 m MSL, respectively. In general, the Krueng Langsa estuary is proper of being used as an international port but the conservation to achieve shipping line standards is needed especially in the estuary mouth of STA A to C. In order to achieve shipping line stability, a model was formed with D_0 = 15 m and D_0 = 20 m, as shown in figure 3 (left section) and table 2.

Table 2. Comparison of model results and measurements of bathymetry.

STASIUN		Width Estuary (m)		Depth Estuary (m)			
		Measured	Model (Wo = 1,869 m)	Measured	Model (Do = 8.263 m)	Model (Do = 15 m)	Model (Do = 20 m)
A	(0 + 000)	1,869.0	1869.00	8.264	8.263	15.000	20.000
B	(1 + 000)	1,002.0	1282.14	9.092	7.533	13.675	18.233
C	(2 + 000)	573.0	879.55	7.913	6.867	12.467	16.622
D	(3 + 000)	456.0	603.37	13.287	6.261	11.365	15.154
E	(4 + 000)	317.0	413.91	21.033	5.708	10.361	13.815
F	(5 + 000)	284.0	283.95	17.763	5.203	9.446	12.594
G	(6 + 000)	383.0	194.79	9.703	4.744	8.611	11.481
H	(7 + 000)	220.0	133.62	17.579	4.324	7.850	10.467
I	(8 + 000)	325.0	91.67	9.296	3.942	7.157	9.542
J	(9 + 000)	295.0	62.88	23.383	3.594	6.524	8.699
K	(10 + 000)	231.0	43.14	8.363	3.277	5.948	7.931
L	(11 + 000)	197.0	29.59	3.683	2.987	5.422	7.230
M	(12 + 000)	94.2	20.30	3.292	2.723	4.943	6.591
N	(13 + 000)	76.0	13.93	2.316	2.483	4.507	6.009
O	(14 + 000)	57.2	9.55	2.263	2.263	4.108	5.478
P	(15 + 000)	45.2	6.55	1.954	2.063	3.745	4.994
Q	(16 + 000)	24.7	4.50	1.836	1.881	3.415	4.553

The tidal model was done by equation (2). The component was obtained from the admiralty method through tide measurement data for 15 days (21st March to 4th April 2019) yielding A_{S2} = 0.143, A_{M2} = 0.326, and A_{M4} = 0.00010 m. Formzal number (F) was equal to 0.117 (F <0.25). This indication classified Krueng Langsa estuary as a semidiurnal tidal type with an average sea level (MSL) of 1,561 m. The highest tide average (HWS) equals to 2.27 m, and the lowest tide average (LWS) is 0.85 m.

The relationship between the model and measurement in figure 4 indicated a good agreement with CC = 0.830. Therefore, the model can be used to predict the occurrence of tides in the Krueng Langsa estuary. The measurement of salinity parameters in the estuary mouth was 38 mg/L where the temperature in STA A (estuary mouth) was 30 °C and in STA Q (Langsa river) was 32.08 °C. So, the temperature of the river (T_R) > temperature of sea (T_s). The distribution of salinity and temperature parameters is formed in numerical models through the gaussian distribution equations (5) and (6). The relationship between the model and measurement data of salinity and temperature parameters shown in figure 5.

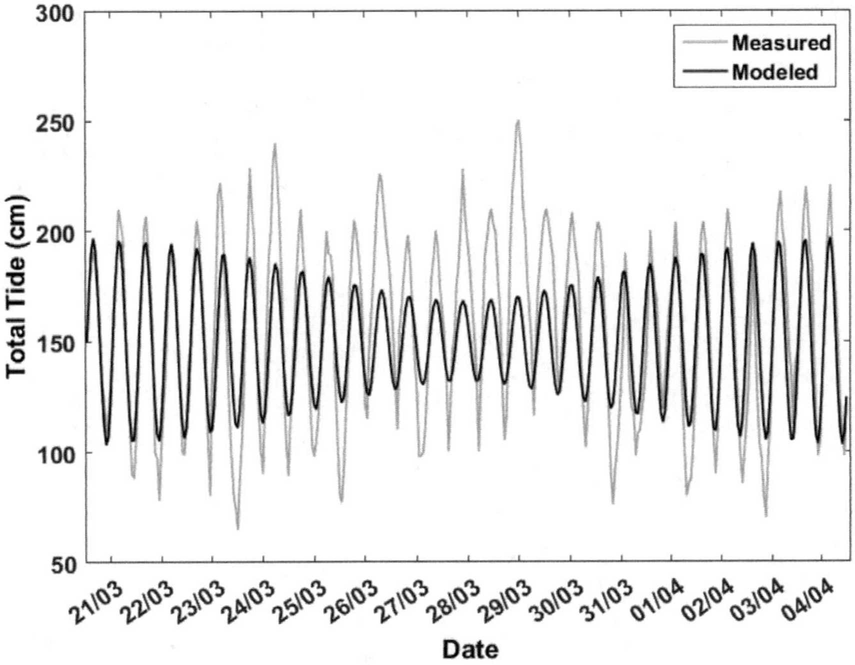

Figure 4. Comparison between the model output and measurement data for tides and lows of Krueng Langsa estuary.

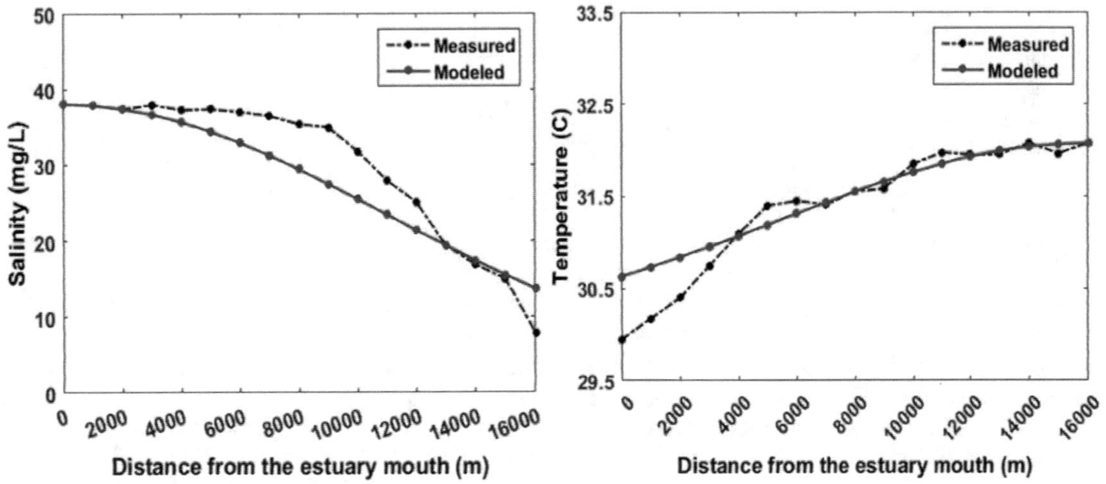

Figure 5. Comparison between the model and measurement data for salinity (left) and temperature (right) of Krueng Langsa estuary.

The σ_x value as the gaussian distribution variance every 1 km is calculated based on T_x and S_x. The model results and the measurement data show good agreement with the value of CCs are 0.96 for temperature and 0.94 for salinity, as shown in table 1. This value expressed the numerical model. In situ measurements are correlated. As a result of the exchange of saltwater (sea) and river freshwater, temperature and salinity parameters in the estuary zone will continue to change in harmony due to tidal currents and river discharge.

The current model was formed using equation (4). From the measurement data of Langsa river, flow velocity was 0.353 m/s at point Q with a width of 24.7 m and depth of 1.836 m resulting in river discharge (Q) of 16 m^3/s. The each-point river flow towards the estuary mouth and tide currents are obtained from the difference of the depth (Δh: hi + 1 - hi) and the upstream volume of (Vut + 1 = Δ_l.Wx / 106 + Vut). The hourly tidal current conditions at STA A state that the peak tidal current is 0.008 m/s, the lowest current is 0.010 m/s, the maximum tide current is 0.028 m/s and ebb-tides are 0.0270, as shown in figure 6.

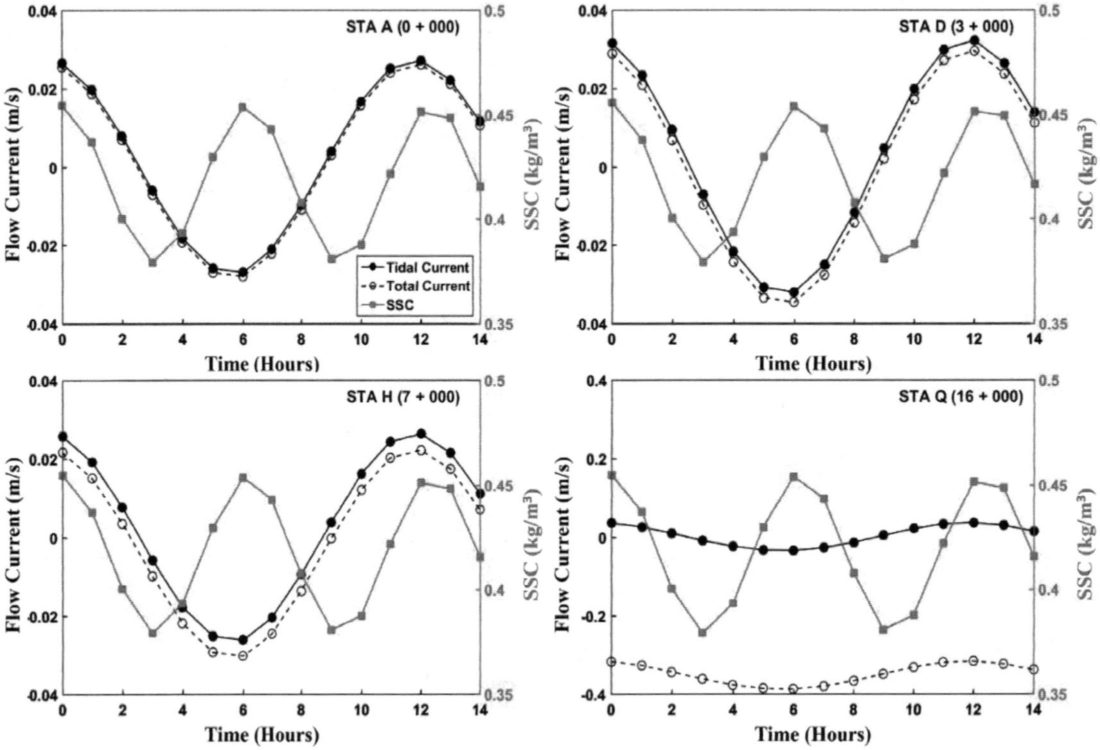

Figure 6. Comparison between tidal currents, total currents, and SSCs of Krueng Langsa estuary.

The modeling of sediment concentration was formed using equation (7). Based on laboratory tests of sediment for each station, C_B is equal to C_B of 379 mg/L. The distribution of sediment grains at the bottom of the Lago Palm estuary is classified as sedimentary sand—u_{cr} = 2 and ω_S = 0.1 m/s. C_{Max} value obtained at STA A is equal to 0.454 kg/m^3. Then, SSC was formed every hour due to the tide of the floating sediment concentration pattern as shown in figure 6. The figure explained that the pattern of SSC during peak tides is 0.379 kg/m^3. During the maximum tides, the value of SSC is greater at 0.453 kg/m^3 and the max (tide) current of the SSC value also increases by 0.451 kg/m^3. These phenomena explain that the higher current will cause an increase in SSC in the Krueng Langsa estuary. By the value of CC = 0.83, it means that the relationship of increasing current is in line with the increase in SSC.

The tidal current pattern and SSC from figure 6 stated that tidal currents in the estuary mouth showed that the river flow upstream of the estuary is 0.353 m/s to 0.001 m/s at STA A. This occurrence is due to greater estuary cross-section so that the river flow does not affect the tidal current as a current total.

Meanwhile, for STA which are located upstream, such as STA D, STA H, and STA Q, river flow affected the tidal current. So that, during low tide conditions, the total current increases together with the tidal current; when a tide occurs, the total current decreases from the tide current. The current pattern also explained the condition of the SSC content in the Krueng Langsa estuary, where during low tide conditions, there is an increase in SSC and slack water conditions during tides and low tide.

4. Conclusion

Numerical modeling was applied to observe the characteristics of the physical properties of the Krueng Langsa estuary, especially in water depth, tidal currents, salinity, temperature, and SSC. Then, numerical modeling was evaluated using measurement data. The observation point consists of 17 stations at every 1 km starting from the estuary mouth (STA A) toward the upstream (STA Q). The results show that the correlations between the model results and observations are quite good for most physical parameters that are modeled where CCs are ranged from 0.42 to 0.696. The correlation between currents and SSCs shows CC of 0.83, which indicates good utility. As a result, it can be concluded that increasing the current will increase the content of SSC in the Krueng Langsa estuary.

Acknowledgments
The author would like to thank the Chancellor and LPPM-PM of Samudra Langsa University for funding the research from the DIPA UNSAM through the "Applied Research" scheme.

References

[1] Hardisty J 2007 *Estuaries: Monitoring and Modeling the Physical System* (Australia: Blackwell Publishing)
[2] Michaelis W 2012 *Coastal and estuarine studies (Monitoring, Modeling and Research)* (Berlin Heidelberg: Springer-Verlag)
[3] Wang J, Li L, He Z, Kalhoro N A and Xu D 2019 Numerical modelling study of seawater intrusion in Indus River Estuary, Pakistan *Ocean Eng.* **184** 74–84
[4] Tarigan A P M, Swandana D and Isma F 2017 Modelling the physical system of Belawan Estuary *Proc. of 1st Annual Applied Sci. and Eng. Conf. (AASEC), in conjunction with The Int. Conf. on Sport Science, Health, and Physical Edu. (ICSSHPE)* (Bandung, Indonesia) p 1–6
[5] Qiao H, Zhang M, Jiang H, Xu T and Zhang H 2018 Numerical study of hydrodynamic and salinity transport processes in the Pink Beach wetlands of the Liao River estuary, China *Ocean Sci.* **14** 437–451
[6] Lintas Atjeh, Akibat perairan dangkal, kapal Rumah Sakit TNI AL tidak bersandar di Pelabuhan Kuala Langsa *accessed July 13, 2019* available at https://www.lintasatjeh.com/2017/12/akibat-perairan-dangkal-kapal-rumah-sakit-tni-al-tidak-bersandar-di-pelabuhan-kuala-langsa.html
[7] Peraturan Menteri Perdagangan RI No. 24 Tahun 2019 tentang Perubahan Ketujuh Atas Peraturan Menteri Perdagangan Nomor 87/M-Dag/Per/10/2015 tentang Ketentuan Impor Produk Tertentu (Jakarta: Sekjen Kemendag)
[8] Keputusan Menteri Perhubungan No KM 53 Tahun 2002 tentang Tatanan Kepelabuhan Nasional (Jakarta: Menhub)
[9] Isma F, Purwandito M and Ardhyan Z 2019 Estimasi erosi dan sedimentasi lahan pada DAS Langsa berbasis sistem informasi geografis (SIG) *Teras Jurnal* **9** 29–41
[10] Nurzahziani and Surussavadee C 2018 Numerical simulations for biomass burning aerosol transport in northern Thailand *Proc. of 2nd Int. Conf. Energy and Environmental Science* (Kuala Lumpur, Malaysia) p 1–6
[11] Temmerman S, Govers G, Wartel S and Meire P 2004 Modelling estuarine variations in tidal marsh sedimentation: response to changing sea level and suspended sediment concentrations *Mar. Geol.* **212** 1–19
[12] Lung W S 1993 *Water Quality Modeling Application to Estuaries, Volume III* (Boca Raton, Florida: CRC Press)

Analysis on the willingness to pay and ability to pay in determining the fares of Sigli - Banda Aceh toll road

Q H Wathan[1], **M Isya**[2] **and Sugiarto**[2]

[1] Civil Engineering Postgraduate Program, Department of Civil Engineering, Universitas Syiah Kuala, Banda Aceh, Indonesia
[2] Department of Civil Engineering, Universitas Syiah Kuala, Banda Aceh, Indonesia

Corresponding E-mail: qasim@mhs.unsyiah.ac.id

Abstract. The Sigli - Banda Aceh toll road, which has a length of 75 km, is part of the construction project of the Banda Aceh - Binjai toll road section. This toll road will become the first toll road in Aceh, so it is needed to do the research in order to plan the toll rates. The purpose of this study is to find out the value of ability to pay (ATP) and willingness to pay (WTP) of the community on toll fares. Observation surveys were carried out for two days, which is on Saturday and Sunday. The types of vehicles reviewed are group i vehicles which is light vehicles namely sedans, passenger car, and small buses, group II which is the truck with two axles, group III which is a truck with three axles, group IV which is the truck with four axles, and group V which is the truck with five axles or more. The results of the ability to pay (ATP) fares analysis for each group are as follows: group I (LV) is Rp. 60,193 which is rounded to Rp. 61,000; group IIA (MHV) is Rp. 174,089 which is rounded to Rp. 174,000; and Group IIB (LT & LB) is Rp. 120,579 which rounded to Rp. 121,000. On the other hand, the results of the willingness to pay (WTP) analysis for each group are as follows: group I (LV) is Rp. 52,305 which is rounded to Rp. 53,000; group IIA (MHV) is Rp. 124,655 which is rounded to Rp. 125,000; and group IIB (LT & LB) is Rp. 110,052 which is rounded to Rp. 110,000.

1. Introduction

The development of infrastructure, such as the construction of Banda Aceh-Binjai toll road, is a logical and strategic choice in order to increase Indonesia's competitiveness. The Aceh toll road consists of four roads with a total length of 455 km, which consists of 75 km for Banda Aceh-Sigli, 135 km for Sigli – Lhokseumawe, 135 km for Lhokseumawe - Langsa, and 110 km for Langsa - Binjai. The construction of this first Aceh toll road started from the Sigli - Banda Aceh Corridor with a length of 75 km. This toll road will become the first toll road in Aceh, so it is needed to do the research in order to plan the toll rates or fares.

Based on the regulation of the Minister of Public Works and Public Housing of the Republic of Indonesia (2018), No 10/PRT/M/2018 [1], the toll road is a public road which is part of the road network system and as the national road; those who use it are required to pay tolls. For toll road users themselves are people who use motorized vehicles by paying the toll. Toll road development will affect economic development, and increase the mobility and accessibility of people and goods.

The purpose of this study is to find out the value of ability to pay (ATP) and willingness to pay (WTP) of the community on toll fares. This research was conducted on the Sigli - Banda Aceh toll road with a length of 75 km. Observation surveys were carried out for two days. The type of vehicles reviewed is the group I vehicles which is light vehicles, namely sedans, passenger car, and small buses, group II which is truck with two axles, group III which is a truck with three axles, group IV which is truck with four axles, and group 5 which is a truck with five axles or more. The analysis of vehicles uses Pacific Consultant International method [2].

2. Literature review

2.1. Toll road

The toll road is a public road as part of a road network system and as a national road. The users of this road are required to pay tolls [3]. Toll road users are people who use motorized vehicles by paying tolls. Toll road fare is a certain amount of money that is paid for the use of toll roads [1]. The operation of the toll road itself is intended to realize the development of equality and balance in the region. The development can be achieved by fostering a road network whose funds come from road users.
The technical requirements for toll roads are:

a. Toll roads have a higher level of security and comfort services compared to the general public roads; it can serve long-distance traffic flow with high mobility.
b. Toll roads used for inter-city traffic are designed based on the plan speed of at least 80 km per hour; the toll road in urban areas are designed with a plan speed of at least 60 km per hour.
c. Toll roads are designed to be able to withstand the heaviest axle loads of at least 8 tons.
d. Each toll road segment must be fenced off and equipped with crossing facilities in the form of bridges or tunnels.
e. In sections that can endanger toll road users, they must have security build, which has the strength and the structure that can absorb the impact energy of the vehicles.
f. Toll roads must be equipped with rules and orders which are stated in the form of traffic signs, road marking, or traffic signaling devices.

2.2. Fare

Fare is the price of transportation services that must be paid by the service users, whether through agreement lease, bargaining, and government provisions [4]. The price of transportation services is determined by following the commonly applied fare system. There are no provisions that bind the transport company and owner of goods or passengers except what is set in the fare books.
Several general principles that underlie the fare determination, which is based on [5]:

- The number and characteristics of passengers,
- Characteristics of public transport passengers, and
- Vehicle operational cost.

In order to handle the fare policies, any purposes that have been made in the end will be made as the decision that consider two things: first, fare rates are the amount charged and has a range from rates of free until the fare rates charged to make a profit for the service; second, the fare rate structure is calculated and the procedure of payment is decided.

2.3. Ability to pay (ATP)

Ability to pay (ATP) is a person's ability to pay for services he or she receives based on ideal income. The method used in the ATP analysis is based on the cost allocation for transportation from the routine income. In other words, ATP is the ability of the community to pay the cost of their trip [5]. The calculation of ATP is done by using these following formulas:

$$\mathrm{ATP} = \frac{\mathrm{Ix}\,.\,\mathrm{Pp}\,.\,\mathrm{Pt}}{\mathrm{Tt}} \tag{1}$$

Information:
ATP : Ability to pay, Rp;
Ix : Respondent income per month, Rp;

Pp : Cost transportation per month, percentage of the total income;
Pt : Allocation percentage cost transportation for toll, %; and
Fr : Frequency of respondent travel time.

2.4. Willingness to pay (WTP)

Willingness to Pay (WTP) is the willingness of service users to pay the fares for the services they obtain. The approach used was based on the perception of the fares of the transportation service [3]. The purpose of WTP is to obtain the optimum and realistic amount of toll fares based on the community's willingness to pay, but still attract investors to invest.

Factors that influence the amount of willingness to pay:

- Products offered or provided by transportation service operators;
- Provided quality and quantity services;
- Users utility for transportation services;
- User characteristics.

The value of WTP is obtained by averaging selected fares perception for each type of work, namely

$$WTP_{\text{type of works}} = \frac{\Sigma(\text{selected fares x number of respondent})}{\text{the total number of respondents for each type of work}} \quad (2)$$

$$WTP_{\text{all works categories}} = \frac{\Sigma(\text{WTP type of works})}{\text{number of work categories}} \quad (3)$$

2.5. The relation between ATP and WTP

In order to determine the fares obstacles between the value of ATP and WTP, these conditions are then illustrated as follows:

a. ATP greater than WTP
   This condition indicates that the ability to pay is greater than the willingness to pay. It occurs when users have a relatively high income, but utility or the needs on services are relatively low.
b. ATP lower than WTP
   This condition shows that the willingness of users to pay the services is greater than the ability to pay. Maybe it occurs because the users have a relatively lower income, but the utility or the needs on service is relatively high so that the user has no options. The willingness of users to pay for these services tends to be influenced by the utility or the high dependency.
c. ATP equal with WTP
   This condition shows that the ability and willingness to pay are equal. This condition occurs when the balance of users' utility with the services cost charge happens.

3. The methodology of the research

3.1. Surveyor and equipment

In this research, several surveyors are in charge of the following jobs:

a) Recording traffic volume.
b) Distributing questionnaires.

The equipment used in this research are:

a) Wristwatch, which is used to review the time in order to record traffic volume.
b) Traffic volume survey form, which is used to record the number of vehicles that passed the Sigli - Banda Aceh road.
c) The questionnaire, which will be filled by passengers.

d) Stationery, which helps both the respondents to fill out the questionnaires and surveyor to fill out the survey form.

3.2. Data collecting method

The data collected in this research are the primary and secondary data. Primary data is the data that directly obtained by surveying the research location, while secondary data is the data obtained in order to support the research. The secondary data is obtained through relevant agencies.

A. Primary data

Primary data Includes

1. Traffic volume that obtained from traffic surveys based on the regulation of Minister of Public Works and Public Housing of the Republic of Indonesia (2007), No. 370/KPTS/M/2007, about the determination of the vehicles type on the operating toll road and the amount of toll fares on the several toll road sections, which are as follows:
   - Group I Vehicles (LV): sedans, jeep, pick up, small bus, small truck, and medium bus;
   - Group II Vehicles (MHV): truck with two axles;
   - Group III Vehicles (LT dan LB): truck with three axles;
   - Group IV Vehicles (LT dan LB): truck with four axles; and
   - Group V Vehicles (LT dan LB): truck with five or more axles.
2. Questionnaire

B. Secondary Data

Secondary data covers:

1. Data of prices and material for vehicle operational cost calculation, including;
   - Premium fuel, in the unit price (Rp/liter);
   - Diesel, in the unit price (Rp/liter);
   - Oil for premium fuel engines, in the unit price (Rp/liter);
   - Oil for diesel fuel engines, in the unit price (Rp/liter);
   - New tire prices;
   - New vehicle prices;
   - Depreciating vehicle prices;
   - Insurance; and
   - Interest rate.
2. Population data and gross regional domestic product data issued by the Central Agency Statistic of Aceh Province in 2018;
3. Planned toll road data and figures of detail engineering drawing (DED);
4. Road geometric data.

3.3. Population and sample

The population taken in this research is the population of Banda Aceh, Aceh Besar, and Pidie, which is in total 564,911 people. There are eight age categories divided in this research, namely: 20 – 24, 25 – 29, 30 – 34, 35 – 39, 40 – 44, 45 – 49, 50 – 54, and 55 – 59. Based on the Slovin Formulas [6], by using 10% error tolerance, the number of samples taken is 100 samples.

3.4. Data processing method

After data from in the field is obtained, the data then being processed based on the formulas that have been stated in section 2. The data was processed by using software Microsoft office excel with MKJI [3] and vehicle operational cost based on the PCI method [2]. From these data, the researcher will plan four scenarios. Then, ATP was analyzed by using the income allocation method, whereas WTP was analyzed using Microsoft excel.

4. Results and discussions

4.1. Respondents characteristics

The most significant percentage for respondents' gender was male (91%); the rest, female, was 9%. The highest percentage of respondents' age is 25-29 years (18%), and the lowest one is 40 - 59 years and 55 - 59 years, each of which is 6%. Then, the highest percentage of respondents' last education is high school (68%), while the lowest percentage of respondents last education is S2 (1%) and junior high school (1%). The most significant percentage of respondent jobs is the driver (40%), whereas the lowest percentage is for the director and notary public, each of which is 1%.

For travel purposes, the largest percentage is the group of people who go for work or business (52%), whereas the group having the smallest percentage is people who go for other activities (1%). For the dependency, respondents who have the most dependence are the person having two family members, with a percentage of 36%, while the smallest percentage is for the family with five members, which is 55%. For the frequency of passing the road, the most dominant group is who passes the existing road route one to two times (81%) and the most minority group is who passes the existing road route more than three times (23%). Regarding the dependents family, the majority group is who have five dependents family (5%).

4.2. Fares scenario

Based on the results of the research, higher the fares proposed in the questionnaire, fewer respondents agree with the fares, and so vice versa. The following is a graph showing the number of respondents who agree and disagree with the fares based on the fares scenario.

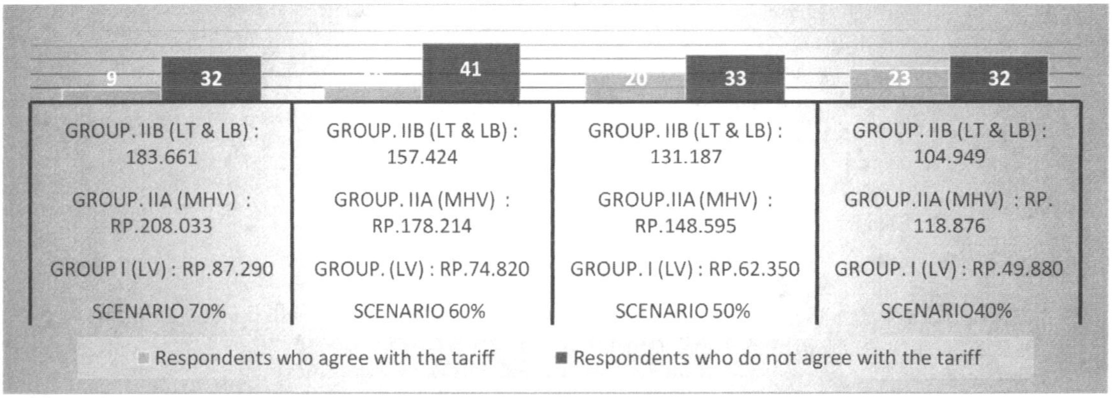

Figure 1. The distribution of respondents based on the reason for choosing the planned toll fares.

4.3. Fare analysis based on ability to pay (ATP)

Table 1. Ability to pay (ATP) for every group of vehicles.

Job	Number of respondents	Average income / month (Rp)	Percentage of transportation cost/ month	Percentage of toll cost/ month	Travel frequency	ATP for each type of jobs (Rp)
Group I	129	2,800,000	26%	25%	3.00	60,193
Group IIA	32	9,050,000	26%	26%	3.44	174,089
Group IIB	39	7,012,821	28%	24%	3.82	120,579
Total	**200**					

Table 2. Recapitulation of ability to pay (ATP) for every group of vehicles.

Job	Number of respondents	ATP for each type of job	Rounded
Group I	129	Rp. 60,193	**Rp. 61,000**
Group IIA	32	Rp. 174,089	**Rp. 174,000**
Group IIB	39	Rp. 120,579	**Rp. 121,000**
Total	**200**		

4.4. Analysis of willingness to pay (WTP)

Table 3. Value of willingness to pay (WTP) each group for vehicle group I (LV).

Type of work	Type WTP of work (Rp)	WTP throughout job categories (Rp)	Rounded up
Retirement	59,233		
Businessman	46,763		
Student	34,293		
Government Employees	81,055	**52,305**	**Rp. 53,000**
Staff/ Private	40,875		
Supervisor	49,881		
Driver	54,037		

Table 4. Value of willingness to pay (WTP) each group for vehicle group IIA (MHV).

Type of work	Type WTP of work (Rp)	WTP throughout job categories (Rp)	Rounded up
Retirement	141,166		
Businessman	111,447		
Student	81,728		
Government Employees	193,174	**124,655**	**Rp. 125,000**
Staff/ Private	97,413		
Supervisor	118,877		
Driver	128,783		

Table 5. Value of willingness to pay (WTP) each group for vehicle group IIB (LB & LT).

Type of work	Type WTP of work (Rp)	WTP throughout job categories (Rp)	Rounded up
Retirement	124,628		
Businessman	98,391		
Student	72,153		
Government Employees	170,543	**110,052**	**Rp. 110,000**
Staff/ Private	86,001		
Supervisor	104,950		
Driver	113,695		

4.5. Analysis of ability to pay (ATP) and willingness to pay (WTP)

Based on the fare of willingness to pay (WTP) of Sigli - Banda Aceh toll road, the fare for vehicles group I is Rp. 53,000, group IIA is Rp. 125,000, and group IIB is Rp. 110,000. The comparison of fares analysis based on a maximum of 70% of the profit of vehicle operational cost, ability to pay (ATP), and willingness to pay (WTP) can be seen on the fares recapitulation on the following table.

Table 6. Recapitulation of the groups II of vehicle fares.

Type of fares	Group I (LV)	Group IIA (MHV)	Group IIB (LT & LB)
Based on the maximum 70% of the profit from vehicle operational cost	Rp. 87,290	Rp. 208,033	Rp. 183,661
Based on ATP approach	Rp. 60,193	Rp. 174,089	Rp. 120,579
Based on WTP approach	Rp. 52,305	Rp. 124,655	Rp. 110,052

Table 7. Recapitulation of the II groups of vehicle fares for a distance of 75 km.

Type of Fares	Plan rates for 75 km (Rp)		
	Group I (LV)	Group IIA (MHV)	Group IIB (LT & LB)
Based on the maximum of 70% of the profit from vehicle operational cost	Rp. 1,164 ≈ 1,200 /1 km	Rp. 2,774 ≈ 3,000 /1 km	Rp. 2,449 ≈ 2,500/1KM
Based on ATP approach	Rp. 803 ≈ 800 /1 km	Rp. 2,321 ≈ 2,300 /1 km	Rp. 1,608 ≈ 1,600/1KM
Based on WTP approach	Rp. 697 ≈ 700 /1 km	Rp. 1,662 ≈ 1,700 /1 km	Rp. 1,467 ≈ 1,500/1KM

5. Conclusions and recommendations

The conclusions, which can be taken in determining vehicle operational cost for Sigli - Banda Aceh toll road fares based on the 70% maximum of the profit from vehicle operational cost, is as follows:

1. The results of the ability to pay (ATP) fares analysis for each group are Group I (LV) is Rp 60,193, which is rounded to Rp. 61,000, group IIA (MHV) is Rp. 174,089, which is rounded to Rp. 174,000, and group IIB (LT & LB) is Rp. 120,579, which is rounded to Rp. 121,000.
2. The results of the willingness to pay (WTP) analysis for each group are the group I (LV) is Rp. 52,305, which is rounded to Rp. 53,000, group IIA (MHV) is Rp. 124,655, which is rounded to Rp. 125,000, group IIB (LT & LB) is Rp. 110,052, which is rounded to Rp. 110,000.

The suggestions proposed based on this research are as follows:

1. It is recommended to conduct financial investment/ financial analysis and economic value to determine the fares of Sigli - Banda Aceh toll road.
2. It is hoped that this research can be a reference and consideration for authorities involved in determining the fares for the Sigli - Banda Aceh toll road.

References

[1] Menteri Pekerjaan Umum dan Perumahan Rakyat Republik Indonesia 2007. *Peraturan Menteri Pekerjaan Umum dan Perumahan Rakyat Republik Indonesia (2018), No 10/PRT/M/2018 Jalan tol merupakan jalan umum yang merupakan bagian sistem jaringan jalan dan sebagai jalan nasional yang penggunaannya diwajibkan membayar tol* (Jakarta)

[2] Metode Perhitungan Biaya Operasional Kendaraan Pacific Consultant International (PCI) 2000

[3] Direktorat Jenderal Bina Marga 1997 *Manual Kapasitas Jalan Indonesia (MKJI)* (Jakarta: Bina Karya)

[4] Warpani S 2002 *Pengelolaan Lalu Lintas dan Angkutan Jalan* (Bandung: ITB)

[5] Tamin O Z 2008 *Perencanaan, Permodelan, dan Rekayasa Transportasi, Contoh Soal dan Aplikasi* (Bandung: Penerbit ITB)

[6] Amirin T M 2011 *Populasi dan Sampel Penelitian 4, Ukuran Sampel Rumus Slovin* (Yogyakarta)

[7] Departemen Pekerjaan Umum Direktorat Jendral Bina Marga 2005 *Peraturan Pemerintah Nomor 15* (Jakarta)

[8] Lembaga Afiliasi Penelitian Dan Industri Institut Teknologi Bandung (LAPI-ITB) 1997 *Studi Perhitungan Nilai Waktu PT. Jasa Marga* (Bandung)

[9] Badan Litbang PU 2005 *Pd T-15-2005-B Pedoman Perhitungan Biaya Operasi Kendaraan* (Jakarta: Badan Litbang PU Departemen Pekerjaan Umum)
[10] Suhartono et al. 2010 *Modul PRJL-III PerencanaanLalu Lintas Dan EkonomiTransportasi. Himpunan Pengembangan Jalan Indonesia* (HPJI) (Jakarta)

IOP Conf. Series: Materials Science and Engineering **933** (2020) 012024 doi:10.1088/1757-899X/933/1/012024

Simulation of road treatment costs based on life-cycle cost analysis

A Mulyawan[1], S M Saleh[2] and R Anggraini[2]

[1] Civil Engineering postgraduate program, Department of Civil Engineering, Universitas Syiah Kuala, Banda Aceh, Indonesia
[2] Department of Civil Engineering, Universitas Syiah Kuala, Banda Aceh, Indonesia

Corresponding e-mail: mulya.mulyawan2@gmail.com

Abstract. Flexible pavement is the most widely used type of road pavement in Indonesia. The main factors in the selection of flexible pavements are the low initial cost and the limited budget for road maintenance. This study aims to examine the level of comparison of road treatment costs based on life-cycle cost analysis (LCCA) rigid and flexible pavement for 40-years analysis period. The road treatment simulation uses road deterioration models based on the equation in the second version of the HDM-4 manual with the International Roughness Index (IRI) and the treatment period as the main variable. Based on the LCCA, The NPV analysis showed that the rigid pavement treatment cost ratio is 0.818 or more economical 18.20% at a 10% discount rate and 0.915 or more economical 8.48% at a 12% discount rate than the flexible pavement. Meanwhile, the EUAC analysis result showed that the rigid pavement treatment cost ratio is 0.847 or more economical 15.31% at a 10% discount rate and 0.862 or more economical 13.79% at a 12% discount rate compared to flexible pavement. The results showed that the rigid pavement treatment cost was more efficient 8.48% - 18.20% compared to flexible pavement.

1. Introduction

Flexible pavement is generally used for road treatment in Indonesia compared to other road pavements. The main factors in choosing this pavement are cheaper initial costs, limited road maintenance budget, relatively faster implementation time, and demand for equitable development. The flexible pavements do have cheaper initial costs, but in the long term period, the flexible pavements maintenance costs will increase annually, so the total cost of flexible pavements will be more expensive and less economical compared to the rigid pavements. Another aspect to be considered in determining pavement type beside maintenance costs is pavement endurance against heavy traffic. There are still many overloading vehicles in Indonesia that will cause an impact on service capability or performance life of the pavement. According to Saleh [1], the excess of the truckload of up to 50% can affect maintenance costs 2.5 times higher than the annual maintenance plan costs in the service life span. In addition, Syafriana [2] also stated that the total number of vehicles violating the allowable weight (JBI) was 31.50% per day of the total number of goods vehicles on the eastern national road cross-section of the Aceh Province.

This study aims to examine the comparison of road treatment cost based on life-cycle cost analysis (LCCA) rigid pavement and flexible pavement for 40-years plan of the analysis period. Road deterioration modeling used the equation in the second version of the HDM-4 manual with the International Roughness Index (IRI) and the treatment period as the main variable. The benefits cost or user cost in this study for the two types of pavement is assumed to be the same so that the value of benefits can be ignored. The total road treatment based on LCCA will also be correlated with the level of road stability in accordance with the standards of the Directorate General of Highways of the Ministry of Public Work and Housing, so the most economical and efficient road treatment costs can be discovered.

2. Literature review and analysis methods

2.1. Road treatment and life cycle cost analysis

Life-cycle cost analysis (LCCA) is an analysis technique, based on well-founded economic principles, used to evaluate the overall long-term economic efficiency between the alternate investment options [3]. LCCA is typically used as a means to evaluate and compare the cost to the agency of any number of alternate pavement alternatives, including variations of concrete and asphalt pavement solutions.

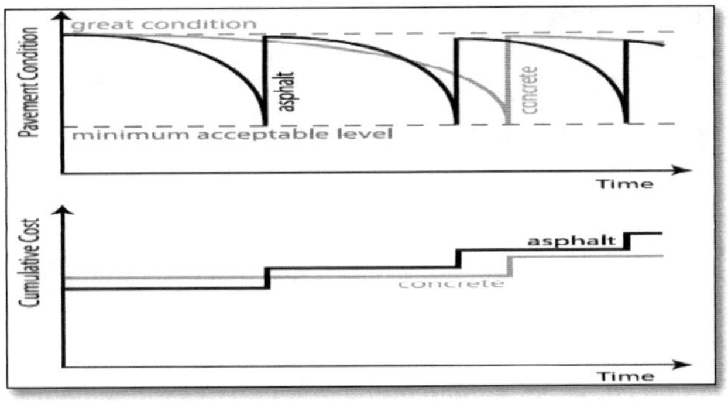

Figure 1. Generalized illustration of pavement condition over time and the financial implications of such through the calculation of total cost [3].

2.2. Pavement design and road deterioration model

Pavement design in this study refers to Directorate General of Highways's Road Pavement Design Manual for 2017. Ditjen Bina Marga [4], states that the service life of flexible pavement plans is 20 years while the rigid pavement is 40 years. The ratio of traffic volume to road capacity (RVK) is a maximum of ≤ 0.85 [5]. The value of RVK or the degree of saturation will be used as an indicator of the need to increase road capacity. Morosiuk [6] states that there are several main components in analyzing the decline in road conditions based on IRI values. The components can be seen in the following equation.

$$\Delta RI = \Delta RI_s + \Delta RI_c + \Delta RI_r + \Delta RI_p + \Delta RI_e \quad (1)$$

Description:

ΔRI = Total incremental change in roughness during analysis year, in m/km IRI.
ΔRI_s = Incremental change due to structural deterioration during analysis year, in m/km IRI.
ΔRI_c = Incremental change due to cracking during analysis year, in m/km IRI.
ΔRI_r = Incremental change due to rutting during analysis year, in m/km IRI.
ΔRI_p = Incremental change due to potholing during analysis year, in m/km IRI.
ΔRI_e = Incremental change due to the environment during analysis year, in m/km IRI.

The following is the equation used to analyze the IRI value on the rigid Joint Plain Concrete Pavement (JPCP) type according to ERES (1995) [6]:

IOP Conf. Series: Materials Science and Engineering **933** (2020) 012024 doi:10.1088/1757-899X/933/1/012024

$$\text{IRI} = IRI_0 + 0{,}00265(TFAULT) + 0{,}0291\,(SPALL) + 0{,}15\text{X}10^{-6}\,(TRACK)^3 \tag{2}$$

Description:

IRI	=	International Roughness Index, in m / km IRI.
IRI_0	=	Initial roughness at construction, in m / km IRI.
TFAULT	=	transverse joint faulting, in mm/km.
SPALL	=	spalled joints, in per cent.
TCRACK	=	transverse cracks, in no/km
	=	10 CRACKING / L CRACKING.
L	=	mean transverse joint spacing, in m.

2.3. Evaluation of economic values based on LCCA

ACPA [3] states that the real discount rate, also known as the real interest rate, is commonly used in engineering economics to refer to the rate of change over time in the true value of money, taking into account fluctuations in both investment interest rates and the rate of inflation. Road treatment costs are calculated using the real discount rate as the discount factor value of road treatment costs during the analysis period. The equation of the real discount rate in the adaptation of Thuesen and Fabrycky is shown below [3].

$$d = \frac{1+i_{int}}{1+i_{inf}} - 1 \tag{3}$$

Description:

d = real discount rate, %.
i_{int} = the interest rate, %.
i_{inf} = the inflation rate, %.

Walls III [7] states that NPV is computed by assigning monetary values to benefits and costs, discounting future benefits and costs using an appropriate discount rate, and subtracting the sum total of discounted costs from the sum total of discounted benefits. Walls III [7] also states that equivalent uniform annual costs represent the NPV of all discounted costs and benefits of an alternative as if they occur uniformly throughout the analysis period. EUAC is a particularly useful indicator when budgets are established on an annual basis. The NPV and EUAC equations used in evaluating project economics based on LCCA are as follow [7]:

$$NPV = Initial\ Cost + \sum_{k=1}^{N} Rehab\ Cost_{\ k} \left[\frac{1}{(1+i)^{n_k}}\right] \tag{4}$$

$$EUAC = NPV \left[\frac{1\,(1+i)^n}{(1+i)^n - 1}\right] \tag{5}$$

Description:

i = discount rate, %.
n = year of expenditure.

2.4. Analysis methods

The research location is the national road section of ‘Bts Bireuen / Aceh Utara - Bts. Lhokseumawe City /Bts. Aceh Utara (N.01.007.2)’. The road has a configuration of 2/2 undivided roads with a total length of 16.32 Km and a road width of 7.0 m. This study uses secondary data from ‘Balai Pelaksanaan Jalan Nasional I (BPJN I)’ and ‘Satker Perencanaan dan Pengawasan Jalan Nasional Provinsi Aceh (Satker P2JN)’. Data collected data includes general information data, traffic volume, CBR subgrade, unit prices in 2017, and routine road maintenance unit prices in 2018. The inflation value used in this study is 3.51% which is an average value inflation rate for the period 2016 - 2018 from the Bank of Indonesia website. The interest rate used in the calculation of treatment costs is 10% during the period analysis using the real discount rate as an economic indicator, while the discount rate used in the economic analysis of LCCA is 10%, 12%, and 15%. Data processing is done with the help of laptops and Microsoft Word and Excel software.

2.5. Types of treatment for sealed roads based on IRI values

The following is the road conditions category index based on IRI values according to Directorate General of Highways which is a policy on road management in Indonesia:

- Good condition : IRI ≤ 4 m/km
- Fair condition : 4 < IRI ≤ 8 m/km
- Poor condition : 8 < IRI ≤ 12 m/km
- Bad condition : IRI > 12 m/km
- Road steadiness : Good condition + Fair condition

Table 1. Simulation model criteria for road treatment on the flexible pavement.

Types of Road Treatment	Type of Simulation Criteria for Road Treatment		
	Model SF01	Model SF02	Model SF03
Routine Maintenance	0 < IRI < 4	0 < IRI < 4	Every year (unless there are other treatments)
1st minor rehabilitation (*overlay* 40 mm)	4 ≤ IRI < 6	4 ≤ IRI < 6	years 5th & 25th
2nd minor rehabilitation (*overlay* 60 mm)	6 ≤ IRI < 7	6 ≤ IRI < 8	years 10th & 30th
1st mayor rehabilitation (*overlay* 100 mm)	7 ≤ IRI < 8	8 ≤ IRI < 10	years 15th & 35th
Reconstruction	IRI ≥ 8	IRI ≥ 10	years 20th

Note: - SF01 and SF02 use treatment trigger criteria based on IRI values.
- The SF03 model uses treatment trigger criteria based on performance life every 5 years.

The work road effect is calculated based on the IRI value. The IRI value after road treatment, especially the major rehabilitation, is 3.0 m/km IRI. While the reduction in IRI value after road treatment, such as road routine maintenance and minor rehabilitation, is 0.01 (the default value from the HDM-4 manual). The reduction was analyzed using the road deterioration analysis results with reduced sensitivity of IRI value,

Table 2. Simulation model criteria for road treatment on rigid pavement.

The handling simulation	Model SR01	Model SR02	Model SR03
Trigger Criteria of Road Treatment	service period every 5 years	service period every 8 years	service period every 10 years
Types of Treatment	- Repairing of *transfer fault* with dowel replacement - Repairing of concrete *spalling* - Repairing of concrete *cracking* - Surface roughness maintenance with diamond grinding for at years 10th and years 30th - Surface drainage maintenance with diamond grooving at year 20th		

IOP Conf. Series: Materials Science and Engineering **933** (2020) 012024 doi:10.1088/1757-899X/933/1/012024

2.6. Data processing methods

The data processing methods used in this study are as follow:

- The manual design of pavement thickness for both types of pavement based on the results of secondary data processing, which are traffic data, soil carrying capacity (CBR) data, analysis of Degree of Saturation (DS), and Cumulative Equivalent Standard Axle (CESA) analysis.
- Analysis of road deterioration model for both types of pavement.
- Calculation of material quantities requirements and road treatment costs for both types of pavement.
- Calculation of the economic value of road treatment costs for each year during the analysis period using a real discount rate economic indicator as a discount factor.
- Analysis of the economic value based on Life Cycle Cost Analysis (LCCA) by using economic indicators Net Present Value (NPV).
- Comparison of LCCA results between rigid pavement and flexible pavement.

3. Results and discussion

3.1. Road pavement thickness design

Based on the calculations and analyzes results that have been carried out, the results of flexible pavement thickness design are as follow:

a. AC-WC : 40 mm
b. AC-BC : 60 mm
c. AC-Base (initial development) : 210 mm
d. AC-Base (for 1st & 2nd reconstruction) : 245 mm
e. Sub-base coarse aggregate : 300 mm
f. Foundation layer : 300 mm

The Rigid pavement of the analysis is JPCP with dowel as the load transfer system. The analysis results of rigid pavement thickness design are as follow:

a. Concrete thickness f'c 35 MPa : 305 mm
b. Dowel size : D 38 mm - distance 300 mm
c. Binding Reinforcement Size (assumption) : 4 D16 and stirrup Ø10–250 mm
d. Concrete shoulder : Not counted in simulation
e. Lean Mix Concrete Foundation Layer : 150 mm
f. Sub Base Course Aggregate : 300 mm (as drainage layer)
g. Foundation layer : 300 mm (CBR 3.0%)

The results of the degree of saturation analysis show that in the 23rd year (in 2042) the value of the degree of saturation has reached 0.88 (DS > 0.85) so that the road capacity is increased from 2/2 un-divided to 4/2 divided. The Degree of saturation value in the 40th year of the analysis is still below the maximum allowable value of 0.81 (DS <0.85) so road widening is not needed.

3.2. Analysis results of flexible pavement simulation

The poor condition on the SF01 only occurred once during the analysis period which is in the year of 23rd (IRI = 9.16 m/km) with the road steadiness level of 98 %. SF01 has a life-cycle with a performance life of 23 years before the reconstruction takes place in the year of 24th. The initial cost of SF01 is the same as SF02 and SF03, which is Rp. 219.90 billion. The total pavement treatment cost during the analysis period is Rp. 2,301.75 billion.

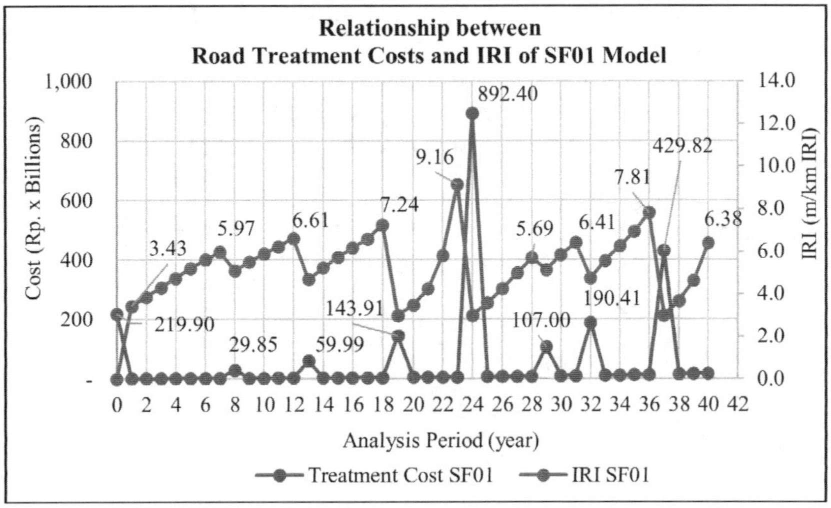

Figure 2. Relationship between road treatment costs and IRI values of SF01 model.

Both SF02 and SF03 have the same analysis. The poor condition on the SF02 occurs twice which is in the year 20th and the year 26th with an IRI value of 8.72 m/km IRI and 11.96 m/km and with the road steadiness level of 95 %. The Total pavement treatment cost during the analysis period of the SF02 model is Rp. 2,652.17 billion. Besides, the poor condition on the SF01 occurs once which is in the year 18th, but the bad condition on the SF02 occurs twice in the year of 39th (IRI = 12.46 m/km). In the year 27th, major rehabilitation was carried out because the IRI value was above 12 m/km. SF03 road steadiness level is 93 % with the requirement of total treatment cost during the analysis period is Rp. 2,475.20 billion. The graph of the relationship between annual treatment costs and the cumulative cost of model SF01 can be seen in figure 3 below.

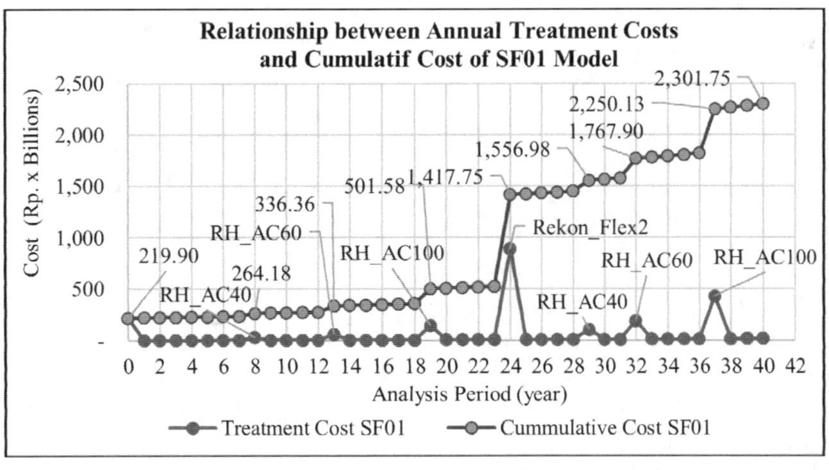

Figure 3. Relationship between annual treatment costs and the cumulative cost of SF01 model.

3.3. Analysis results of rigid pavement simulation

The results of the analysis on the SR01 model indicate that poor conditions occurred at the end of the analysis period of 8.06 m/km IRI with 98% of road steadiness level during the 40-year analysis period.

The initial cost for all of the rigid pavement model is Rp. 293.84 billion. The life-cycle age of rigid pavement is considered the same as that of the analysis period of 40 years.

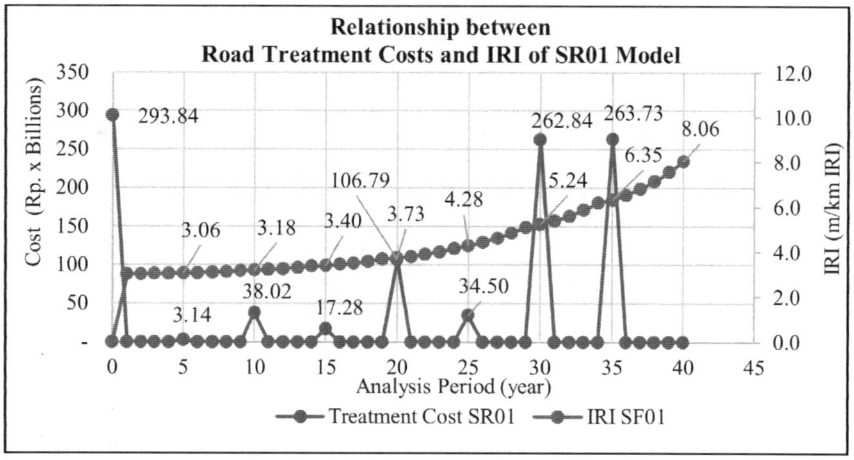

Figure 4. Relationship between Road Treatment Costs and IRI Values of SR01 Model.

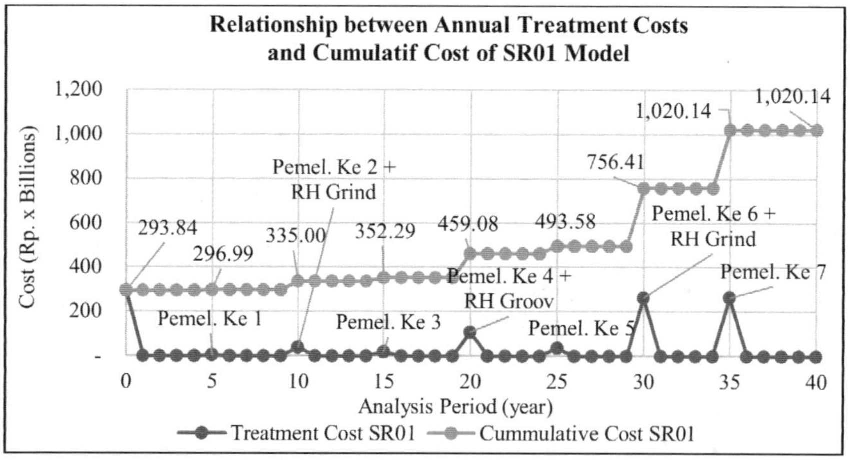

Figure 5. Relationship between Annual Treatment Costs and Cumulative Cost of SR01Model.

The same analysis was carried out on SR02 and SR03. The poor condition on the SR02 occurs three times which is in the year 38th (8.31 m/km), in the year 39th (8.88 m/km), and in the year 40th (9.53 m/km). SR02 road steadiness level is 95 % with the total treatment cost during the analysis period of Rp. 841.44 billion. The poor IRI conditions on the SR05, occurring five times starting from the year 36th to the end of the analysis period, are 8.02 m/km, 8.54 m/km, 9.31 m/km, 9.79 m/km and 10.53 m/km. The total treatment cost of SR03 during the analysis period is Rp. 759.73 billion.

3.4. Results of rigid and flexible pavement base on life cycle-cost analysis

Table 3 below is a recapitulation of the results of LCCA analysis using NPV and EUAC indicators to evaluate the value of actual pavement treatment costs over the 40-year analysis period. Based on the NPV, SF03 has the highest pavement treatment costs at a discount rate of 10% and 12% with the total cost of Rp. 455.83 billion and Rp. 388.01 Billion. While SR03 also has the lowest pavement treatment costs at both discount rates of 10% and 12% with the total costs of Rp. 345.07 billion and Rp. 329.73 billion.

Furthermore, based on the analysis results of EUAC indicators, SF01 has the highest pavement treatment costs at a discount rate of 10% and 12% with the total cost of Rp. 498.09 billion and Rp. 478.23 billion. While SR03 has the lowest pavement treatment costs also at both discount rates of 10% and 12% with total costs of Rp. 325.91 billion and Rp. 321.90 billion.

Table 3. Results of LCCA Analysis with NPV and EUAC.

No.	Model	***Net Present Value* (NPV)**		***Equivalent Uniform Annual Cost* (EUAC)**	
		***Discount Rate* 10 %**	***Discount Rate* 12 %**	***Discount Rate* 10 %**	***Discount Rate* 12 %**
1	SF01	422.60	359.09	498.09	478.23
2	SF02	408.79	345.96	408.56	392.29
3	SF03	455.83	388.01	333.84	322.55
4	SR01	358.09	337.89	376.15	365.09
5	SR02	349.80	332.74	348.56	341.54
6	SR03	345.07	329.73	325.91	321.90

3.5. Discussion

Based on the table of the relationship between road treatment costs and the road steadiness below, the SR01 rigid pavement is the best model with a total treatment cost of Rp. 1,020.14 billion with 98% of the road steadiness level. Even though SR01 costs are not the cheapest costs, but in terms of cost and road steadiness level, the SR01 model is the best choice. SF01 has the same steadiness level as SR01 but has a much more expensive cost which is Rp. 2,301.75 billion; however, the SR01 road treatment costs are more efficient at 1.26 times compared to the SF01. The total average of road treatment costs of rigid pavement model saves more 1.83 times than the flexible pavement model. However, the average road steadiness of rigid pavement is lower 2% compared to the flexible pavement, which is 98%.

Table 4. Relationship between road treatment costs and road steadiness.

No.	Model	Road Steadiness	Road Steadiness	Total Cost (Rp. Billion)	Average of Treatment Cost (Rp. Milyar)	Comparison of Rigid and Flexible Pavement
1	SF01	98 %		2,301.75		
2	SF02	95 %	95 %	2,652.17	2,476.37	
3	SF03	93 %		2,475.20		183 %
4	SR01	98 %		1,020.14		
5	SR02	93 %	93 %	841.44	873.77	
6	SR03	88 %		759.73		

Figure 6 below is an LCCA analysis result using economic indicators of NPV and EUAC on both types of pavement. Based on figure 6, the treatment cost of rigid pavement based on NPV saves Rp. 78.09 billion (18.20%) at a discount rate of 10% and Rp. 30.90 billion (8.48%) at a discount rate of 12% compared to flexible pavement. Whereas based on EUAC calculation results, the treatment cost of rigid pavement saves Rp. 63.29 billion (15.31%) at a discount rate of 10% and Rp. 54.84 billion (13.79 %) at a discount rate of 12% compared to flexible pavement.

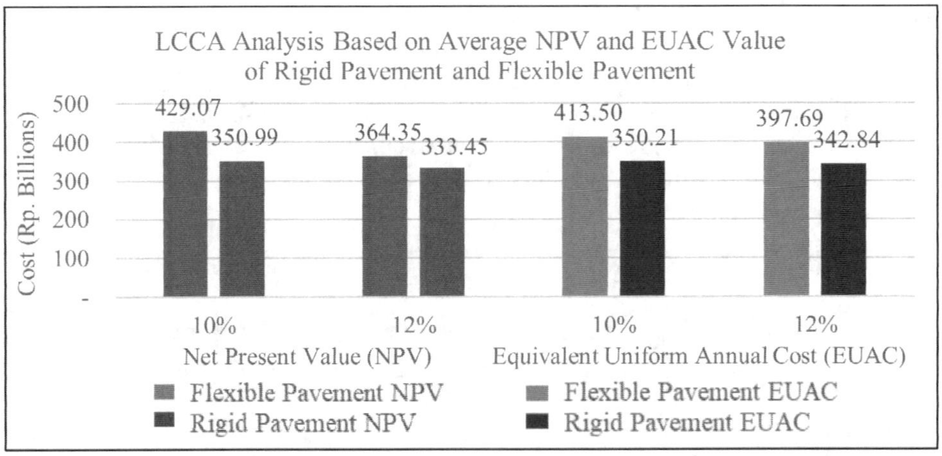

Figure 6. LCCA analysis based on average NPV and EUAC value of rigid pavement and flexible pavement.

4. Conclusion

Based on the results of the analysis and discussion it can be concluded that the SR01 model is the most economical and efficient pavement model in terms of road treatment cost and road stability during the 40-year analysis period with a cost of Rp. 1,020.14 billion with 98% road steadiness level. The total average of rigid pavement road treatment cost is 1.83 times or 183% more efficient than the flexible pavement, but the average of rigid pavement road steadiness is lower 2 % compared to the flexible pavement's, which is 98 %. Based on the LCCA analysis with NPV indicators, the road treatment cost of a rigid pavement ratio is 0.818 or more economical 18.20% at a discount rate of 10% and 0.915 or more economical 8.48% at a discount rate of 12 % than the flexible pavement. Besides, the EUAC indicator shows that the road treatment cost of rigid pavement ratio is 0.847 or more economical 15.31% at a discount rate of 10% and 0.862 or more economical 13.79% at a discount rate of 12% compared to flexible pavement.

Acknowledgments

I would like to express my deep gratitude to Dr. Ir. Sofyan M. Saleh, M.Sc.Eng. IPM and Dr. Renni Anggraini, S.T, M.Eng, my research supervisors, for their patient guidance, enthusiastic encouragement, and useful critiques of this research work. I would also like to thank Dr. Eng. Sugiarto, S.T, M.Eng, Dr. Ir. M. Isya, M.T., Dr. Teuku Budi Aulia, S.T. Dipl, Ing, and Dr. Noer Fadhly, S.T, M.T., for the advice and assistance in data analysis. Finally, I wish to thank my parents, my wife, and my Son for their support and encouragement throughout my study.

References

[1] Saleh S M, Sjafruddin A, Frazila R B and Tamin O Z 2009 Pengaruh muatan truk berlebih terhadap biaya pemeliharaan jalan *Transportation J. of FSTPT* **9** 1 79

[2] Syafriana, Saleh S M and Anggraini R 2015 Evaluasi umur layan dengan memperhitungkan beban berlebih di ruas jalan lintas timur Provinsi Aceh *Transportation J. of FSTPT* **15** 2 115

[3] American Concrete Pavement Association 2012 *Life Cycle Cost Analysis: A Tool for Better Pavement Invesment and Engineering Decicions* (Rosemont-Illinois: US) vii 1 p 6

[4] Direktorat Jenderal Bina Marga 2017 *Manual Desain Perkerasan Jalan Nomor 02/M/BM/2017, Kementerian Pekerjaan Umum dan Perumahan Rakyat* (Jakarta: Direktorat Jenderal Bina Marga Kementerian Pekerjaan Umum dan Perumahan Rakyat) 2-1

[5] Kementerian Pekerjaan Umum dan Perumahan Rakyat 2011 *Peraturan Menteri PUPR Nomor:*

19/PRT/M/2011 tentang Persyaratan Teknis Jalan dan Kriteria Perencanaan Teknis Jalan (Jakarta)

[6] Morosiuk G, Riley M J and Odoki J B 2004 *Highway Development & Mangement (HDM -4) Volume Six: Modelling Road Deterioration and Works Effect Version 2.* (Wahington DC : The World Road Assosiation (PIARC) on behalf of the ISOHDM Sponsors Paris and The World Bank) B 10-11 C 5-2

[7] Walls III J and Smith M R 1998 *Life-Cycle Cost Analysis in Pavement Design-Interim Technical Bulletin* (Washington: FHWA US) pp 4-5

IOP Conf. Series: Materials Science and Engineering **933** (2020) 012025 doi:10.1088/1757-899X/933/1/012025

Small-sized contractors' strategies of construction material purchasing in road rehabilitation projects

F S C S Maisarah[1], R Irfanto[2] and M Abduh[3]

[1] Doctoral Student in Civil Engineering, Faculty of Civil and Environmental Engineering, Bandung Institute of Technology, Indonesia
[2] Master Student in Civil Engineering, Faculty of Civil and Environmental Engineering, Bandung Institute of Technology, Indonesia
[3] Associate professor in construction engineering and management research group, Faculty of Civil and Environmental Engineering, Bandung Institute of Technology, Indonesia

Corresponding e-mail: fauziah.scs.maisarah@gmail.com

Abstract. Material management in construction project consists of material requirements planning, selecting suppliers, purchasing process, managing inventories, utilizing material, and managing waste materials. Suppliers selection and purchasing materials are very important stage related to suppliers and contractor relationship that needs to be considered. This paper discusses strategic issues in managing the purchasing of construction materials for road projects accomplished by small-sized contractors in The Greater Bandung areas. Kraljic's portfolio matrix approach is used to identify strategic construction materials. Initial data were obtained from three projects of road rehabilitation to identify seven strategic materials which further validated by gathering more data from the other 10 small-sized contractors. Based on the result of the identifying stage, it was found six strategic materials subsequently its strategic purchasing method was identified based on the characteristics of each material; building a strategic partnership was proposed for all materials.

1. Introduction

Many infrastructure projects are currently being developed in Indonesia, such as roads, bridges, railways, tunnels, dams, airports, and irrigation channels. The infrastructure projects are expected to be completed at the end of 2024, but the construction project cycle will not stop even after the construction project is completed, it must be continued with the maintenance process so that the infrastructure facilities can reach the optimum building age as planned.

The routine maintenance process of infrastructure buildings will involve small-sized contractors, since the value and complexity of the work following the qualifications of small-sized contractors. Road infrastructure projects carried out by small-sized contractors are projects to restore and maintain roads' infrastructure and its support elements, such as drainage channels [1]. Construction methods for road infrastructure maintenance are relatively simple and can be done by almost all contractors with the field of road infrastructure works, such as those included in Indonesian Construction Services Development Board (LPJKN) qualifications. Therefore, the technical aspects, in general, are not a serious problem, nevertheless, from the aspect of material costs, some potential things that can be improved therefore resources efficiency are can be achieved.

The value of material costs in a construction project is in the range of 60% to 80% of the total construction costs [2, 3]. In consequence, the aspect of purchasing construction materials needs to be considered by the contractor not only to achieve cost efficiency but also to attain a significant impact of construction operation productivity. Therefore, the step of selecting suppliers and determining purchasing strategies might reduce the risk of material-availability on the project. The activity of procuring construction materials must be started from the planning of material requirements that are integrated with a construction work schedule so that delivering materials are appropriate with project needs hence it will reduce the cost of material-storage in the site. Besides, determining purchasing strategies is not only deciding on selecting suppliers but also coordinating them throughout the project implementation [4].

The construction materials required for road rehabilitation are very diverse, therefore there is a need to identify construction materials that classify into the strategic commodities category hence purchasing strategy will be a focus on the construction materials in this category. The strategic commodities category is a group of materials that has great value in the importance of purchase aspects and supply risk aspects. The assessment of strategic commodities is conducted in terms of quantity and quality analysis and its availability in the long term that will be affected production productivity [5]. In the context of construction, construction material is stated as a strategic commodity if the amount in monetary value is large and meets the assessment of the level of purchase and supply risk importance so that its availability will affect the construction operation process in the site.

This paper discusses strategies in purchasing construction materials for road rehabilitation projects carried out by small-sized contractors, by taking research samples in Bandung and surrounding areas (The Greater Bandung areas). The results of the study are expected to be a reference for contractors, especially small-sized contractors who work on road infrastructure maintenance projects in formulating a strategy for purchasing construction materials.

2. Methodology

The research was conducted with a survey of small-sized contractors working on road rehabilitation projects in Bandung and surrounding areas. In addition, three road rehabilitation projects were applied to validate the result of Kraljic's matrix in materials value aspect to total material cost for each project.

The determination of strategic commodities starts from determining eligible construction materials to be selected as test commodities taking the Kraljic's portfolio matrix. The selection of test commodities is carried out by evaluating contract documents related to material needs and costs. From the results of the Pareto diagram with the highest percentage of material costs for overall material costs, seven test commodities were obtained [6].

Furthermore, the seven test commodities are assessed on the supplier's risk aspects and the level of profit impact. Tests for commodities were carried out by 10 representatives of small-sized contractors as research samples. Analysis of the results of the questionnaire regarding supplier's risk and the level of purchase interest is plotted on the Kraljic's diagram, and 6 out of 7 test commodities are included in the leverage category, namely sand, aggregate, reinforcing steel, cement, asphalt, and concrete. Analysis of building purchasing strategy for strategic commodities which is defined from the previous stage was conducted in a qualitative approach based on literature review.

2.1. Identifying strategic commodities

Identifying strategic commodities in construction materials contexts for road rehabilitation projects are conducted in two studies, assessment by small-sized contractors and calculation of construction materials consumption. Construction materials used in road rehabilitation projects carried out by small-sized contractors are sand, aggregate, landfill, asphalt, reinforcing steel, cement, and concrete. Those materials are will be analyzed through small-sized contractor responses in terms of assessment from supply risk and profit impact aspects. Besides, materials consumption in monetary value for the total material cost is analyzed based on the results of contract documents for three road rehabilitation projects as case studies.

Based on the results of contract documents analysis in terms of material cost, each material had. The project description and the percentage of each material's cost to the total construction costs are shown in table 1 and table 2.

Table 1. Description of project case.

Project Case	Total project cost	Total material cost
Case A: public road works for rural area in Bandung Regency	1,521,588,00 IDR	1,038,617,398 IDR
Case B: public road works, including pedestrian and drainage in Bandung City	783,276,001 IDR	502,201,980 IDR
Case C: public road works, including pedestrian and drainage in Bandung City	461,813,000 IDR	294,267,357 IDR

Table 2. Materials consumption of each case study.

Commodities	Case A	Case B	Case C	Means
Asphalt	34.53%	28.09%	34.68%	32.43%
Concrete	13.31%	19.98%	15.45%	16.25%
Sand	4.00%	4.27%	3.81%	4.03%
Reinforced steel	4.90%	2.68%	2.81%	3.46%
Aggregate	4.56%	2.11%	1.65%	2.78%
Cement	1.62%	2.83%	0.91%	1.79%
Landfill	1.23%	0.22%	0.17%	0.54%

The study of three road rehabilitation projects carried out by small-sized contractors in Bandung and surrounding areas shows that asphalt has the highest percentage compared to the others. This is finding due to the specifications of the road being studied are flexible pavement roads. Asphalt and concrete are commodities for the road overlay process, which is very in accordance with the scope of the project. Project scope area of road rehabilitation project generally does not attain the subbase works. In addition, the unit price for asphalt and concrete is also relatively higher compared to other commodities, therefore, it will increase the monetary value of them.

The determination of strategic commodities is conducted with Kraljic's portfolio matrix approach which consists of four quadrants as can be seen in figure 1. The attributes of each quadrant describe the level of importance of commodities in its business process. Kraljic's matrix classifies four-level in strategic commodities namely, strategic, leverage, bottleneck, and non-critical items. The category levels assess in two dimensions i.e. profit impact and supply risk (low and high level) [5].

In this matrix, the value of each indicator is shown at range 1 to 10, which 1 is for the condition of there's no impact for supply risk and profit impact, and 10 is for vice versa. The midpoint in range (5 to 5 point) is cutting line to divide four areas in the matrix.

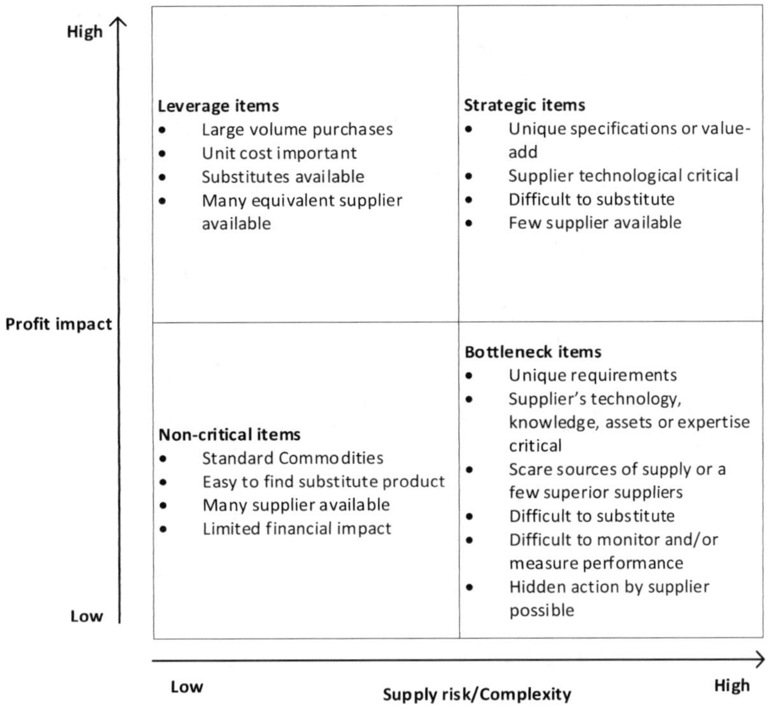

Figure 1. *Kraljic portfolio matrix (KPM).*

The dimensions in Kraljic's portfolio matrix are based on the complexity of commodity supply risk and commodity profit impact. Kraljic's portfolio matrix is a generic theory that has been used in various industries, ranging from manufacturing, agriculture, and construction. The set of indicators is built as a tool to determine the strategic level of a commodity [7], which can be seen in table 3.
Based on the research conducted by Irfanto [8], the results of plotting respondents' assessment of supply risk variables and profit impact are obtained from all respondents for each test commodities. Kraljic's portfolio matrix results of each commodity are shown in table 4.

The results of commodity testing on cost percentage and assessment of profit impact and supply risk attributes indicate that the landfill can be eliminated because based on the cost of purchasing material the percentage is very low at 0.54%, and the supply risk aspect is only 3.90 and profit impact level is only 0.75 above the mid-quadrant threshold.

Based on the results of the supply risk assessment all test commodities are at a low threshold (less than 5.00), this indicates that commodities for road rehabilitation projects have no risk to supplier availability or supplier performance. However, based on the profit impact of commodity purchases, the material used in infrastructure projects is at the middle of the upper threshold of Kraljic's portfolio matrix, which means that purchasing commodities will have an effect on the value of profits related to overall project costs, absence of substitute commodities, and commodity value is very high even without being affected by the costs beyond the purchasing cost the commodity itself, such as delivery costs and other costs.

Table 3. Indicators of strategic commodities determination [7].

Dimension	Variables	Indicators
Supply risk	Market risk	Availability of suppliers
	Performance risk	Supplier's performance in delivering
	Complexity risk	Difficulties in finding suppliers
Profit impact	Impact on profitability	Profit on purchasing commodities
	Critically of purchase	Substitute commodities
	Value/cost of purchase	Including tangible and intangible cost for the commodities

Table 4. The result of each-commodities [8].

Commodities	Importance of purchase	Supply risk/Complexity
Asphalt	8.55	4.47
Concrete	6.98	4.55
Sand	7.20	3.92
Reinforced steel	8.15	4.48
Aggregate	7.90	4.12
Cement	7.98	4.07
Landfill	5.75	3.90

Based on the results of testing of project contract documents from three case studies as well as an assessment of the dimensions of supplier risk and the profit impact, the test commodities fall into the leverage category in Kraljic's portfolio matrix. Furthermore, the purchasing strategies to be analyzed are asphalt, concrete, cement, reinforcing steel, aggregate, and sand.

2.2. The strategy of purchasing materials

By using the Kraljic's approach, an organization can develop a purchasing strategy by focusing more on strategic commodities, because controlling all commodities requires enormous effort. By controlling the commodities that have been identified as strategic commodities, the purchasing strategies formulated can be implemented properly.

The contractor carries out a strategy to purchase construction materials used in the project not only to guarantee the availability of material for the project but also to reduce the risk of material available in the projects that are carried out later. Hence, the sustainability aspect in terms of material supply will not be hampered by the change in the project location or the characteristics of the different projects. The purchasing strategy was conducted based on commodity categories from Kraljic"s portfolio matrix by Caniëls and Gelderman [9] as given in table 5.

IOP Conf. Series: Materials Science and Engineering **933** (2020) 012025 doi:10.1088/1757-899X/933/1/012025

Table 5. Purchasing strategy related to Kraljic's matrix [9].

Categories	Purchasing strategy/scenario	
Strategic terms	Maintain a strategic partnership	Strategic commodities that have a supply risk must be controlled by building partnerships with suppliers so that sustainable production can be maintained without any obstacles to the availability of commodities
	Accept locked-in partnership	This happens if the supplier's bargaining position is higher than the buyer
	Terminate a partnership	If the supplier's performance can be accepted again by the buyer, the buyer must start reducing the reliance on suppliers and even stop the partnership contract and start looking for alternative suppliers
Bottleneck items	Accept dependence, reduce negative consequences	Risk analysis needs to be done and determine the most important commodities, so that dependence on suppliers can be minimized
	Reduce dependence and risk, find other solutions	Looking for a solution to the use of commodities in the bottleneck category, it can be done by finding new substitute commodities by expanding the required specifications.
Leverage items	Exploit buying power	The bargaining position of the buyer is higher than the supplier so that the buyer can enter into a cooperation contract with a supplier that can provide more benefits
	Develop a strategic partnership	If the buyer has found a supplier in accordance with the criteria, it is necessary to build a strategic partnership with suppliers to ensure supply availability and can increase the efficiency and effectiveness of the buyers
Non-critical items	Pool purchasing requirement	For routine commodity needs, steps need to be taken to reduce the complexity of administration and logistics, namely by establishing a system of contracts with fixed suppliers
	Individual ordering, efficient purchasing	If the steps to build a system of contracts with suppliers cannot be done, then individual purchases can be made using a purchase card to reduce indirect costs, such as booking administration fees and billing.

Based on the results of commodity testing for construction materials in road infrastructure maintenance projects, the test commodities are stated in the leverage category, hence based on the results of the study of Caniëls and Gelderman [9], strategic steps in commodity purchases can be carried out by exploiting purchases and building strategic partnerships with suppliers.

IOP Conf. Series: Materials Science and Engineering **933** (2020) 012025 doi:10.1088/1757-899X/933/1/012025

3. Results and discussion

The purchasing strategy for a commodity can not only be determined by referring to the strategic level of the project, the characteristics of the material or commodity is also needed to be seen. In the context of construction, the construction material used is part of the construction design specifications, therefore the determination of the purchasing strategy must also be adjusted to the characteristics of the construction project, related to the technical specifications of the design and project conditions in the field.

Infrastructure projects, in this case, are road maintenance projects, are projects that may be carried out in urban and rural areas, with normal environmental conditions surrounding the project taking place. This is due to the nature of maintenance, so the project location cannot be completely closed from access to community activities. Therefore, it is possible for the location of the placement of material stocks to be very limited so that the amount of material imported to the project location must be adjusted to the project implementation schedule.

Determination of strategic commodity purchasing strategies must also look at the material characteristics or commodities to be purchased. Construction materials used in road infrastructure maintenance projects included in the leverage category based on Kraljic's portfolio matrix are sand, aggregate, asphalt, concrete, reinforcing steel and cement. Each material must be assessed based on its material properties, whether it is bulk or manufacturing material, whether the material sent is just in time or stock can be carried out in the field, and whether the material storage location in the field requires a safe place and protected from unexpected weather, theft, and whether the material storage location requires large space or not. These things must be assessed before establishing a commodity purchasing strategy. In addition, the external environmental context of material supply must also be considered, whether there are certain times when material supply is difficult or in general the material supply is always stable. Based on the concept of commodity purchasing strategies provided by Caniëls and Gelderman [9], in commodities in the leverage category two purchase strategies can be carried out, namely:

- Exploiting purchasing power, which is when the bargaining position of buyers is higher than the bargaining position of suppliers, in the sense that material suppliers are still available in sufficient quantities compared to the needs of buyers. With such a buyer can choose suppliers that provide more benefits;
- Develop a strategic partnership, although there is no problem in terms of supply risk. However, in certain conditions, the supply in the field must be maintained when the project implementation time occurs during the peak season of the project, especially in road infrastructure maintenance projects. The government generally issues project budgets in time at the same time hence, there is a very possible supply shortage at certain times. Consequently, the step of building strategic intimacy with material suppliers is a necessity to maintain supply availability and the timely delivery and specifications of the ordered materials.

Table 6 shown a set of strategies to purchase construction materials which identified as strategic commodities for road projects that should be conducted, small-sized contractors. Sand and aggregate material is a bulk material category that requires extensive storage in the project location but does not need to be protected from the weather and is a potential crime such as stealing. The quality of the two materials is quite varied, so to maintain the quality of the material that is in line with what is expected by the contractor, a strategic partnership needs to be built so that the contractor can gain confidence in the quality of the material sent. Moreover, because the project location may be far from the supplier's location, delivery costs may arise. To reduce delivery cost it is necessary to do a strategy in terms of economies of scale, where if partnerships have been established with suppliers, indirect costs can be reduced.

Table 6. Strategy of purchasing materials.

Commodities	Characteristic delivery and stock material	Purchasing Strategy
Asphalt	Just in time	Develop a strategic partnership
Concrete	Just in time	Develop a strategic partnership
Reinforcing Steel	Stocking available	Develop a strategic partnership
Cement	Stocking available	Exploiting Purchasing Power
Sand	Stocking available	Develop a strategic partnership
Aggregate	Stocking available	Develop a strategic partnership

Asphalt and concrete materials are materials that cannot be stored at the project location and must be used immediately if they have arrived at the project location (just in time). Such supply must be maintained by suppliers so that contractors do not need to take risks in terms of availability, thus a strategic partnership needs to be built.

Cement material, according to a study from the Ministry of Public Works, is a large manufacturing material that is available in Indonesia and it could be exploited. However, due to the location of material storage that is prone to weather and theft crimes, it is better to ship cement material as needed on the material implementation schedule. Although the contractor can store stock of material in other locations; to make storage cost efficiency, cement material should also be carried out a strategy by building strategic partnerships.

Reinforced steel is a material that is still available in its stock but based on the results of the Ministry of Public Works study, at certain times the stock of materials for reinforcing steel is quite difficult to find. Moreover, the storage of reinforced steel material in the project location requires a large area, and this can hamper work activities in the field. Therefore, building a strategic partnership is the right step that can be done by contractors in buying reinforced steel material.

4. Conclusion

In this paper, we conclude that building strategy in construction materials could not only determined by considering the final result of material plotting in Kraljic's portfolio matrix but also its should be conceived characteristics of construction materials, project site conditions, project engineering design, material storage and handling, and external conditions that influenced materials supply as a whole that is associated with the time of project implementation. Moreover, even though some commodities declared in strategic quadrants based on Kraljic's portfolio matrix for road rehabilitation projects can be carried out strategies with the exploitation of purchases. As an anticipation step, contractors should build strategic partnerships with material suppliers to ensure the availability of materials, so that the construction process can be more productive and efficient. This purchasing strategy is expected to help small-sized contractors in managing commodity purchases so that they can increase the efficiency of purchasing construction commodities.

References

[1] Johannessen B 2009 *Training Guidelines: Contracting Local Infrastructure Works ILO* (International Labor Organization)

[2] Eksekär A and Rudberg M 2015 Third-party logistics in construction: perspective from supplier and transport providers *22nd Int. Annual EurOMA Conf.* (Switzerland, July 2015)

IOP Conf. Series: Materials Science and Engineering **933** (2020) 012025 doi:10.1088/1757-899X/933/1/012025

[3] Sobotka A and Czarnigowska A 2005 Analysis of supply system models for planning construction project logistics *J. of Civil Eng. and Management* **11** 1 73-82
[4] Bildsten L 2016 *Purchasing in Construction Companies* (Netherland: University of Twente)
[5] Kraljic P 1983 Purchasing Must Become Supply Management Harvard Business Review
[6] Irfanto R, Maisarah F S C S and Abduh M 2018 Metode identifikasi komoditas strategis untuk kontraktor kecil di Bandung Raya *Prosiding Konferensi Nasional Pascasarjana Teknik Sipil (KNPTS)* pp 19-26
[7] Wagner S M, Padhi S S and Bode C 2013 The Procurement Process - Refining inputs for Kraljic matrix yields objective purchasing portfolios and strategies *Industrial Engineer* **45** 34-39
[8] Irfanto R 2019 *Kajian kebutuhan komoditas strategis untuk kontraktor kecil di Bandung Raya.* (Bandung: Unpublished Thesis Bandung Institute of Technology)
[9] Caniëls M C J and Gelderman C J 2005 Purchasing strategies in the Kraljic matrix—a power and dependence perspective *J. of Purchasing and Supply Management* **11** 141-155

IOP Conf. Series: Materials Science and Engineering **933** (2020) 012026 doi:10.1088/1757-899X/933/1/012026

Optimization of water distribution for channel irrigation networks Krueng Baro Kiri scheme (Aceh-Indonesia)

I Yanita[1], A Azmeri[2] and E Fatimah[2]

[1] Civil Engineering Postgraduate Program, Department of Civil Engineering, Universitas Syiah Kuala, Banda Aceh, Indonesia
[2] Department of Civil Engineering, Universitas Syiah Kuala, Banda Aceh, Indonesia

Corresponding e-mail: irmayanita202@gmail.com

Abstract. The standard area of Krueng Baro irrigation scheme is 11.950 Ha, divided into *Baro Kanan* irrigation (8,920 Ha) and *Baro Kiri* irrigation (3.030 Ha). This irrigation focuses on increasing the productivity of food crops in Pidie District. Previous research reported that the distribution efficiency of the *Baro Kiri* primary channel (74,1%) was smaller than the *Baro Kanan* (86,8%). The decreasing efficiency level of the channel results in a large number of unirrigated paddy fields that which in turn decrease the rice productivity. This study aimed to determine the optimum area of the rice fields that can be irrigated and examine the amount of discharge distributed to each channel by minimizing water loss. The method used was the optimization linear program. The optimization had the objective function of minimizing water loss. The land area was a decision variable, while the water loss in each channel and the water availability in each weir were the constraint function. The optimization results created an area of 2.810 Ha out of 3.030 Ha total area of rice fields in *Krueng Baro Kiri* by minimizing water loss in each channel based on the efficiency value of each channel. The rice field terrace that had highly reduced from the standard area in the *Krueng Baro Kiri* were the ones in the channel section of R4 and R5, 10 Ha of the standard rice field area. This occurred because these two channel sections (R4 and R5) had the smallest efficiency value of 47,26%, and 58,03%, respectively.

1. Introduction

The *Krueng Baro* irrigation scheme focuses on increasing the productivity of food crops in Pidie District, which has a total area of 11.950 Ha divided into the *Baro Kanan* (8.920 Ha) and the *Baro Kiri* irrigation scheme (3.030 Ha). The location of the *Krueng Baro* irrigation scheme is presented in figure 1. Some efforts to increase the productivity of food crops include expanding the planting area and implementing the intensive method on the existing land. One of the efforts to increase food production is establishing good water allocation to evenly distribute the water to each rice field terrace [1]. However, various current water allocation systems require to be reviewed because of the water supply is not optimal for each rice field terrace. One of the factors influencing this condition is a large amount of water loss, causing the rice fields not to obtain enough water in the irrigation scheme. A large amount of water loss can be resulted from the damaged and leaking primary and secondary channels. Thus, it is necessary to optimize the irrigation water allocation spatially (between the terrace).

The irrigation water distribution is problematic if the available discharge is smaller than the water requirement (especially during the dry season). Thus, the efficient use of irrigation water is required.

The harvest is not only influenced by the fulfillment level of water needs but by way of providing water [2].

Figure 1. *Krueng Baro* irrigation scheme.

Efficiency is an examination of water provision compared to the needs. From an engineering point of view, the definition of irrigation efficiency is based on the fact that not all water supplied or tapped into the canal can be channeled to the tapping place/ irrigated land; some are lost [3].

Agricultural production, especially the paddy fields in the upstream of Krueng Baro irrigation scheme, has met the agriculture's target, around 5 to 6 tons/ha. However, those in the downstream areas did not reach the goal because of the lack of water. The planting intensity in Baro irrigation scheme is currently twice a year, Paddy-Paddy. The planting realization in MT I and MT II are 9.015 Ha and 8.030 Ha, respectively. The total planting realization is 17.046 Ha. The Planting Index (PI) in Baro irrigation scheme is 143%, while the percentage of planting realization area is 71% [4].

According to Azmeri et al. [5] the distribution efficiency of the Baro Kiri primary channel is smaller than the distribution efficiency of Baro Kanan channel. The actual efficiency level of Baro Kiri primary channel is 74,1%, meaning that the water loss along the Baro Kiri primary channel is 15,9% from the efficiency of the design conditions (90%). On the other hand, Baro Kanan primary channel has a distribution efficiency of 86,8%, indicating that the water loss along the channel is 3,2% of the efficiency of design conditions (90%). A large amount of water loss in the irrigation channel of Baro Kiri is due to the damage and leakage leading to the number of unirrigated rice terrace. Thus, it is necessary to optimize the water allocation in each channel by minimizing water loss.

Azmeri et al. [6] stated that an optimization model is usually formed by combining the system behavior and the objective function of the system. The constraint system objective is created by being the objective function. Next, suitable mathematical algorithms will be selected to be applied to the optimization. Sometimes, the optimization model of a system requires to be simplified the form of the optimization algorithm used, to meet the requirements of one optimization algorithm.

The results of a research conducted by Azmeri et al. [5] showed that the implementation of cropping patterns in Krueng Baro Kiri irrigation scheme is not optimal. The cropping pattern consists of only two planting seasons yearly, namely MT I and MT II. The best and ideal planting schedule in Baro Kiri irrigation scheme is October II as it produces the largest planting area.

Based on the problem statement above, this study aimed to analyze the optimum area of the Baro Kiri Irrigation scheme that can be irrigated, the discharge distributed to each channel by optimizing the water loss minimizing.

2. Methodology

Problems concerning irrigation water allocation often occur if the amount of available discharge is smaller than the water required (especially during the dry season); thus, efficient use of irrigation water is needed. This study conducted the water allocator process employing the Linear Program using Solver in Microsoft Excel for optimization and its stages. The optimization was carried out to obtain the land area and the discharge distributing into each primary and secondary channel. The limitation, an obstacle in optimization, is the water availability that must be greater or equal to the water requirements for each rice field terrace.

The optimization using linear program is undertaken by creating mathematical equation model based on the objective function, constraint function, and non-negative mathematical model function built according to the objective function to achieve. The stages of optimization are:

2.1. Initial optimization variable

The creation of the optimization algorithm must consider the variables of the objective function and the constraint function. This study had several primary variables that were used to solve the optimization, as presented in table 1.

2.2. Optimization using linear program

The optimization solutions of several alternatives and scenarios were obtained using the Solver program. The solver was activated through the Manage Excel Add-Ins feature and displayed in the Data Analysis menu. This study analyzed basic solutions in linear programs to find the best combination of resources and constraints to obtain the optimal results/conditions so that they can be interpreted and used for decision-making. The mathematical model in this linear program was created based on the objective function to achieve. The formulation of the optimization analysis consisted of:

2.2.1. The objective function. The objective function to achieve was to minimize the water loss in the irrigation channel, aiming to irrigate all rice field as needed. This objective function was an equation consisting of the independent variable which will be optimized, and the form of its function is to minimize water loss.
Minimum water loss:

$$\sum_{i=1}^{n}(Q_{Ri} * E_i) \tag{1}$$

Where:
Q_{Ri} = Distribution discharge of each channel in Baro Kiri irrigation scheme (m^3/sec);
E_i = The water loss in each irrigation channel (%); and
n = The number of irrigation channels.

2.2.2. The constraint function. This constraint function was an equation limiting the main use, and the form of this function is the water availability in the Krueng Baro river and the sufficient amount of water reaching the related place.
Constraint:

- Irrigation water demand:

$$\sum_{i=1}^{n} Q_{Ri} \le Qsj \tag{2}$$

IOP Conf. Series: Materials Science and Engineering **933** (2020) 012026 doi:10.1088/1757-899X/933/1/012026

➢ The water distributed:

$$\sum_{i=1}^{n} (Q_{Ri} * Ei) + Q_{Ri} \leq Qsj \tag{3}$$

Where

$$Q_p = DR \times A_i \tag{4}$$

Decision Variable: the land area (A)

Where:

QRi = distribution discharge of each channel in *Baro Kiri* irrigation scheme (m³/s);
Qsj = the availability discharge in the weir (m³/s);
Ei = the water loss in each channel (%);
Ai = the coverage of each channel (Ha); and
DR = the water demand at the intake valve (m³/s).

Table 1. Initial variables in each channel in Krueng Baro Kiri irrigation scheme.

No	Channel Section	Efficiency (%)	Initial
1	BBrkr 1 - BBrkr 2	88.83	R1
2	BBrkr 2 - BBrkr 3	70.25	R2
3	BBrkr 3 - BBrkr 4	67.39	R3
4	BBrkr 4 - BBrkr 5	47.26	R4
5	BBrkr 5 - BBrkr 6	58.03	R5
6	BBrkr 6 - BBrkr 7	84.92	R6
7	BBrkr 7 - BBrkr 8	84.42	R7
8	BBrkr 8 - BBrkr 9	78.3	R8
9	BBrkr 9 - BBrkr 10	87.26	R9
10	BBrkr 10 - BRb 1	80.03	R10
11	BRb 1 - BRb 2	64.27	R11
12	BRb 2 - BRb 3	85.51	R12
13	BRb 3 - BRb 4	77.95	R13
14	BBrkr 10 - BLk 1	78.24	R14
15	BLk 1 - BLk 2	88.27	R15
16	BLk 2 - BLk 3	75.72	R16
17	BLk 3 - BLk 4	88.44	R17
18	BLk 4 - BLk 5	87.83	R18
19	BLk 5 - BLk 6	81.77	R19
20	BBrkr 6 - Bla 1	90.69	R20
21	Bla 1 - Bla 2	91.29	R21
22	Bla 2 - Bla 3	88.53	R22
23	Bla 3 - Bla 4	89.54	R23
24	Bla 4 - Bla 5	89.13	R24
25	Bla 5 - Bla 6	89.33	R25
26	Bla 6 - Bla 7	61.03	R26
27	Bla 7 - Bla 8	67.89	R27

IOP Conf. Series: Materials Science and Engineering **933** (2020) 012026 doi:10.1088/1757-899X/933/1/012026

3. Results and discussion

The optimization process and stages in this study employed the Linear Program using Solver in Microsoft Excel. The optimization was used to obtain the optimum land area by minimizing the water loss in each channel in the *Krueng Baro Kiri* irrigation scheme. The constraints in the optimization calculation were the land area and the water availability in the weir that serves the *Baro Kiri* irrigation scheme. Table 2 presents the optimization results of optimum rice field area by minimizing water loss in each channel in the Baro Kiri irrigation scheme.

Table 2 shows that the planned rice field is smaller than the standard area of the rice fields served by *Krueng Baro Kiri* Dam. The discharge available at this weir was 4,469 m^3/sec, while the discharge required was 4,814 m^3/sec. The total standard area of rice fields that should be irrigated by *Keumala Kiri* Dam was 3,030 Ha, while the total of rice fields that can be irrigated based on the water availability in the optimization results was 2.810 Ha. This finding indicates that 220 ha of rice field terrace that cannot be served due to lack of water available in the weir.

The channel section of R4 had the smallest efficiency value in *Krueng Baro Kiri* irrigation scheme (47.26%), with the water loss of 52,74%. The optimization results showed that R1, R20, R21, R23, R24, and R25 channel section were reduced by 7 Ha of the total standard area, while R2, R3, R11, R26, and R27 channel sections were decreased by 9 Ha of the standard area. The rice field terrace with the highest area reduction compared to the standard area in the *Krueng Baro Kiri* irrigation scheme were R4 and R5 channel section, 10 Ha decline of the standard area. This occurs because R4 and R5 were the channel sections with the smallest efficiency values (47, 26% and 58.03% respectively), indicating that theses channel section had the highest water loss.

Table 2. The optimization results of optimum rice field area by minimizing water loss in each channel in the *Baro Kiri* irrigation scheme.

No	Channel Section	Initial	Efficiency (%)	Water Loss (%)	Water Demand (m^3/s)	Total Standard Area (Ha)	Optimization of Total Area (Ha)
1	BBrkr 1 - BBrkr 2	R1	88.83	11.17	0.193	154	147
2	BBrkr 2 - BBrkr 3	R2	70.25	29.75	0.229	160	151
3	BBrkr 3 - BBrkr 4	R3	67.39	32.61	0.080	70	61
4	BBrkr 4 - BBrkr 5	R4	47.26	52.74	0.236	147	137
5	BBrkr 5 - BBrkr 6	R5	58.03	41.97	0.301	189.5	180
6	BBrkr 6 - BBrkr 7	R6	84.92	15.08	0.146	119	111
7	BBrkr 7 - BBrkr 8	R7	84.42	15.58	0.180	141	133
8	BBrkr 8 - BBrkr 9	R8	78.3	21.7	0.027	39	31
9	BBrkr 9 - BBrkr 10	R9	87.26	12.74	0.262	200	192
10	BBrkr 10 - BRb 1	R10	80.03	19.97	0.500	247	239
11	BRb 1 - BRb 2	R11	64.27	35.73	0.383	210	201
12	BRb 2 - BRb 3	R12	85.51	14.49	0.017	40	32
13	BRb 3 - BRb 4	R13	77.95	22.05	0.040	49	41
14	BBrkr 10 - BLk 1	R14	78.24	21.76	0.013	37	29
15	BLk 1 - BLk 2	R15	88.27	11.73	0.053	55	47
16	BLk 2 - BLk 3	R16	75.72	24.28	0.163	103	95
17	BLk 3 - BLk 4	R17	88.44	11.56	0.038	49	41
18	BLk 4 - BLk 5	R18	87.83	12.17	0.116	81	73
19	BLk 5 - BLk 6	R19	81.77	18.23	0.323	170	162
20	BBrkr 6 - Bla 1	R20	90.69	9.31	0.136	89	82
21	Bla 1 - Bla 2	R21	91.29	8.71	0.005	36	29
22	Bla 2 - Bla 3	R22	88.53	11.47	0.053	55	47
23	Bla 3 - Bla 4	R23	89.54	10.46	0.018	41	34
24	Bla 4 - Bla 5	R24	89.13	10.87	0.131	87	80
25	Bla 5 - Bla 6	R25	89.33	10.67	0.309	160	153
26	Bla 6 - Bla 7	R26	61.03	38.97	0.274	160.5	151
27	Bla 7 - Bla 8	R27	67.89	32.11	0.241	141	132

4. Conclusion

Based on the optimization results of the land area by minimizing the water loss in *Krueng Baro Kiri* Irrigation scheme, it can be concluded that The total area of optimized rice field by minimizing the water loss in each channel section based on the efficiency value of each channel is 2810 Ha out of 3.030 Ha total area of rice fields in *Krueng Baro Kiri.*

References

[1] Azmeri, Meilianda E, Ifrayaski and Mirza I 2019a Solusi kekurangan air irigasi pada musim gadu melalui optimasi pengoperasian embung *Konferensi Nasional Teknik Sipil 13* (Banda Aceh: Indonesia)

[2] Ahmad M, Masih I and Turral H 2004 Diagnostic analysis of spatial and temporal variations in crop water productivity: A field scale analysis of the rice-wheat cropping system of Punjab, Pakistan *J. of Applied Irrigation Science* **39** 1 43-63

[3] Chow V T 1989 *Hidrolika Saluran Terbuka (Open Channel Hydraulics)* (Jakarta: Erlangga, Indonesia)

[4] Balai Wilayah Sungai Sumatera BWS S – I 2017 *Laporan Akhir Pekerjaan Penyusunan Penilaian Kinerja dan Aknop Jaringan Irigasi Kewenangan Pemerintah Pusat (D.I. Krueng Baro)* (Banda Aceh: Indonesia)

[5] Azmeri, Yulianur A, Zahrati U and Faudli I 2019b Effects of irrigation performance on water balance: Krueng Baro irrigation scheme (Aceh-Indonesia) as a case study *J. of Water and Land Development* **42** VII-IX 12-20

[6] Azmeri, Hadihardaja I K, Shaskia N and Admaja K S 2017 Completion of potential conflicts of interest through optimization of Rukoh reservoir operation in Pidie District, Aceh Province *AIP Conf. Proc.* vol 1903 Iss 1 (Palembang: Indonesia) 10.1063/1/5011611

IOP Conf. Series: Materials Science and Engineering **933** (2020) 012027 doi:10.1088/1757-899X/933/1/012027

Rainfall interception loss by corn crop

T Erlangga[1], A Yulianur[2], Sugianto[3] and L Aditya[4]

[1] Civil Engineering Postgraduate Program, Department of Civil Engineering, Universitas Syiah Kuala, Banda Aceh, Indonesia
[2] Department of Civil Engineering, Universitas Syiah Kuala, Banda Aceh, Indonesia
[3] Agriculture Departement, Universitas Syiah Kuala, Banda Aceh, Indonesia
[4] Environmental Engineering Departement, Gdańsk University of Technology, Gabriela Narutowicza 11/12, 80-233 Gdańsk, Polandia

Corresponding e-mail: erlanggats0809@gmail.com

Abstract. One factor that influences the water balance on earth is the presence of the rain factors. The Hydrological cycle is a process of water that is evaporated and if the atmosphere allows, water will fall as rain. Interception is an important factor in the hydrological cycle. With the process of interception, rain water that reaches the ground surface becomes reduced. The amount of interception is influenced by land use in a sub-watershed. This research was conducted in the Krueng Jreu sub-watershed, located in Aceh Besar District with an area of 23,189.53 Ha. Land use at this location is generally intended for plantations, wetland agriculture, protected forests, and rural settlements. Most of the citizen around the research location, work as farmers, such as farmers of rice, corn, peanuts, oil palm, and others. In comparison to all types of plants, corn is one of the dominant plants developed in this area. The purpose of this research is to find out the value of interception in maize plants. The value of interception in maize plants aged 1-3 months was analyzed by making a Mini Hydrological Watershed model to obtain a regression equation, where this equation is used to see the interception of maize intercepts on rainwater rainfall observations of 19 variants of observational data from 3 months of research. The amount of interception in maize plants obtained based on the Mini Hydrology Watershed model with logarithmatic regression is 17.62 mm from 205.950 mm of total rainfall.

1. Introduction

One factor that influences the water balance on earth is the presence of the rain factors. Rain is the event of drops of water droplets from the sky and is the cycle of water on earth. In the hydrological cycle there is a process of water travel caused by solar thermal energy which is also called the evaporation process, both on the surface of sea water and other bodies of water. The water vapor will be carried by the wind across the mountainous and flat land, if the atmosphere allows some of the water to rain. Before reaching ground level, the rainwater will be held back by the vegetation canopy. Some of the rainwater will be stored on the surface of the canopy/leaf during the canopy wetting process, and some will fall to the ground through the sidelines of the leaf (throughfall) or flow down through the surface of the tree trunk (stemflow). Rainwater that reaches the ground surface, some will enter (absorbed) into the ground (infiltration). Whereas rainwater that is not absorbed into the ground will be temporarily accommodated in surface detention basins (surface detention) to then flow above the surface to a lower place (runoff), and then enter the river and some others will be intercepted [1].

Rain interception is the process of retaining rain water on the surface of vegetation which is then re-evaporated into the atmosphere during and after rain. According to Bruijnzeel in Asdak [1], the amount of rain interception in primary forests ranges from 10-35% of total rainfall and changes in land cover stands from one type of vegetation to another can affect the annual water balance. Several studies have been conducted to determine the amount of interception in a type of plant. From the results of Benara's research [2], the magnitude of Lamtoro plant interception was 70.27% of total rainfall and the amount of interception of Arabica Coffee plant was 62.33% of total rainfall. Rao [3] also conducted a rain interception study on guava plants, where the percentage of rain that was intercepted by this plant was 31%.

This research was conducted in the Krueng Jreu Sub-watershed, located in Aceh Besar District, this Sub-watershed has an area of 23,189.53 Ha. Land use at the study site is generally intended for plantations, wetland agriculture, protected forests, and rural settlements. Most of the people around the study site work as farmers, be it farmers of rice, corn, peanuts, oil palm and others. Of the various types of plants, maize is one of the dominant plants developed in this area, where maize is a food crop sub-sector which is the second staple food in Indonesia after rice and the third staple food in the world after wheat and rice.

In this study, interception of maize research was carried out by creating a Mini Hydrology Watershed model. The making of this Mini Hydrology Watershed aims to obtain a regression between the relationship between the value of interception and rainfall. The expected results of this study are to find out the magnitude of interception of the Krueng Jreu Sub-watershed with land cover on corn plantations.

2. Methodology

The amount of rain interception due to maize in the Krueng Jreu Sub Watershed was observed directly in the field by making Mini Hydrology Watershed as a model. The process of making Mini Hydrology Watershed functions to estimate or as and approach from rain data in units of mm/day or called the intensity of rain into runoff and then reconfigured into units of mm/day. Broadly speaking the research was conducted through four stages, namely the type and source of data, the method of rain interception research, analysis of research data and the infiltration estimation formula.

2.1. Types and data's sources

The data used in this study consisted of primary data and secondary data. Primary data in this study were obtained from the manufacture and observation of the Mini Hydrology Watershed. The tools used to obtain several parameters in this study are water meters, storage containers, pipes, elbo, zinc, sanyo, hose and dron. Parameters observed include measurements of rainfall, runoff and infiltration. Sketch of the Mini Hydrological Watershed can be seen in figure 1 below, without corn-crop *(left)* and with corn-crop *(right)*.

Rainfall data used is to collect daily rainfall or every 24-hour period measured every morning. Based on these daily data, weekly, ten daily, monthly, annual rainfall data can be collected and so on. Furthermore, it can also be taken into account the rainy day. Rainfall is measured using the Automatic Rain Measuring Tool (ARG). Automatic rain gauge or Automatic Rain Gauge (ARG) is equipment used to calculate the amount of rainfall in a certain time unit automatically with the help of a battery as a source of energy. The ARG component consists of two parts, namely the ARG body that functions to accommodate and measure rainfall and the ARG logger which functions to calculate and record rainfall.

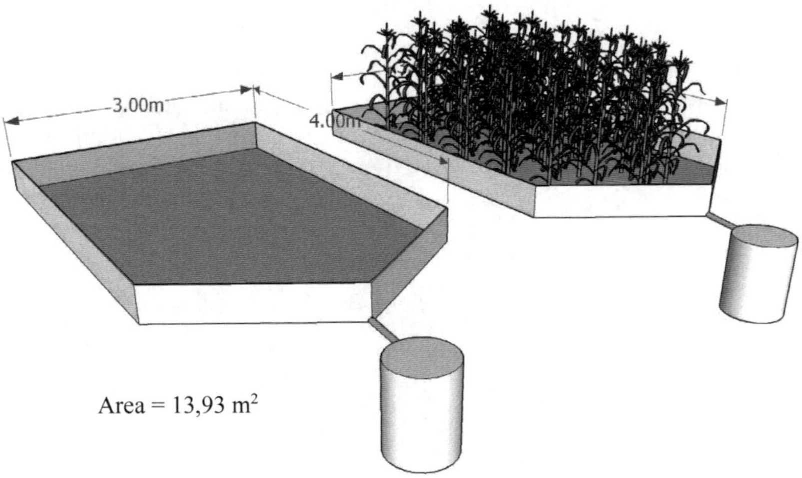

Figure 1. A mini hydrological watershed plan, without corn-crop (left) and with corn-crop (right).

During the process of rain, surface runoff water flowing in the Mini Hirology Watershed will gather at the observation outlet point which is then accommodated in a storage container. The water from the storage is then calculated by using a water meter. Furthermore, direct runoff discharge calculations are based on the length of time the rain occurs. Based on figure 1 we can see that there are two types of Mini Hydrological Watersheds where the first is the Mini Hydrological Watershed with corn plants which serves to see the amount of interception/water in the tree stand and return to the atmosphere assuming the water that passes as runoff. Whereas the second Mini Hydrology Watershed without corn plants is used for infiltration measurement. While secondary data used in this study is hydrological data. Hydrological data in the form of rainfall data recorded on the Rainfall Recording Station from 2009 to 2018 [4]. The data was obtained from the Station of the Meteorology, Climatology and Geology Agency (BMKG) Indrapuri District of Aceh Besar.

2.2. Rain interception research method

The research was conducted on 1-3 months old corn plants. Corn plants 1-3 months old have a height of 40-150 cm, canopy diameter 50-100 cm and stem diameter of 2-5 cm. determining the amount of rain interception value is determined by the method of volume balance approach as follows:

$$R = I + Q + I_f \tag{1}$$
$$I = R - Q - I_f \tag{2}$$

Where:

I = interception (mm/day);
R = rainfall (mm/day);
Q = runoff Discharge (m^3/s); and
I_f = infiltration (mm/day).

In this study the R rain data in the equation above is the total rainfall (gross participation) as in equation 2.19. Runoff discharge data Q represents the sum of the values of T_f and S_f in equation 2.19 after deducting the value of I_f. The I_f values are obtained by two methods by directly measuring using an infiltration meter (ring infiltrometer) and by using a Mini Hydrological Watershed model without land cover assuming zero evaporation.

IOP Conf. Series: Materials Science and Engineering **933** (2020) 012027 doi:10.1088/1757-899X/933/1/012027

2.3. Data analysis

Data analysis was performed to determine the value of rain interception in maize plants that have been measured in the field. The data analysed consisted of runoff data which is the sum of flow through fall and stemflow in volume units and infiltration data in volume units and rainfall data at the time of observations in the field as many as 19 events of rainfall in units of high rainfall. Based on the results of the rainfall that occurred at the time of the study observation, runoff data and infiltration data in volume units were converted into data in the highest rainfall unit after considering the canopy area of corn plants. The unit change is explained in the following equation:

$$\alpha = \frac{A_{corn-area}}{A_{total-area}} \quad (3)$$

$$Q_{vol.of\,corn} = \alpha\ x\ Q_{total\ vol} \quad (4)$$

$$Q_{mm\ corn} = \frac{Q_{vol.of\ corn}}{A_{corn}} \quad (5)$$

$$R = Q + I_f \quad (6)$$

$$I_{f\,vol.T} = Q_{vol.T} - R_{vol} \quad (7)$$

$$R_{vol} = R_{rain}\ \mathrm{x}\ A_T \quad (8)$$

$$I_{f\,vol.\ of\,corn} = \alpha\ x\ I_{f\,vol.T} \quad (9)$$

$$I_{f\ mm\ corn} = \frac{I_{f\ vol.of\ corn}}{A_{corn}} \quad (10)$$

After the runoff and infiltration data are converted into rain height units (mm), then using equation 3.2 can be obtained the value of interception in maize plants as follows:

$$I = R - (Q_{mm\ corn} + I_{f\,mm\ corn}) \quad (11)$$

Where:

$A_{corn-area}$	=	area planted with corn (m^2),
$A_{total\ area}$	=	total area (m^2),
α	=	comparison of the area of land planted with corn and the total area,
$Q_{vol.of\,corn}$	=	runoff in corn plants (m^3),
$Q_{mm\ corn}$	=	discharge due to corn plants (mm/day),
R	=	rain (mm/day),
Q	=	runoff (m^3),
I_f	=	infiltration (mm/day),
$I_{f\,vol.T}$	=	total volume infiltration (m^3),
$Q_{vol.T}$	=	total runoff (m^3),
R_{vol}	=	actual rain (m^3),
$I_{f\,vol.corn}$	=	infiltration affected by corn-crops (m^3),
$I_{f\,mm\ corn}$	=	infiltration affected by corn-crops (mm/day), and
I	=	interception (mm/day).

2.4. Infiltration estimation

Infiltration estimation formula is calculated by using the regression formula of the relationship between interception data and rain data from field measurements measured within 3 months of research or as much as 19 times of the rain incidence, interception and rain relationships can be seen in the equation below:

$$I = F(R) \quad (12)$$

Where:

I = Interception (mm/day);
R = Rain (mm/day);

3. Results and discussion

3.1. Mini hydrological watershed of corn plants

At the research location, preliminary preparations were carried out, such as the tools used and the selection of seedlings of corn which were used as the object of research and made adjustments in the selection of the types of corn plants that were dominantly used by the community at the study site. Land preparation and installation of research tools according to land preparation procedures from agriculture, with a spacing of 20 cm x 40 cm and the creation of a mini hydrology watershed, the watershed can accommodate the entire rain water that falls and flows at the outlet point which is then accommodated in a storage container or referred to as runoff discharge. The process of making Mini Hydrology Watershed and Mini Hydrology Watershed that has been planted with corn can be seen in figure 2 and 3.

Figure 2. The making of Mini Hirology Watershed.

Figure 3. Mini Corn Hydrology Watershed.

3.2. Runoff and rainfall discharge

The flow measurement of runoff and water runoff discharge is carried out by assuming or grouping in one flow, i.e. runoff discharge in the Mini Hydrology Watershed which is recorded manually every time a rain event occurs. Rain water that falls directly hit by tree trunks which then falls hit by the ground gathered at the outlet point is inserted into the reservoir to obtain the volume. In the Mini Hydrology Watershed without corn plants, the runoff volume is also measured by the same method. The runoff discharge data is first converted to get this value in height (mm). While the high rainfall results in the field are obtained directly on the reading in the rainfall gauge. If there is no rain, no data is collected. Research data collection activities can be seen in figure 4.

3.3. Infiltration analysis

The results are obtained through the amount of water that falls to the ground after hitting the tree trunk and also accommodated in the leaves of the tree. This value is also subtracted from the results of the study area in the mini watershed that is not covered by corn stalks. The area covered by corn stems in the mini watershed is 85% of the entire research area. Some of the rainwater that falls on the stems and leaves is evaporated and some of it is infiltrated. This process is carried out so that later the amount of interception due to maize in the Krueng Jreu Sub-watershed can be determined later.

IOP Conf. Series: Materials Science and Engineering **933** (2020) 012027 doi:10.1088/1757-899X/933/1/012027

Figure 4. The process of collecting runoff discharge data.

3.4. Rainfall interception estimation

From the time span used for this study, there were 19 rainy days. The process of analysing rainy day data is carried out to obtain a regression equation that is in accordance with the rainfall data that occurs. Where this regression is used to see the relationship between interceptions of rainfall data. Based on the analysis of rainfall interception estimation formula obtained logarithmic regression, logarithmic graphs can be seen in figure 5. Based on logarithmic regression, the height of rainfall that varies, ranging from 1.3 mm to 33.8 mm, with the characteristic rainfall that is light rain/drizzle to light. The highest rain interception value of 1.51 mm occurred on the first rainy day on January 8, 2019, with a rain height of 33.8 mm, while the lowest rain interception value of 0.25 mm occurred on the 10th rainy day on March 3, 2019 with high rain of 1.3 mm. Based on the calculation of rain interception on corn plants, the amount of intercepted rainwater is 10.16 mm from the total rainfall of 154.8 mm.

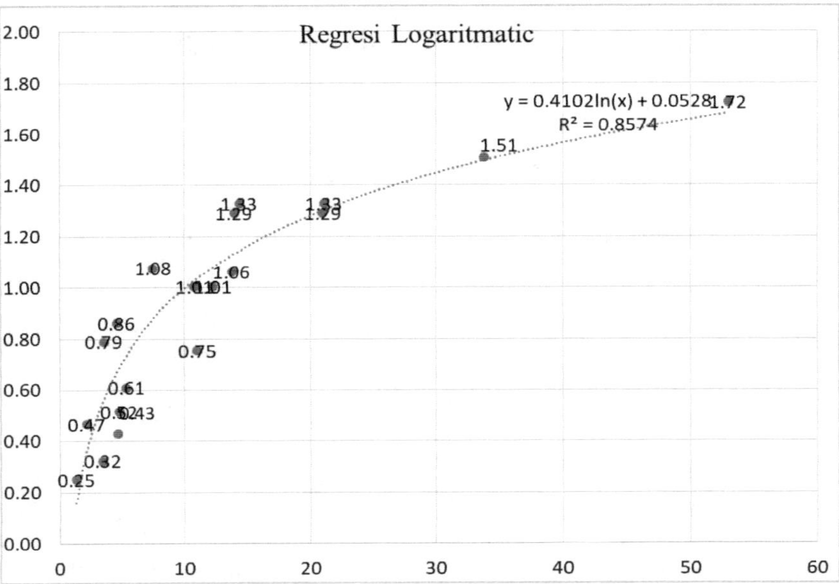

Figure 5. The relationship between rainfall and interception.

4. Conclusions

Corn plants have the ability to intercept smaller rainfall when compared with Arabica Coffee Plants and Palm Oil. The Arabica Coffee Plant studied by Benara was able to intercept rainwater by 62.33% and the Oil Palm plant studied by Meylis had a rainwater interception value of 63.43%. While the results of this study, the interception value of corn planting is 6.56%. Even though the corn plant has leaves which are arranged overlapping one another with each other, this does not make the absorption of water from the base of the stem to the stem become even greater. So, the interception value of the rainwater of the corn plants is much lower than that of Arabica Coffee and Palm Oil. This is also influenced by the size of the stems that are still small so as to minimize the absorption of water into the stems of plants. Thus, the runoff and infiltration discharge become greater.

References

[1] Asdak C 1995 Hydrology and watershed management (Yogyakarta: Gadjah Mada University Press)

[2] Pandey A, Mishra R M, Sahu D, Benara S K, Biswas M and Sengupta U et al 2012 Heterosexual risk behaviour among long distance truck drivers in India: role of marital status *Indian J. of Medical Research.* **136** 7 44

[3] Rao S M, Aubin-Faubert P S and Leo G J 1989 Information processing speed in patients with multiple sclerosis *J. of Clinical and Experimental Neuropsychology* **11** 4 471-7

[4] Yanto M, Rajagopalan B and Zagona E 2016 Space–time variability of Indonesian rainfall at inter-annual and multi-decadal time scales *Climate Dynamics* **47** 9-10 2975-89

IOP Conf. Series: Materials Science and Engineering **933** (2020) 012028 doi:10.1088/1757-899X/933/1/012028

A simple and safe brick masonry infill construction for high seismic-prone area

J Tanjung[1], F A Ismail[1], M Maidiawati[2] and Y Hayati[3]

[1] Civil Engineering Department, Universitas Andalas, Padang, Indonesia
[2] Civil Engineering Department, Padang Institute of Technology, Padang, Indonesia
[3] Civil Engineering Department, Universitas Syiah Kuala, Banda Aceh, Indonesia

Corresponding e-mail: jafriltanjung@eng.unand.ac.id

Abstract. West Sumatera Province has been known as one of the high seismic-prone areas in Indonesia. In this area, peoples commonly use the brick masonry as infill in the Reinforced Concrete (RC) frame structures. Field investigation after the earthquake informed that there were many bricks masonry infill collapsed caused by the earthquake motion. In this study, a simple and safe brick masonry infill construction in the RC frame structure is proposed for applying in the high seismic-prone area such as West Sumatera. The proposed construction method was defined based on the experimental works of the three 1/4 reduced-scale singe-bay and single-story RC frame specimens. The specimens include one specimen of RC frame infilled with brick masonry, one specimen of RC frame infilled with strengthened brick masonry by using chicken wire mesh, and one specimen of RC frame infilled with brick masonry infill and anchored to the columns with plain steel rebars. The wire mesh was sewn on both sides of the diagonal area of the brick masonry infill. All the specimens have subjected to lateral static reversed loading. The structural test results have shown that the strengthening using the chicken wire mesh increase the lateral strength of the RC frame about 22.%, while the use of the plain rebars anchor decreases the lateral strength about 1.8% and both methods can delay the failure of the brick masonry infill.

1. Introduction

The burned clay bricks masonry has commonly used as the construction material for wall or infills in RC frame structures in seismic-prone areas such as West Sumatra, Indonesia. Field investigation after the earthquakes showed that one of the causes of fatalities was the collapse of the brick masonry wall [1]-[2]. Improper brick masonry construction has been believed to be one of the causes of the collapse of its brick masonry. In this study is proposed a simple and safe construction for the brick masonry infill in RC frame structures. The construction is proposed based on the experimental results of laboratory tests of three RC frame specimens.

This simple and safe brick masonry construction is the brick masonry infills with was strengthening the weak areas in the brick masonry infill. That weak area is the diagonal region of the wall and the interface area between the wall and column. Indeed, the research on the strengthening of the brick masonry infill has been carried out by several researchers. Tanjung [3], for instance, has used the embedded plain steel on bed mortar joint to strengthen the brick masonry infill in the RC frame structure, while Ismail [4] uses wire mesh banded (ferrocement) to strengthen the unconfined brick masonry housing in Pariaman, West Sumatera. Leeanansaksiri [5] also used the ferrocement for straightening the brick masonry infilled in RC frame structures. The strengthening methods by using modern materials

IOP Conf. Series: Materials Science and Engineering **933** (2020) 012028 doi:10.1088/1757-899X/933/1/012028

such as textile-reinforced-mortar, welded wire mesh, and Carbon Fiber Reinforced Polymer (CFRP) have also been conducted and proposed by the researchers as were well summarized in [6]-[9]. Although their test results show that the strengthening materials and methods gave better brick masonry infill construction, unfortunately, the strengthening methods require the relative expensive materials and specialized labor for applying the strengthening.

2. Methodology

The experimental works which are explained in the paper were conducted by utilizing the structural testing facilities at Structure and Construction Material Laboratory of Syiah Kuala University, Banda Aceh, Indonesia. Three 1/4 reduce-scaled one-bay and one-story RC frame specimens were prepared. The specimens represent the first story of the typical construction of low-rise RC frame structures in West Sumatra. The materials for constructing these specimens were collected from local markets in Banda Aceh, Indonesia. All the specimens were subjected to lateral static reversed cyclic loading.

2.1. RC frame specimens

Figure 1 shows the details of the geometry and reinforcement applied for all RC frame specimens. The columns of the RC frame were detailed to yield in flexure before shear failure. The dimension of the cross-section of columns was 125 mm x 125 mm and reinforced with 4D10 longitudinal bars and $\phi 4@50$ transverse hoops. The cross-section of the columns was designed by considering the scale reduction. The clear height of the columns was 750 mm. The dimension of top-beam was 200 mm wide, 200 mm deep and 1550 mm long and reinforced with 4D13 longitudinal bars and $\phi 6@50$ transverse stirrups. The dimension of the lower-beam was 700 mm wide, 150 mm deep and 1650 mm long and reinforced with 12D16 longitudinal bars and $\phi 6@50$ transverse stirrups.

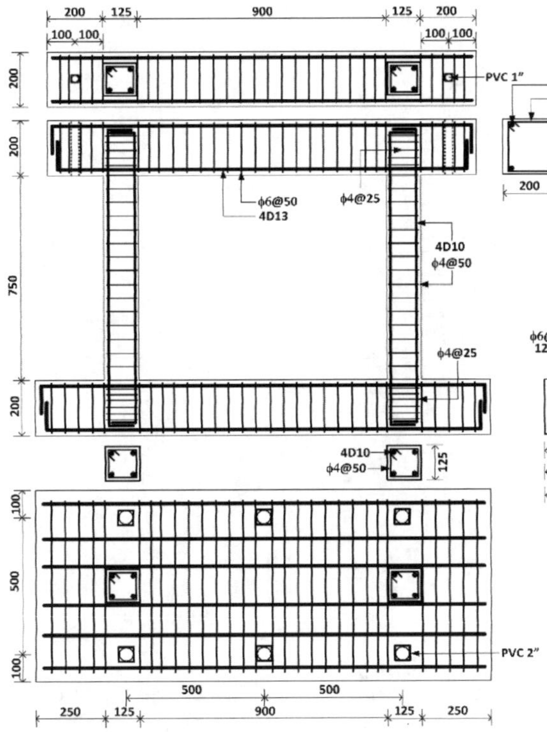

Figure 1. Reinforcements detail of the RC frame specimen (unit: mm).

IOP Conf. Series: Materials Science and Engineering **933** (2020) 012028 doi:10.1088/1757-899X/933/1/012028

Three 1/4 reduced-scale singe-bay and single-story RC frame specimens were built for experimental works. The specimens were a specimen of RC frame infilled with brick masonry (IF-SW), a specimen of RC frame infilled with strengthened brick masonry by using chicken wire mesh (IF-SM-A), and a specimen of RC frame infilled with brick masonry and anchored to the columns with plain steel rebars (IF-SM-C). The wire mesh was sewn on both sides of the diagonal area of the brick masonry wall. The chicken wire mesh was then fastened to columns and beams of RC frame by using chemical epoxy adhesive. The schematic of these RC frame specimens is shown in figure 2.

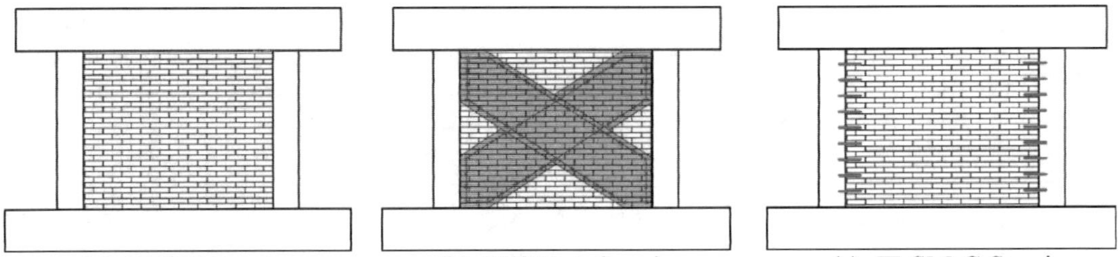

(a) IF-SW Specimen (b) IF-SM-A Specimen (c) IF-SM-C Specimen

Figure 2. Design the RC frame specimens.

2.2. The experimental setup and instrumentations

As was mentioned above, during experimental works, the prepared RC frame specimens were subjected to static lateral reversed cyclic loading. A schematically image of the experimental setup for the current study is shown in figure 3.a. For testing the specimen, at first, its specimen was placed on the rigid-floor. The lower-beam of the specimen was fastened to the rigid-floor by using six post-tensioning rods to keep the specimen remain in its position during the test. A double-action lateral actuator force equipment was attached and fastened to the strong wall by using four post-tensioning rods.

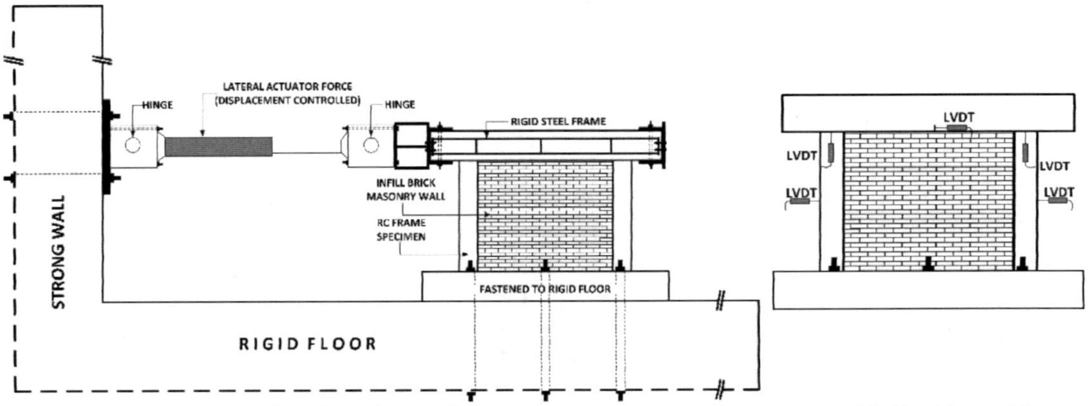

(a) Experimental Setup (b) Position of the LVDT

Figure 3. Experimental setup and instrumentations.

Two horizontal steel beams were used to restrain top-beam of the specimen from preventing the applied force on its top-beam causes out-of-plane deformation occurs during testing. These two horizontal steel beams were connected to the actuator force, which mounted on the strong wall. The displacement transducers (LVDTs) were installed at several points to measure the deformation of the

IOP Conf. Series: Materials Science and Engineering **933** (2020) 012028 doi:10.1088/1757-899X/933/1/012028

RC frame specimen, as is shown in figure 3.b. A displacement transducer which was placed in the middle of top-beam was used as a displacement-control point. The amount of lateral movement of the top-beam was defined based on the drift ratio of the column R= δ /H, where δ is the lateral displacement at a tip of the top-beam measured by the displacement transducer and H is the distance between the transducer and the bottom of the column. The loading program was R=1/800, R=1/400, R=1/200, R=1/100, R=1/50, R=1/25, R=1/12.5 rad., and followed by a pushover to R=+1/10. Except for the first drift ratio R=+1/800, two cycles were applied for each drift ratio [10].

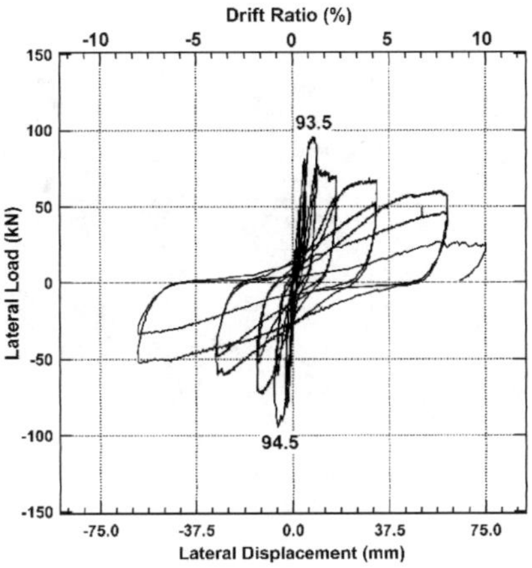

(a) IF-SW Specimen

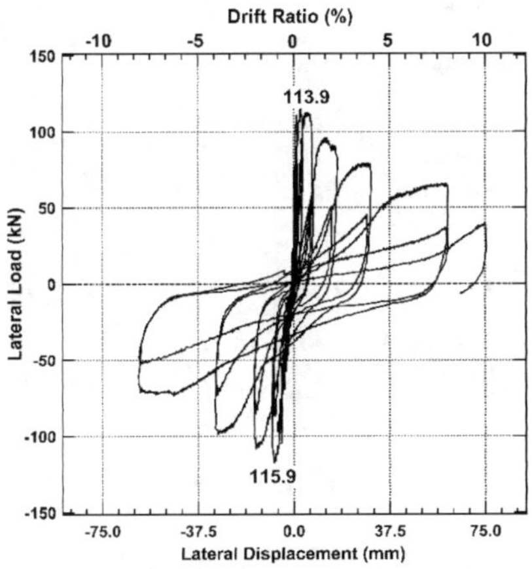

(b) IF-SM-A Specimen

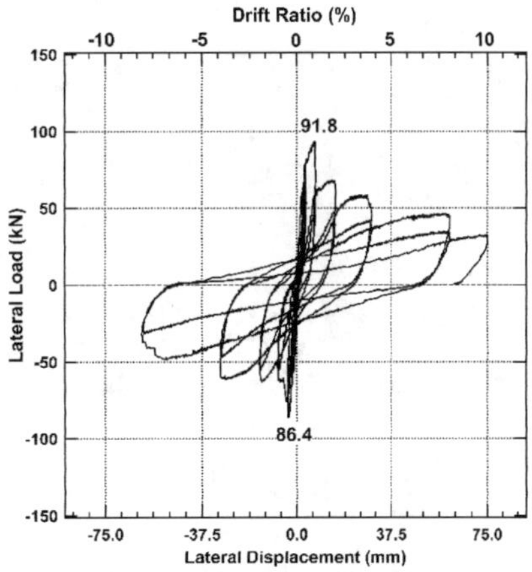

(c) IF-SM-C Specimen

Figure 4. Comparison of the load-displacement hysteresis curves.

3. Results and discussion

Figure 4.a shows the test results of the IF-SW specimen in the load-displacement hysteresis curve, while the crack pattern of its specimen is given in figure 5.a. The ultimate lateral strength was reached about 93.5 kN at the first cycle of R =+1/100 and 94.5 kN at R=-1/100. An initial crack on the interface between columns and brick masonry infill firstly appeared at R = +1/400 at a load of 30 kN and the initial flexural crack in the column has also been seen at it R= +1/400 due to the lateral load of 47.6 kN. Meanwhile, the brick masonry infill started to crack on it diagonal at the first cycle of R =+1/200 caused by the lateral loading of 78 kN.

Furthermore, the diagonal crack on the brick masonry infill spreads and widens when the specimen was loaded to 92 kN at cyclic R = +1/100. At the same time, the flexural cracks in the left column also increased. The shear crack on the upper of the left column occurred at the first cycle of R = +1/50, and the diagonal cracks on the brick masonry infill increased become wider than 5 mm. The surface plastering on the brick masonry infill started to peel off at R = +1/50 as it is shown in figure 5.a., then the diagonal crack on the brick masonry infill increased more than 10 mm wide when R = +1/25. Finally, the brick masonry infill collapsed at the second cycle of R = +1/12.5, and the transverse hoop in the left column ruptured when the specimen has been subjected a pushover loading to R = +1/10.

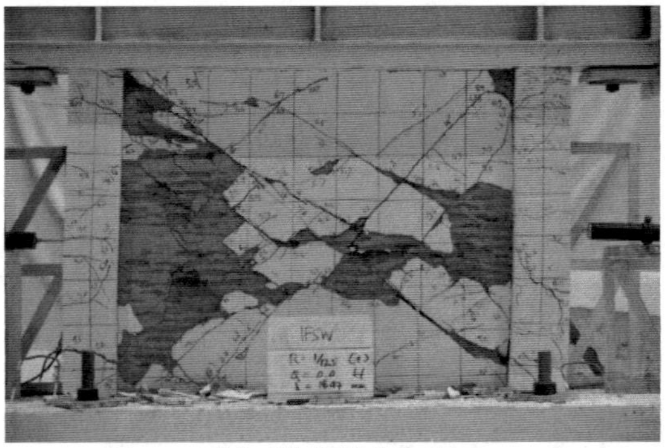

(a) IF-SW Specimen

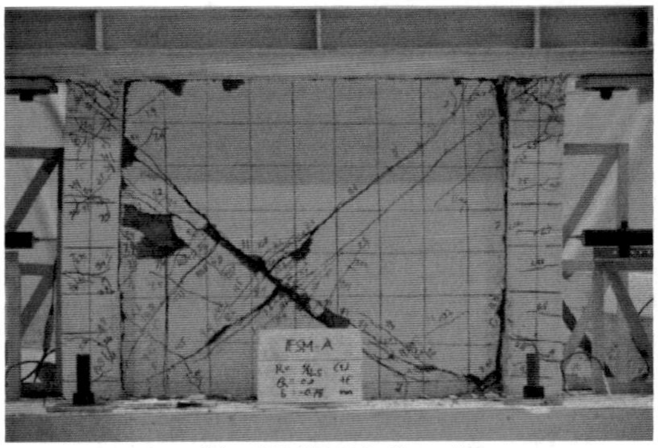

(b) IF-SM-A Specimen

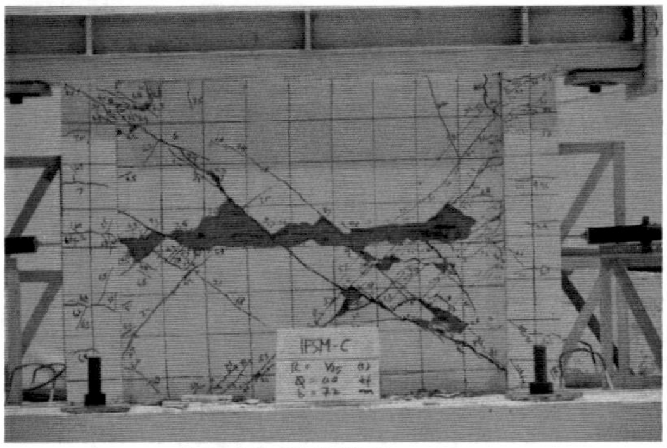

(c) IF-SM-C Specimen

Figure 5. Comparison of the crack pattern of the RC frame specimens.

The comparison of the load-displacement hysteresis curves based on the experimental results for all three brick masonry infill specimens is shown in Fig. 4. Compare to the test results of the IF-SW specimen, increasing of the lateral strength due to the strengthening of brick masonry infill of IF-SM-A specimen was 22.6%, while the IF-SM-C specimen decreased about 1.8%. Although the increase or decrease in its lateral strength is not significant, however, the presence of the strengthened brick masonry infill and anchors in the RC frame specimen have significantly changed the deformation behavior, crack pattern, and the progressive failure of the RC frame specimens.

The experimental results have shown that strengthen by using the chicken wire mesh succeeded in reducing cracks that occur in the brick masonry infill and delay the collapse occurred in the RC frame specimens. A flexible material of the wire mesh chicken works effectively against the tensile forces on the diagonal area of the brick masonry infill such that cracks in the brick masonry infill only appeared outside area where the chicken wire mesh was installed. Its cracks are clearly shown in figure 5.b. The diagonal cracks on the IF-SM-A specimen occurred at the cycle of R=+1/100; compare to the IF-SW specimen at the cycle of R=+1/200. The use of chicken wire mesh has confirmed to increase the strength of the brick masonry infill without reducing its ductility. The use of the plane steel anchors to reduce the weaknesses at the interface between the brick masonry infill and the columns have also shown the encouraging results. The shear cracks on the column for both specimen IF-SM-A and IF-SM-C occurred at the cycle of R=+1/25; compare to the IF-SW specimen at the cycle of R=+1/50. As it is shown in figure 5.c., almost no cracks occurred on the interface. On the other hand, there was almost no change in lateral forces due to the presence of the anchors.

4. Conclusion

The experimental results presented in this paper have shown that strengthening the brick masonry infill by using the stitched chicken wire mesh can increase the lateral strength of the RC frame specimens, increase the strength and delay the collapse of the brick masonry infill. The use of the anchors between brick masonry infill and columns effectively increase the strength of the interface between the brick masonry infill and the columns. As a result, the connection between the brick masonry infill and the columns become monolithic when receiving the lateral loads. The experimental results concluded that the combination of strengthening by using chicken wire mesh and anchoring the plain steel rebars between brick masonry infill and the columns make the brick masonry infill construction becomes simple and safe.

Acknowledgments

This research work is financially supported by the Institute of Research and Community Development of Andalas University through the Research Grand of Program Penelitian Skim Klaster Riset Percepatan Guru Besar (KRP2GB) with the contract number T/45/UN.16.17/PP.IS-KRP2GB/LPPM/2019. The authors also sincerely acknowledge to Professor Katsuki Takiguchi, Takiguchi Laboratory, Department of Mechanical and Environmental Informatics, Tokyo Institute of Technology, Japan; Dr. Abdullah, Civil Engineering Department, Syiah Kuala University; head and staffs of Structure and Construction Material Laboratory, Engineering Faculty, Syiah Kuala University, for supporting during prepared the specimens, provided the structural testing facilities and conducted the structural tests.

References

[1] Maidiawati and Sanada Y 2008 Investigation and analysis of buildings damaged during the September 2007 Sumatra, Indonesia Earthquakes *J. of Asian Arch. Building Engg. JAABE* **7** 2 371-8

[2] Tanjung J 2017 *Post-earthquake Investigation Report of Damaged Reinforced Concrete Buildings after Pidie-Jaya Earthquake 2016* (Padang: Civil Engineering Department, Andalas University, Unpublished Report)

[3] Tanjung J, Maidiawati and Nugroho F 2017 Experimental investigation of the seismic performance of the R/C frames with reinforced masonry infills *AIP Conf. Proc. 1892 020009* doi: 10.1063/1.5005640

[4] Ismail F A, Tanjung J, Hakam A, Fauzan and Boen T 2015 Plastered wire-mesh bandaged: an effective alternative technique for seismic strengthening of the unconfined brick masonry housing in Pariaman City, West Sumatera, Indonesia, *Int. J. Civil Eng. Tech. IJCIET*) **6** 7 44-52

[5] Leeanansaksiri A, Panyakapo P and Ruangrassamee A 2018 Seismic capacity of masonry infilled RC frame strengthening with expanded metal ferrocement *Eng. Struct.* **159** 110-127

[6] Akhoundia F, Vasconcelos G, Lourenço P, Silva L M, Cunha F and Fangueiro R 2018 In-plane behavior of cavity masonry infills and strengthening with textile reinforced mortar *Engi. Struct.* **156** 145-60

[7] Fagone M and Ranocchiai G 2018 Experimental investigation on out-of-plane behavior of masonry panels strengthened with CFRP sheets *Composites Part B* **150** 14-26

[8] Can Ö 2018 Investigation of seismic performance of in-plane aligned masonry panels strengthened with carbon fiber reinforced polymer *Const. Build. Mats.* **186** 854-62

[9] Guerreiro J, Proença J, Ferreira J G and Gago A 2018 Experimental characterization of in-plane behaviour of old masonry walls strengthened through the addition of CFRP reinforced render *Composites Part B* **148** 14–26
[10] FEMA-461 2007 *Interim Testing Protocols for Determining the Seismic Performance Characteristics of Structural and Nonstructural Components* Federal Emergency Management Agency.

IOP Conf. Series: Materials Science and Engineering **933** (2020) 012029 doi:10.1088/1757-899X/933/1/012029

Effects of Gondorukem addition on AC-WC pavement containing reclaimed asphalt pavement

E E Putri[1], T Kurniati[1], Y Yosritzal[1] and A D E Putra[1]

[1] Civil Engineering Department, Engineering Faculty, Universitas Andalas, Padang, Indonesia

Corresponding e-mail: elsaeka@eng.unand.ac.id

Abstract. Increasing the level of the road due to overlay, by adding additional layers continuously, will thicken the pavement layer and utilize the virgin materials. The dredged asphalt pavement materials can be reused, for example for new pavements, commonly called Reclaimed Asphalt Pavement (RAP). However, adding RAP on the new pavement result in a decrease in the physical properties of the pavement. Gondorukem is a sap that was chosen as an asphalt concrete additive ingredient that can absorb heat. The use of Gondorukem as additive was expected to increase the stability of the asphalt pavement. In this study, the Marshall characteristics of the mixture of asphalt concrete - wearing course (AC-WC) containing RAP and Gondorukem additives were tested. Asphalt concrete containing 35% of RAP and Gondorukem which content varied of 0%, 3%, 5%, 7%, and 10% was tested. The obtained stability value increased by 28.48%. The optimum Gondorukem addition into the asphalt was 3.2%; the optimum bitumen content was 6.45%.

1. Introduction

The use of recycled asphalt pavement nowadays become a popular issue. The usage of RAP may prevent the exploitation of the virgin aggregate. RAP is a stripped pavement or a pavement repacking material containing asphalt and aggregate. This material arises when asphalt pavement is peeled for reconstruction, resurfaced, or dug to access the embedded utility network underneath. When shelled, and filtered properly, RAP can contain high-grade and well-aggregated aggregates [1].

There are many benefits obtained from recycling asphalt pavement. Based on estimation, the asphalt pavements recycle saves the US taxpayer more than 2.5 billion USD annually, saves more than 60 million yd^3 of landfill space annually. In addition, it reduces the needs for natural resources and material transportation process [2].

RAP contains the aggregate material that has been used in the old pavement service life. Hence, it is recommended that the RAP be mixed with the new material to produce the new pavement structure. RAP can be used for 20% - 60% of the total pavement material. By mixing the RAP and the new material, the new pavement material still has the appropriate quality to serve in the next service life.

The use of RAP for AC-WC, AC-BC, and HRS-WC has been investigated by Putri et al. [3]. RAP was varied by 35%, 50%, and 60% for the three types of pavement. The use of 35% RAP on AC-WC pavements results in high stability values compared to the stability of the asphalt concrete containing 55% RAP and 60% RAP. Unfortunately, the performance of asphalt pavement containing RAP and virgin material is not as strong as the pavement containing the virgin material only [3].

Moreover, it is important to investigate the improvement of the performance of the pavement mixture containing RAP. Gondorukem is chosen as an additive material because its physical properties like rubber: soften when the temperature increases and harden when the temperature decreases. The use of Gondorukem for porous asphalt pavement containing all new virgin aggregates has been investigated by Putri et al. [4]. Based on the investigation, the Marshall stability increased by 63.3% [4].

Hence, this study aims to improve the AC-WC pavement mixture containing 35% RAP by adding Gondorukem. This study also aims to determine optimum Gondorukem content.

2. Methodology

2.1. Aggregate properties

The properties of aggregate used have been investigated in the Civil Engineering Department, Andalas University, which is presented in table 1.

Table 1 Aggregate parameter.

No	Parameter	Method	Value	Requirement as per SNI
1	Coarse aggregate density			
	a. Bulk density	SNI 1969-2008	2.53	
	b. Apparent density	SNI 1969-2009	2.59	2.5 – 3.0
	c. SSD density	SNI 1969-2010	2.56	
	d. Water absorption	SNI 1969-2011	0.83	max. 2
2	Fine aggregate density			
	a. Bulk density	SNI 1969-2008	2.47	
	b. Apparent density	SNI 1969-2009	2.81	2.5 – 3.0
	c. SSD density	SNI 1969-2010	2.59	
	d. Water absorption	SNI 1969-2011	4.98	
3	Mass volume			
	a. Lose mass volume	PB-0204-76	1.41	
	b. Shaking	PB-0204-77	1.55	
	c. Stabbing	PB-0204-78	1.60	
4	Aggregate adherence to asphalt, %	SNI 06-2439-1991	96	min. 95
5	Los Angeles test abrasion			
	a. New aggregate, %	SNI 247 : 2008	29.09%	max. 35
	b. RAP aggregate, %	SNI 247 : 2009	44.30%	max. 40
6	*Aggregate impact value* (AIV)			
	a. New aggregate, %	BS : 182 Part 111: 1990	7.18%	max. 35
	b. RAP aggregate, %	BS : 182 Part 111: 1990	31.72%	max. 40
7	*Aggregate crushing value* (ACV)			
	a. New aggregate, %	BS : 182 Part 111: 1990	26.02%	max. 35
	b. RAP aggregate, %	BS : 182 Part 111: 1990	35.70%	max. 40

Based on table 1, the properties of aggregate meet the requirements. Meanwhile, the first parameter property until the fourth one is to characterize their physical properties. The properties were suitable for pavement mixture. Meanwhile, based on the fifth until the seventh properties, it was observed that RAP aggregate has lower quality in terms of strength, such as abrasion test, impact test and crushing test. As a consequence, the usage of RAP to substitute the aggregate in the pavement can reduce the pavement quality.

Thus, the addition of Gondorukem in the asphalt was investigated. It is expected to increase the adhesion between aggregate and asphalt in order to protect the aggregate from oxidation, impact load, etc. during its service life. As a result, adding Gondorukem is expected to increase the Marshall properties, especially the Stability value.

2.2. Gondorukem

Gondorukem (resina colophonium), as shown in figure 1, is a product of sap from the lead in the tusam (pine) stem. Gondorukem is the result of the cleaning process of the residue obtained from the steam distillation of tusam sap. The result of the distillation is turpentine [1-2].

Figure 1. Gondorukem.

The result of the distillation of the sap becomes turpentine with the chemical element of C10H16. Gondorukem is obtained from the processing of pine sap from the tapping process. Table 1 shows the properties of Gondorukem.

Table 2. Properties of Gondorukem (colophony) [1-2].

No.	Characteristics	Colophony
1	Form	Liquid
2	Colour	White/ pale yellow
3	Solubility	Hydrophobic
4	Softening point	65 - 75 °C
5	Density (gr/cm^3)	1.11 - 1.12
6	Other specification	Adhering
7	Tensile strength (MPa)	0.077 ± 0.06
8	Tensile modulus (GPa)	0.009 ± 0.001
9	Max strain (%)	1.71 ± 0.03
10	Moisture (%)	2.42 - 3.93
11	Ash (%)	0.01 - 0.03
12	Fixed carbon (%)	0.01 - 0.05

Gondorukem is a product of the pinewood processing industry. Pine trees at a certain age can be tapped into the sap. When the pine sap is distilled, it will produce Gondorukem and turpentine. Generally, Gondorukem is used in the batik industry, while the turpentine oil is used as paint solvents. However, by the development of pine resin processing technology, it provides an opportunity to utilize products produced from pine sap as adhesives raw material, one of the ingredients in papermaking, printing ink, chewing gum, and others [5].

Gondorukem can be added to asphalt to produce the modified binder [6]. It has been investigated that Gondorukem can improve the conventional properties of the base bitumen such as penetration, softening point, and temperature susceptibility. Moreover, Gondorukem is applicable to road construction. The optimum percentage of modified binder Gondorukem in bitumen was achieved at 7%.

2.3. Asphalt content theories

To calculate the theoretical asphalt content needed, firstly the area of the aggregate particle of AC-WC pavement is determined in order to calculate the area of the surface aggregate to be coated by asphalt for sample testing. Then, the area of the aggregate surface obtained is used to calculate the theoretical asphalt content.

2.4. Optimum asphalt content

The optimum asphalt content is the asphalt content that yields maximum stability based on Marshall test, on condition that the other Marshall parameters, such as flow, VMA, VIM, and Marshall stability are within the standard. The optimum asphalt content for AC-WC pavement containing 35% RAP was used to produce the samples with various percentages of Gondorukem addition, between 0% until 10%.

2.5. Optimum Gondorukem content

Optimum Gondorukem content in the mixture is the content of Gondorukem to obtain the maximum stability and to fulfill the requirement of the other Marshall test parameters, such as flow, VIM, and VMA.

In general, the research is recapitulated as the following steps:

- Conducted the sieve analysis investigation for aggregate that contains 35% RAP;
- Calculated the theoretical optimum asphalt content using Surface Area Methods for aggregate containing 35% RAP;
- Investigated the optimum asphalt content by conducting Marshall testing for AC-WC pavement mixture containing 35% RAP;
- The samples were produced using the optimum asphalt content by adding various percentages of Gondorukem in asphalt in order to determine the optimum Gondorukem to the 35% RAP AC-WC mixture.

3. Results and discussion

3.1. Theoretical asphalt content

The surface area of the aggregate was determined in order to obtain the asphalt needed to coat the aggregate. In this study, 35% of the coarse aggregate was substituted with the RAP. The total surface area of the aggregate that was used can be seen in table 3.

Table 3. Surface area calculations.

Sieve No	Specification	Passing cumulative (%)	Retain of sieve (%)	% of each sieve	Surface Area (mm^2)
¾"	100	100	0	0	0
1/2"	90 – 100	95	5	5	7
3/8"	77 – 90	83.5	16.5	11.5	36.5
#4	53 – 69	61	39	22.5	72
#8	33 – 53	43	57	18	113
#16	21 – 40	30.5	69.5	12.5	205
#30	14 – 30	22	78	8.5	309.5
#50	9 – 22	15.5	84.5	6.5	530
#100	6 – 15	10.5	89.5	5	910
#200	4 – 9	6.5	93.5	4	728
pan	0	0	100	6.5	3997.5
Total					6908.5

The asphalt binder content can be calculated by using the empirical formula as follows:

$$P = S \, x \, K \, x \, T \tag{1}$$

$$SG\ for\ the\ used\ aggregates = \frac{100}{\frac{Coarse\ Agg.}{SG\ Coarse\ Agg.} + \frac{Fine\ Agg.}{SG\ Fine\ Agg.} + \frac{Filler}{SG\ Filler}} \quad (2)$$

Where:
P = Asphalt binder content before aggregate gradation determination;
S = The ratio of SG from the material standard, with the material used as seen on equation (2), determined by (2. 65/ (SG for the aggregates used);
K = Surface roughness index; and
T = The amount of asphalt based on total surface area method.

With a total surface area determined from aggregate gradation testing at 6908.5 cm^2 shown in table 3, a value of T equal to the amount of asphalt of 7.832% can be obtained from the total surface area method. The value of S of 1.049 and K of 0.95 are chosen because the surface type is considered to be sheathed slightly irregularly. Hence, the value of P calculated based on equation (1) is equal to 7.035%. It is required for the surface layer to have an air void of 0.3% - 0.5% to prevent the flow of the pavement or the loss of stability. Therefore, the optimum bitumen content theoretically is 7.4%.

After obtaining the theoretical optimum asphalt content of 7.4%, the asphalt content used was varied by 6.4%, 6.9%, 7.4%, 7.5%, and 8.4%, namely 1% below and 1% above the theoretical optimum asphalt content. For this stage, the Marshall sample is made for 15 pieces, where 3 pieces for each asphalt content to obtain optimum bitumen content by Marshall test.

The optimum bitumen content is then used for making samples to determine the optimum levels of Gondorukem for AC-WC pavements, which results in maximum Marshall stability. So, the AC-WC pavements containing the RAP have enough strength and durability during service life.

3.2. Optimum asphalt content obtained by Marshall parameter

To determine the optimum asphalt content for AC-WC pavements containing 35% RAP, Marshall tests were carried out for 15 samples. The results obtained can be seen in figure 2(a) until figure 2(f).

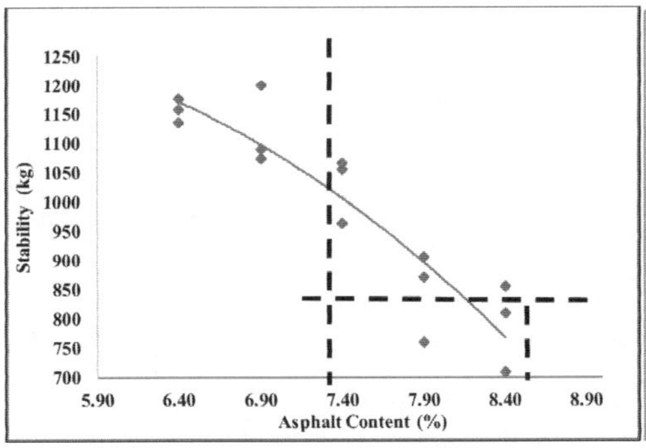

(a) Stability vs. Asphalt content

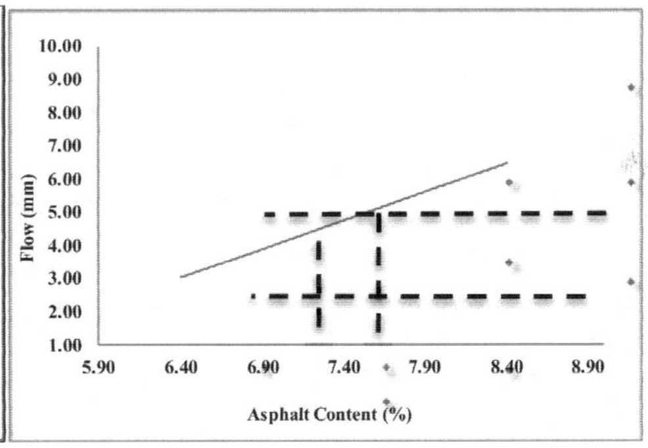

(b) Flow vs. Asphalt content

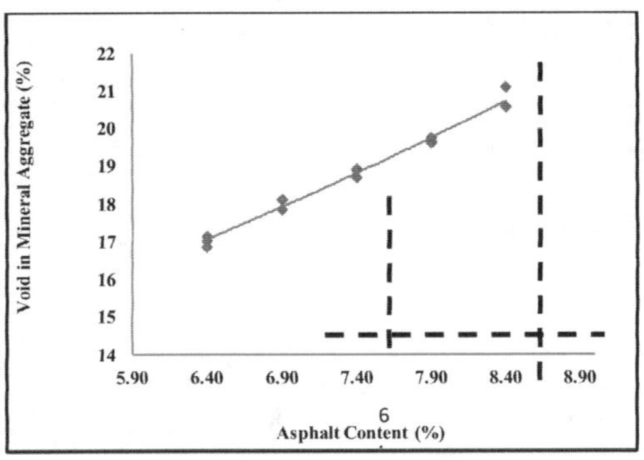

(c) VMA vs. Asphalt content

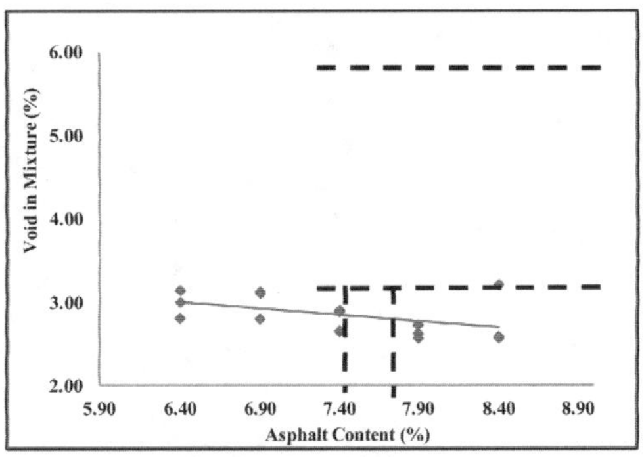

(d) VIM vs. Asphalt content

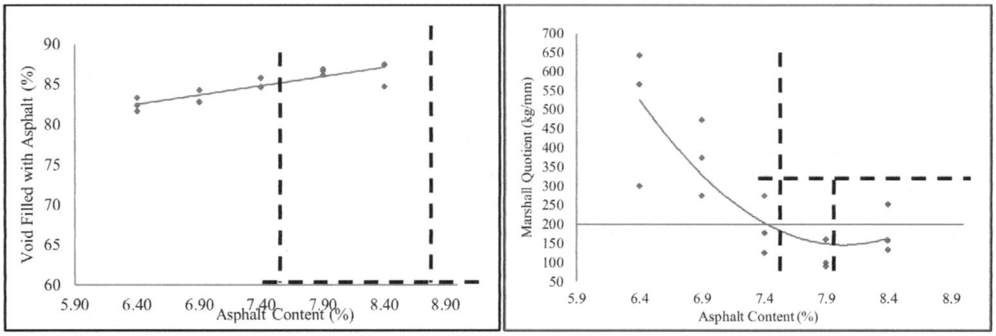

(e) VFA vs. Asphalt content

(f) MQ vs. Asphalt content

Figure 2. The Marshall test results of 35% RAP AC-WC for asphalt content optimum determination.

The stability test results were shown in figure 2(a). The results show that the stability increases with increasing asphalt content up to the optimum asphalt content and after that, it decreases. The stability for all samples was above the minimum standards, except for samples with overwhelming asphalt content. The asphalt content having maximum stability was 6.4%. Moreover, the flow values of asphalt mixtures against asphalt content are shown in figure 2(b). The asphalt content that resulted in Flow value within the standard was 6.4% to 6.92%.

Figure 2(c), 2(d) and 2(e) show the average VMA value, void in the mixture, void filled with asphalt and the above-mentioned parameters' relationship to the asphalt binder content. Figure 2(f) shows Marshall quotient of the sample which is the ratio of stability and flow.

Based on the Marshall test results from Fig 2(a) until Fig 2(f), the optimum asphalt content was found to be 6.4% for AC-WC pavement mixture containing 35% RAP. It is recommended because it provides maximum stability and sufficient VIM and VFB values. Hence, this optimum asphalt content will be used to prepare the sample to determine the optimum Gondorukem content for the 35% RAP of the AC-WC pavement.

3.3. Optimum Gondorukem content

To determine the optimum content of Gondorukem addition to the asphalt, the Marshall test was conducted. The Marshall parameter test results for 35% RAP AC-WC pavement for various Gondorukem addition were presented from figure 3(a) to 3(f).

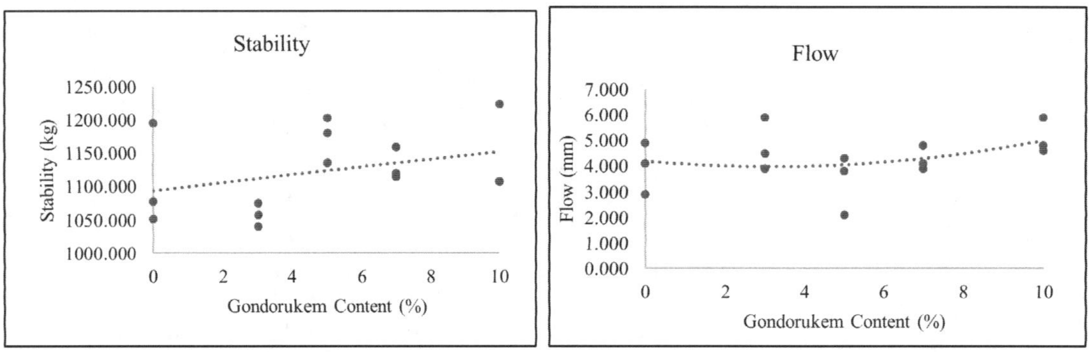

(a) Stability vs. Gondorukem content

(b) Flow vs. Gondorukem content

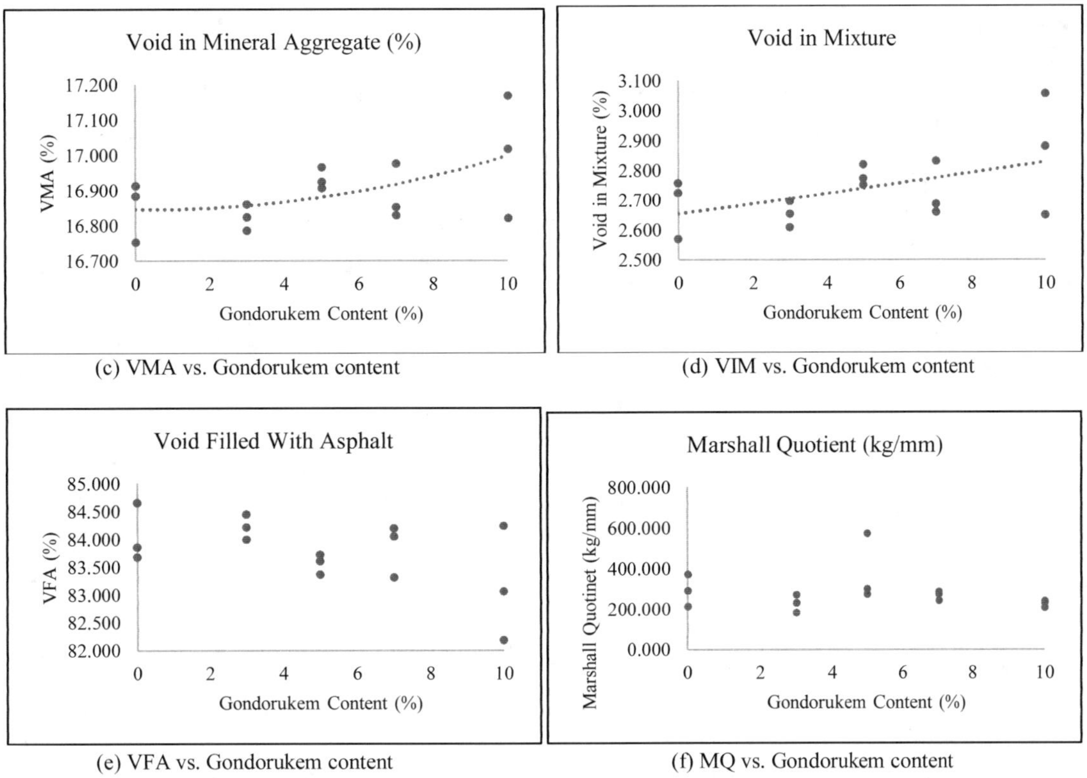

Figure 3. The Marshall test result vs Gondorukem content.

Figure 3(a) until figure 3(f) shows the results of various Gondorukem content in the Asphalt versus the Marshall Parameters, such as stability, flow, VIM, VMA, VFA, and MQ. The stability test results, as seen in Fig. 3(a), increases with the increasing of the Gondorukem content until the optimum content; and thereafter the value decreases. Moreover, it can be seen that the stability values for all the asphalt concrete containing the Gondorukem with various contents met the general specification from the Department of Public Works Indonesia, which is equal to or higher than 800 kg [7].

Meanwhile, in figure 3(b), it can be seen from the results that flow values are in the range of 3 to 5mm. Some data falls under the standard flow requirement values of about 2 to 4 mm.

As can be seen from figure 3(c), some of the voids in the mixture of the various Gondorukem value was not within the standard of the 2010 General Specification, which is 3 - 5%. Furthermore, the value of void in mineral aggregate falls under the standard requirement (figure 3 (d)).

Figure 3(e) shows the various Gondorukem content in the asphalt versus void filled with asphalt. The results show that all the VFA values were within the standard requirement, which is less than 65%. It should be noted that an increase in VFA means that there are more spaces to be filled with asphalt. This condition results in less strength of the asphalt concrete.

The 2010 General Specification requires that the Marshall quotient must be more than 250 kg/mm. As a result, the Gondorukem content must be between 0 – 9,1%, as shown in figure 5(f). From the overall results of the Marshall parameter investigation, the Gondorukem content in the asphalt was assessed. The Gondorukem content is chosen to fulfill the 2010 specification standard. The optimum value of Gondorukem contained in the AC-WC pavement mixture is 3.2% with the asphalt content of 6.45%. Thus, from all the Marshall parameter results, it can be concluded that the Gondorukem addition is strongly recommended for asphalt modification.

4. Conclusion

Based on the results of the laboratory investigation, it can be concluded as follows:

- The optimum asphalt content for 35% RAP for AC-WC pavement is 6.45%;
- The addition of Gondorukem tends to increase the strength and quality of the asphalt mixture. Using the asphalt mixed with the addition of the Gondorukem which percentage of total asphalt concrete weight is 2% to 10%, improve the Marshall Stability, flow, bulk density, void in the mix, and the Marshall quotient of the asphalt concrete. Results showed that Gondorukem rubber may increase stability by 28.48%;
- The optimum content of Gondorukem is 3.2% with 6.45% of asphalt content for AC-WC pavement mixture.

With the result obtained it can be concluded that the use of Gondorukem rubber as an additive in the asphalt for 35%RAP AC-WC pavement mixture is recommended as the Marshall properties results show good properties in their asphalt binder.

Acknowledgments
This research was supported by the Engineering Faculty of Andalas University grant under contract no. 025/UN.16.09.D/PL/2019.

References

[1] Wiyono B, Tachibana S and Tinambunan D 2006. Chemical compositions of pine resin, rosin, and turpentine oil from West Java. *Indonesian J. of Forestry Research* **3** 1 7-17
[2] Kencanawati C I P K, Suardana N P G, Sugita I K G and Suyasa I W B 2017 Characterization physical, mechanical, thermal and morphological properties of Colophony. *Intern. J. of Eng. and Emerging Techn.* **2** 2 17-19
[3] Putri E E, Purnawan, Triandila M A, Pratama A and Rindi 2018 Experimental study on use of reclaimed asphalt pavement as aggregate substitution for flexible pavement. *Matec Web of Conf.* p 229
[4] Putri E E, Idral M, Gungat L and Makinda J 2019 Marshall properties of porous asphalt with Gondorukem rubber addition, *J. of Eng. Science and Techn.* **14** 167-80
[5] Sjostrom E 1998 *Kimia Kayu: dasar-dasar dan penggunaannya* 2nd Edition *transl.* Hardjono Sastrohamidjojo H (Yogyakarta: Gadjah Mada University Press)
[6] Putri E E and Perdana P 2016 Properties of bitumen containing powdered Gondorukem rubber additives *Intern. J. of Earth Sciences and Eng.* **09** 03 149-53
[7] Directorate of Highways Indonesia *General Specifications, Revision 3* 2010

IOP Conf. Series: Materials Science and Engineering **933** (2020) 012030 doi:10.1088/1757-899X/933/1/012030

Predicting of land logistics travel demand in Aceh Province Indonesia

M Isya[1] and N Fadhly[1]

[1] Civil Engineering Department, Universitas Syiah Kuala, Darussalam Banda Aceh, Indonesia.

Corresponding E-mail: m_isya@unsyiah.ac.id

Abstract. In the process of planning a transportation system, it is important to predict transportation demand in the next few years. Predicting transportation demands for a wide area requires large resources including funding. Transport demand is usually done for the scope of the study area, such as formulating documents for the national transportation system, regional transportation system, local transportation system, and another similar one. One method of forecasting transportation demand is through transportation modeling. This paper is an analysis of land logistic transportation demand in Aceh Province Indonesia. Modeling is done based on data from the Origin-Destination survey of the movement of logistics in 2011, which has been calibrated to 2018. Finally, the prediction of the logistic movement is determined for land transportation mode in Aceh Province.

1. Introduction

Transportation is the activity of moving people or logistics from one place to another. Transportation infrastructure is one of the main infrastructures to improve the economy of a region [1]. Freight or logistics is also very important to support economic growth. Transportation has characteristics and attributes that show specific meaning and function. The main function is to connect humans with land use. Related to the transportation demand in a city and region, it is necessary to have good transportation planning to achieve efficiency and optimization of existing conditions [2].

Transportation of logistics in Aceh Province is still dominated by road transport, especially by trucking. The choice of shipping expedition logistics using truck mode is chosen by many freight forwarding service companies because it is not bound by time considering the delivery of logistics can be done at any time. Thus, truck transportation is the main choice of shipping logistics. More than 95% of the movement of logistics in Aceh Province uses road transportation modes [3]. Demand for land transportation for logistics can be seen based on the number of special vehicles in Aceh Province. Until now, the number of logistic vehicles in the form of 3,579 large trucks, while the types of small trucks are 24,489 units [4].

In the process of planning a transportation system, it is necessary to predict transportation demand in the next few years. Predict transportation demand in the next few years in a wide area requires large resources including funding. The scope of planning transportation systems is like the formulation of documents for the national transportation system, regional transportation system, local transportation system, and another similar one. One method of forecasting transportation demand is through transportation modeling [1]. In this paper, an analysis of the transportation demand for land logistics in the province is carried out, namely Aceh Province. Based on data from the Origin-Destination (O-D)

IOP Conf. Series: Materials Science and Engineering **933** (2020) 012030 doi:10.1088/1757-899X/933/1/012030

Matrix of the movement of logistics in 2011 that has been calibrated to 2018, projections of the distribution of trips in Aceh Province can be carried for land transport mode.

2. Transport demand modeling and predicting methods

Prediction of transport demand (traffic) is carried out through two approaches, namely the macro approach and the micro approach. The macro approach aims to obtain flows in general that enter or exit the study area and their influence on other parts of the network, while the micro approach aims to obtain micro-currents in the study area [1]. A four-stage transportation planning model will be used for the macro approach, because in addition to the ease of its ability to describe various interactions between the road transportation system and spatial planning in the study area.

Transport network data and zone system data are the main inputs in the four-stage transportation demand model. Transportation network data represents the supply and performance of transportation networks in the study area, while zone system data represents spatial characteristics in the study area and the socio-economic characteristics of the population in the spatial layout. The interaction between the two systems will be the main part analyzed in the four-stage transportation model.

Trip generation models (people and logistics) are a form of mathematical equations that represent the correlation between socioeconomic variables in the study area with the current reality of transportation or traffic (people/logistics), which can be obtained from National Origin-Destination (O-D) survey data. Based on the correlation of the relationship and predictions of the development of the region that is expected to occur, future travel demand can be predicted.

The trip generation model that is most often used in regional transportation studies is a multilinear regression analysis model, where travel demand (trip generation/attraction) as the dependent variable will be correlated with some socioeconomic data as independent variables, for example, population per zone, number office/trade / industrial floor area, and so on. The flow chart of the trip generation modeling process is presented in figure 1.

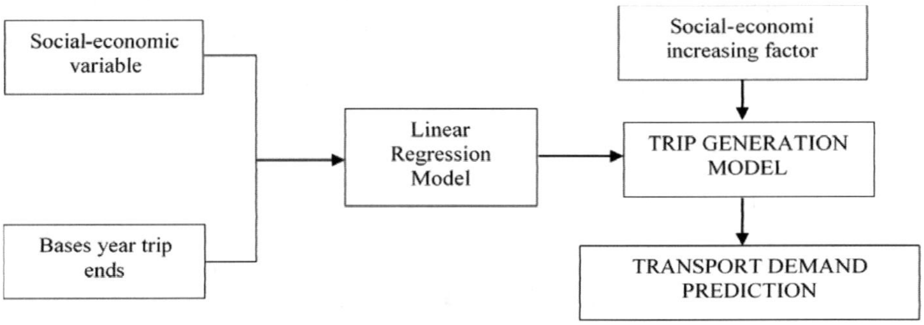

Figure 1. Trip generation and prediction process.

The trip distribution model is carried out to obtain the Origin-Destination (O-D) Matrix of the trip for every pair of zones in the study area. The trip data for each zone (trip end) has been obtained from the previous trip generation process. Preliminary data on O-D for the study area can be predicted based on population and employment data. This predicted data, then used as the base matrix (prior matrix) and are assumed to reflect travel patterns in the study area.

To get an O-D Matrix at year review, predictions are made according to the latest traffic data collected from the primary survey for road transportation modes. Furthermore, with the O-D data, this year the O-D prediction model was formed in the coming year with a Gravity or Furness model approach which is likely to be suitable for the condition of the study area.

Mode choice model is generally not established in this study, given its review in the city and the influence of modes other than roads can be ignored. In the subsequent analysis process, other modes other than roads will be seen qualitatively and quantitatively through the determined performance magnitude.

Trip assignment model also called route choice model will be done by using the software. Origin-destination Matrix will be distributed to the road network. The structure step of the trip assignment model is shown in figure 2.

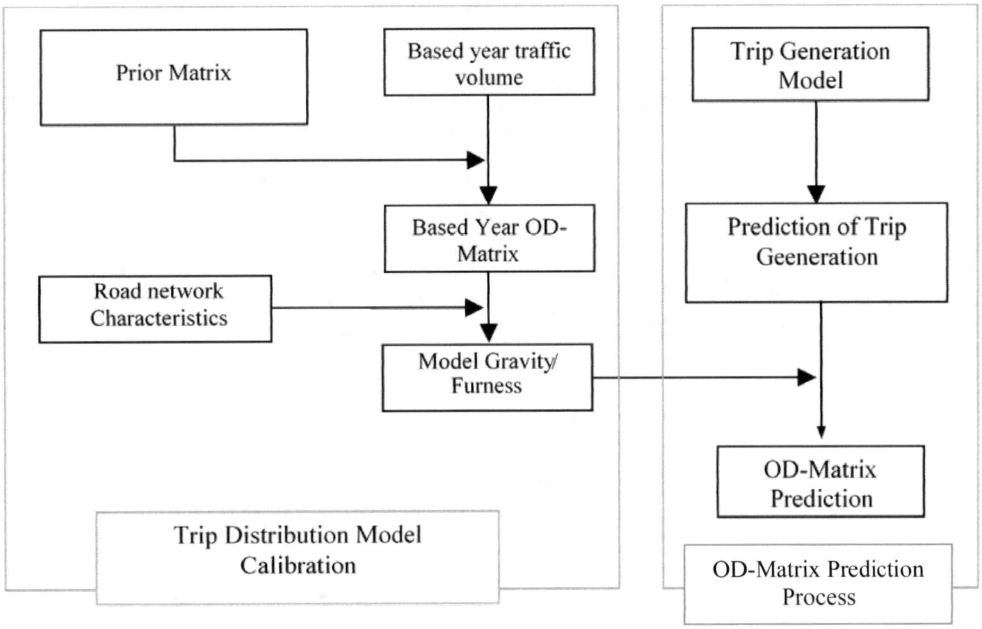

Figure 2. Trip distribution model and prediction process.

Trip Zone Modeling. The division of trip zones established at the location of activities is a zone with internal and external analysis units. The study area is divided into 23 internal zones and 13 external zones. The zone of internal movement between the center of activity consists of the activity center of each district/city while the external zone consists of the nodes of movement of the seaport. For more details about zones can be seen in figure 3 and related to table 1.

3. Results and discussion

3.1. Network system modelling

The developed road network system is pursued according to the level of development demand for multimodal transportation. In planning this multimodal transportation, the road network that is used as modeling is a road network with primary functions or a road network that connects activity centers. Based on the existing base map, a road network model was created at the location of the activity. The road networks in this plan are several roads that connect between zones of movement, both internal and external trips. In this road network modeling, there were 446 review links spread across 36 zones. By considering the possible impacts, the road network system model is formed to include the national and provincial roads in Aceh Province. The attribute data on this road network system consists of road length, basic capacity, speeds, and travel time. The road networks system can be seen in figure 4.

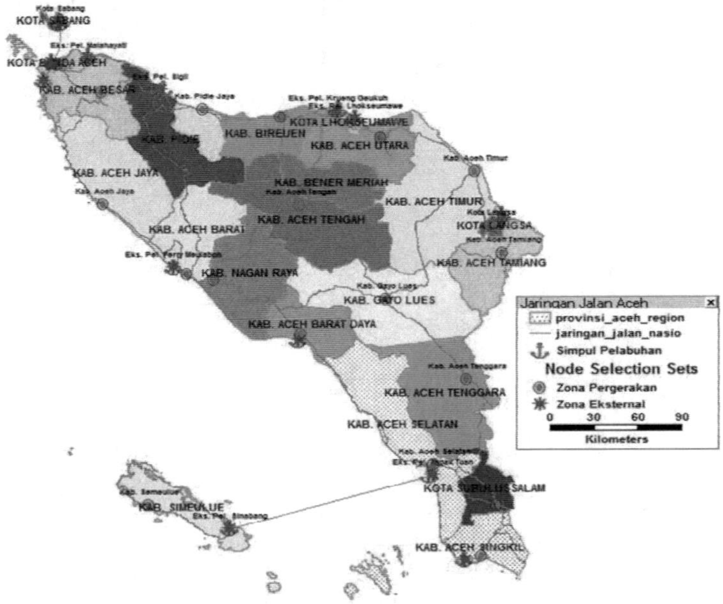

Figure 3. Zones in the studied area.

Table 1. The name of the internal and external zone.

No	Internal Zone	No	External Zone	Note
1	Kab. Aceh Barat	1	Eks. Pel. Ferry Meulaboh	The ports that are actively carrying out logistic loading and unloading activities are at the Ports of Pel. Lhokseumawe, Kuala Langsa, Ulee Lheue, Krueng Geukuh, and Malahayati
2	Kab. Aceh Barat Daya	2	Eks. Pel. Ferry Susoh	
3	Kab. Aceh Besar	3	Eks. Pel. Krueng Geukuh	
4	Kab. Aceh Jaya	4	Eks. Pel. Kuala Langsa	
5	Kab. Aceh Selatan	5	Eks. Pel. Lhoknga	
6	Kab. Aceh Singkil	6	Eks. Pel. Lhokseumawe	
7	Kab. Aceh Tamiang	7	Eks. Pel. Malahayati	
8	Kab. Aceh Tengah	8	Eks. Pel. Sabang	
9	Kab. Aceh Tenggara	9	Eks. Pel. Sigli	
10	Kab. Aceh Timur	10	Eks. Pel. Sinabang	
11	Kab. Aceh Utara	11	Eks. Pel. Singkel	
12	Kab. Bener Meriah	12	Eks. Pel. Tapak Tuan	
13	Kab. Bireuem	13	Eks. Pel. Ulee Lheue	
14	Kab. Gayo Lues			
15	Kab. Nagan Raya			
16	Kab. Pidie			
17	Kab. Pidie Jaya			
18	Kab. Semeulue			
19	Kota Banda Aceh			
20	Kota Langsa			
21	Kota Lhokseumawe			
22	Kota Sabang			
23	Kota Subulussalam			

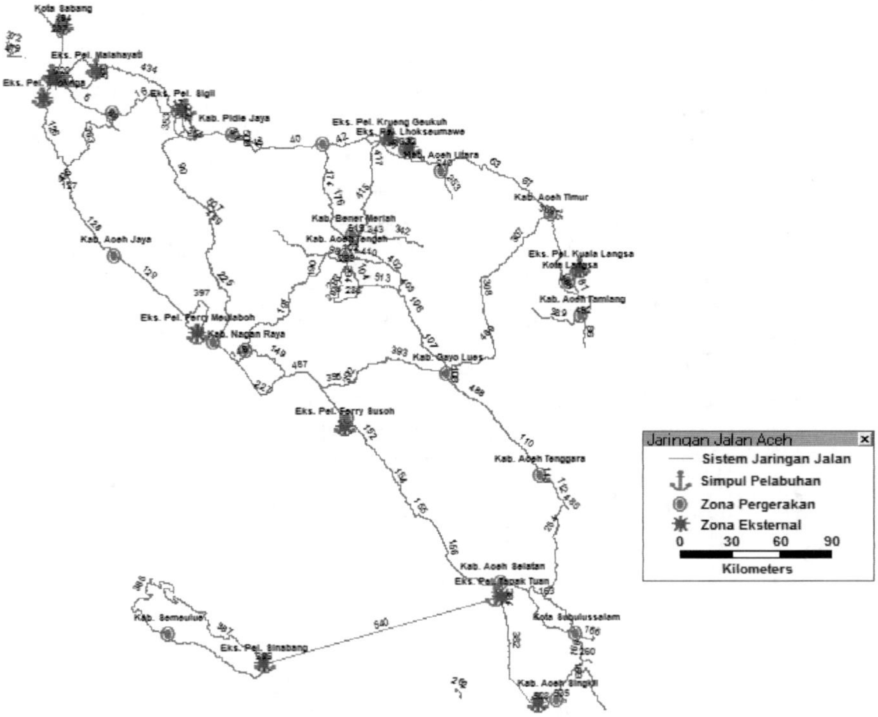

Figure 4. The model of the road network system.

3.2. Trip generation modelling

To determine the number of trips in the future at the location of the activity, the current model of logistic movement is determined in advance. Production and attraction trip number at the activity site using the 2011-year Origin-Destination (O-D) matrix data that has been calibrated to 2018 by using vehicle growth rates and linear regeneration methods. The results of trip generation modeling of logistic based on O-D Matrix can be seen in table 2 and table 3. Table 2 shows the trip generation of the internal zone and table 3 shows that the trip generation of the external zone.

Table 2 Generating internal zone of logistic in activity locations in 2018 (tons/year).

No	Internal Zone	Trip Generation	
		Trip Production	Trip Attraction
1	Aceh Selatan	1,526,554	1,521,277
2	Aceh Tenggara	1,050,573	997,261
3	Aceh Timur	4,963,349	4,428,763
4	Aceh Tengah	4,149,464	3,682,232
5	Aceh Barat	2,993,889	3,158,994
6	Aceh Besar	1,580,270	1,765,923
7	Pidie	3,877,621	3,622,706
8	Aceh Utara	5,003,986	5,002,193
9	Simeulue	459,193	492,081
10	Aceh Singkil	380,728	404,656
11	Bireun	4,205,993	4,125,737

12	Aceh Barat Daya	1,194,508	1,278,604
13	Gayo Lues	823,213	729,721
14	Aceh Jaya	957,515	901,242
15	Nagan Raya	1,659,697	1,597,003
16	Aceh Tamiang	330,032	296,360
17	Bener Meriah	1,207,076	1,378,343
18	Pidie Jaya	1,260,548	1,265,311
19	Kota Banda Aceh	2,039,463	2,271,011
20	Kota Sabang	218,510	265,923
21	Kota Lhokseumawe	1,798,491	2,216,090
22	Kota Langsa	1,396,892	1,647,480
23	Kota Subulussalam	250,624	279,277

Table 3 Generating external zone trips of logistic in activity locations in 2018 (tons/year).

No	External Zone	Trip Generation		Note
		Trip Production	Trip Attraction	
1	Eks. Pel. Ferry Susoh	-	-	The ports that are actively loading and unloading logistic are the Ports of Pel. Lhokseumawe, Kuala Langsa, Sabang, and Malahayati
2	Eks. Pel. Lhoknga	-	-	
3	Eks. Pel. Sigli	-	-	
4	Eks. Pel. Sinabang	-	-	
5	Eks. Pel. Tapak Tuan	-	-	
6	Eks. Pel. Sabang	25,228	49,775	
7	Eks. Pel. Fery Meulaboh	-	-	
8	Eks. Pel. Krueng Geukuh	-	-	
9	Eks. Pel. Kuala Langsa	6	20,661	
10	Eks. Pel. Lhokseumawe	257,544	873,336	
11	Eks. Pel. Malahayati	1,338,065	979,856	
12	Eks. Pel. Singkil	-	-	
13	Eks. Pel. Ulee Lheue	-	-	

3.3. Trip distribution modelling

Trips distribution describes the pattern of travel from the origin and to a destination zone. Analysis of trip distribution is made based on the results of data processing from the destination between the movement zones that have been adjusted to the center of activity. Modeling the trip distribution of people and logistics in the current conditions uses the gravity model method. This model takes into account the magnitude of accessibility between movement zones. The higher the accessibility, the greater the movement between the zones. The results of modeling the movement distribution with the gravity model method are mentioned in table 4 as an example. Predictions of road network loading had been completed for 2023, 2028, 2033 and 2038.

Based on table 4 on can be seen briefly that the number of movements of the O-D matrix in the location of the activity, it can be seen that the movement patterns are mostly still towards the activity centers (provincial and district capitals cities). Most of the trips go to Banda Aceh, Lhokseumawe City, Meulaboh, Pidie, East Aceh and North Aceh). While the high movement to the seaport transportation nodes (Sinabang, Sigli, Singkil, Ulee Lheue, Krueng Geukuh, and ferry ports is still relatively low).

3.4. Trip assignment modelling

The modeling of road network assignment was used as the equilibrium method. The state of equilibrium is assumed that the traffic flow will stand itself in the road network in such a way that no one driver will be able to reduce their travel cost by changing to another route. In other words, in equilibrium conditions, all selected routes have the same costs, while non-selected routes have the same or greater costs.

Table 4 The result of trip distribution modeling for the 2023 year as an example (tons/year) (1/2).

The Name of Zone	Eks. Pel. Ferry Maulaboh	Eks. Pel. Ferry Susoh	Eks. Pel. Krueng Geukuh	Eks. Pel. Kuala Langsa	Eks. Pel. Lhoknga	Eks. Pel. Lhokseu mawe	Eks. Pel. Malahaya ti	Eks. Pel. Sabang	Eks. Pel. Sigli	Eks. Pel. Sinabang	Eks. Pel. Singkil	Eks. Pel. Tapak Tuan	Eks. Pel. Ulee Lheue	Kab. Aceh Barat	Kab. Aceh Barat Daya	Kab. Aceh Besar	Kab. Aceh Jaya	Kab. Aceh Selatan
Eks. Pel. Ferry Maulaboh	-	2	2	2	2	11	2	2	2	2	2	2	2	2	2	2	2	2
Eks. Pel. Ferry Susoh	2	-	2	2	2	2	2	2	2	2	2	2	2	2	2	2	2	2
Eks. Pel. Krueng Geukuh	2	2	-	2	2	12	2	2	2	2	2	2	2	2	2	2	2	2
Eks. Pel. Kuala Langsa	2	2	2	-	2	14	2	2	2	2	2	2	2	2	2	2	2	2
Eks. Pel. Lhoknga	2	2	2	2	-	3	2	2	2	2	2	2	2	2	2	2	2	2
Eks. Pel. Lhokseumawe	11	2	12	14	2	-	2	9	2	6	18	8	20	20	21	23	24	26
Eks. Pel. Malahayati	3	2	2	2	2	2	-	3	3	3	2	3	3	3	3	3	3	3
Eks. Pel. Sabang	2	2	2	2	2	9	2	-	2	2	2	2	2	2	2	2	2	2
Eks. Pel. Sigli	2	2	2	2	2	5	2	2	-	2	2	2	2	2	2	2	2	2
Eks. Pel. Sinabang	2	2	2	2	2	6	2	2	2	-	2	2	2	2	2	2	2	2
Eks. Pel. Singkil	2	2	2	2	2	2	2	2	2	2	-	2	2	2	2	2	2	2
Eks. Pel. Tapak Tuan	2	2	2	2	2	8	2	2	2	2	2	-	2	2	2	2	2	2
Eks. Pel. Ulee Lheue	2	2	2	2	2	2	2	2	2	2	3	2	-	2	2	2	2	2
Kab. Aceh Barat	2	2	2	2	2	2	2	2	2	2	3	2	3	-	38	58	36	26
Kab. Aceh Barat Daya	2	2	2	2	2	2	2	2	2	2	3	2	3	38	-	3	7	28
Kab. Aceh Besar	2	2	2	2	2	2	2	2	2	2	3	2	3	73	3	-	6	20
Kab. Aceh Jaya	2	2	2	2	2	2	2	2	2	2	3	2	3	32	7	6	-	6
Kab. Aceh Selatan	2	2	2	2	2	2	2	2	2	2	3	2	3	27	42	24	8	-
Kab. Aceh Singkil	2	2	2	2	2	2	2	2	2	2	3	2	3	6	5	4	2	12
Kab. Aceh Tamiang	2	2	2	2	2	2	2	2	2	2	3	2	3	4	3	3	3	4
Kab. Aceh Tengah	2	2	2	2	2	2	2	2	2	2	3	2	3	87	40	72	23	30
Kab. Aceh Tenggara	2	2	2	2	2	2	2	2	2	2	3	2	3	15	13	15	4	36
Kab. Aceh Timur	2	2	2	2	2	2	2	2	2	2	3	2	3	57	25	51	15	38
Kab. Aceh Utara	2	2	2	2	2	2	2	2	2	2	3	2	3	61	30	14	18	45
Kab. Bener Meriah	2	2	2	2	2	2	2	2	2	2	3	2	3	17	9	3	6	8
Kab. Bireuen	2	2	2	2	2	2	2	2	2	2	3	2	3	60	28	21	20	29
Kab. Gayo Lues	2	2	2	2	2	2	2	2	2	2	3	2	3	14	15	3	3	20
Kab. Nagan Raya	2	2	2	2	2	2	2	2	2	2	3	2	3	166	16	4	12	12
kab. Pidie	2	2	2	2	2	2	2	2	2	2	3	2	3	79	22	120	37	29
Kab. Pidie Jaya	2	2	2	2	2	2	2	2	2	2	3	2	3	25	7	37	12	9
Kab. Simeulue	2	2	2	2	2	2	2	2	2	2	3	2	3	10	7	9	3	11
Kota Banda Aceh	2	2	2	2	2	2	2	2	2	2	3	2	3	27	12	17	16	14
Kota Langsa	2	2	2	2	2	2	2	2	2	2	3	2	3	18	8	3	5	11
Kota Lhokseumawe	2	2	2	2	2	2	2	2	2	2	3	2	3	25	8	4	6	9
Kota Sabang	2	2	2	2	2	2	2	2	2	2	3	2	3	3	3	2	2	2
Kota Subulussalam	2	2	2	2	2	2	2	2	2	2	3	2	3	4	3	3	3	8

Table 4 The result of trip distribution modeling for the 2023 year as an example (tons/year) (2/2).

The Name of Zone	Kab. Aceh Singkil	Kab. Aceh Tamiang	Kab. Aceh Tengah	Kab. Aceh Tenggara	Kab. Aceh Timur	Kab. Aceh Utara	Kab. Bener Meriah	Kab. Bireuen	Kab. Gayo Lues	Kab. Nagan Raya	Kab. Pidie	Kab. Pidie Jaya	Kab. Simeulue	Kota Banda Aceh	Kota Langsa	Kota Lhokseu mawe	Kota Sabang	Kota Subuluss alam
Eks. Pel. Ferry Maulaboh	2	2	2	2	2	2	2	2	2	2	2	2	2	2	2	2	2	2
Eks. Pel. Ferry Susoh	2	2	2	2	2	2	2	2	2	2	2	2	2	2	2	2	2	2
Eks. Pel. Krueng Geukuh	2	2	2	2	2	2	2	2	2	2	2	2	2	2	2	2	2	2
Eks. Pel. Kuala Langsa	2	2	2	2	2	2	2	2	2	2	2	2	2	2	2	2	2	2
Eks. Pel. Lhoknga	2	2	2	2	2	2	2	2	2	2	2	2	2	2	2	2	2	2
Eks. Pel. Lhokseumawe	27	29	30	32	33	35	36	38	40	41	47	46	15	53	52	55	50	3
Eks. Pel. Malahayati	4	4	4	4	4	4	5	5	5	5	6	2	7	7	7	6	6	3
Eks. Pel. Sabang	2	2	2	2	2	2	2	2	2	2	2	2	2	2	2	2	2	2
Eks. Pel. Sigli	2	2	2	2	2	2	2	2	2	2	2	2	2	2	2	2	2	2
Eks. Pel. Sinabang	2	2	2	2	2	2	2	2	2	2	2	2	2	2	2	2	2	2
Eks. Pel. Singkil	2	2	2	2	2	2	2	2	2	2	2	2	2	2	2	2	2	2
Eks. Pel. Tapak Tuan	2	2	2	2	2	2	2	2	2	2	2	2	2	2	2	2	2	2
Eks. Pel. Ulee Lheue	2	2	2	2	2	2	2	2	2	2	2	2	2	2	2	2	2	2
Kab. Aceh Barat	7	4	75	14	39	60	17	63	15	151	73	25	15	34	17	27	4	5
Kab. Aceh Barat Daya	5	3	32	10	22	28	9	25	15	19	22	8	7	12	12	9	4	4
Kab. Aceh Besar	5	3	53	12	52	10	3	21	3	5	85	29	9	21	3	6	2	3
Kab. Aceh Jaya	2	3	16	4	14	19	7	30	3	14	38	13	4	21	6	9	2	3
Kab. Aceh Selatan	12	3	33	29	29	43	8	27	13	16	22	8	9	14	19	12	3	8
Kab. Aceh Singkil	-	3	9	11	7	9	2	6	3	3	5	2	4	3	3	3	3	3
Kab. Aceh Tamiang	3	-	7	5	6	14	1	7	2	3	5	2	3	2	19	3	3	3
Kab. Aceh Tengah	7	6	-	21	99	148	116	159	17	42	72	24	13	41	29	53	4	5
Kab. Aceh Tenggara	12	4	17	-	17	34	4	16	12	6	18	6	5	10	20	8	3	8
Kab. Aceh Timur	7	9	99	27	-	436	35	163	21	21	72	24	7	31	40	143	4	5
Kab. Aceh Utara	13	13	131	32	383	-	37	143	26	26	83	28	14	36	80	109	5	9
Kab. Bener Meriah	3	3	109	5	19	23	-	55	3	8	19	7	2	7	6	12	3	3
Kab. Bireuen	6	5	143	14	156	146	48	-	17	38	130	44	7	59	34	108	5	4
Kab. Gayo Lues	3	3	27	10	16	28	5	22	-	7	12	4	4	7	11	5	3	2
Kab. Nagan Raya	2	2	47	9	22	30	12	35	6	-	23	8	5	11	8	10	2	2
kab. Pidie	7	4	62	14	73	76	26	114	11	25	-	13	9	189	79	33	8	5
Kab. Pidie Jaya	2	3	19	4	23	24	8	36	4	8	19	-	3	59	25	10	3	3
Kab. Simeulue	4	3	11	6	6	11	2	7	3	5	9	3	-	5	3	3	3	3
Kota Banda Aceh	4	3	26	9	32	37	6	54	5	13	163	56	4	-	10	14	23	3
Kota Langsa	4	15	20	16	36	82	5	27	9	6	58	20	4	10	-	12	3	3
Kota Lhokseumawe	2	2	40	6	125	69	14	84	6	9	29	10	3	14	13	-	2	3
Kota Sabang	3	3	3	2	2	5	1	4	3	3	9	3	3	17	3	2	-	3
Kota Subulussalam	3	2	6	3	5	6	1	4	2	2	4	3	2	2	2	2	3	-

3.5. Forecasting roads loading

Forecasting road network loading due to the movement of goods in Aceh Province is obtained from the modeling of road loading using the equilibrium (balance) method. Modeling predictions for road loading were carried out in 2023, 2028, 2033, and 2038. Changes in road network load due to the travel logistics in Aceh Province can be seen in figure 5.

IOP Conf. Series: Materials Science and Engineering **933** (2020) 012030 doi:10.1088/1757-899X/933/1/012030

Forecasting road network loading due to the movement of logistics in Aceh Province is obtained from the modeling of road network assignment using the equilibrium (balance) method. Modeling predictions for road network assignments were carried out for 2023, 2028, 2033, and 2038 year of prediction. The changes in travel demand at road network assignment due to travel of logistics in Aceh Province can be seen in figure 5.

The pattern of distribution of goods movement in Aceh Province from 2018 until the end of the planning year did not experience much change, this is because the center of movement (transportation node) did not change much. The change that occurred was an increase in the amount of logistic distribution. As a result of the increasing number of goods distributions, the traffic load on the road network has also increased. The development of road network loading in Aceh Province can be seen in the following sequential figure 5.

As seen at that picture that the road load due to the travel of logistics in Aceh Province is increasing in the eastern lane section. At the end of the planning year (2038) the logistic vehicle load in the eastern, middle, and western lanes is higher than any other road network. This means that the road network running parallel from south to north or opposite direction must be the focus of attention to facilitate logistics transportation to and from port outlets. Thus, the road routes that connect between the regional development cities must be the next focus to support the main lines that have been mentioned.

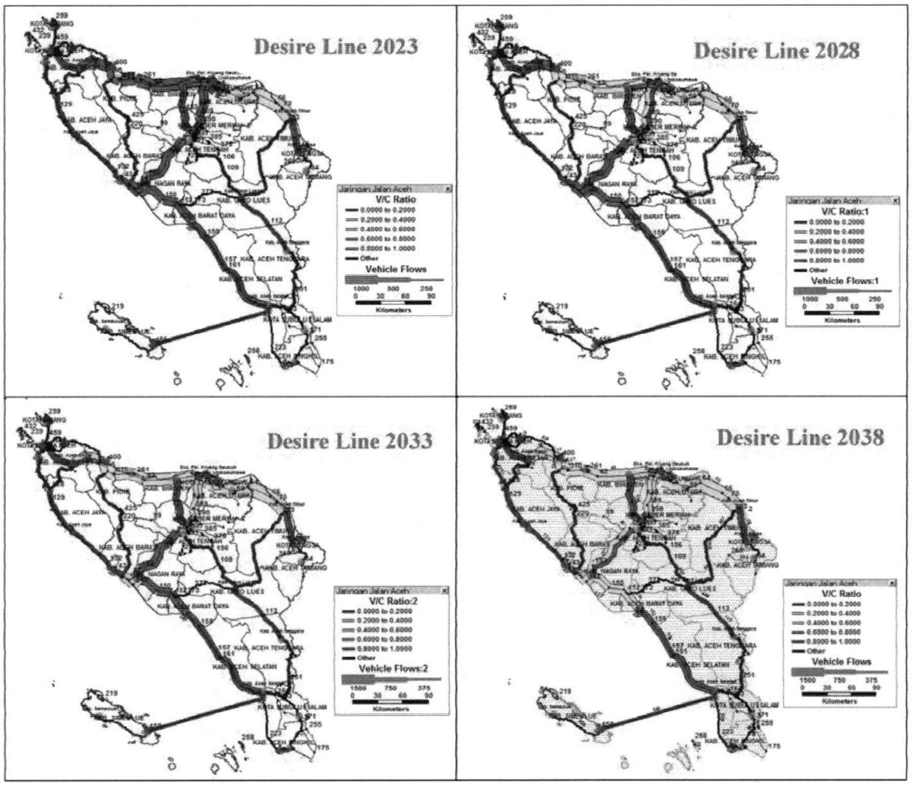

Figure 5. The traffic condition in the ratio of road volume to capacity at prediction years.

4. Conclusions

Based on the steps that have been done to predict the transport demand of logistic and the prediction of road performance, several points can be concluded and suggested, are:

1. Logistic travel demand at basic and prediction year can be determined by modeling techniques using data of origin and destination survey, socio-economic, socio-economic growth rates, transportation networks, road traffic by using four-steps modeling methods.
2. Analysis of travel distribution shows the potential path of Origin - Destination that can be the focus of attention in handling both the starting point and destination and on the path.
3. In the predicted years, the logistic movement in the east lane of Aceh will be more congested. For this reason, on the eastern lane, it is necessary to immediately construct a new transportation modal such as the railway and toll roads.
4. Similar studies with this study at the same location need to be carried out for air and sea transportation needs.

Acknowledgments
The author would like to thank the Aceh Provincial Transportation Agency and all colleagues who have contributed to this paper.

References
[1] Tamin O Z 2008. *Perencanaan, Permodelan dan Rekayasa Transportasi* (Bandung: Institut Teknologi Bandung)
[2] [Dephub RI] Departemen Perhubungan Republik Indonesia.1997. Keputusan Menteri Perhubungan Nomor KM.15 tahun 1997 tentang Sistem Transportasi Nasional (Sistranas), Dephub RI, Jakarta.
[3] Saleh, S M, *et al.* 2010." *Kebijakan Sistem Transportasi Barang Multimoda Di Provinsi Nanggroe Aceh Darussalam*", http://journal.unpar.ac.id/index.php/journaltransportasi/article/view/381/
[4] Jinca, M Y, 2010. "Keterpaduan Sistem Jaringan Antar Moda Transportasi Di Pulau Sulawesi" *Jurnal Transportasi* 9 (1): [29 November 2016] http://download.portalgaruda.org/article.php?article=95682&val=3917
[5] [Dishub Aceh] Dinas Perhubungan Aceh, 2018, *Laporan Akhir Cetak Biru Transportasi Antarmoda/Multimoda Aceh*, Dishub Aceh, Banda Aceh.
[6] [Kemenhub RI] Kementerian Perhubungan Republik Indonesia. 2012. Peraturan Menteri Perhubungan Republik Indonesia Nomor 8 tahun 2012. Penyelenggaraan & Pengusahaan Angkutan Multimoda. , Kemenhub RI., Jakarta.
[7] [Kemenhub RI] Kementerian Perhubungan Republik Indonesia. 2010. Peraturan Menteri Perhubungan Nomor : 15 Tahun 2010 Tentang Cetak Biru Transportasi Antar moda/Multimoda, Tahun 2010-2030. Kemenhub RI., Jakarta.
[8] Direktorat Jenderal Bina Marga, 1997, *Manual Kapasitas Jalan Indonesia*, Kementerian Pekerjaan Umum Direktorat Jenderal Bina Marga, Jakarta.

IOP Conf. Series: Materials Science and Engineering **933** (2020) 012031 doi:10.1088/1757-899X/933/1/012031

Builder perception on the implementation of earthquake resistant non-engineered house construction practice in Padang City

Y Hesna[1], **J Sunaryati**[1] **and P A Marlin**[2]

[1] Lecturer, Department of Civil Engineering, Faculty of Engineering, Andalas University, Padang, Indonesia
[2] Student, Department of Civil Engineering, Faculty of Engineering, Andalas University, Padang, Indonesia

Correspondence e-mail: yervi@eng.unand.ac.id

Abstract. Disaster preparedness from the builder is crucial due to the increase of earthquake vulnerability in Padang City. To ensure that the house is built following the concept of an earthquake-resistant non-engineered house must be followed in building the house to secure and minimizes the risk of casualties and losses caused by the earthquake. Previous experience shows that technical factors are not enough to increase the awareness of the builders. It requires non-technical factors that may play a significant role in disaster reduction in the community. That is why this study intends to identify builder perception about indicators of earthquake-resistant non-engineered house development and its implementation in the field. Questionnaires are used to gather the data to 30 housing projects in Padang City. In general, all statements about the concept of earthquake-resistant houses are strongly considered by the builders. However, it does not implement in the field run smoothly. Many construction practices found did not meet the principles. Government and stakeholder participation are needed to enhance the competency and knowledge of the homeowners and builders.

1. Introduction

People living in a dangerous city/location are becoming more vulnerable to natural hazards just by still living in that location. But the option to move the community to a new safe area is not easy because of the social and economic impact factor of the community. Hence the disaster mitigation program should be implemented to the target community. Some studies investigated that disasters can make a disturbance to communities and cause physical, social, economic, and environmental damage [1].

An effective disaster mitigation program cannot be achieved and implemented effectively without the participation of the community in general. Get involved by the vulnerable community in the disaster mitigation process is the prime component [2]. Newport et al. [2] also argued disaster mitigation also requires the involvement of government institutions and the NGOs the participation, not only the individual or group within the community.

With the increase of earthquake vulnerability, demanding preparedness from people in Padang City to ensure the houses built are following the concept of earthquake resistant. Research from Hesna and Hidayat [3] states that one of the factors that play a role in increasing community participation in the implementation of an earthquake-resistant non-engineered house is social capital found in the community. In this case, social bonding, linking, and networking between homeowners and builders.

But involving local communities in disaster mitigation and preparedness program cannot be realized easily [4]. Khan et al. [4] summarized that many interventions by the authorities, ignoring local input, and lack of coordination are the source of failure in a community-based disaster management program. The community members are also being excluded from the process of decision making [5]. Disaster experience and perception of risk are crucial in disaster preparedness, mitigation, and prevention program [4]. The perception and practices of builders need to be taken into account. However, it is not yet known how the influence of perceptions of builders on indicators of earthquake-resistant non-engineered house development and its implementation in the field. This research is carried out to fill this gap.

2. Theoretical background of non-engineered house

As it functions as a community residence, it is important to the non-engineered house to secure the inhabitant. The earthquake damage to the non-engineered house is of particular concern because it can harm the inhabitant [3]. From the observation in many experiences, the earthquake damage causes the largest number of infrastructure damage [6].

Non-engineered house is defined as those that are informally constructed in the traditional manner [7]. The owner and the builder work together without a formal contractual relationship. Another characteristic of the non-engineered house is that the house is built by a simple construction method, without any intervention from the engineer's structural analysis and technologies [6]–[8]. Boen [9] divided the non-engineered house into two main categories. The non-engineered house usually uses local material and follow local tradition and culture, generally called indigenous or vernacular house. The second categories are urban-type masonry construction. It is designated as the first category, but not adopting the traditional craftsmanship and materials. Such buildings involve fieldstone, fired brick, concrete blocks, wood, or a combination of these materials informing the wall construction [7]. Cement, lime, or clay mud are used for the mortars. This type of masonry house will include load-bearing masonry wall buildings, either confined or unconfined [9].

In Indonesia, confined masonry is the most common construction method used for single-story houses. This type of construction method has many advantages, such as lower construction cost, less engineering input required, and lower skill of worker needed. Basic concepts of confined masonry are all four sides of masonry walls confined by reinforced concrete members – bond beams and tie columns. Masonry walls constructed first, tie columns, and bond beams are poured in afterward to confine the wall. Last, walls assumed to carry all vertical and lateral loads.

3. Research method

A questionnaire is an information-gathering technique to study the attitudes, beliefs, behaviors, and characteristics of the respondents concerned about the problems raised. The questionnaire was addressed to the builders in the construction of earthquake-resistant non-engineered houses in Padang City.

The questionnaires about the perception of builders on an earthquake-resistant non-engineered house are compiled and adapted from the reference book [10], [11]. The questionnaires consist of three parts. First, it is about information about builders. The second, builder perception on Basic Principles of Earthquake Resistant Non-Engineered House. The third, builder perception of Specification and Main Structure of Earthquake Resistant Non-Engineered House. Complete indicators for builder perception can be seen in table 1 in the appendix.

Data collection began in July 2018 until December 2018, based on a purposive sampling method. The number of samples in this study is 30 residential development projects whose distribution of location can be seen in the following figure.

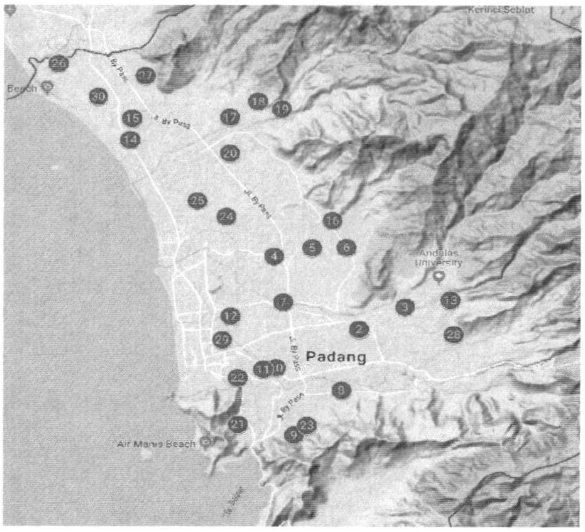

Figure 1. Project location distribution.

Generally, the houses built are categorized as small houses, with the size of the house can be seen in figure 2.

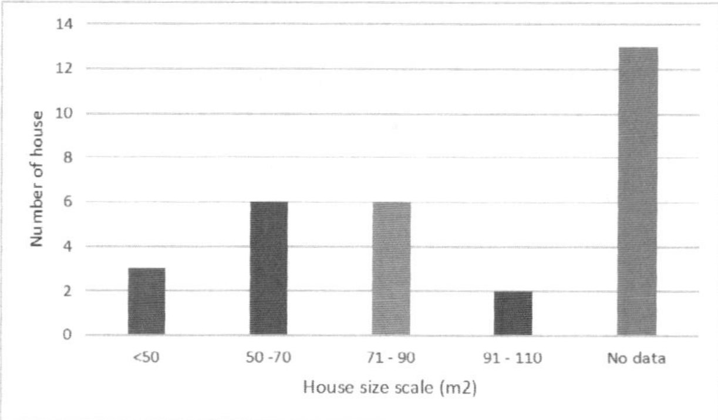

Figure 2. The number of houses based on the area.

An assessment of the perceptions of homeowners on the implementation of earthquake-resistant non-engineered house construction was carried out on a Likert scale with the following scores:

- Very Considerable (VC) = 4
- Consider (C) = 3
- Don't Know (DK) = 2
- Not considered (NC) = 1
- Very Not Considered (VNC) = 0

The respondent's answers were processed with descriptive statistics.

4. Respondent profile

In the field area, there are three positions of the builder. These are the head of builders, builders, and labor. We conducted this survey to the head of builders who are functioned as the director of builders. Most of the head of builders are in the age of 41 - 50 years old. The proportion of respondent's age can be seen from figure 3.

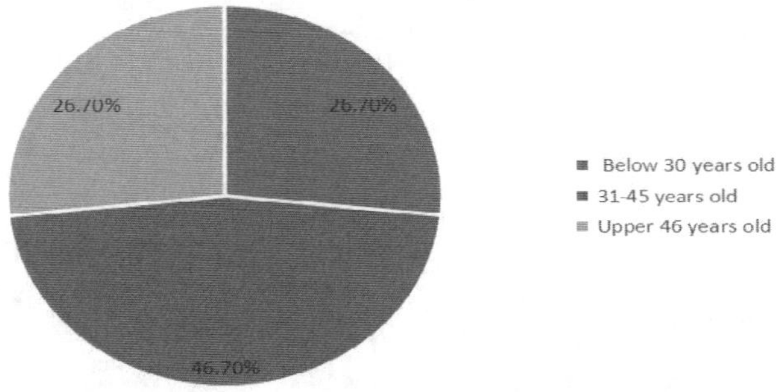

Figure 3. Builders age (in years).

The distribution of builders based on the educational background can be found in figure 4 below. Most of the builder has their education in Junior High School (JHS) and Senior High School (SHS). Only 6.7% of builders were in engineering vocational senior high (VHS).

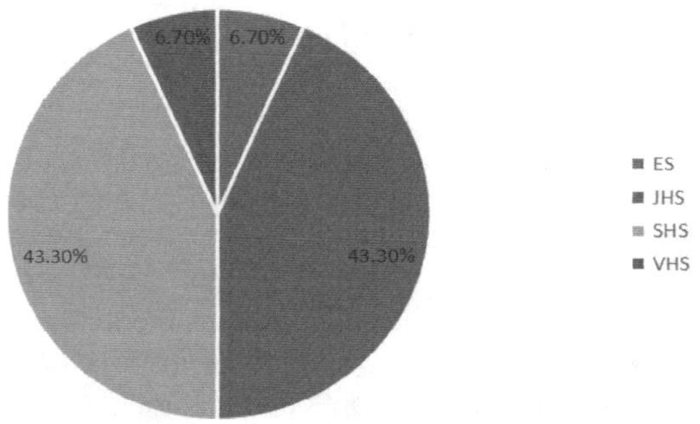

Figure 4. Educational background distribution of builders.

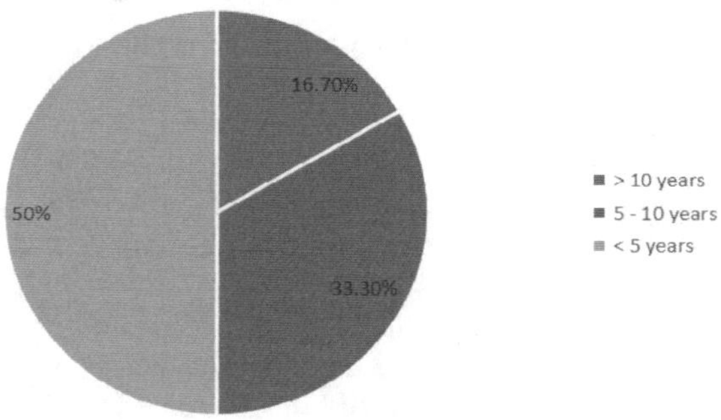

Figure 5. Work experience of builders.

Figure 5 presents the data of the work experience of builders. Overall, most of the builder has less experience as a builder in the construction of a non-engineered house. Only 16.7% of builders experienced with the construction of a non-engineered house.

5. Perception of the builder on earthquake resistant non-engineered house

For the basic principle of the concept of earthquake-resistant houses, get an average score of 4.89. Likewise, the use of building materials has an average score of 4.95. The main structural part has an average score of 4.85. So, it is concluded that all of the indicators of the non-engineered house get strongly considered by the builder. The builders consider the requirement of the basic principle, building material, and the main structure are important to implement at project locations to fulfill the principles of earthquake-resistant non-engineered house.

All projects were built on stable land, none of which were on sloping land. The houses made with a simple house plan. However, the window openings have not been placed systematically. And the relationship between the column-sloof is not very good because it is not given anchorage for the seismic hook.

However, in certain locations, there are discrepancies found, as shown in figure 6. The perception of the builders is different from the implementation. What found in the fields do not prove the builder's perception on the implementation of earthquake-resistant non-engineered house.

It found at the location the available gravel is irregular in size, and some gravels are more than 20 mm in size. Also, there was no apparent anchor between the pairs of bricks and columns. Likewise, in the implementation of casting, there are not many that impose casting each one meter gradually and are punctured so that casting is evenly distributed. So that it cannot be ascertained highly considered of builder's perceptions on an earthquake-resistant non-engineered house will be carried out well in the field.

Figure 6. Examples of problems on non-engineered house construction practice.

There are many mistakes in practice due to the lack of knowledge from the builders. This can be explained from the source of the builder's knowledge. All of the respondents have a different answer to the source of knowledge. Some of the respondents said that self-experience in developed his own house, but the majority get knowledge from the guidance of senior colleagues.

All respondents did not have attended competency training and did not have certain certificates of expertise. None of the respondents joined either certain institutions or associations in the construction sector. And they also rarely participate in socialization/workshop related to guidelines for earthquake-resistant non-engineered houses. All respondents did not use technical guidelines for implementation. According to respondents 4, specifications used usually follows the request of the homeowner. Whereas according to respondents 18, the specifications used are not far away different on each project, so it doesn't need specific guidelines at work. 90% of respondents knew the concept of earthquake-resistant houses from socialization; only 10% said they did not know.

This research confirmed the result from Juliafad et.al.[12], that the low quality of Reinforced Concrete building due to cultural ethic work. Because most builders do not have an engineering educational background, so their skills and knowledge are transferred by their senior. Juliafad et.al. [12] stated that the influence of previous experience is dominant compared to the contribution of knowledge from construction workshop. The builders tend to learn and adopt the same mistakes from their seniors.

6. Conclusion and recommendation

Based on the responses of the respondent, it was found that the builders highly consider all statements about the concept of the earthquake-resistant non-engineered house. Except for comments about gradual casting, which get mixed responses. However, the high consideration given by the builder's perception of earthquake-resistant house indicators does not implement in the field run smoothly. Many construction practices found did not meet the principles. This is due to the lack of understanding and knowledge of the builders. For this reason, socialization and education activities must continue.

This research concludes that the success of the implementation of an earthquake-resistant non-engineered house in Padang City is far away. Research by Bempah and Olav [8] is often attributed to the awareness of the house owners and builders, training, and transferring the proper technical skills to the builders as an effective method for non-engineered construction.

Besides that, it is interesting to explore in the next research about builder motivation and builder risk perception to take pre-disaster risk reduction action. Khan et al. [4] also argued that community risk perception has a significant impact on disaster preparedness as Alam et al. [13] also said that lack of motivation is the problem of the earthquake-resistant non-engineered house, even though the additional cost incurred are small.

Acknowledgment
The publication of this article is supported by the Engineering Faculty of Andalas University Grant No. 005/UN.16.09.D/PL/2019. Program Bantuan Seminar Dalam Negeri supports conference attendance for this publication.

References

[1] Martono M, Satino S, Nursalam N, Efendi F and Bushy A 2019 *Chinese J. Traumatol.* - English Ed. **22** 1 41–6
[2] Newport J K and Jawahar G G P 2003 *Disaster Prev. Manag.* **12** 1 33–6
[3] Hesna Y and Hidayat B 2019 *MATEC Web Conf.* vol 276 p 02004
[4] Khan M G and Baba A N 2017 *Jurnal Kemanusiaan* **26** 1
[5] Pearce L 2003 *Nat. Hazards* (Kluwer Academic Publishers) **28** 2–3 211–28
[6] Putra R R 2016 *Int. J. Geomate.* **7** 2 1076–83, 2016
[7] Arya A S 2000 Non-engineered construction in developing countries-an approach toward earthquake risk preduction *12th World Conf. on Eeartq. Eng.* 2824 pp 1–22
[8] Bempah S A and Øyhus A O 2017 *Int. J. Disaster Risk Reduct.* **23** 104–8
[9] Boen T and Pribadi K S 2007 *Engineering the Non Engineered Houses for Better Earthquake Resistance in Indonesia* Pt 8 pp. 202–8
[10] JICA and PU 2009 *Persyaratan Pokok Rumah Yang Lebih Aman*
[11] D. PU 2006 *Pedoman Teknis Rumah dan Bangunan Gedung Tahan Gempa*
[12] Juliafad E, Rani I G and Rifwan F 2019 *Int. J. on Adv. Sci., Eng. and Inf. Technol.* **9** 1 300-6
[13] Alam M J, Ahsan R, Akhter F and Paul A 2008 Earthquake resistant non-engineered building construction for rural area in Bangladesh *Proc. 14th World Conf. on Earthq. Eng. (Beijing)*

Appendix

Table 1. Indicators of Earthquake Resistant Non-Engineered House.

I	Basic Principles of Earthquake Resistant Non-Engineered House	VC	C	DK	NC	VNC	Score	Mean Score	Dev. St
1	The house is built on stable land	29	1	-	-	-	4.97	4.89	0.1
2	Simple building plan	25	3	2	-	-	4.77		
3	Placement of insulation walls and door/window openings are made symmetrical	26	3	1	-	-	4.83		
4	All building components (foundations, columns, beams, walls, roofs and roof trusses) are joined together	30	-	-	-	-	5.0		
II	**Building Materials**								
A	*Concrete*							4.95	0.05
1	Concrete mixture has a composition of 1: 2: 3: 0.5 (comparison of cement, sand, gravel, and water)	30	-	-	-	-	5.0		
2	The maximum size of gravel is 20 mm	25	4	1	-	-	4.77		
B	*Mortar*								
1	Mortar mixture has a composition of 1: 4 (cement and gravel)	30	-	-	-	-	5.0		
C	*River stone*								
1	The stone size is uniform (not too varied)	28	2	-	-	-	4.93		
2	The surface of the stone is rough, not smooth	30	-	-	-	-	5.0		
D	*Wood*								
1	The wood used is straight and dry	30	-	-	-	-	5.0		
2	There are not many wood eyes	29	1	-	-	-	4.97		
III	**Main Structure**								
A	*Foundation*							4.85	0.14
1	The arrangement of stones is made alternately	30	-	-	-	-	5.0		
2	The foundation is given anchor every 1 m from Sloof	23	7	-	-	-	4.77		
B	*Rebar*								
1	Curved bends form an angle of 45° at least 5 cm	29	1	-	-	-	4.97		
2	Connection of column, beam, and sloof	30	-	-	-	-	5.0		
C	*Casting*								
1	The concrete is watered periodically, and even the formwork is removed	30	-	-	-	-	5.0		
2	Column casting is carried out in stages every 1 m height	14	5	5	6	-	3.9		
3	Concrete casting is pierced with iron or bamboo, so the concrete is evenly distributed and not porous	30	-	-	-	-	5.0		
4	Column, formwork, sloof, and beams are opened after three days	29	1	-	-	-	4.97		
5	Hanging beam formwork is minimally opened after 14 days	30	-	-	-	-	5.0		
D	*Roof Structures*								
1	Each connection of the easel is given a steel clamp/plate clamp	23	6	-	1	-	4.7		
2	The connection of easel and beams is given iron anchor planted	30	-	-	-	-	5.0		
3	Wind ties bind mountains with horses	30	-	-	-	-	5.0		
E	*Walls*								
1	Installation of 1/2 alternating brick wall and anchored every six layers of brick	23	7	-	-	-	4.77		
2	The maximum wall area is 9 m2, and the maximum column distance is 3 m, give a practical column if more	30	-	-	-	-	5.0		

Joint operation of parallel reservoir with interconnected tunnel

K Iqbal[1], L M Limantara[2], W Soetopo[2] and U Andawayanti[2]

[1] Doctoral Program on the Civil Department, Faculty of Engineering, Universitas Brawijaya, Malang-Indonesia
[2] Department of Water Resources, Faculty of Engineering, Universitas Brawijaya, Malang-Indonesia

Corresponding e-mail: khairuliqbal@gmail.com

Abstract. Reservoir operation is strongly influenced by water balance conditions in the catchment area. The amount of inflow, reservoir capacity, and water requirement are major factors in arranging the reservoir operation rules. The purpose of this research is to obtain the extent of the parallel reservoir operation reliability that performed simultaneously by utilizing water transfer from wet to dry watershed. Simulation method used to operate the parallel reservoir with interconnecting tunnel between. From simulated results on three seasons, there are obtained that the water potential to be transferred to the dry watershed. By utilizing this water potential, further simulation for 25 years will come to see the reliability of the joint operating rule in this location.

1. Introduction

Management of Water Resources requires a comprehensive study indeed. The involvement of all sectors in the Watershed area is needed. Problems that often occur in management are flood prevention, water supply for irrigation, raw water availability, and energy. In its implementation, social problems also often arise related to the water allocation process. The solution to overcoming problems in management is by managing the water resources such as constructed dams. The practical application scenarios of reservoir operation are extremely complex and involve multiple time scales and multiflow regimes, often accompanied by occasional emergencies [1].

Multireservoir operating policies are usually defined by rules that specify either individual reservoir desired (target) storage volumes or desired (target) releases based on the time of year and the existing total storage volume in all reservoirs [2]. Reservoir Operation is largely determined by the function of the reservoir. Besides that, the management rule becomes a reference in decision making in the event of extreme conditions such as floods or droughts. Join operating rules are also proposed based on a water diversion rule, a hedging rule based on an aggregated reservoir, and a storage allocation rule [3].

To obtain the appropriate operating rule, a new set of release rules for each reservoir is made in the form of an analytical consideration of the optimal conditions of water balance and reservoir value in each reservoir [4]. Water Transfers between Watersheds are usually considered as one of the most effective facilities for balancing non-uniform temporal and spatial distributions between water resources and water requirements. This was done with the intention of diverting water from surplus areas to less areas. Several methods of operating reservoirs with additional water from other watersheds have long been developed. The development of a multilevel model of the pattern of multi-reservoir operations with the aim of minimizing runoff and maximizing the pattern of water availability with the transfer of water between watersheds to minimize the risk of water shortages [3]. A set of water transfers optimization between watersheds modeled through three stages to maximize profits from agricultural products on land and minimize water transfer costs [2].

The purpose of this paper is to show the reliability of joint operation rules from parallel reservoir with interconnected tunnel. Potential discharge that could be transferred becoming one basis of the utilization of two reservoirs through the interconnected tunnel. The potential discharge volume simulated from three seasons wet, normal and dry. Furthermore, the level of reservoir reliability shown by simulation for the next 25 years.

2. Case study

This study took the case of Rukoh and Tiro Dam that located in Pidie Regency Aceh province Indonesia. Tiro Dam has a Catchment Area of 174,24 km^2 with the main river has 46 km length. While the Rukoh Dam has a Catchment Area of 19,63 km^2 with the main river length was 6,87 km. the purpose of each dam was to provide raw water requirement for irrigation and cities downstream. The location of these two reservoirs is shown in figure 1.

The Baro Irrigation area of 11,950 ha where the irrigation demand at Intake 14.50 m^3/s taken from the Baro River. While the water demand of Tiro Irrigation area (6,330 ha) sourced from Tiro River. Dependable discharge of Baro River is 2.52 m^3/s, Tiro River is 1.89 m^3/s and Rukoh River 0.77 m^3/s. The condition of dependable discharge the condition caused only 6417 hectares of Baro Irrigation land covered. Likewise, the area of Tiro irrigation can only be served 2463 hectares. The condition is decreasing in the dry season [5].

By considering the rules to minimize potential conflicts of interest between the raw water and irrigation water, it would require suppletion from Krueng Tiro amounted to 52.30% [6]. With the potential of Rukoh Reservoir and interconnected with the potential discharge from Tiro River obtained a reservoir system that has a large potential to be developed. Schematic of the system shown in figure 2. Hopefully as a supply of irrigation water Kr. Baro covering 11,950 ha and irrigation Kr. Tiro Area of 6,330 ha. are expected to supply the irrigation water requirement of the Baro Irrigation Area of 11,950 ha and Tiro Irrigation Area of 6,330 ha. In addition, to fulfill the needs of the raw water, river maintenance flows, and power plants.

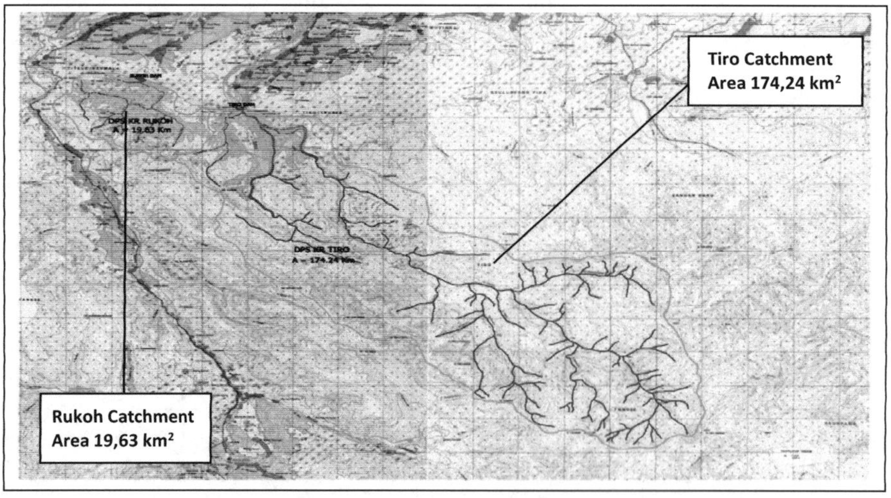

Figure 1. Maps of Rukoh and Tiro catchment area.

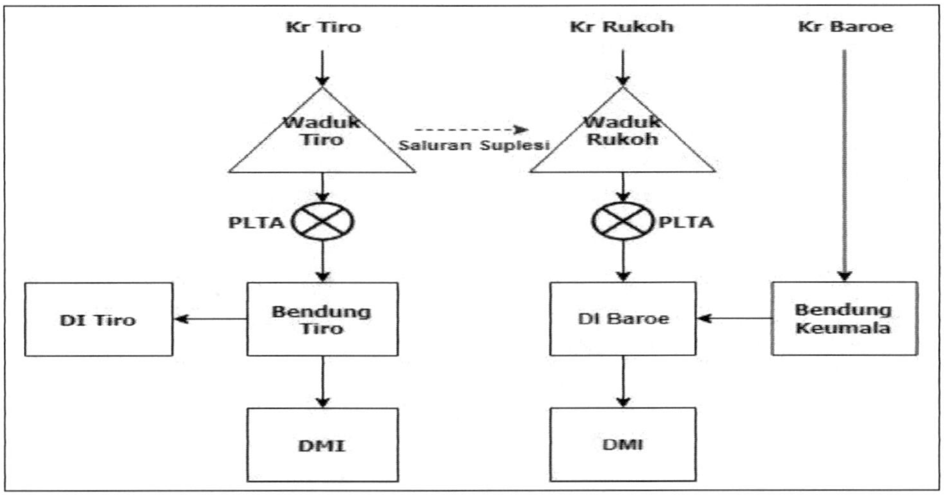

Figure 2. Scheme of Rukoh Tiro dam.

3. Simulation model

Simulation in water resources is a modeling technique that used to simulate and move the behavior of a system into a model with the help of a computer, to spread all the characteristics of the system widely with mathematical or algebraic description [7]. The simulation model could predict and show what will happen to a system at a certain moment when the systems are given certain inputs. Thus, management rule system could decide and define by studying the reaction of various system management scenarios without needed to have the system itself in real.

Application of the simulation model in reservoir system is a problem-solving technique for complex systems, in addition to analyze the model, the simulation can also be used for decision making process. The modelling of the reservoir operation for conservation purposes is based on the water balance for the time interval [8].

$$V_t = V_{t-1} + Total\ Inflow - Total\ Outflow \quad (1)$$

$$STr_t = STr_{t-1} + ITr_t - LTr_t - DivTr_t(x) - \sum DTr_t(y) - SpillTr_t\ (x,y) \quad (2)$$

$$SRk_t = SRk_{t-1} + IRk_t - LRk_t + DivTr_t(x) - \sum DRk_t(y) - SpillRk_t\ (x,y) \quad (3)$$

Where V_{t-1} is Reservoir volume in period t-$_1$ (m^3) and V_t is Reservoir volume in period t (m^3), *IR* is irrigation water requirement, S_t is initial storage in period t (m^3) and S_{t-1} is storage in the end of period t (m^3), I_t is Inflow period t (m^3), L_t is water loses in period t (m^3), Div_t is volume of water transfer in periode t (m^3), D_t is water demand in period t (m^3), $Spill_t$ is Volume of Spillway in period t (m^3), *Tr* is Tiro and *Rk* is Rukoh.

3.1. Boundaries condition

The main boundary of reservoir simulation is the reservoir capacity. At each stage of operation, the reservoir capacity is limited by the maximum volume of operation and minimum operating volume. The water Transfer from Tiro Reservoir is an additional inflow to Rukoh Reservoir. While the constraints on water transfer from the Tiro reservoir can only be done after fulfilling the water requirements in the downstream [9].

3.1.1. Storage capacity. The reservoir volume corresponds to the technical data of the Tiro Reservoir and the Rukoh Reservoir. The lower limit is the dead storage while the upper limit is the cap on the normal water level.

$$S_{i,min} \leq S_{i,t} \leq S_{i,max};\ S_t = T \quad (4)$$

Where $S_{i,min}$ Dead Storage in Reservoir i and $S_{i,max}$ Maximum Storage in Reservoir i.

In its application, the capacity of these two reservoirs is considered to be a system, so that the principle used as associated vessel. The minimum and optimum limit of the reservoir is the number of two reservoirs capacity. While the normal and maximum water level is equal. The volume of water that exceeds the maximum capacity will be released.

3.1.2. Water release. The amount of water released from the reservoir must meet the water demand in the downstream. The Volume of water requirement includes irrigation, raw water and river maintenance.

$$Q_R \geq D_t \quad (5)$$

Where Q_R is Release Debit from storage, D_t is Water demand

The volume of water requirement for irrigation is the total needs of both irrigation areas, Tiro irrigation area and Baro irrigation area. Land area of both irrigation areas that will be aggregated as a whole amount of water requirement.

3.1.3. Water Transfer in Joint Operation. Water transfer in joint operation connected with 120 meters long and 4,5 meters width tunnel. The maximum discharge value that can be transferred through a tunnel depends on the water elevation of the Tiro Reservoir.

$$0 \leq Div_{tr}(x) \leq Div_{max} \quad (6)$$

Where Div_{Tr} is Water Transfer and Div_{max} is Maximum Water Transfer Discharge.

3.2. Simulation operation

The simulation is done using Standard Operating Rule method [10]. In this simulation, the water requirement was divided into the needs of raw water and irrigation water demand. It is intended to be the need for raw water being the main priority to be fulfilled and the irrigation water demand to be the second priority of water release. Simulated conducted monthly on three conditions of the year wet (20%), normal year (50%), and dry year (80%). Figure 3 shows the simulated scheme performed.

4. Results and discussion

4.1. Potential of water transfer

To see the potential for the discharge of Kr Tiro that can be transferred to the Rukoh Reservoir, simulated the Tiro reservoir. The simulation is done in three seasons, dry year, normal year, and wet year. The simulation was conducted on a monthly basis by looking at the amount of discharge from the Tiro reservoir in all three seasons. The potential of transferred water based on simulation of three-season shown in figure 4.a to 4.c.

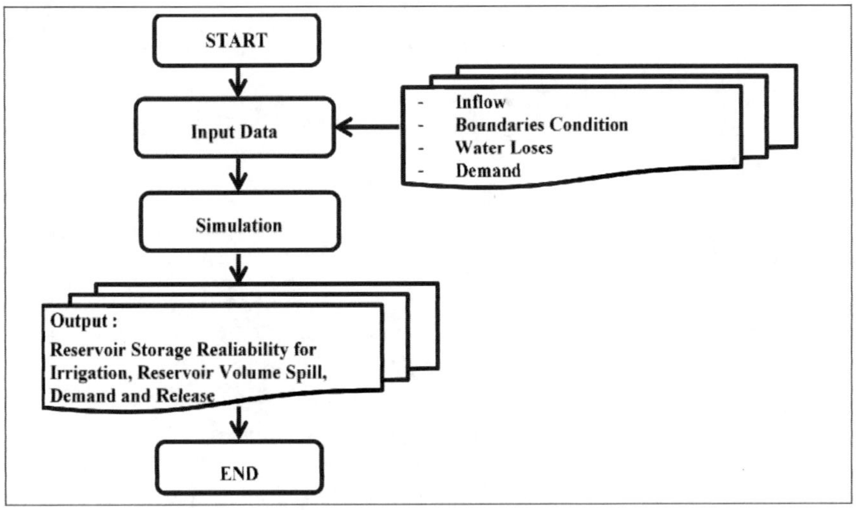

Figure 3. Simulation procedure.

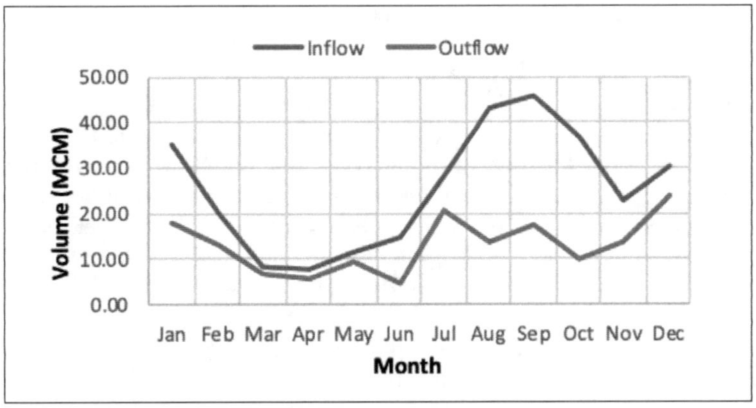

Figure 4.a. Simulation of Tiro reservoir in dry season.

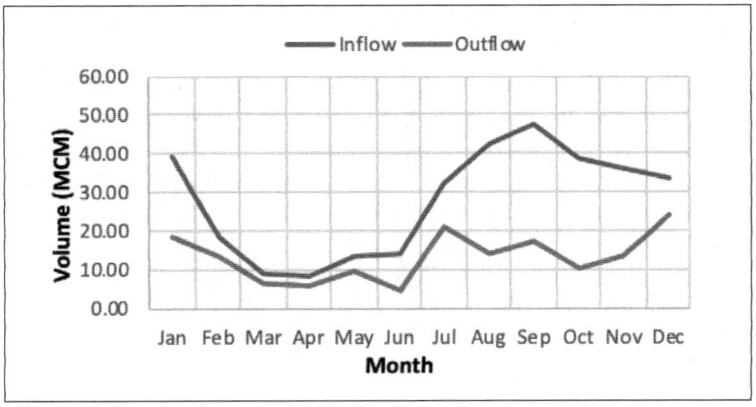

Figure 4.b. Simulation of Tiro reservoir in normal season.

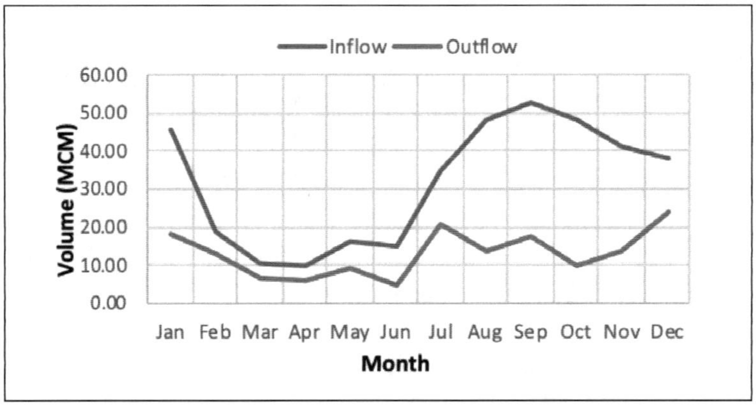

Figure 4.c. Simulation of Tiro reservoir in wet season.

The amount of potential discharge of Kr. Tiro can be transferred as a Rukoh Reservoir inflow can be seen from the simulation results of the Tiro Reservoir. Utilization of two potential of the dam of Rukoh Reservoir and Tiro Reservoir and inflow of both rivers is carried out as a joint operation rule. Then from the simulation, both reservoirs can be seen the reliability of fulfillment of the needs of irrigation water and raw water occurring. From the results of the simulation done on the reservoir of Tiro for all three conditions of the year, it appears that the water potential can be transferred very large. The greatest volume of water that can be utilized is in the wet season between July and November. In the dry year, the amount of water volume from the Tiro reservoir reaches 147.31 MCM. In the normal year, the water volume would reach 176.61 MCM. As for the wet year, the amount of water volume is 221.17 MCM. With a large amount of flow can be used as a limitation in subsequent simulations. The next simulation was done utilizing both the capacity of the Rukoh Reservoir and the Tiro Reservoir as a joint operation rule.

4.2. Joint operation reservoir simulation

In the simulating process of parallel reservoir with the interconnecting channels between the reservoir, the two reservoirs become a unified operating rule. Inflow reservoirs are sourced from both rivers. Reservoir capacity is the total of the two reservoirs. The capacity of Tiro Reservoir is 37.7 MCM while for Rukoh Reservoir of 124.4 MCM. The area of irrigation that must be served for Tiro irrigation area amounted to 6,330 ha and Baro irrigation area of 5,803 ha which cannot be served from the Bendung Keumala on the Baro River. The results of simulation in three season condition shown in figure 5.a to 5.c.

Figure 5.a shown that the joint operating rules of dry season. In May, July and December Outflow greater than inflow. This led to the water level being below the elevation of 150 meters in August and October. However, due to the high inflow in the wet months, the water surface at the end of the year returns above 150 meters. While in the figure 5.b and 5.c which is the result of simulated joint operation in the normal and wet season, there are no significant constrains.

From the simulation chart above, shows that by merging the operations of the two reservoirs through interconnection tunnel could reached high levels of service fulfillment. The lowest water reservoir in May in dry year is 143.42 MCM at 120.10 meters. While in other conditions the position of the water surface is almost always full.

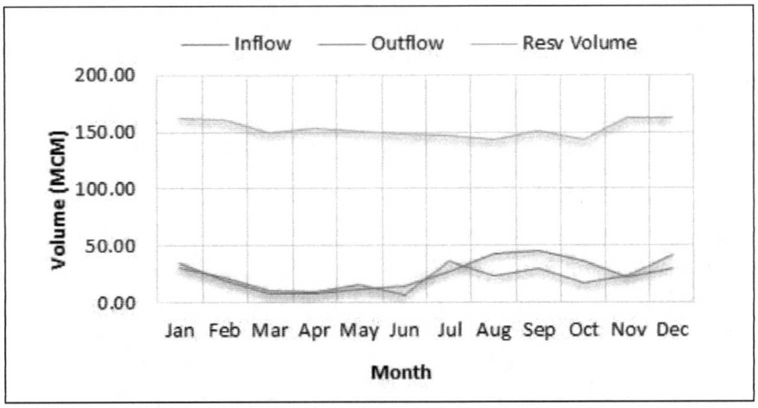

Figure 5.a. Simulation of parallel reservoir in dry season.

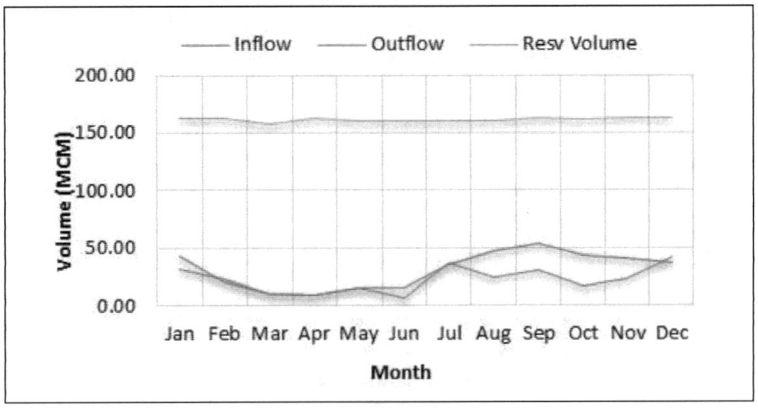

Figure 5.b. Simulation of parallel reservoir in normal season.

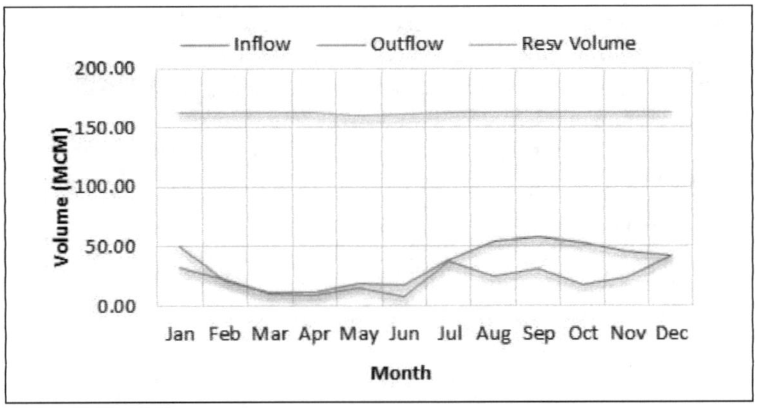

Figure 5.c. Simulation of parallel reservoir in wet season.

4.3. Reservoir reliability
Reliability was measured by the probability of a system being in a satisfactory state [11]. Simulation was done in series for the next 25 years. Where it does not result in any failure of the 600-simulation period performed. Thus, simulated Rukoh-Tiro joint operating of parallel reservoirs with inter-connecting channels can provide 100% reliability for the fulfillment of raw water and irrigation.

5. Conclusion
Based on the analysis above, it can be concluded that with operating the two reservoirs simultaneously with diversion tunnel between could increase the water requirement fulfillment services. Two reservoir capacity will be operating as one system. According to that, tunnel capacity being an important constraint on the reservoir operating model besides the reservoir capacity. With managing the diversion discharge, multi reservoir operation model could give optimize result. In addition, it can also be seen that the operation of two parallel reservoirs still has the potential to serve more extensive irrigation. Further optimization can be done to get a wide range of services that can be covered.

References
[1] Zhang et al 2018 Modeling and simulating of reservoir operation using the artificial neural network, support vector regression, deep learning algorithm *J. of Hydrology;* **565**; 720–36
[2] Rani et al 2015 A Set of linked optimization models for an inter-basin water transfer *Hydrological Sciences J.*
[3] Gu et al 2017 Simulation and optimization of multi-reservoir operation in inter-basin water transfer system *Water Resources Management*
[4] Zeng et al 2015 Derivation of operation rules for reservoirs in parallel with joint water demand *Water Resources* **51** 9539–63
[5] Anonymous 2015 *Dokumen Teknis Pembangunan Waduk Rukoh dan Tiro* (Balai Wilayah Sungai Sumatera I)
[6] Azmeri et al 2017 Completion of potential conflicts of interest through optimization of Rukoh reservoir operation in Pidie District, Aceh Province, Indonesia *AIP Conference Proceedings*
[7] Gunawan G and Kurniawandi A 2010 Penerapan teknik optimasi dan simulasi dalam penyusunan pola operasi waduk untuk pemenuhan kebutuhan energi listrik *Seminar Nasional Fakultas Teknik UR* pp 1-22
[8] Montarcih L 2011 *Manajemen Sumber Daya Air* (Bandung, Lubuk Agung)
[9] Iqbal et al 2020 Multi-reservoirs with inter-reservoir water transfer operation rules *IOP Conf. Series: Earth and Env. Sci.* **437** p 012021
[10] Maass A et al 1962 Design of water resource system *Soil Sci.* **94** 135
[11] Chen et al 2018 Stochastic linear programming for reservoir operation with constraints on reliability and vulnerability *Water* **10** 175

IOP Conf. Series: Materials Science and Engineering **933** (2020) 012033 doi:10.1088/1757-899X/933/1/012033

Study of potential implementations for motorcycle lanes in Padang city

T Kurniati[1], E Afyendo[2] and E E Putri[1]

[1] Associate Professor, Civil Engineering Department, Andalas University
[2] Graduate, Civil Engineering Department, Andalas University

Corresponding e-mail: titi@eng.unand.ac.id

Abstract. Currently, two-wheeled vehicles or motorcycle dominate more than 80% of motorized vehicles in Indonesia. In addition to low prices, these motorized vehicles are fuel efficient, have easier and faster mobilization. However, the large number of motorcycles results in complicated traffic operations and reduced levels of safety. Lanes separation is the best engineering solution to overcome this problem. The purpose of this study was to review the potential application of motorcycle lanes in the city of Padang, a case study on the Raya Ampang and By Pass roads. Collecting the primary data were conducted by Video Image Processor. Each road was carried out recording for 1 day during peak hours in the morning, afternoon, and evening. Analysis of road performance is based on the Highway Capacity Manual of Indonesia (MKJI) 1997 with the degree of saturation (DS) parameter. Road performance analysis is reviewed by comparing the existing conditions and the conditions in which the inclusive motorcycle lanes are applied. The results of the study show that both the Raya Ampang and By Pass streets is feasible to provide an inclusive lane or shared lanes for motorcycles—with a lane width of 3 meters. With motorcycle lane application scenarios, on the Raya Ampang street for non-motorcycles lanes, there is a 10.7% increase in road performance, while for By Pass road, the increase in road performance is 31.5%.

1. Introduction

Motorcycle is the most favorite vehicle and as one of transportation mode in some countries in Asia, especially ASEAN countries [1-2]. Indonesia is the top three countries in trading motorcycle following China and India, respectively. In 2017, data indicated that total of motorcycles reached 113.000.000 units, which was equivalent to 81,6% of total number of vehicles in Indonesia [3]. Growth of motorcycle proportion in traffic flow of Indonesia increased from 25-45% in 1997 to 50-85% in 2011 [4]. The result of research in Padang in 2017 was that proportion of motorcycle volume was around 53-76% in traffic flow where the location of study on street with its type 4/2 D (4 lanes, 2 divided direction) and 4/2 UD (4 lanes, 2 undivided direction) [5]. It is predicted that it is going to develop in line with android application based on vehicle which is well known by its name online motorbike taxi. Various effects occur because of the growth of the total number of motorcycles, which most of them are negative. Among the effects are decreasing space mean speed [5-6], traffic congestion [7] and accident [7-9]. Some efforts are conducted to decrease the negative effects of the growth of motorcycle volume in traffic flow. The result of study proved that separation of lane is the best engineering solution to save motorcycle rider [1-2, 8, 10-12]. This research observes the potentiality in applying particular lane for motorcycle in Padang. The locations of study are Raya Ampang and By Pass km 8 street. The feasibility of applying motorcycle lanes is assessed from existing road performance compared to performance after application in those locations.

2. Literature review

2.1. Definition and design of motorcycle lanes
According to Act number 22 Year 2009 regarding traffic and transportation in Indonesia, "motorcycle is a two-wheeled motor vehicle with or without accessories or a three-wheeled vehicle without accessories" [14]. Ministry of transportation ordinance of the Republic of Indonesia number 34 year 2014 regarding road signs defines that "Lane is a long part of road with or without road signs which has enough width in order that one vehicle can run, except motorcycle"[15]. Motorcycle lane sign is another sign which is marked by a picture of white motorcycle which is placed on the left of traffic. Motorcycle lane sign is placed on the same lane of other vehicles.
In Malaysia, there are two types of motorcycle lanes; they are exclusive motorcycle lanes and inclusive motorcycle lanes [5]. Exclusive motorcycle lanes are constructed specially to separate motorcycle and other vehicles, which is separated by fence or median. Meanwhile, inclusive motorcycle lane or shared lane [10] separates motorcycle with other vehicles which is bordered by road sign. Applying exclusive motorcycle lane in Indonesia has been done on some highways in Jakarta; Darmo street in Surabaya; Lingkar Utara street, Sleman, Yogyakarta; Bypass Sukarno-Hatta arteri street, Bandung.
Directorate General of Highways, 2004 made a rule that determination of the need of exclusive or inclusive motorcycle lane is decided if percentage of motorcycle volume > 40 than volume of total vehicles [12].

Lane widths of 2.0m, 2.5m, and 3.0m were recommended for motorcycle volumes ranging from 1000-1500, 1500-2000, and above 2000 per hour, respectively. A motorcyclist physically spans 0.8m wide, 2.0m in length. An area of 1.6m2 represents the physical space occupied by static motorcyclist. The mean operating speed required by single motorcyclist is 1.3m [1].

2.2. Road performance
Analysis of road performance is based on MKJI 1997 [16] with degree of saturation parameter. Degree of saturation (DS) is defined as ratio of traffic flow Q (pcu/hour) on capacity C (pcu/hour), which is used as the main factor in determining level of road performance. DS score shows whether the road has a capacity problem or not.
Capacity is defined as the maximum traffic flow on a section of road under fluent traffic conditions per hour. To estimate road capacity in the city area, the formula is: [16]

$$C = C_O \times FC_W \times FC_{SF} \times FC_{SP} \times FC_{CS} \quad (1)$$

Where : C_O: basic capacity,
FC_W: lane width adjustment factor,
FC_{SF}: side friction adjustment factor,
FC_{SP}: separation of directions adjustment factor, and
FC_{CS}: city size adjustment factor.

According to [16], the basic capacity is determined as a table 1. and adjustment factor as tables in [16].

Table 1. Basic capacity of urban roads.

Roads type	Basic capacity (pcu/hour)	Note
Four lanes divided or one way street	1650	per lane
Four lanes undivided	1500	per lane
Two lanes undivided	2900	total of 2 directions

IOP Conf. Series: Materials Science and Engineering **933** (2020) 012033 doi:10.1088/1757-899X/933/1/012033

3. Methodology

3.1. Data collection

To analyze road performance, some primary data were collected by doing a survey. The survey is a road geometric survey by measuring the road using tape-measure.

Collecting the data of traffic volume and side barrier were conducted by Video Image Processor. Collecting the data on By Pass street was conducted on Wednesday, March 20th 2019 and on Raya Ampang street on was conducted on Monday, March 25th 2019. Data of traffic volume were taken in peak hours; they are the peak hour from 6:30 a.m. to 8:30 a.m., the peak hour from 11:30 a.m. to 1:30 p.m., and the peak hour from 4.30 p.m. to 6:30 p.m.

3.2. Data processing

Data of recorded video were read by using software *Avidemux*. The observation result showed traffic volume and side barrier. Data of traffic volume were recapitulated per 15 minutes and classified as a motorcycle (MC), light vehicle (LV) and heavy vehicle (HV). The volume of classified vehicles is equivalent to passenger cars unit (pcu). Volume of peak hour (1 hour) was determined based on the highest volume. Based on types of the road, volume of peak hour on Raya Ampang street was taken from two directions. Meanwhile, volume on By Pass street was taken from each direction (North and South) and volume of peak season was the highest volume from one of them. Flow chart of data processing is shown in figure 1.

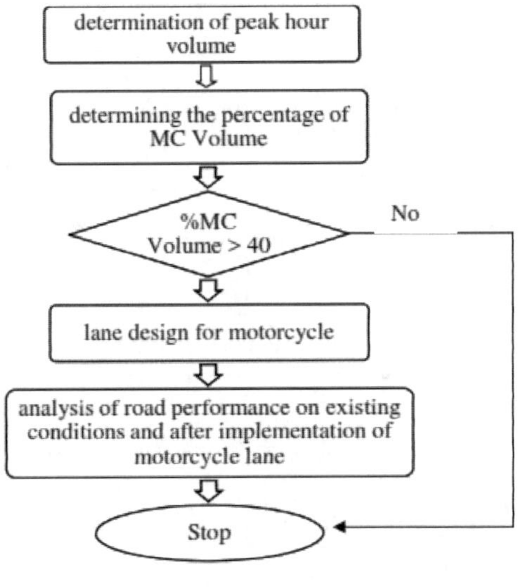

Figure 1. Flow chart of data processing.

4. Results and discussion

4.1. The existing conditions

The geometric data of Raya Ampang and By Pass street and their cross sections are shown in figure 2 and figure 3.

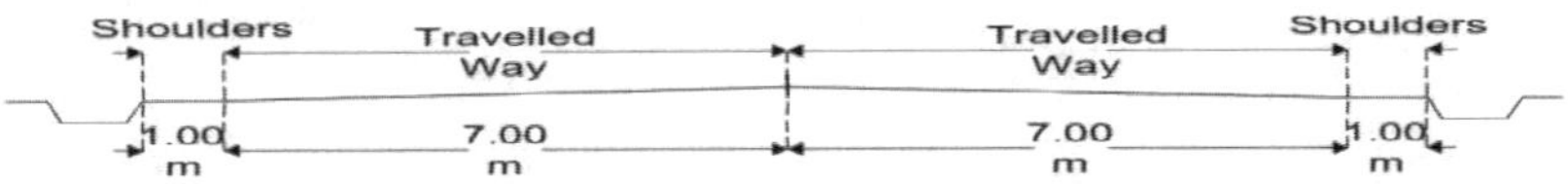

Figure 2. Cross section of Raya Ampang street.

Raya Ampang street is a 4-lane 2-way undivided road type with a width of 7 m pavement for travelled way and shoulder width on the left and right sides of 1 m. On this road side frictions are medium.

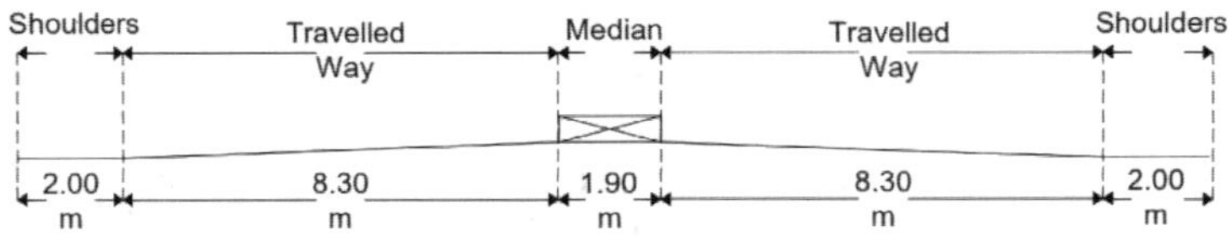

Figure 3. Cross section of By Pass street.

By Pass street is a 4-lane 2-way divided road type with a width of 8.3m pavement for travelled way and shoulder width on the left and right sides of 2 m. On this road side frictions are low.

Data processing result of traffic flow volume on Raya Ampang street is that volume of peak hour is 6623 vehicles/hour (2936 pcu/hour) from 6:45 a.m. to 7:45 a.m. The highest volume of motorcycle occurs at the same time that is 4920 vehicles/hour (1230 pcu/hour) with its percentage 74.3 %. Based on this percentage, it is concluded that the Raya Ampang road requires a special lane for motorcycle. On By Pass street, volume of peak hour is 3961 vehicles/hour (1837.5 pcu/hour) from 7:00 a.m. to 8.00 a.m. toward north direction. The highest volume of motorcycle (MC) is 2842 vehicles/hour (710.5 pcu/hour) with its percentage 71.8. The By Pass street need a special lane for motorcycle too. Result of analysis of road performance and the need for special lane for motorcycle on Raya Ampang street are summarized in table 2, and on By Pass street in table 3.

Table 2. Raya Ampang street performance analysis.

Time	Volume (veh/hr)			Total volume veh/hr (pcu/hr)	% MC	C pcu/hr	DS	Suggestion
	MC	LV	HV					
6.45-7.45 a.m	4920	1690	13	6623 (2936)	74.3	5224.6	0.56	Need of motorcycle lane

Table 3. By Pass street performance analysis.

Time	Volume (veh/hr)			Total volume veh/hr (pcu/hr)	% MC	C pcu/hr	DS	Suggestion
	MC	LV	HV					
7:00-8:00 a.m.	2842	1690	13	3961 (1837.5)	71.8	3417	0.54	Need of motorcycle lane

With a DS value of 0.5, the traffic conditions are stable flow, vehicle speed and movement controlled by traffic volume. Both roads require special lane for motorcycle. Based on the geometric condition of the road, the specific types of lanes applied are inclusive motorcycle lane or shared lane. Inclusive motorcycle lane or shared lane separate motorcycle with other vehicles which are bordered by road sign and marking, placed on the left of traffic.

4.2. The application of motorcycle lanes conditions

Raya Ampang and Bypass roads require a shared lane with 3 metre in width. Lane widths of 3.0m were recommended for motorcycle volumes above 2000 per hour [1]. Motorcycle lane is designed by using road shoulder and part of the road body as it is shown in figure 4 and figure 5.

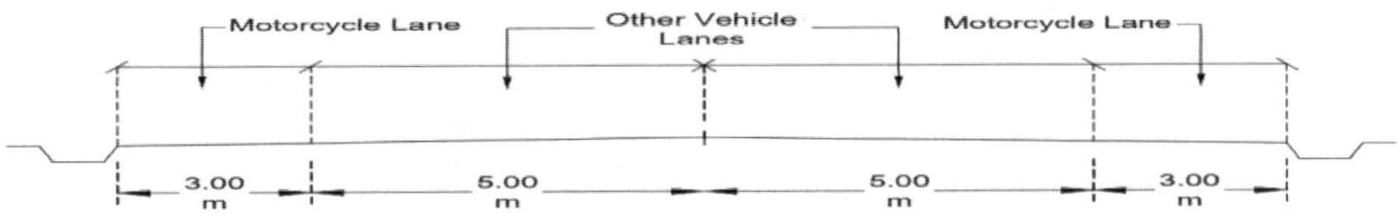

Figure 4. Cross section of the Ampang street after the application of motorcycle lane.

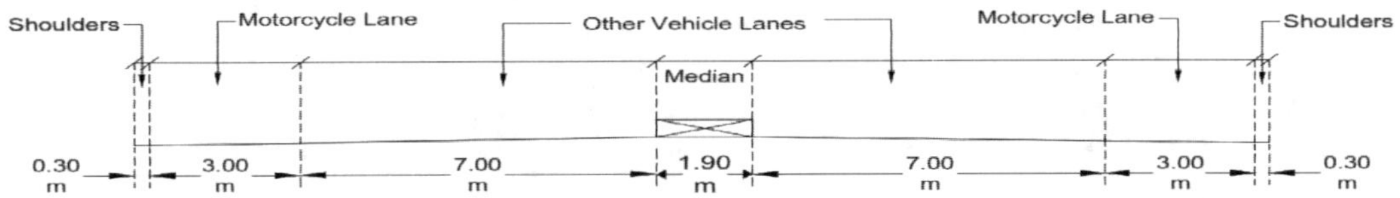

Figure 5. Cross section of the By Pass street after the application of motorcycle lane.

Analysis of road performance with scenario of shared lane is shown in table 4.

Table 4. Road performance analysis after shared lane.

	$Q_{non\ MC}$ (pcu/hr)	C (pcu/hr)	DS
Raya Ampang	1705	3481	0.50
By Pass	1127	3070	0.37

On Raya Ampang street, road performance increases 10.7 % on non motorcycle lane and DS score decreases from 0.56 to 0.50. The same thing, on By Pass street, there is an increase, 31.5% and the decrease of DS score from 0.54 to 0.37. Road performance improvement is quite significant so that traffic conditions are stable with moderate traffic volumes.

5. Conclusion

From the results of the analysis of the two road sections that are the locations of the research objects, Raya Ampang and Bypass road have the potential to apply motorcycle lanes. The right type of motorcycle lane applied is inclusive lane or shared lane, separates motorcycle with other vehicles which is bordered by road sign and marking. Motorcycle lane with 3 metre in width is designed by using road shoulder and part of the road body. Analysis of road performance with scenario of motorcycle lane implementation shows that there is an increase in performance on non-motorcycle lanes for both roads.

Acknowledgments

Thanks to Edo Afyendo who gave data of survey so that this article could be composed. Data of this research were taken from Edo Afyendo's undergraduate thesis under the writer supervision. We are grateful for Program Bantuan Seminar International Dalam Negeri Andalas University.

References

[1] Hamid H, Sohadi R U R, Sadullah A F M and Ma'soem D M 2005 *IATSS Research* **29** **1** 50-6
[2] Le T Q and Nurhidayati Z A 2016 *Procedia Engineering* 142 292-8
[3] Central Bereau of Statistics 2018 *Statistik Transportasi Darat* https://www.bps.go.id/publication/2018

IOP Conf. Series: Materials Science and Engineering **933** (2020) 012033 doi:10.1088/1757-899X/933/1/012033

[4] Putranto L S, Suardika G P, Sunggiardi R, Munandar A S and Lutfi I 2011 *Proc. of the Eastern Asia Soc. for Transportation Studies* vol 8

[5] Kurniati T and Fajriati R 2019 *MATEC Web of Conf.* vol 276 03016

[6] Kusnandar E 2010 The effect of the proportion of motorcycle on the speed of traffic flow *J. Puslitbang Jalan dan Jembatan* (in bahasa)

[7] Lubis H A R S 2009 *Proc.of the Eastern Asia Soc. for Transp. Studies* vol 7

[8] Law T H and Sohadi R U R 2005 *J. of the Eastern Asia Society for Transp. Studies* **6** 3372-85

[9] Khaidir N M, Manan M M A and Johari N M 2015 *Proc. of Conf. of ASEAN Road Safety* pp 80-5

[10] Mama S and Taneerananom P 2016 *Engineering Journal* **20** 3 113-21

[11] Sulistio H 2018 *The Open Transp. J* **12** 1-7

[12] Zukhruf F, Frazila R B dan Wibowo S S 2010 *J. Transportasi* **10** 1 23-32

[13] Ibrahim M K A, Hamid H, Hua L T and Voon W S (2018) *Accident Analysis and Prevention J.* **111** 63-70

[14] Anonymous 2009 *Act of the Republic of Indonesia number 22 Year 2009 Traffic and Transportation*

[15] Anonymous 2014 *Ministry of Transportation Ordinance of the Republic of Indonesia number 34 year 2014 Road Markings*

[16] Directorate General of Highways (Bina Marga) 1997 *Indonesian Highway Capacity Manual Bandung Indonesia*

Large deformation of building frame structure due to December 7, 2016 Pidie Jaya earthquake

M Hasan[1], M Afifuddin[1] and R Putra[1]

[1]Department of Civil Engineering, Universitas Syiah Kuala, Banda Aceh, Indonesia

Corresponding e-mail: muttaqin@unsyiah.ac.id

Abstract. Pidie Jaya earthquake on December 7, 2016 caused the damage of many buildings. This paper presents the case study on a building frame structure that has undergone large deformation during Pidie Jaya earthquake without structural damage. The building has 3 stories. The permanent lateral displacement and inter-story drift of the structure in Y direction at the third story after the earthquake is 420 mm and 4.2%, respectively, which exceed the requirement of Indonesian standard for design building and other structures under earthquake load (SNI 1726:2012). The building structure was modeled as a space frame. A response spectral analysis with design response spectral provided in SNI 1726:2012 was conducted. For comparison, a time history analysis based on real time ground acceleration during Pidie Jaya earthquake was also conducted. The maximum lateral displacement based on response spectral analysis is 469 mm with maximum inter-story drift of 4.79%. Based on time history analysis, the maximum lateral displacement and inter-story drifts are 550 mm and 5.81%, respectively. Based on those results, it is necessary to revise the design response spectrum in SNI 1726:2012.

1. Introduction

Indonesia is known as the ring of fire because it is located at the confluence of three world tectonic plates, namely the Indo-Australian plate which moves northward, the Pacific plate which moves northwest and the Eurasian plate. The existence of the interaction between those plates makes Indonesia as a region that is very vulnerable to earthquakes [1]. Ring of fire is a series of active volcanoes in the world that causes Indonesia to experience the frequency of earthquakes quite often.

Aceh Province, as one of the most west provinces in Indonesia, occupies a very active tectonic zone, because in addition to the three large plates of the world, there are also nine other small plates that meet each other and form a complex meeting point of the plates [2]. Earthquakes have occurred in Aceh both in vertical and horizontal directions, such as in December 2004 with 9.2 SR which is followed by huge tsunami, in April 2012 with 8.6 SR, in July 2013 centered in Bener Meriah District with 6.2 SR and on December 7, 2016, centered in Pidie Jaya District with 6.5 SR. For the purpose of seismic vulnerability assessment, the spectral displacement and spectral acceleration of those earthquake events have been developed [3-6].

The principal reasons for building failure during the earthquake are due to soft storeys, floating columns, mass irregularities, poor quality of construction materials and faulty construction practices, inconsistent earthquake response, soil and foundation effect and pounding of adjacent structures due to large lateral deformation [7]. This large deformation also initiates cracks and under the subsequent load, the crack propagates and causes the structural damages [8]. Types of structural failure due to earthquake can be in the form of shear failure on reinforced concrete beams, destruction of one floor due to insufficient strength between floors and very large inter-storey drift during an earthquake, failures related to strong beam-weak column, beam-column joint failures and failures related to structural system selection errors during design [9-11].

According to the National Disaster Management Agency (BNPB), the Pidie Jaya earthquake caused 104 deaths, 700 injured, 16,238 buildings damaged and 85,161 people displaced [12]. Damage to the building structures has different levels depending on the magnitude of the earthquake that occurred, the distance of the location to the epicenter, the foundation system, the quality of the building, the mass and geometry of the building and the type of soil at the building site. The behavior of building structures to retain earthquake loads needs to be known to evaluate the strength of structural elements under earthquake loads. With the occurrence of major earthquakes, the Indonesian Government through the National Standardization Agency (BSN) has issued the design code of building and other structures under earthquake load (SNI 1726:2012) [13].

In this paper, the analysis of building frame structure that has undergone large deformation during December 7, 2016 Pidie Jaya earthquake is presented. The building structure is modelled as a space frame and response spectral analysis based on SNI 1726:2012 is conducted. In addition, the time history analysis based on ground acceleration during Pidie Jaya earthquake is also conducted. The lateral displacement and inter-storey drift of structure is presented and compared with that was measured on the site.

2. Methodology

The analytical model used in this study is the frame structure of the Multipurpose Hall Building of Pidie Jaya. The building has 3 stories with the height of the first, second and third story is 4 m, 4 m and 2 m, respectively. The plan of the building is shown in figure 1. In the second floor, there is a cantilever slab and beam with the span of 2 m, as shown in figure 1. The columns and beams of the structure are steel-concrete composites, except cantilever beam which is reinforced concrete. The columns and beams cross sections are shown in figure 2. The slab of the building is reinforced concrete. During the Pidie Jaya earthquake, the structure has undergone a large permanent deformation. The lateral displacements in Y direction of first, second and third stories are 150 mm, 330 mm and 420 mm, respectively. The inter-story drift ratio in Y direction of first, second and third stories are 3.75%, 4.50% and 4.50%. This drift ratio has exceeded the requirement in SNI 1726:2012 which allows the maximum drift ratio of 2%. In X direction there is no permanent deformation was found.

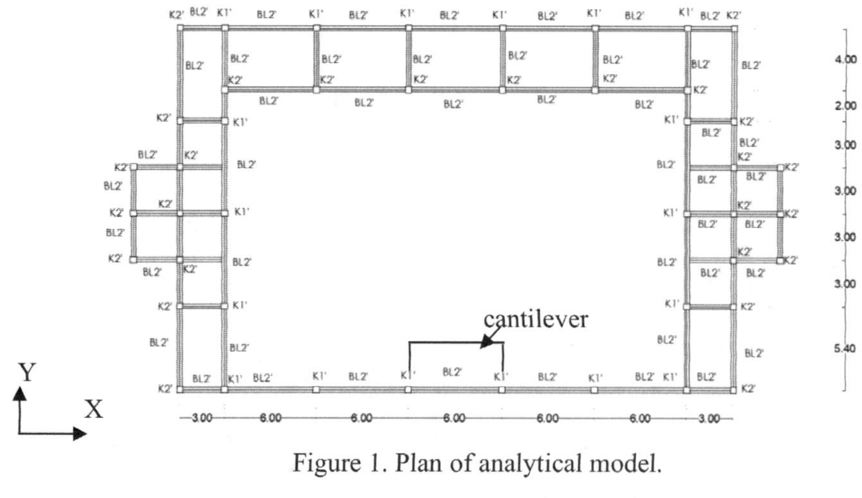

Figure 1. Plan of analytical model.

IOP Conf. Series: Materials Science and Engineering **933** (2020) 012034 doi:10.1088/1757-899X/933/1/012034

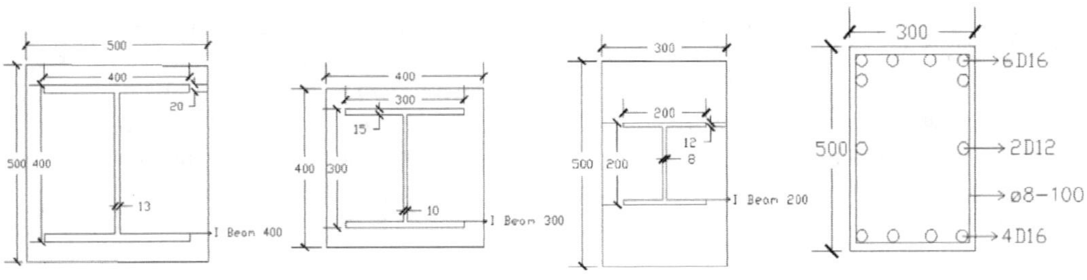

Column K1' Column K2' Beam BL2' Cantilever Beam

Figure 2. Columns and beams cross section.

The building structure is modelled as space frame as shown in figure 3 using structural analysis software, namely SAP 2000. The material properties used in the analysis are shown in table 1.

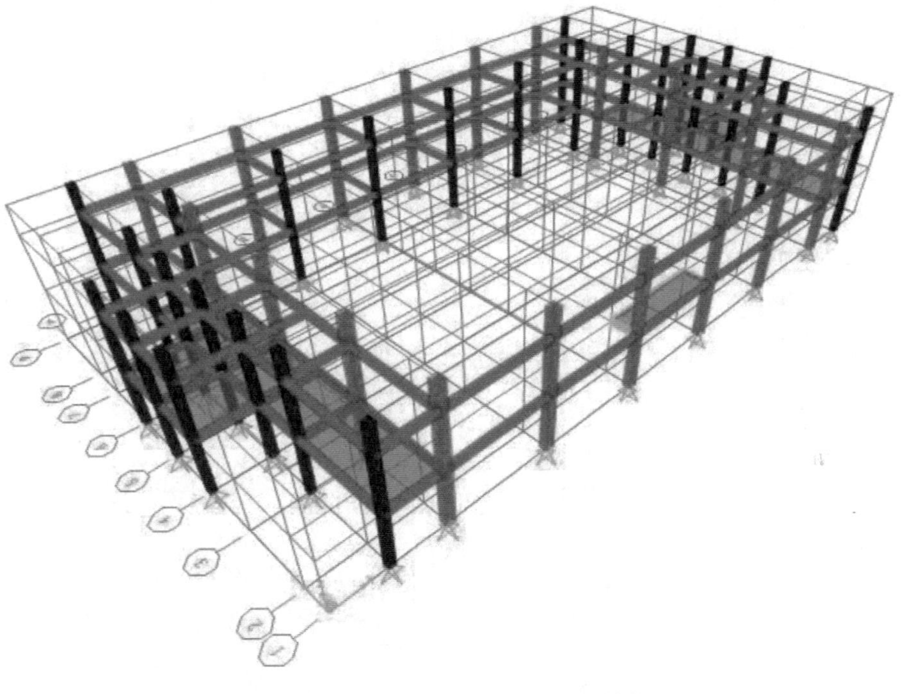

Figure 3. Space frame model of analysed building structure.

Table 1. Material properties.

Concrete compressive strength (f_c')	25 MPa
Steel yield strength (f_y)	240 MPa
Steel ultimate strength (f_u)	370 MPa
Concrete elastic modulus (E_c)	23.5 GPa
Steel elastic modulus (E_s)	200 GPa
Reinforced concrete density	2400 kg/m^3
Concrete density	2200 kg/m^3
Steel density	7850 kg/m^3

The loads considered in this study are dead load, live load, wind load and earthquake load. Dead load, live load and wind load are calculated based on minimum load requirement as specified in Indonesian standard [14]. For earthquake load analysis, a response spectral analysis with design response spectral specified in SNI 1726:2012 [13] was conducted. In addition, a time history analysis based on ground acceleration during December 7, 2016 Pidie Jaya earthquake was also conducted. The ground acceleration of Pidie Jaya earthquake can be found in the reference [15]. The data for earthquake load analysis is shown in table 2. The following load combinations (LC) were calculated:

LC 1 : 1.4 D
LC 2 : 1.2 D + 1.6 L
LC 3 : 1.2 D + 0.5 W_x
LC 4 : 1.2 D + 0.5 W_y
LC 5 : 1.2 D + 1.0 W_x + L
LC 6 : 1.2 D + 1.0 W_y + L
LC 7 : 1.2 D + 1.0 L + 1.0 E_x
LC 8 : 1.2 D + 1.0 L + 1.0 E_y
LC 9 : 0.9 D + 1.0 W_x
LC 10 : 0.9 D + 1.0 W_y
LC 11 : 0.9 D + 1.0 E_x
LC 12 : 0.9 D + 1.0 E_y

Where D = dead load, L = live load, W_x = wind load in X direction, W_y = wind load in Y direction, E_x = earthquake load in X direction and E_y = earthquake load in Y direction.

Table 2. Data for earthquake load analysis.

Building location	Mayang Lancok, Pidie Jaya
Coordinate	5.229 NL; 96.245 EL
Soil classification	Soft soil
Seismic design category	D
Seismic importance factor (I)	1.25
Response modification coeffecient (R)	8
Bedrock acceleration at short period (S_S)	0.802 g
Bedrock acceleration at period of 1 second (S_1)	0.398 g
Design spectra acceleration parameter at short period (S_{DS})	0.639 g
Design spectra acceleration parameter at period of 1 second (S_{D1})	0.608 g

3. Results and discussion

3.1. Lateral displacement

The lateral displacement in X and Y directions obtained from response spectral analysis for all load combinations are shown in figures 4 and 5, respectively. For comparison, the real lateral displacement in Y direction measured in the site is also plotted in figure 5. The displacements in X direction meet the requirement of SNI 1726:2012. However, the displacement in Y direction exceeds the SNI 1726:2012 requirement. The higher displacement in Y direction compared to in X direction is presumably due to the presence of cantilever plate and beam. The presence of cantilever plate results in high bending moments in the columns which cause the large displacement. Another reason for high lateral displacement in Y direction is the stiffness of the structure in Y direction because there is only one raw of columns in the front of the building.

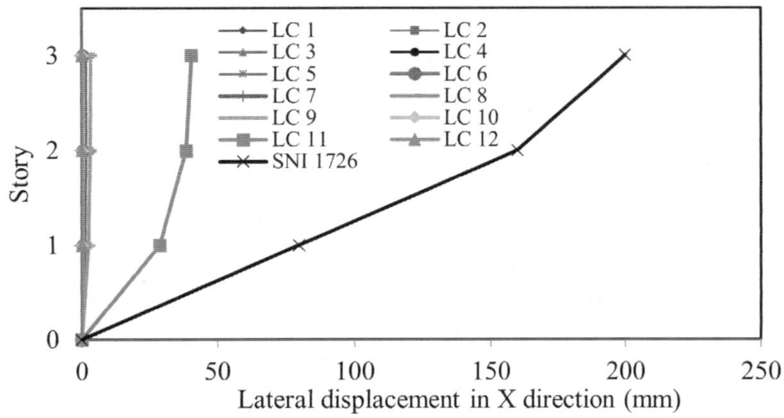

Figure 4. Lateral displacement in X direction obtained from response spectral analysis.

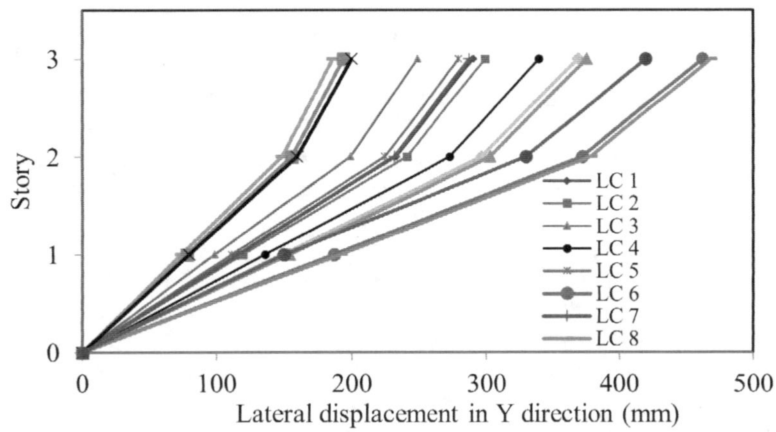

Figure 5. Lateral displacement in Y direction obtained from response spectral analysis.

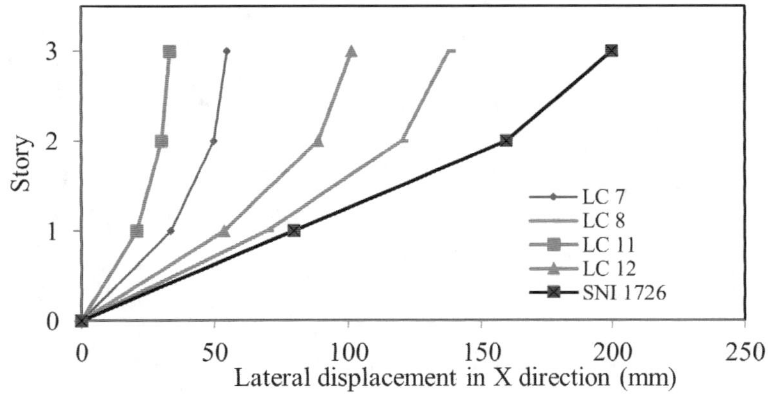

Figure 6. Lateral displacement in X direction obtained from time history analysis.

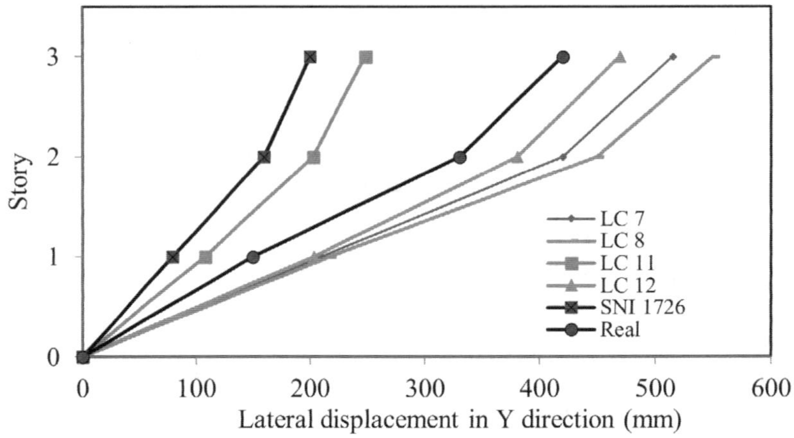

Figure 7. Lateral displacement in Y direction obtained from time history analysis.

The lateral displacement in X and Y directions obtained from time history analysis are shown in figures 6 and 7, respectively. Since the difference from the response spectral analysis is only for load combination with earthquake load, the lateral displacement shown in figures 6 and 7 is only for load combination with earthquake load (LC 7, LC 8, LC 11 and LC 12). The real lateral displacement in Y direction measured in the site is also plotted in figure 7. The lateral displacements obtained from time history analysis are higher than those obtained from response spectral analysis.

3.2. Inter-story drift

The inter-story in X and Y directions obtained from response spectral analysis for all load combinations are shown in figures 8 and 9, respectively. For comparison, the real inter-story drift in Y direction measured in the site is also plotted in figure 9. The inter-story drifts in X direction meet the requirement of SNI 1726:2012. However, the inter-story drifts in Y direction exceed the SNI 1726:2012 requirement.

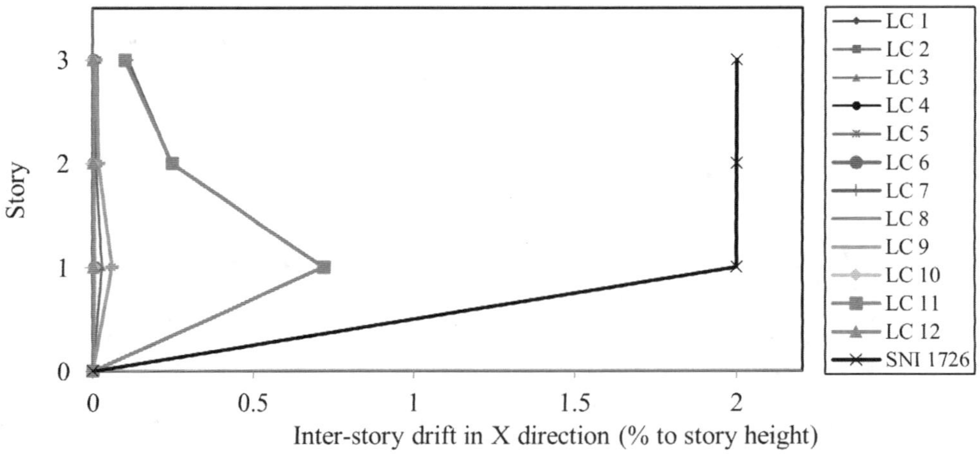

Figure 8. Inter-story drift in X direction obtained from response spectrum analysis.

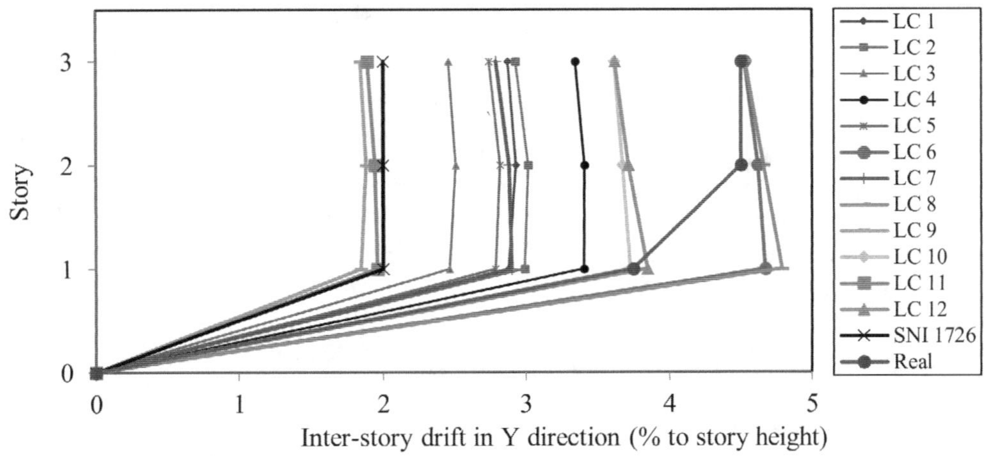

Figure 9. Inter-story drift in Y direction obtained from response spectrum analysis.

The inter-story in X and Y directions obtained from time history analysis are shown in figures 10 and 11, respectively. Since the difference from the response spectral analysis is only for load combination with earthquake load, the inter-story drifts shown in figures 10 and 11 are only for load combination with earthquake load (LC 7, LC 8, LC 11 and LC 12). The real inter-story drift in Y direction measured in the site is also plotted in igure 11. The inter-story drift obtained from time history analysis is higher than that obtained from response spectral analysis. The higher lateral displacements and inter-story drift obtained from time history analysis and what found in the real structure at the site compared to those obtained from response spectral analysis confirms that the design response spectral in SNI 1726:2012 cannot result in the safe structures in the earthquake-prone areas. Therefore, to have the safe structures in the future, the revision of design response spectral in SNI 1726:2012 is needed.

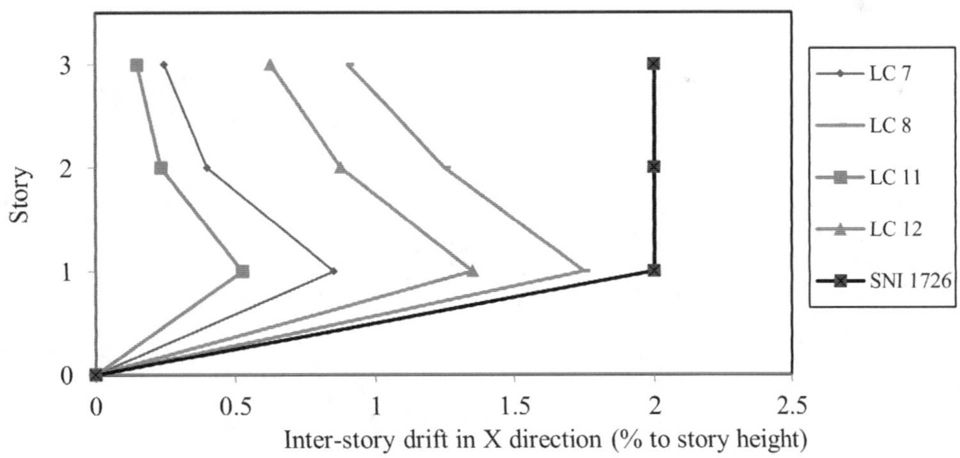

Figure 10. Inter-story drift in X direction obtained from time history analysis.

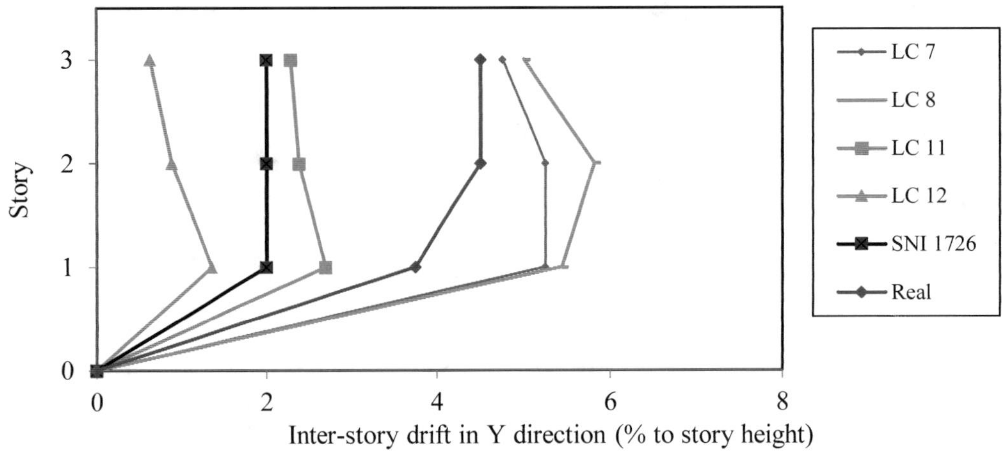

Figure 11. Inter-story drift in Y direction obtained from time history analysis.

4. Conclusion

The response spectral and time history analysis was conducted to study the deformation of building structure due to December 7, 2016 Pidie Jaya earthquake. Base on the analysis, the following conclusions can be drawn:

1. The deformation of structure in X direction meets the requirement of SNI 1726:2012. In the site, there is no permanent deformation in X direction.
2. The deformation of structure in Y direction exceeds the requirement of SNI 1726:2012. The large deformation in Y direction is presumably due to the presence of cantilever in the front row of the column.
3. The deformation of the structure obtained from time history analysis using real time ground acceleration during December 7, 2016 Pidie Jaya earthquake is higher than that obtained from response spectral analysis using the design response spectral provided in SNI 1726:2012. Therefore, the design response spectral in SNI 1726:2012 needs to be revised.

Acknowledgments

This research was supported by the grant provided by Syiah Kuala University Research and Community Service Center (LPPM) under contract number of 12/UN11.2/PP/SP3/2019.

References

[1] Milson J, Masson D, Nicholas D, Sikumbang N, Dwiyanto B, Parson L and Kallagher H 1992 *Tectonics* 11 145-53
[2] Bird P 2003 *Geochemistry, Geophysics, Geosystems* **4** 3 1027
[3] Saidi T, Aulia T B, Setiawan B, Abdullah N and Hasan M 2018 *Matec Web of Conference* vol 197 10001
[4] Saidi T, Hasan M and Fakhrurrazi 2019 *IOP Conf. Series: Mater. Sci. and Eng.* vol 523 012039
[5] Saidi T, Muttaqin and Sarana D 2015 *Proc. KoNTekS* vol 9 741-8
[6] Sarana D, Saidi T and Muttaqin 2015 *J. Teknik Sipil Univ. Syiah Kuala* **4** 3
[7] Agarwal P, Thakkar S K and Dubey R N 2002 *ISET J. Earthquake Technol.* **34** 3 195-217
[8] Hasan M, Husaini and Abdullah N M 2019 *IOP Conf. Series: Mater. Sci. and Eng.* vol 523 012037
[9] Celep Z, Erken A, Taskin B and Ilki A 2011 *Eng. Failure Anal.* **18** 868-89
[10] Cosgun C, Dindar A A, Seckin E and Onen Y H 2013 *Gradevinar* **65** 743-52
[11] Damci E, Temur R, Bekdas G and Sayin B 2015 *Case Studies in Constr. Mater.* **3** 112-31

IOP Conf. Series: Materials Science and Engineering **933** (2020) 012034 doi:10.1088/1757-899X/933/1/012034

[12] Supartoyo, Subandriyo and Amalfi O 2017 *Tanggap Darurat Gempa Bumi Pidie Jaya (Frekuensi 1) 6.5 Sr-7 Desember 2016* (Bandung: Badan Geologi, Kementerian Energi dan Sumber Daya Mineral)

[13] SNI 1726:2012 *Tata Cara Perencanaan Ketahanan Gempa untuk Struktur Bangunan Ggedung dan Non Gedung* (Badan Standarisasi Nasional)

[14] SNI 1727:2013 *Beban Minimum untuk Perancangan Bangunan Gedung dan Struktur Lain* (Badan Standardisasi Nasional)

[15] Hasan M, Afifuddin M, Putra R, Asyifa C N and Maulana M I 2019 *J. Teknik Sipil ITB* (submitted)

IOP Conf. Series: Materials Science and Engineering **933** (2020) 012035 doi:10.1088/1757-899X/933/1/012035

Slope stability analysis using Bishop and Finite Element Methods

M Sungkar[1], M Munirwansyah[1], R P Munirwan[1] and D Safrina[1]

[1] Department of Civil Engineering, Universitas Syiah Kuala, Banda Aceh, Indonesia

Corresponding e-mail: munira.sungkar@unsyiah.ac.id

Abstract. Analysis of slope stability in this study aims to determine the safety factor (SF) of the landslide for the existing and the reinforcement condition. The study case was in STA 84+910 of national road Meulaboh-Geumpang, West Aceh Regency in Aceh Province. The condition of the Meulaboh-Geumpang road often causes landslides due to the topography of the area and is also triggered by rainfall. The analysis of safety factor for the existing condition was calculated by using the Bishop method that assumes a circular failure surface and the calculation was also validated by the finite element method using 2D Plaxis version 8.6 software. However, computation requires iterative procedures because of the nonlinear relationship for the safety factors. The result obtained from calculations using the Bishop method for the existing condition is 1.08, while by using Plaxis software, the result is 1.10, which means the slopes are in unstable conditions and need to be reinforced to prevent landslides. Alternative reinforcement was conducted by using sheet pile and calculations performed using 2D Plaxis version 8.6 software. Variations in the placement of sheet piles are at the upper and middle of the slope. The results obtained from the calculation of the safety factor after reinforcement are 1.57 (with a 12m sheet pile depth at the upper slope) and 1.48 (with a 10m sheet pile depth in the middle of the slope).

1. Introduction

The development of transportation encourages humans to make the best use of existing land, such as in hilly and sloped areas, whose topography tends to be diverse. Slope stability analysis is required from the planning or design stage to create a safe, comfortable, and durable construction in the slope area. Object location for this study is Meulaboh-Geumpang national road, which often causes landslides due to the topography of the area and is also triggered by rainfall. The landslide that occurred on Wednesday (10/5/2017) night around 11.00 p.m caused the disruption to the Meulaboh-Geumpang national road, which experienced a landslide and failure. The case study located in Lancong Village, Sungai Mas District, West Aceh Regency (figure 1).

Natural factors causing landslides are topography or slope, rock conditions and slope subgrade, as well as hydrological conditions or groundwater conditions on slopes. The heavy rainfall, which flushed most of the high area of Lancong Village, Sungai Mas District, West Aceh Regency, applies landslides and erosion of most of the main roads that have quite steep cliffs on STA 84 + 910 as shown in figure 2.

The objectives of slope stability analyses are: (1) Understanding the development and formation of natural slopes and processes that occur under different natural conditions; (2) Determine the slope stability in the short-term and long-term conditions; (3) Determine the possibility of failure on natural or artificial slopes; and (4) Analyzing and understanding the mechanism of failure and the factors that cause it [1].

Slopes can occur naturally or be formed by humans for certain purposes. If the surface forms a slope, the soil mass component above the slip plane tends to move downward due to gravity. If the gravity component that occurs is large enough, it can cause landslides on the slope [2].

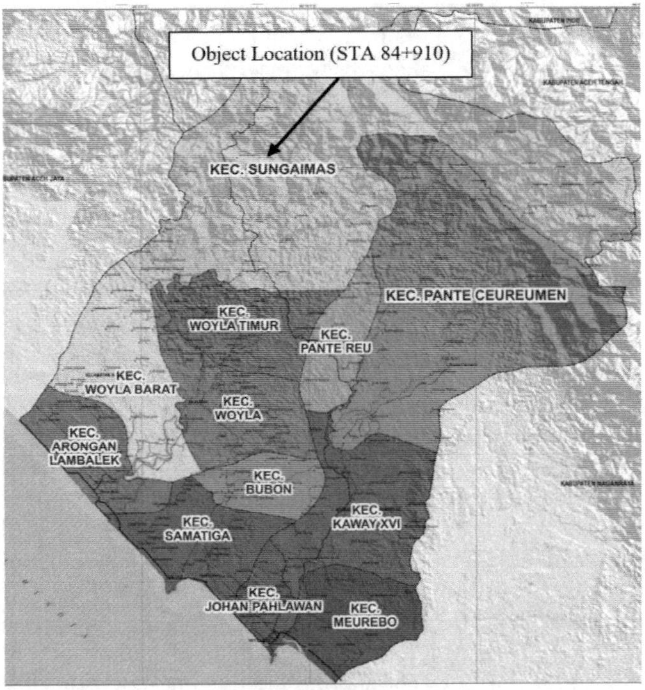

Figure 1. Map of case study locations in West Aceh Regency.

Figure 2. The latest condition of Meulaboh-Geumpang Road (December 2018).

The concept of slope stability is very much needed in developing the current use of slopes. Because the use of slopes for the benefit of humans increases, it is necessary to develop the concept of slope stability aimed to overcome the problem of slope landslide [3]. There are many solutions for landslide remediation. Common principles used are unloading, drainage, reinforcement, retaining structures, and surface vegetation [4]. However, [1] and [4] indicate that not all stabilization methods are appropriate for every type of slope failure; the most expensive treatment may not always be the most effective. Therefore, the determination of treatment for landslide appears to have no definitive rule to follow.

Reinforcement by installing sheet pile has been used for decades to meet engineering purposes: considering safety, economic, and uncomplicated approach [5]. Many similar studies using sheet pile reinforcement have been done before, as in research [5] using sheet pile wall for bridge pile foundation, and [6] describe the effect of seismic reinforcement of sheet pile quay wall using the ground anchor. This study describes the result of the safety factor that has been calculated to anticipate slope failure by the installation of the sheet pile.

The purpose of the slope stability analysis is to obtain the value of safety factor (SF) from the landslide area in the existing and reinforcement condition. Calculations for the existing condition is performed using the Bishop method and validated by the finite element method using 2D Plaxis version 8.6 software. Whereas for conditions with reinforcement, calculations are only performed using Plaxis software for 2 (two) variations of the placement of sheet piles, namely at the upper slope and in the middle of the slope. Then, the next calculation is varying depth of the sheet pile.

2. Methodology

This study was based on experimental works in Soil Mechanics Laboratory in University of Syiah Kuala, Indonesia. The soil sample was taken from two test pits in STA 84+910 namely B1 and B2. Some physical and mechanical properties of the soil were then performed; the tests were unit weight of soil, water content, Atterberg limit test, grain size distribution test, and specific gravity. Mechanical tests of soil were also conducted to obtained soil cohesion and friction angle. All of the soil testing to obtain physical and mechanical properties of soil were follow the ASTM standard [7]. Soil classification is done by two methods specifically the AASHTO and USCS soil classification system [8]. Physical and mechanical soil properties are used for designing or calculating reinforcement by the installation of sheet pile and as input parameters for 2D Plaxis software.

Sheet pile is a relatively thin, long flat vertical wall, usually made of steel or concrete material that serves to restrain the soil and also water entering the digging hole [9]. This study uses the steel sheet pile, as shown in figure 3.

Figure 3. Steel sheet pile.

The type of sheet pile can be divided in terms of construction (sheet pile without anchor and anchored sheet pile), and type of sheet pile according to the material (wood, steel, and reinforced concrete sheet pile). [10] Sheet pile without anchor is used for differences in soil height (h) that are not too large and are often used for semi-permanent work. Anchored sheet pile is used for quite a large difference in the height of the retained soil (h). The stability of construction is obtained apart from soil clamps in the sheet pile construction section embedded underground.

In general, the value of safety factor (SF) $\geq$ 1.25 was used for normal design in slope stability analysis. This is important to ensure that the slope design is safe and prevents unexpected factors during analysis, construction such as incorrect data, analysis errors, work skills, and the lack of field supervision

[10]. Safety factor values for design [11] and based on slope failure studies [12] are shown in Table 1 and Table 2.

Table 1. Value of safety factors for design [11].

No.	Safety Factor (SF)	Description
1	< 1.0	Unsafe
2	1.0 – 1.2	Query safety
3	1.25 – 1.4	Safe for excavation, embankment, query for dam
4	1.5 – 1.75	Safe for dam/reinforcement

Table 2. Safety factors based on study of slope landslide [12].

No.	Safety Factor (SF)	Description
1	< 1.07	Unstable
2	1.07 – 1.25	Critical
3	> 1.25	Stable

The safety factor (SF) is defined as the comparison between the maximum shear resistance (τ) and the shear stress due to gravity of the soil will failure (τd) as shown in equation (1).

$$SF = \frac{\tau}{\tau d} \qquad (1)$$

Where:

τ = maximum shear resistance (kN/m^2); and
τd = shear stress due to the gravity of the soil that will failure (kN/m^2).

2.1. Simplified Bishop Method

The Bishop method assumes that the forces acting on the slice have zero resultant in the vertical direction. The system of forces in the Bishop method is shown in figure 4 [9].

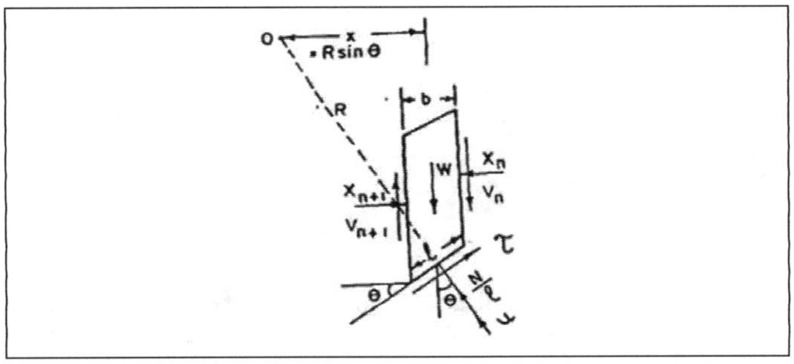

Figure 4. Forces acting on segments by the Bishop method [12].

By calculating the entire balance of forces, the formula for the safety factor (SF) of the Bishop method is obtained as follows [13].

$$SF = \frac{\sum_{i=1}^{i=n}[c' b_i + (W_i - u_i b_i)\tan\varphi'](\frac{1}{\cos\alpha_i\left(1+\tan\alpha_i.\frac{\tan\varphi'}{F}\right)})}{\sum_{i=1}^{i=n} W_i \sin\alpha_i} \quad (2)$$

Where:
SF = safety factor;
c' = effective soil cohesion (kN/m^2);
ϕ' = effective friction angle (°);
b_i = width of the-i slice (m);
Wi = γ.bi.h, weight of the-i slice (kN);
γ = unit weight of soil (kN/m^3);
h = average slice height (kN/m^3);
α_i = slope angle of the tangent on point in the middle of the base of the slice with respect to the horizontal plane;
The slope angle is positive if it is in the direction of the slope and negative if it is in the opposite direction to the slope (°); and
ui = pore pressure of the-i slice (kN/m^2).

2.2. Finite element method

The finite element method (FEM) represents a powerful alternative approach for slope stability analyses. This method is accurate, versatile, and requires fewer prior assumptions, especially regarding the failure mechanism. The FEM is powerful in solving problems with irregular boundaries and complex variation of potential and flow lines [14].
The reduction in cohesion parameters (c) of the friction angle (ϕ) and SRF (strength reduction factor) can be stated in Equation (3) and Equation (4) below [15].

$$c_f = \frac{c}{SRF} \quad (3)$$

$$\varphi_f = \tan^{-1}\frac{\tan\phi}{SRF} \quad (4)$$

Where:
Cf = certain factor;
c = cohesion;
ϕ = friction angle; and
SRF = strength reduction factor.

With SRF (*strength reduction factor*) is a shear strength reduction factor, so the safety factor (SF) is the same as the SRF value at the time of failure.

2.3. Plaxis software

The Plaxis software for Windows was first released in 1998 at Delft Technical University, the Netherlands. According to [16] the Mohr-Coloumb model is highly recommended to be used as an initial analysis of a problem. Mohr-Coloumb can be used to calculate realistic supporting stresses on tunnel faces, foundation boundary loads, and others. This model can also be used to calculate the safety factor with the ' ϕ -c Reduction' approach.

The Mohr-Coloumb model is included in the elastoplastic model with two shear strength parameters namely cohesion (c) and friction angle (ϕ). 2D Plaxis version 8.6 software can handle models with parameter values of cohesion (c) of zero, friction angle (ϕ), dilatation angle (Ψ), where in the sand soil, the angle of dilation depends on the density and friction angle, Young modulus/modulus elasticity (E), and Poisson's ratio (ν).

In general, Poisson's ratio values range from 0.3-0.4 depending on the type and condition of the soil. In soil mechanics, two types of analysis are used, namely undrained analysis and drained analysis. In 2D Plaxis software, the calculation used undrained analysis. The modulus of elasticity of the soil in an undrained condition, according to [17], can be calculated by Equation (5).

$$E_u = \frac{15.000 \text{ x } C_u}{IP} \tag{5}$$

Where:
E_u = modulus of elasticity for *undrained* condition (kN/m^2);
C_u = undrained shear strength (kN/m^2);
IP = index of plasticity (%).

The calculation of safety factors (SF) in the 2D Plaxis software can be done by selecting ϕ-c reduction. This option is only available in plastic calculations in the manual setting procedure or load advancement number of steps. 'ϕ-c reduction' approaches the shear angle parameters (ϕ) and cohesion (c) by reducing shear parameters to failure. The total multiplier (ΣMsf) is used to determine the parameters given at the analysis stage [16] can be calculated by Equation (6).

$$\Sigma Msf = \frac{\tan\varphi_{input}}{\tan\varphi_{reduced}} = \frac{c_{input}}{c_{reduced}} \tag{6}$$

The advantages of the 2D Plaxis version 8.6 software in analyzing slope stability are:
1. Provides a clear picture, both at the time of failure and slope stability and provides an effective safety chart according to safety criteria;
2. Able to simulate construction in stages, as is usually done in embankment construction;
3. Can model reinforcement elements, such as retaining walls, geogrids, geotextiles, anchors, and their interfaces

According to [9], the research method is the steps carried out systematically with a clear frame of reference in solving problems starting from the stage of collecting information on landslides, looking for literature studies related to landslides, primary and secondary data collection. Then proceed with analyzing the stability of the existing slope conditions using the Bishop method and the 2D Plaxis software. If these conditions do not meet the safety factor (SF), proceed with analyzing the stability of the slope with steel sheet pile reinforcement to obtain a safety factor (SF) ≥ 1.5. Furthermore, analyzed using the 2D Plaxis software has three stages, namely the data input stage, the calculation stage, and the data output stage.

In analyzing slope stability with the 2D Plaxis software requires modeling the slope in accordance with the actual conditions so that accurate results are obtained to determine the value of the safety factor of these conditions. However, if after analyzing the natural slope conditions but the safety factor value is not required, slope stability is reinforced by installing sheet piles.

3. Results and discussion

The results of soil laboratory testing and slope stability analyses for the study case in Meulaboh–Geumpang national road located in Lancong Village, Sungai Mas District, West Aceh Regency are shown below.

3.1. Results of soil laboratory testing and slope geometry

Results of soil laboratory testing regarding the physical and mechanical properties of the soil are shown in Table 3 and slope geometry in STA 84+910 is shown in figure 5.

Table 3. Results of soil laboratory testing.

No.	Description	B1	B2
1	Unit weight of soil (gr/cm³)	1.645	1.65
2	Dry soil weight (gr/cm³)	1.235	1.10
3	Water content (%)	33.15	49.43
4	Liquid Limit, **LL %**	89.85	65.79
5	Plastic Limit, **PL %**	35.95	40.79
6	Index of Plastiicity, **IP %**	53.90	25.01
7	Specific gravity, Gs	2.44	2.46
8	Friction angle, ϕ (°)	25.6	27.9
9	Cohesion, c (kg/cm²)	0.175	0.14
10	AASHTO Classification System	A-7-6	A-7-6
11	USCS	CH	MH

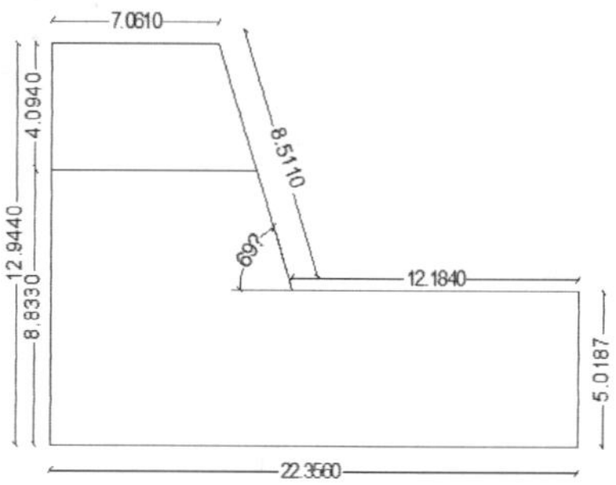

Figure 5. Slope geometry in STA 84+910.

The obtained soil properties values are used for calculation of safety factor by using Bishop method for the existing condition. Specification of steel sheet pile is used for slope reinforcement and as the input parameter for 2D Plaxis calculation as shown in Table 4. In addition, the physical and mechanical properties of soil (input parameters) also used in calculation by using 2D Plaxis software, as shown in Table 5.

Table 4. Specification of Steel Sheet pile [14].

	Sheet pile specification (Steel)
EA	1.200E + 07 KN/m
EI	1.200E + 05 KNm²/m
d	0.346 m
w	8.300 KN/m
v	0.15
Mp	1.000E + 15 KNm/m
Np	1.000E + 15 KNm/m

Table 5. Input Parameters on the Meulaboh-Geumpang STA 84 + 910 Road Section.

Description	Unit	STA 84+910	
		B1	B2
Material model	-	MC	MC
Type of behavior	-	*Drained*	*Drained*
Dry soil weight (*γdry*)	kN/m^3	12.99	10.80
Wet soil weight (*γwet*)	kN/m^3	16.13	16.15
Saturated Soil (*γsat*)	kN/m^3	23.95	24.08
Horizontal permeability (k_x)	m/day	0.00864	0.00864
Vertical permeability (k_y)	m/day	0.00864	0.0086
Young's modulus (E_{ref})	kN/m^2	6376.5	6376.5
Poisson's ratio (ν)	-	0.3	0.3
Cohesion (c)	kN/m^2	17.27	14.13
Friction angle (ϕ)	°	25.6	27.9
Dilatancy angle (Ψ)	°	0	0

3.2. Slope stability analyses results

The results of the safety factor for the existing condition as shown in figure 6 is used to see the location of the slope failure pattern which is then used to calculate the Bishop method and also the calculation for reinforcement using the sheet pile in the 2D Plaxis software.

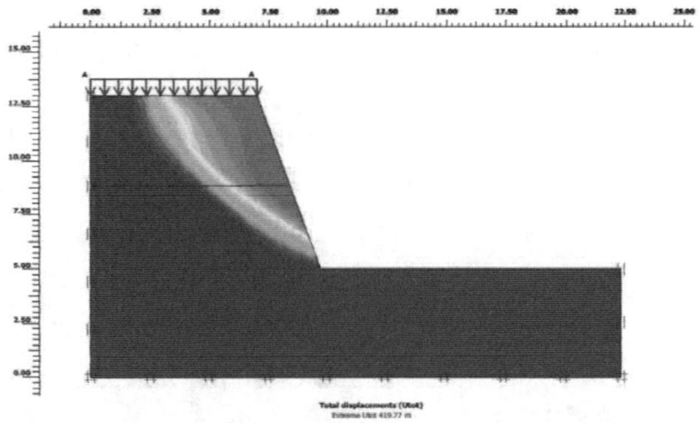

Figure 6. Slope failure pattern for existing conditions in STA 84 + 910 (SF = 1.10).

The results of slope stability analyses obtained by reviewing 2 (two) slope conditions, namely the existing conditions and the condition of sheet pile reinforcement with different sheet pile depths and placement. The existing conditions are calculated using the Bishop method and also using the 2D Plaxis software (as shown in Table 6) while the sheet pile reinforcement is performed by simulating the 2D Plaxis software to obtain a safety factor that matches the safe and stable conditions.

Table 6. Results of Analysis of Safety Factors in Existing Conditions Using the Bishop's Method and the Plaxis Software.

No.	Location (STA)	Safety Factor (Bishop)	Description	Safety Factor (2D *Plaxis* 8.6)	Description
1	STA 84+910	1.08	Unstable	1.10	Unstable

In the existing condition, two calculation models are used: the Bishop method, which obtains a safety factor value of 1.08, and the 2D Plaxis software, which obtains a safety factor of 1.10. In addition to the influence of the slope angle, the depth of the sheet pile also has a great influence on the value of the safety factor obtained after the reinforcement is performed. STA 84 + 910 has a slope angle of 69° so it has the different safety factors for various conditions of placement and depth of sheet pile.

Slope stability analysis using sheet pile reinforcement was carried out by changing the sheet pile depth from 8m, 9m, 10m, 11m, 12m, for the location of the upper slope, and 6m, 7m, 8m, 9m, 10m, for the location of the middle of the slope. Results of the safety factor obtained from 2D Plaxis software for various placement and depth of the sheet pile are shown in Table 7.

Table 7 shows that at STA 84 + 910, reinforcement by sheet pile installation gets the highest safety factor value of 1.57 for 12m sheet pile depth for upper slope. It can be seen very clearly from figure 7, the influence of the pattern of slope failure, if the sheet pile is installed, does not pass through the failure plane, then it does not increase the value of the safety factor. Slope stability analysis using sheet pile reinforcement in the 2D Plaxis software shows that the results obtained will be better if the sheet pile is installed at a depth that passes through the failured plane. If the installation of sheet pile on the soil layer located in the failured plane, so it does not have the contribution of reinforcement due to the weight of the sheet pile itself.

Table 7. Results of Safety Factor for Variation in Placement and Depth of Sheet Pile (STA 84 + 910).

No.	Location of *Sheet Pile*	Position	Slope angle	*Sheet Pile* depth (m)				
1	Upper slope	A	69°	8	9	10	11	12
	Safety Factor (SF)			1.19	1.12	1.33	1.42	1.57
2	Middle of slope	B	69°	6	7	8	9	10
	Safety Factor (SF)			1.18	1.25	1.34	1.42	1.48

On the other hand, figures 8 shows the results of the slope failure pattern after reinforcement by installing sheet pile with various placement and depth of the sheet pile. The results obtained from the calculation of the safety factor after reinforcement is 1.48 with a 10m sheet pile depth in the middle of the slope.

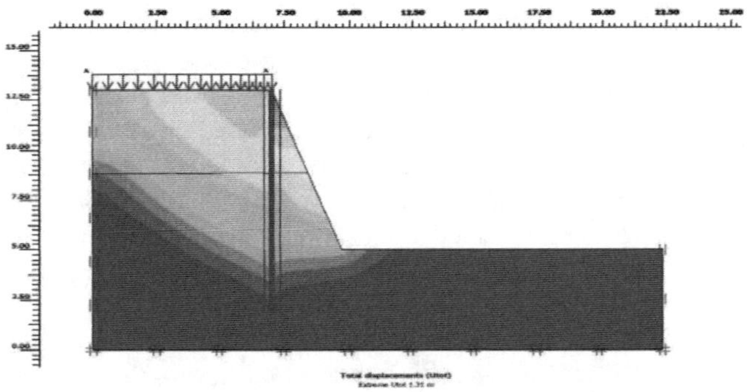

Figure 7. Analysis of Slope Stability at STA 84 + 910 with Sheet Pile Installation of 12m depth (upper slope), SF = 1.57.

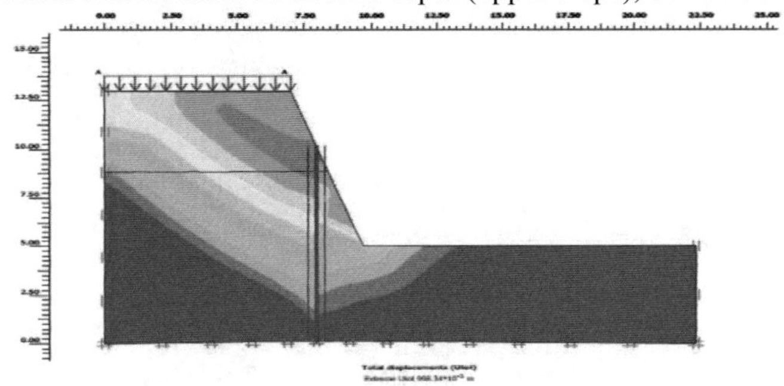

Figure 8. Analysis of Slope Stability at STA 84 + 910 by Installing a Sheet Pile Depth of 10m (the middle of the slope), SF = 1.48

4. Conclusion

Based on the results of slope stability analyses using the Bishop method and 2D Plaxis software on the Meulaboh-Geumpang road section, Sungai Mas District, West Aceh Regency, it can be concluded that the safety factor obtained for existing condition by using Bishop method for is 1.08 while using 2D Plaxis software the safety factor is 1.10.

The results of the safety factor values for after reinforcement condition vary greatly, the position of sheet pile installation also greatly influences the value of the safety factor. The results of the safety factors that meet the safety requirements for STA 84 + 910 are for the position of upper slope, sheet pile with a depth of 12m SF is 1.57, on the other hand, the placement of sheet pile in the middle of the slope with a depth of 10m SF is 1.48. Those indicate that sheet pile reinforcement for a 12 m sheet pile depth meets the required safety factor (SF $\geq$ 1.5).

It is generally obtained that the longer the sheet pile, the value of the safety factor in the upper slope will increase. However, the addition of a certain depth of the sheet pile will decrease the value of the safety factor. Placement of sheet piles must be located below the slope failure plane if the sheet pile has not influenced the safety factor or if the sheet pile only increases the soil overburden pressure resulting in the landslide even worse.

References

[1] Abramson L W et al 1996 *Slope Stability and Stabilization Methods* (New York: Wiley and Sons Inc)

[2] Susi and Yohan 2007 *Program Analisis Stabilitas Lereng* (Universitas Diponegoro)

[3] Azmeri and Sundary D 2013 Stability Analysis of Edge River Lawe Liang Pangi at Leuser Sub-District, Southeast Aceh Regency Towards Flash Flood *J Inersia* **5** 73-85
[4] Wu J Y, Huang K and Sungkar M 2017 Remediation of Slope Failures by Compacted Soil-Cement Fill *J. of Performance of Constructed Facilities* **1** (ASCE) 1 1-9
[5] Tikanta T, Matsumoto T, Vu A, T Kobayasi S, Shimono S and Bamrungwong C 2017 Fundamental Experiments on Reinforcement Method using Sheet Pile Wall for Bridge Pile Foundation Subjected to Pile Embedment Reduction and Numerical Validation Geotechnical Engineering *J. of the SEAGS & AGSEA* **48** 3 25-39
[6] Yoshida M, Mitou M, Kiyomiya O, Tashoro S and Gouda K 2012 *Effect of Seismic Reinforcement of Sheet Pile Quay Wall Using Ground Anchor 15 WCEE* (Lisboa)
[7] Wesley L D 2010 *Mekanika Tanah untuk Endapan dan Residu* (Yogyakarta: Andi Offiset)
[8] Bowles J E 1993 *Sifat-sifat Fisis dan Geoteknis Tanah* terjemahan J.K. Hainim Edisi Kedua (Jakarta: Penerbit Erlangga)
[9] Das B M 2000 *Fundamentals of Geotechnical Engineering* (USA: Brooks/Cole Thomson Learning)
[10] Hardiyatmo H C 2007 *Mekanika Tanah II* Edisi Keempat (Yogyakarta: Gadjah Mada University Press)
[11] Hasyim A 2007 *Slope Stability Analysis in Saturated Slope* Faculty of Civil Engineering Universiti Teknologi Malaysia, inside.mines.edu (Malaysia)
[12] Bishop AW 1955 *The Use of Slip Surface in The Stability of Analysis Slopes Geotechnique* **5** (London) 1 7-17
[13] Anderson M G and Richard K D 1987 *Slope Stability Geotechnical Engineering and Geomorphology* (John Wiley and Sons)
[14] Rashed A 2014 *A new prediction model for slope stability analysis* (Bologna: PhD Dissertation Department of Civil Engineering University of Di Bologna, Italy)
[15] Griffiths D V and Lane P A 1999 *Slope stability analysis by finite elements* Geotechnique **49** 3 387-403
[16] Brinkgreve R B J and Vermeer P A 1998 *Finite Element Code for Soil and Rock Analyses* version 7 Balkema (Rotterdam)
[17] Plaxis 2007 *Manual Plaxis 2D* – Versi 8 (Amsterdam)

IOP Conf. Series: Materials Science and Engineering **933** (2020) 012036 doi:10.1088/1757-899X/933/1/012036

The effect of fishbone fiber and rice husk ash additive on the mechanical properties of normal concrete

S Damayanti[1], T B Aulia[2] and Y Hayati[2]

[1] Civil Engineering Postgraduate Program, Department of Civil Engineering, Universitas Syiah Kuala, Banda Aceh, Indonesia
[2] Department of Civil Engineering, Universitas Syiah Kuala, Banda Aceh, Indonesia

Corresponding e-mail: sri.d@mhs.unsyiah.ac.id

Abstract. Concrete is a material that strong in holding compression force but weak in holding tension force so reinforcement is needed to prevent cracks in the pull region. Tuna bone fibers work as reinforcement, which is expected to overcome the problem of brittle concrete and produce ductile concrete. To improve concrete performance, fibers and additives are used. Tuna bone fibers and rice husk ash additives were selected to be able to utilize local natural materials. The purpose of this study is to examine the mechanical properties of normal concrete, which include compressive strength and tensile strength, by adding fiber and ash additives. The percentage of fibers added is 0%, 0.5%, 1%, and 1.5% of volume of concrete, and the percentage of additives is 0%, 10%, and 15% of weight of cement. Compression tests were performed to a total of 90 pieces of objects at 7, 28, and 56 days of age. Split tensile tests were performed to a total of 60 pieces of objects at the 28 and 56 days of age. The maximum compressive strength at 7, 28, and 56 days of age are 31.45 MPa, 36.81 MPa, and 41.16 MPa, respectively which were produced by variation BNS1 (BN+fiber 0.5%). The maximum tensile strength at 28 and 56 days of age are 4.08 MPa and 4.12 MPa; respectively which were produced by variation of BNS2 (BN+fiber 1%). All specimens at 7, 28, and 56 days of age produce concrete compressive strength on average 20 MPa of concrete above so that it is categorized into structural concrete.

1. Introduction

Concrete has advantages in strong holding force/compressive stress but also has specific disadvantages that it is not strong holding force/tensile stress. The tensile strength of concrete is only one of twenty of its compressive strength [1]. To improve this performance, fiber concrete is used. The idea of using fiber is to reinforce concrete with fiber that is spread evenly to prevent cracks in the tensile area. The addition of fiber overcomes the problem of brittle concrete (becoming more ductile). Adding fiber to concrete not only increase the concrete's tensile strength, but it also can improve other concrete's mechanical properties [2]. Also, fiber concrete has advantages compared to fiber without concrete in several structural properties, among which are ductility, resistance to shock loads (impact resistance), tensile and flexural strength, resistance to fatigue, resistance to shrinkage, and resistance of abrasion.

Tuna bone fiber is a natural fiber derived from animals. Tuna bone is part of the body of fish. It has the highest calcium content compared to other fish species, the calcium content ranges from 12.9% - 39.24% [3]. Bone has relatively hard and light tissue; it is formed from a portion of calcium phosphate. Bone has a relatively high compressive strength of about 170 MPa, a tensile strength of 104-121 MPa, and flexural strength of 51.6 MPa. Bone has a significant level of elasticity contributed by collagen [4]

To improve the performance of concrete, in addition to fiber concrete, the added ingredients is also often used. Rice husk ash is classified as one of the additives. Rice husk ash is the result of burning rice husk which has a dominant silica content of 93%—almost the same as the silica content found in factory-made micro silica [5].

2. Methodology

2.1. Fiber concrete

Fiber concrete is a composite material consisting of ordinary concrete or other material in the form of fiber (in the form of rods with diameters between 5 and 500 μm with a length of about 2.5 mm to 10 mm). The addition of fiber to the intended concrete is to improve the weaknesses of the concrete that has a low tensile strength [6]. Fiber concrete is concrete made from a mixture of cement, aggregate, water, and some fibers that are distributed randomly [2]. The principle of adding fiber into concrete is to give reinforcement to concrete. The fiber is spread evenly into the concrete mix with a random orientation to prevent concrete cracks that are too early in the tensile area due to heat hydration or due to loading.

Fiber concrete has the concrete's mechanical properties which are influenced by several factors, including the type of fiber, fiber aspect ratio, fiber fraction volume, the strength of concrete, geometry, and manufacture of test specimens and aggregates. Adding fiber to the concrete mixture will reduce the elasticity of the affected mixture, including fiber aspect ratio and percentage amount of fiber added to the concrete mix (those will be explained in the following sentences). First, the fiber aspect ratio is the ratio between the length and diameter of the fiber. The maximum limit of fiber aspect ratio that still allows mixing can be easily achieved is $lf/df < 100$ (lf and df are the length and diameter of the fiber). The high aspect ratio causes the tendency of the fiber to clot (balling effect) and the difficulty to spread evenly. Second, the percentage amount of fiber added to the concrete mix (fiber volume fraction) [7]. The decrease in workability can be overcome by increasing the cement water factor, reducing the maximum aggregate diameter, or using additives. Fibers that are often used in concrete mixtures have a length of 10 - 60 mm [2].

2.2. Tuna bone fiber

Tuna in the Province Aceh Indonesia is often processed into 'keumamah' fish. This processed food is closely related to the culture of Aceh, so it is in demand and continues to be produced. 'Keumamah' fish production reaches 2.725 tons/year. To utilize the processing of 'keumamah' fish, tuna bones are used as fiber.

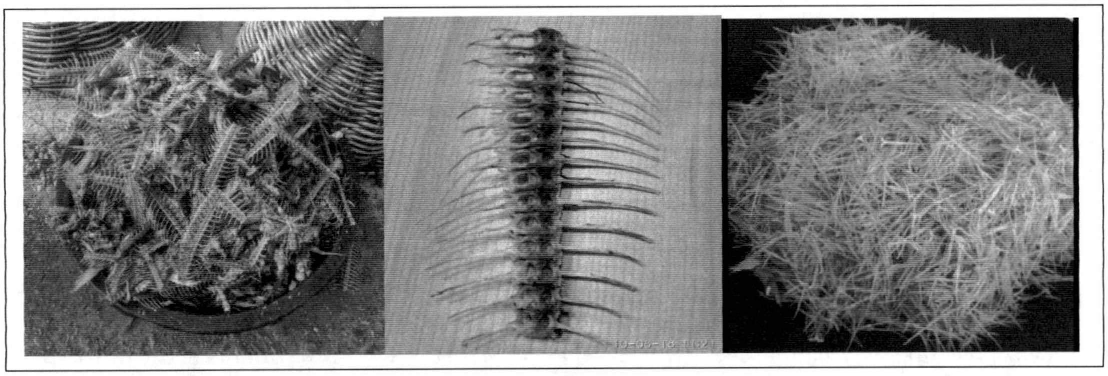

Figure 1. Tuna bone.

2.2.1. Mechanical characteristics of bones. The main tissue of the bone, the osseous network, is relatively hard and light, formed mostly from Calcium Phosphate in a chemical arrangement called Calcium Hydroxylapatite (this is the tissue which gives rigid properties to the bone). This network has a relatively high compressive strength of about 170 MPa (1800 Kgf / cm) but has a low tensile strength from 104 MPa to 121 MPa and a very low flexural strength of 51.6 MPa, which means it holds the push well, but not good in traction or moment. Bones are brittle but have a significant level of elasticity, which is donated by collagen. All bones are composed of living and dead cells that are in the organic matrix of minerals that make up the tissues [4].

Table 1. Composition of calcium phosphate as a mineral in bone [8].

Mineral content	Percentage
Calcium phosphate	85%
Calcium carbonate	14%
Magnesium	1%

Table 2. Content of tuna bones in the form of bone flour [9].

Content	Percentage
Water content	5.60 - 8.30%
Ash	77.54 - 84.22%
Protein	0.48 - 1.29%
Fat	1.70 - 4.13%
Calcium	23.72 - 39.24%
Phosphor	11.34 - 14.25%

Tuna bones have the highest calcium content compared to other fish, which ranges from 12.9% - 39.24%. The main elements that form tuna fish bones are calcium oxide, phosphate, and carbonate [3]. Tuna bones contain calcium oxide (CaO) of 31.36%.

2.3. Rice husk ash

Rice husk ash or RHA has very high pozzolanic activity superior to other SCM, such as fly ash, slag, and silica fume, because they contain silica at 87% -97%. Rice husk ash has a particle size of 5-10 μm, cellular and irregular particle shape, the specific gravity of 2.0-2.4 kg/m^3, the surface area of 20-50 m^2/kg and has an amorphous silica content of 85-90%, so that it can act as a potential additive to cement [10].

Rice husk is a cellulosic material like other biomass but contains high silica. The chemical content of rice husk consists of 50% cellulose, 25-30% lignin, and 15-20% silica [11]. Rice husk ash can be classified as pozzolanic because it contains SiO2 + Al2O3 + Fe2O3 greater than 70% according to the required pozzolanic quality [12]. The use of rice husk ash in cement composites can provide several benefits such as increasing strength and durability, reducing material costs, reducing the environmental impact of waste materials, and reducing carbon dioxide emissions [13].

2.4. Mechanical properties of concrete

2.4.1. Concrete compressive strength. Compressive strength is one of the main performances of concrete. The compressive strength is the ability of concrete to accept compressive force per unit area [14]. Concrete compressive strength testing refers to the ASTM C 234 standard by using a compression testing machine. The compressive strength of concrete can be calculated by equation (1).

$$f_c' = \sigma = \frac{P}{A} \tag{1}$$

IOP Conf. Series: Materials Science and Engineering **933** (2020) 012036 doi:10.1088/1757-899X/933/1/012036

Where:
f_c' = compressive strength of concrete cylinders (kg/cm^2),
P = maximum compressed load/ crushed (kg), and
A = the cross-sectional area of the test specimen (cm^2).

2.4.2. Concrete tensile strength. The tensile strength is determined by the tensile strength of a concrete cylinder (pressed on its long side) [11]. Of the several methods, the most frequently used is split cylinder testing. According to ASTM C496, the test is carried out by giving load to the side of the cylinder until it breaks or splits. The tensile stress that arises when the test object is split is called the split cylinder strength. The tensile strength is obtained from equation (2).

$$f_t = \frac{2.P}{\pi.L.D} \tag{2}$$

Information:
f_t = split tensile strength (kg/cm^2),
P = load time split (kg),
L = the length of the cylindrical test specimen, and
D = diameter of cylindrical test specimen (cm).

Based on the research of several experts, the value of tensile strength to concrete compressive strength varies. The tensile strength of concrete is around 20% of the compressive strength. The tensile strength of concrete ranges from 10-15% of the compressive strength [7]. The value of the tensile strength approach obtained from the split cylinder test reaches strength ($0.5\sqrt{f_c'} - 0.678\sqrt{f_c'}$ MPa) [7].

2.5. Equipment and materials
The equipment used is available at the Construction and Building Materials Laboratory, Department of Civil Engineering, Faculty of Engineering, Syiah Kuala University. The types of equipment are oven, a set of filters, measuring cups, scales, containers, cornering sticks, glass plates, concrete mixer machines with a capacity of 90 liters, slump gauges, rubber hammers, cement spoons, cement spoons, wheelbarrows, shovels and molds cylindrical specimens of 15 cm diameter, 30 cm height and 30 cm x 10 cm x 30 cm of mold prism specimens dimensions. The planned material used can be seen in table 3 below.

Table 3. Plans of material used.

No	Material	Goals to Be Achieved	Requirements	Size Plan Used
1	Aggregate	Fiber can be spread evenly	Aggregate diameter < 19 mm	≤ 4.75 mm (fine sand)
				4.76 - 9.5 mm (rough sand)
				< 19 mm (gravel)
2	Tuna Fish Fiber	A measure often used	Long 10 mm - 60 mm	Diameter 1 mm, length of 20 – 60 mm
		Easy stirring/ workability	Aspect ratio lf/df < 100	20 - 60 mm
		Good speediness	Fiber content < 2%	Percentage 0%, 0.5%, 1%, 1.5%
3	Rice Husk Abu	As a partial substitution	Percentage of fiber (10-15%)	Percentage 0%, 10%, 15%
4	Cement	Slump limits fiber concrete	Slump value; 25 - 100 mm (tolerance ± 20 mm)	The weight of cement is adjusted accordingly water used

No	Material	Goals to Be Achieved	Requirements	Size Plan Used
5	Water	WCR of normal concrete	WCR (0.4 - 0.65)	WCR 0.4
*	Concrete	Medium quality concrete	f'c; 20 - < 35 MPa	f'c Plans: 30 MPa

Material inspection carried out includes checking specific gravity (ASTM C.127-93), absorption (ASTM C.128-93), bulk volume (ASTM C.127-88), and sieve analyzer (ASTM C.136- 93).

3. Results and discussion

3.1. Aggregate physical examination results

Table 4. Results of calculation of physical properties of aggregates.

Material Check Type						
I	**Bulk Density Aggregate**					
No	**Aggregate Type**	**Volume Weight (kg/l)**			**Permit Limits Orchard (1979)**	
1	Gravel	1.829			> 1.445	
2	Rough sands	1.778				
3	Fine sands	1.632				
II	**Specific Gravity and Absorption**					
No	**Aggregate Type**	**Specific Gravity SG (SSD)**	**Specific Gravity SG (OD)**	**Limit on Specific Gravity Permit (SNI 03-1970-1990)**	**Absorption (%)**	**Absorption Permit Limit (%) (SNI)**
1	Gravel Ø19 mm	2.757	2.730	> 2.5	0.988	< 3
2	Rough sand Ø9.52 mm	2.660	2.597		2.421	
3	Fine sand Ø4,76 mm	2.548	2.502		1.837	
III	**Fineness Modulus Aggregate**					
No	**Aggregate Type**	**Fineness Modulus**			**Permit Limits (SNI 03 1970 1990)**	
1	Gravel	6.222			6.0 - 8.0	
2	Rough sands	2.788			1.5 - 3.8	
3	Fine Sand	2.646				

From table 4, it can be concluded that the material used in the study as a concrete forming material has fulfilled the requirements.

3.1.1. Physical properties of other materials. Other physical properties data include fishbone fibers. The result of the calculation of fishbone volume weight is 0.813 kg/l, and the specific gravity is 0.639 g/ml.

3.2. Aggregate physical examination results

The calculation of a concrete mix design uses the ACI method. The results of the calculation of concrete mix designs for each variation with a composition of 1 m³ concrete can be seen in the table below:

Table 5. 1 m^3 Concrete mixture design for all fiber variations and additives.

Variations	Composition per m^3 of concrete (kg)		
	Material	**Total**	**Unit**
1	**BN (Normal Concrete)**		
	Water	205.00	Kg
	Cement	512.50	Kg
	Rough Aggregate	1220.63	Kg
	Fine Aggregate	406.88	Kg
2	**BNS1 (BN + 0.5% ST)**		
	0.5% Bone Fiber	3.20	Kg
3	**BNS2 (BN + 1% ST)**		
	1% Bone Fiber	6.40	Kg
4	**BNS3 (BN + 1.5% ST)**		
	1.5% Bone Fiber	9,60	Kg
5	**BS1A1 (BN + 0.5% ST + 10% AS)**		
	0.5% Bone Fiber	3.20	Kg
	10% Rice Husk Ash	51.25	Kg
6	**BS1A2 (BN + 0.5% ST + 15% AS)**		
	0.5% Bone Fiber	3.20	Kg
	15% Rice Husk Ash	76.88	Kg
7	**BS2A1 (BN + 1% ST + 10% AS)**		
	1% Bone Fiber	6.40	Kg
	10% Rice Husk Ash	51.25	Kg
8	**BS2A2 (BN + 1% ST + 15% AS)**		
	1% Bone Fiber	6.40	Kg
	15% Rice Husk Ash	76.88	Kg
9	**BS3A1 (BN + 1.5% ST + 10% US)**		
	1.5% Bone Fiber	9.60	Kg
	Rice Husk Ash 10%	51.25	Kg
10	**BS3A2 (BN + 1.5% ST + 15% US)**		
	1.5% Bone Fiber	9.60	Kg
	Rice Husk Ash 15%	76.88	Kg

3.3. Concrete mechanical properties test results

3.3.1. Compressive strength. The compressive strength of the concrete with 0.5% fiber addition on all variations of age tests which were 7, 28, and 56 days increased 16.78% (31.45 MPa), 7.57% (36.81 MPa), and 18.00% (41.16 MPa) compared to normal concrete, respectively.

The average compressive strength of concrete with 0.5%, 1%, and 1.5% fiber addition at 7, 28, and 56 days of age is higher than normal concrete (see figure 2). This shows that 31.36% of calcium oxide (CaO) found in the bones of tuna fish which CaO is also found in cement as the main constituent component (60-67%) affects to increase the hardening speed and the strength of concrete.

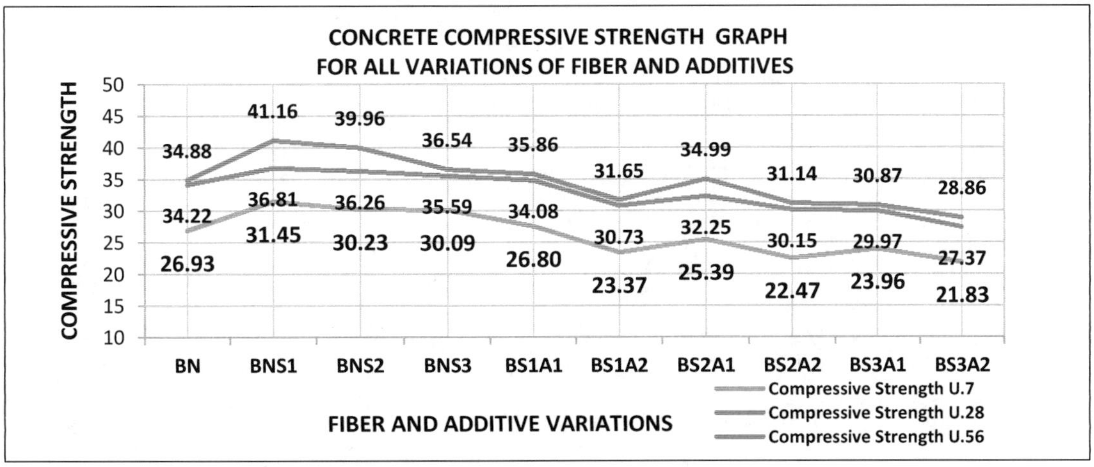

Figure 2. Concrete Compressive Strength Graph for All Variations of Fiber and Additives.
Note: S1 (0.5% Fiber), S2 (1% Fiber), S3 (1.5% Fiber), A1 (Additive 10%), A2 (Additive 15%)

The compressive strength of concrete with 0.5% and 1% fiber addition and 10% additives at 28 days of age is below normal concrete however at 56 days of age the strength is above normal concrete (see figure 2). This shows that the decomposition of rice husk ash particles with carbon above 45 μm affects the rate of hydration to increase the strength of the concrete. The silica SiO_2 in the rice husk ash that reacts to decompose $Ca(OH)_2$ produces calcium silicate hydrate (CSH) and the silica also affects the rate of strength increment.

The average compressive strength of concrete with 15% additives is below the strength of normal concrete. This demonstrates that the addition of fiber and additives in the concrete mixture reduces the workability which results in decreasing the concrete strength.

The test results of BNS1, BNS2, and BNS3 variations at 7 days achieved the minimum compressive strength target which was 30 MPa. For 28 days test, almost all variations reached the minimum compressive strength target (30 MPa) except BS3A1 and BS3A2 variations. For 56 days test, only BS3A2 variation did not reach the strength target. Concerning the structural quality of concrete, all specimens produce the average compressive strength above 20 MPa, so that they can be categorized as structural concrete.

3.3.2. Split tensile strength. The average split tensile strength of all sample variations at 28 and 56 days of age is higher than the split tensile strength of normal concrete (see figure 3). The average split tensile strength of concrete with 1% fiber addition at 28 and 56 days of age increase 7.33% (4.08 MPa) and 6.06% (4.12 MPa) from normal concrete, respectively.

The split tensile strength of concrete with addition of fiber and 10% additives at 28 days of age is lower than the split tensile strength of normal concrete however at 56 days of age the strength is higher than the strength of normal concrete. This shows that the decomposition of rice husk ash particles containing carbon above 45 μm is believed to affect the hydration rate which increases the strength of the concrete. The silica SiO2 in the rice husk ash decomposes $Ca(OH)_2$ to produce calcium silicate hydrate (CSH) and also affects to increase the rate of strength.

The average tensile strength of all variations of concrete with fiber addition and 15% of additives is lower than the tensile strength of normal concrete. This shows that the effect of fiber and additives in the concrete mixture reduces workability and clarity and decreases the strength of concrete produced.

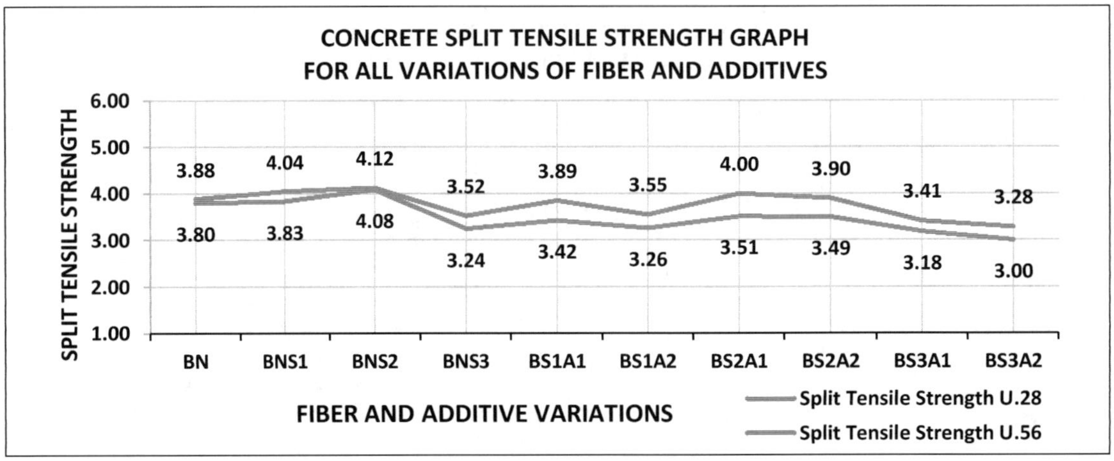

Figure 3. Concrete Split Tensile Strength Graph For All Variations Of Fiber And Additives.

Note: S1 (0.5% Fiber), S2 (1% Fiber), S3 (1.5% Fiber), A1 (Additive 10%), A2 (Additive 15%)

4. Conclusion

From the results of the study, the addition of tuna bone fibers and rice husk ash as additives to the normal concrete obtained several conclusions regarding mechanical properties as follows:

- The compressive strength of concrete with 0.5% fiber addition at 7, 28, and 56 days of age is higher than other variations. The compressive strength of concrete with 0.5%, 1%, and 1.5% addition of the fiber is higher than the strength of normal concrete. The concretes with fiber addition and 10% of additives at 28 days of age have compressive strength lower than normal concrete however at 56 days of age, the concretes have the strength higher than normal concrete. All variations of concrete with fiber addition and 15% of additives have the average compressive strength lower than normal concrete.
- The tensile strength of concrete with 1% of fiber addition at 28 and 56 days of age is higher than other variations. The concrete using 0.5% and 1% of fiber addition can increase the tensile strength. Concrete with fiber addition and 10% of additives at 28 days of age has the tensile strength lower than the normal concrete however at 56 days of age this concrete has the tensile strength higher than the normal concrete. Concrete with fiber addition and 15% of additives, all variations produce lower average tensile strength than the normal concrete.
- Regarding the structural quality of concrete, all specimens produce concrete compressive strength values above 20 MPa, so they can be categorized as structural concrete.

The results of this study are expected to be useful in the development of science, especially materials technology, and it can be taken into consideration that the local natural materials should be used as an alternative material to substitute synthetic fibers and factory-made additives.

Research on the addition of tuna fishbone fibers and rice husk ash additives in normal concrete mixes is expected to be continued by other researchers; therefore, it is suggested as follows:

- In further studies, it can be done by adding the percentage of tuna bone fibers with another percentage to see its effect on the tensile strength of concrete and other mechanical properties.
- In subsequent studies, the addition of rice husk ash additives can be done with other percentages to see the effect on the compressive strength of fiber concrete along with other mechanical properties.
- In subsequent studies, variations in the addition of tuna fishbone fibers with other types of additives at different percentages can be done.

IOP Conf. Series: Materials Science and Engineering **933** (2020) 012036 doi:10.1088/1757-899X/933/1/012036

References

[1] Murdock L J and Brook K M 1986 *Concrete Materials and Practices* 4th Edition (Jakarta: Erlangga)
[2] Soroushian P and Bayasi Z 1987 *Concept of Fiber Reinforced Concrete.* In proceeding of the international seminar on fiber reinforced concrete (Michigan: USA, Michigan State University)
[3] Detiknews 2011 *Thanks to Fish Bones; UGM Student Wins Food Innovation Competition* (Yogyakarta)
[4] Anam M S et al. 2013 *The Effect of Use of Bone Ash and Rice Husk Ash on the Mechanical Properties of Cement Paste*
[5] Swamy R N 1986 *Concrete Technology and Design: Cement Replacement Materials* Volume 3 (University of Sheffield: Reader in Civil and Structural Engineering)
[6] Tjokrodimulyo K 1996 *Concrete Technology* (Yogyakarta: Department of Civil Engineering, Faculty of Engineering, Gadjah Mada University)
[7] Sudarmoko 1991 *Flexural Strength of Fiber Concrete with Full-Scale Model* Seminar mechanics of materials in various aspects, inter-university center for science and engineering (Yogyakarta: Gadjah Mada University)
[8] Morrison F B 1959 *Feeds and Feeding* 9th (New York: The Morrison Publishing Company)
[9] Nabil M 2005 *Utilization of Tuna Bone Waste as a Source of Calcium Using Protein Hydrolysis Method* (Thesis, Faculty of Fisheries and Marine Sciences, Bogor Agricultural University)
[10] Nugraha P and Antoni 2007 *Concrete Technology of Materials, Manufacture, to the High-Performance Concrete* (Yogyakarta: Publisher ANDI)
[11] Ismail M S and Waliuddin A M 1996 Effect of rice husk ash on high strength concrete *Construction and Building Materials*
[12] Putra D 2006 Addition of husk ash to concrete in anticipation of damage due to Magnesium Sulfate from Sea Water *Scientific J. of Civil Engineering*
[13] Bui D D, Hu J and Stroeven P 2005 Particle size effect on the strength of rice husk ash blended gap-graded Portland cement concrete *Cement and Concrete Composites*
[14] Mulyono T 2006 *Concrete Technology* (Yogyakarta: Andi Publisher)
[15] Anonymous 2002 *Concrete Technology Practicum Handbook* Physics of Semiconductor Devices (New York: Wiley–Interscience, Concrete Technology Laboratory, Medan, USU FT-DPSSze S M 1969)

Analysis of reinforced concrete beam-column joint structures retrofitting

Z Araby[1], A Abdullah[1] and M Affifuddin[1]

[1] Department of Civil Engineering, Universitas Syiah Kuala, Banda Aceh, Indonesia

Corresponding e-mail: zardan_araby@unsyiah.ac.id

Abstract. There are many earthquake-related damages among Indonesian buildings, especially old buildings constructed based on SNI T-15-1993 standard. This research aims to study the comparison of the reinforced concrete beam-column joint structures' ability in resisting cyclic loads according to SNI T-15-1991-03 and SNI T-15-1991 codes to the ability of the joint structure that had been improved to the newer SNI 2847-2013 code. The beam-column joints structures designed by SNI T-15-1991-03 then retrofitted according to newer SNI 2847-2013 code in real-life situations refers to the damaged old building structures that fixed without the need to construct new buildings thus this will impact to budget efficiency. The joint specimens were made into three types of variation, which are the specimen with the reinforcement distribution not following the SNI T-15-1991-03 (BBK-1), retrofitting the specimen with the addition of woven wire (BBK-2), and the specimen with the reinforcement detailing according to SNI 2847-2013 (BBK-3). The beam dimension is 30 × 40 ×120 cm, and the column dimension is 30 × 30 × 200 cm. A test was carried out by applying a cyclic load in stages to a specimen through a hydraulic jack connected to a portable data logger for load recording. The specimen was equipped with strain gauges installed on steel reinforcements to detect strains that occur. The expected results in this study are to find out the most efficient and effective method used to reinforce the beam-column joints in resisting cyclic loads, both in the construction of new or old buildings.

1. Introduction

Indonesia is one of the countries where earthquakes frequently happen. Until now, there is not much that can be done to prevent earthquakes, which also directly cause damage to buildings. However, governments continue to update the codes regarding the structure of earthquake-resistant buildings to prevent damage to building structures so that there are no fatalities and material losses due to the impact of the earthquake [1].

Currently, the building structure failures related to the earthquake are caused by the design that does not follow the standard or by the design that refers to the older standard. The structure that frequently fail is the beam-column joint. Beam-column joint in buildings is one of the essential parts to receive loads; however the joint also has limited capacity. When an earthquake hits the structure, there will be a very large shear force at the beam-column joint; this shear force results in collapsing the beam-column joint because of the over shear-capacity force, the bonding breakdown, or the result of both [2].

The strength of the joint must not be less than the maximum requirement, which the strength can enlarge the mechanism of the plastic joint of the structure. The well design joint can relatively reduce the need for repairing the joint and the energy in a joint decreases when repetitive loads hit it in the elastic phase and the capacity of the column should not be affected by the possibility of a strength decreased in the joint (SK SNI T-15-1991). The joint area must be calculated accurately as part of the column. So that during an earthquake, a joint is expected in a plastic phase, and joint deformation may not increase the occurrence of a joint. Strengthening the joints needed to increase the capacity should not cause difficulties in the construction process. This research aims to find out and to study the comparison of the strength ability of beam-column joint designed according to SNI T-15-1991-03 and SNI 2847-2013 codes to resist shear forces from cyclic loads and also capability of the beam-column joint designed according to SNI T-15-1991-03 code which then the joint strengthened by wiremesh to resist shear force from cyclic loads [3].

External joint failure Internal joint failure

Figure 1. Destruction beam-column joint.

The compressive strength of concrete is the ability of concrete to accept the compressive force of broad unity. Concrete compressive strength identifies the quality of a structural plan, where the higher the level of strength of the desired structure, the higher the quality of the concrete was produced. Factors that influence the compressive strength of concrete include the proportion of constituent materials, methods of implementation, maintenance, and circumstances at the time of casting [4].

The value of compressive strength of concrete is obtained through standard testing using a compressive testing machine with a loading speed of 0.15 MPa / second up to 0.34 MPa / second until the concrete cylinder specimen is destroyed [5]. The commonly used test procedure is the American Society for Testing and Materials (ASTM) C39 standard. Compressive strength testing of each specimen done at 7, 28, and 56 days.

The compressive strength that arises according to the American Society of Testing Materials (ASTM) C-39 was calculated using the equation (1):

$$f_c' = \frac{P}{A} \tag{1}$$

Annotation:

f'_c = Concrete cylinder compressive strength (MPa);
P = Maximum press load/ press (Newton); and
A = Cross-sectional area of the specimen (mm^2).

1.1. Beam-column joint

1.1.1. Beam-column joint according to SNI T-15-1991-03 Code. The force in the reinforcement of the longitudinal beam on the surface side of the joint must be determined by assuming that the tension inside the flexural tensile reinforcement is 1.25 fy. Flexural moments and column shear forces, as well as horizontal shear and vertical forces passing through the joint, must be evaluated with rational analysis, which will count all the forces forming a balance on the joint column reviewed (6). The nominal horizontal shear force in the beam-column joint is given by the following equation:

$$V_c = \left(\frac{\sqrt{f_c'}}{6}\right).\ b_w\ .d \tag{2}$$

Annotation:

bw = board width, or circle cross-section diameter (mm);
d = effective cross-section; and
The shear strength provided by the shear tension is given by the following equation.

$$V_s = \frac{A_v.\ f_y.\ d}{S} \tag{3}$$

Annotation:

Av = Total cross-sectional area of shear reinforcement (mm^2);
Fy = Steel yield tension (MPa);
d = Distance from the outer press fiber to the tensile reinforcement (mm); and
S = Reinforcement space distance (mm).

The requirements for shear reinforcement planning are as follows:

$$V_u \leq \phi\, V_n \tag{4}$$

$$V_n = V_c + V_s \tag{5}$$

$$V_u \leq \phi\,(V_c + V_s) \tag{6}$$

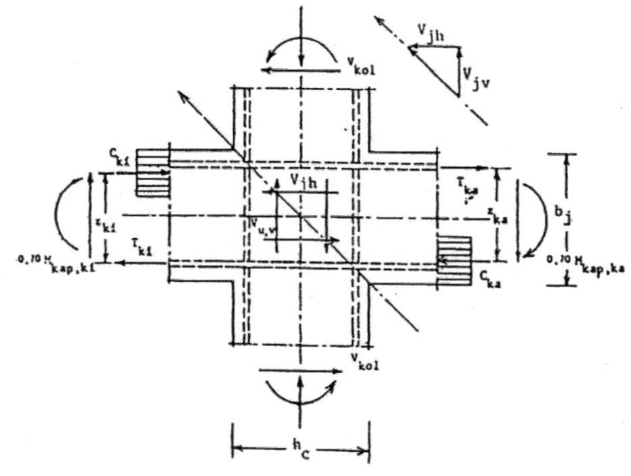

Figure 2. Beam-column joint according to SNI-T-15-1991 code.

1.1.2. Beam-column joint according to SNI 2847-2013 Code. SNI 2847:2013 explains that the shear force design (Ve) must be determined from an observation of the static force on the joint component structure of the joint. It must be assumed that torques with opposite signs are related to the strength of

IOP Conf. Series: Materials Science and Engineering **933** (2020) 012037 doi:10.1088/1757-899X/933/1/012037

the bending torque (Mpr) that might be working on the surfaces of the joint and that the structural components which deducted with factored gravity loads along the beam [7]. The picture of the beam-column joint, according to SNI 2847: 2013, can be seen in figure 3 below:

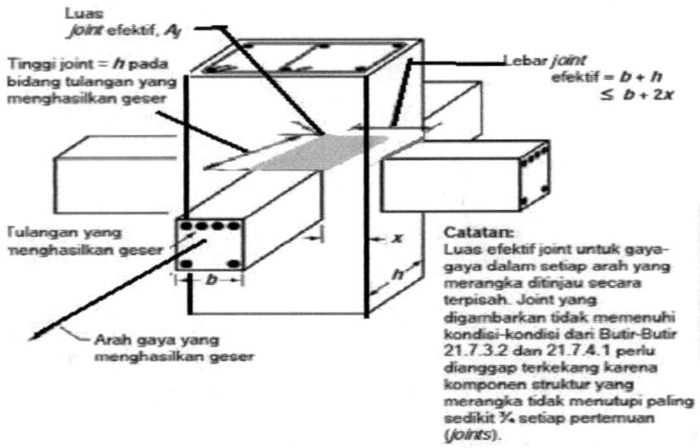

Figure 3. Illustration of beam-column joints from SNI 2847: 2013 code.

There are two types of factored shear forces that work on beam-column relations (Vu), i.e., for interior joints in equations (7) and exterior joints in equations (8) and (9). For the exterior joint, the biggest value is taken from those two equations.

$$Vu=1{,}25(As+As)fy-V_{kol} \tag{7}$$

$$Vu=1{,}25\,.\,As\,.fy-V_{kol} \tag{8}$$

$$Vu=1{,}25\,.\,As\,.fy'-V_{kol} \tag{9}$$

The shear force in the column can be calculated based on the values of Mpr- and Mpr+ divided by half of the upper column height (h1), plus the lower half of the column height (h2). The equation is shown below:

$$V_{kol} = \frac{Mpr^{-} + Mpr^{+}}{\frac{h1}{2}+\frac{h2}{2}} \tag{10}$$

Nominal shear tension (Vn) in beam-column joints can be seen from the equation below:

$$V_n = \frac{V_u}{bj\,.\,hc} \tag{11}$$

Annotation:

V_n = Nominal shear tension joint;
V_u = Factored shear force;
bj = Effective width of beam-column relations; and
hc = Effective height in beam-column relations.

The value of the shear force (Vn), for the confined beam-column relationship, can be seen from the following equation:

$$V_n \leq 1.7 \sqrt{f_c'.A_j} \tag{12}$$

The shear tension borne by concrete is calculated from the following equation:

$$V_c = \frac{2}{3}\sqrt{\left(\left(\frac{N_n.K}{A_g}\right) - 0.1f_c'\right)} \tag{13}$$

Annotation:

Vc = Shear stress borne by concrete;
Nn,k = Axial force column;
Ag = Cross-sectional area of the column; and
f'c = Concrete compressive strength.

Transverse reinforcement in beam-column joints is needed to provide sufficient restraints on the concrete, so it is capable of showing the ductile behavior and of carrying vertical gravity due to exfoliation on the concrete covers. Planning a shear reinforcement must meet the requirements as follows:

1. If Vn ≤ Vc, the minimum shear reinforcement will be used, and
2. If Vn > Vc, it will need shear reinforcement.

1.2. Retrofitting on reinforced concrete

According to [8], damage to unstable structures may be caused by external forces such as earthquakes, wind loads, and vibrations. Earthquakes can cause buildings to sway and cause the structure to crack where the damage level of building structures is based on the width of the cracks. Building damages that often occur are cracks, faults, debris, arches, and deflections. As a result of the damage to the structure, there is a need for improvements in the structure of the building. Durable building retrofitting requires the selection of suitable materials, including the retrofitting that require chemical resistance. The selection of retrofitting materials must be based on low-cost ease, the availability of skilled workers, and the work equipment needed.

1.3. Ferrocement

According to [9], ferrocement itself is a composite material with materials made from a mixture of sand and cement using mesh wires that can form a unified whole and form a strong construction. Generally, ferrocement thickness ranges from 10 mm to 50 mm, with reinforcement volumes ranging from 6% to 8% of all construction contents.

According to [10], the shear strength of ferrocement's contribution can be calculated by the following equation:

$$V_j = \frac{0{,}125\pi^2 d_w{}^2 f_{yj} n D\prime}{g_w} \tag{14}$$

Meanwhile, to calculate the number of wire mesh layers needed for shear reinforcement, it is calculated by equation (15) for round columns and equation (16) for square columns.

$$n = \frac{0{,}81 g_w V_j}{d_w{}^2 f_{yj} D\prime} \tag{15}$$

$$n = \frac{0{,}785 g_w V_j}{d_w{}^2 f_{yj} D\prime} \tag{16}$$

Annotation:

V_j = Jacket slide strength / ferrocement (kg/cm^2);

n = The amount of wiremesh layers (unit);
g_w = Distance/space wiremesh (mm);
f_{yj} = Jacket pull strength/ ferrocement (kg/cm^2);
d_w = Wiremesh diameter (mm); and
D' = Colomn diameter (mm).

2. Methodology

To achieve good research results, it must be done with a good and correct method and procedure. The stages of this research began with a literature study, followed by preparation of equipment and materials, materials testing, concrete mixture proportions planning, specimen preparation and maintenance, specimen testing, and data analysis.

2.1. Study objectives

The specimens were varied into three types, the specimens with the length of the reinforcement distribution were not in accordance with the SNI T-15-1991 code (BBK-1), the retrofitting specimens with the addition of woven wire (BBK-2), and the specimens with reinforcement length according to SNI 2847-2013 code (BBK-3), with the quality of concrete for each specimen is 18.68 MPa and the steel quality is 240 MPa. The dimensions of beams were 30 x 40 x 120 cm, and the dimension of columns were 30 x 30 x 200 cm, both beams and columns used 8Ø14 mm for main reinforcement and Ø10-100 mm for stirrup reinforcement with yield strength (fy) 240 MPa and tested at 28 days.

Cylindrical specimens were also prepared in size of 15 x 30 cm. The mix design was obtained by the trial method based on the concrete volume weight. The specimen is made for testing the cyclic load capacity in reinforced concrete beam-column joints. The planned specimen consists of normal reinforced concrete with smooth aggregates, coarse aggregates, reinforcing iron, and water. In this research, planning for reinforced concrete will use normal aggregates such as gravel, sand, and water. The plan and dimensions of the specimen can be seen in table 1 and table 2 below.

Table 1. Specimens production plans.

Specimens Production	Total of Specimens		Specimens Code	Reinforcement	
	Reinforced Concrete	Cylinders Concrete		Main	Sliding
SNI T-15-1991 Code	1	4	BBK-1	8 ∅ 14	∅ 10 – 100
Retrofitting SNI T-15-1991 Code	1	4	BBK-2	8 ∅ 14	∅ 10 – 100
SNI 2847-2013 Code	1	4	BBK-3	8 ∅ 14	∅ 10 – 100

Table 2. Specimens dimension production.

Specimens Production	Specimens Code	Specimen Dimensions (cm)	
		Beams	Columns
SNI T-15-1991 Code	BBK-1	30 x 40 x 120	30 x 30 x 200
Retrofitting SNI T-15-1991 Code	BBK-2	30 x 40 x 120	30 x 30 x 200
SNI 2847-2013 Code	BBK-3	30 40 x 120	30 x 30 x 200

The details of the planning and cross-section of the specimens can be seen in figure 4 below.

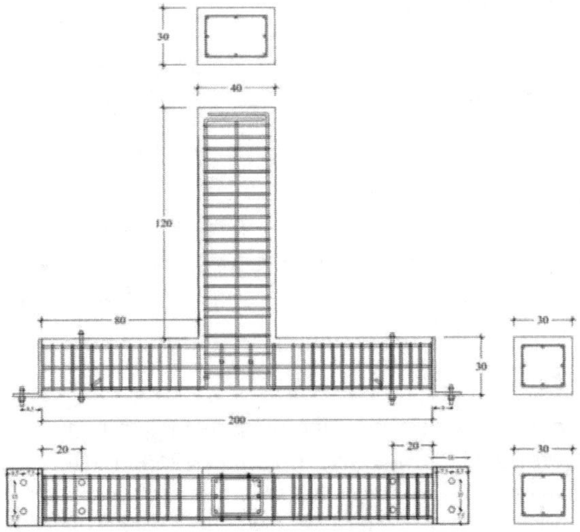

Figure 4. The details of planning and cross-section of the specimens

3. Results and discussion

3.1. Concrete compressive strength testing

The method used for the concrete press test is the ASTM C39-86 method. A concrete cylinder compressive strength test is done to determine the desired compressive strength of the concrete. A compressive strength test carried out when the specimen is at 28 days old. Concrete cylinder compressive strength test was used to represent beam casting as a control specimen. The set up of the compressive strength of cylindrical specimens test can be seen in figure 5. This compressive strength test was also carried out according to SNI 03-1974-1990, which concerns the concrete press strength testing method by placing the specimen on a press machine centrically. Then, the press machine was run with constant loading (ranging from 2-4 kg/cm^2 per second) until the specimen was destroyed. The maximum load was recorded.

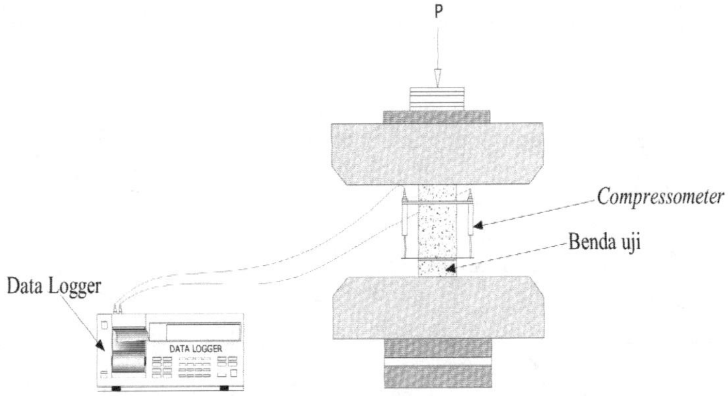

Figure 5. Concrete compressive strength test set up

3.2. Beam-column joint specimens test

The test will be carried out when the specimen is at 28 days old. The beam-column specimen test was carried out together with a cylindrical specimen. The cyclic load capacity that will be count is the compressive strength resulting from the compressive strength test of the cylindrical specimen. Before testing, the surface of the specimen was painted in white and the grid is drawn first in order to make it easier to draw the resulting crack pattern. The specimen is rigidly mounted on the beam frame. On the surface of the end of the beam area, the steel plate was placed as the foundation for the load cell in applying the compression load. The load was given horizontally at the end of the beam. The horizontal load was given by a hydraulic jack connected to the load cell and forwarded to the specimen. The load given was controlled by reading the dial on the hydrant jack. Loading was carried out in stages, and at each step, the load was read and the crack pattern that arises was illustrated. The load was given until the specimen was destroyed. On all sides of the joint, a transducer was installed to read the deflection.

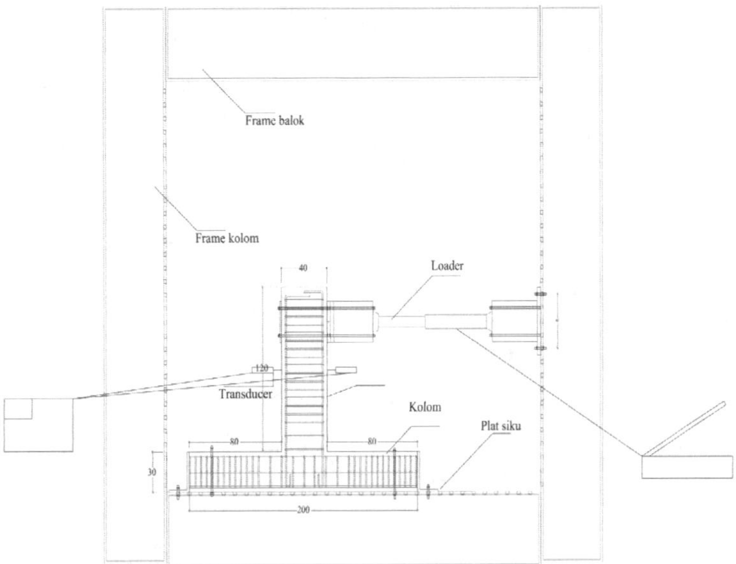

Figure 6. Beam-column joint specimens set up

Strains that occur in beam-column joints are read by a Portable Data Logger that has been connected to a strain gauge. Loading is stopped when the load is no longer increases due to the specimen being unable to accept the load so that the specimen is cracked and destroyed. The pattern of crack development is monitored at all times by making a crack image that occurs in the column according to the amount of load given.

3.3. Specimens Retrofitting

After the experiment was carried out with a cyclic load on the beam, the failure that occurred will be cleaned and then repaired by coating the failed area, which is in the column joint and beams. The reinforced layer material to be used is a layer made of ferrocement. The number of layers and the length of the layer to be used are calculated by equations 15 and 16.

4. Conclusion

After the test completed in the laboratory, it is expected that the conclusions can be drawn regarding the cyclic load capacity of beam-column joint specimen designed according to SNI T-15-1991 code, the joint designed according to SNI T-15-1991 code retrofitting with woven wire, and the joint designed according to SNI-2847-2013.

References

[1] Wijaya dan Saputra 2016 Repairing and retrofitting of earthquake-affected exterior beam-column connection by using resin concrete *J. of Civil Engineering*

[2] Venkatesan B and Ilangovan R 2016 Structural behaviour of beam column joint retrofitted with ferrocement laminates *Int. J. of Advanced Engineering Technology* **VII** 1272-80

[3] Ghobarah A and Said A M 2002 Shear strengthening of beam-column joints *Engineering Structures* **24** 881-8 (Canada: Elsevier, Department of Civil Engineering, McMaster University, Hamilton, Ontario L8s L7)

[4] Mourad S M and Shannag M J 2012 Repair and strengthening of reinforced concrete square columns using ferrocement jackets *Cement and Concrete Composite* **34** 2012 288-96 (Saudi Arabia: Elsevier, Department of Civil Engineering, King Saud University)

[5] Ristanto E, Suyadi S and Irianti L 2015 Analisis joint balok kolom dengan metode SNI 2847-2013 dan ACI 352R-2002 pada Hotel Serela Lampung *J. Rekayasa Sipil dan Desain (JRSDD)* **3** 3 521-40

[6] Widyawati R 2009 Keruntuhan lentur balok pada struktur balok kolom beton bertulang eksterior akibat beban siklik *J. Rekayasa* **13** 3 (Universitas Lampung)

[7] Bechtoula H, Sakashita M, Kono S, Watanabe F and Eberhard M O 2006 Cyclic performance of lower stories of mid-rise reinforced concrete frame buildings *Structural J.* **103** 4 513-21

[8] Achmad K, Agoes A and Tavio T 2012 Peningkatan kekuatan akibat beban siklik pada kolom beton bertulang persegi dengan pengekangan eksternal FRP *Jurnal Sains Terapan* **4** 1 845-56

[9] Artiningsih T P 2017 Kajian penggunaan ferrocement untuk retrofit kolom beton bertulang dengan variasi tingkat pembebanan *Prosiding Seminar Nasional Sains dan Teknologi* (Jakarta: Fakultas Teknik Universitas Muhammadiyah)

[10] Abdullah and Takiguchi K 2003 An investigation into the behavior and strength of reinforced concrete columns strengthened with ferrocement jackets *Cement and Concrete Composites* **25** 233-42 (Jepang: Elsevier, Department of Mechanical and Environmental Informatics)

IOP Conf. Series: Materials Science and Engineering **933** (2020) 012038 doi:10.1088/1757-899X/933/1/012038

Coconut fiber strength test as a potential to prevent landslides on badlands

M Suhendra[1], H Yunita[2] and S M Saleh[2]

[1] Civil Engineering Magister, Department of Civil Engineering, Universitas Syiah Kuala, Banda Aceh, Indonesia
[2] Department of Civil Engineering, Universitas Syiah Kuala, Banda Aceh, Indonesia

Corresponding e-mail: halida_yunita@unsyiah.ac.id

Abstract. Landslide is one of the natural disasters that often occur on natural slopes. Landslides and erosion of slopes mostly happen during the rainy season, resulting in a decrease in soil shear strength. One method of reinforcing slopes is to use coconut fibers which have the features of fibers. Coconut fiber is not only relatively more economical but also can last a long time as a substitute for geosynthetics and even as a place for vegetation growth. The purpose of this study was to determine the tensile strength value of coconut fiber and to identify the shear strength of barren slopes and bush slopes on Takengon-Bintang-Blangkejeren Road, Bur Retak Area, Kampung Mendale, Central Aceh. The tensile strength testing of coconut fiber was carried out by taking into account the diameter size and the number of fibers. The tensile stress testing of coconut coir fibers was done by clamping both sides, with the diameter and the number of fibers varied. The results showed that the tensile strength (tr, MPa) of coconut coir fibers decreased with the increase in diameter and the number of the fibers. In the application, coconut coir fibers placed on a barren slope in one square meter area has the potential to produce additional shear strength.

1. Introduction

The failure of a slope can be related to several factors, such as weather conditions, soil type, slope angle, topography, or a combination of these. Slope failure is due to a decrease in the shear strength of the soil where the soil is unable to resist the shear stress that occurs when burdened. Soil shear failure happens not because of the destruction of the soil grains but the relative motion between the soil grains [1]. Landslides and erosion are examples of natural disasters related to slope failure.

Landslides are the displacement of slope-forming materials in the form of rocks, rubble, soil, or mixed materials, moving down or out of the slope. Shallow landslides often occur in the rainy season. Shallow avalanche incidents generally are on steep road cliffs, with the depth of the slide field is no more than two meters, and more often less than one meter. The shallow landslide usually links to a fairly high frequency of rain. Landslides are classified [2] into six types, namely translational landslides, rotational landslides, falls, topple, lateral spread, and the flows.

Landslides often occur worldwide, as well as in Indonesia, especially Aceh. Aceh Province is one of the areas where landslides frequently happen, such as in the Bur Retak area of Kampung Mendale, Kebayak, Aceh Tengah District, on March 31, 2014. This landslide caused the Takengon - Bintang - Blangkejeren causeway to be covered by large rocks. This incident also killed two university students due to being hit by the rocks. This tragic incident encouraged the authors to research the method of slope strengthening.

One method of slope strengthening is using coconut mesh or coconut coir fiber. This technique is relatively inexpensive compared to other slope reinforcement methods, such as spray concrete, a term used for concrete that is shot and hardened in a relatively fast time (shotcrete).

Coconut coir fiber is widely used because of its long-lasting characteristics. It is very resilient, strong against friction, water-resistant, mold and pests resistant, as well as not easily broken, rot, nor inhabited by termites and rats (Rumokoi, 1990). Coconut coir fiber is also an alternative to composites development because it is not only cheap but also abundance. The area of coconut trees is around 3,712 million ha (Soba, 2003) or 3.88 million ha [3]. The islands of Sumatra, Java, Sulawesi and East Nusa Tenggara are the central areas for growing coconut trees, with an area of 2.841 million ha (76.5% of the total area of Indonesia). With the coconut area of 3.29 million ha [4], the potential for coconut production can reach an average of 15.5 billion per year. The total coconut fiber produced can reach up to 1.8 million tons. One coconut produces approximately 0.4 kg of coir, consisting of 30% fiber [5].

Coconut coir fiber consists of the chemical composition: cellulose, lignin, pyroligneous acid, gas, charcoal, tar, tannin, and potassium (Rindengan et al., 1995). Cellulose is the main organic constituent of plant cell walls [6]. Hartati et al. (2013), in their research, reported that cellulose can increase the strength of coconut coir fiber. Lignin, the largest part of cellulose, has a major role in forming layers between the fibers that function as a binding between cellulose fibers in wood or non-wood [7]. From this composition, coconut coir fiber can also be used as a suitable medium for plant growth. Plants growing around coconut coir fibers help maintain the slope stability by increasing soil shear strength, estimated by measuring the root length density and soil shear strength at depths of 0 - 5 cm [8].

The combination of coconut coir fibers with native soil on the slope has the potential to increase the soil shear strength. The results of Lastiko's research [9] explained that coconut coir fibers could increase and increase the soil shear strength, as indicated by the variation of the soil mixture with 1.5% coconut coir fibers increasing the shear strength by 10.30%. Another study also [10] showed that the shear strength of the soil added coconut coir fibers was enhanced with an average increase of 3.32%. The shear strength of the original soil without coconut coir fibre was 53.50 kN/m^2 while for the soil with 5%, additional coconut coir, the shear strength was 64.05 kN /m^2.

This study aimed to determine the tensile strength of coconut coir fibers and identify the shear strength of the barren slope and bush slope overgrown with plants as a comparison of the shear strength of the original soil. The barren slopes studied were on Jalan Takengon-Bintang-Blangkejeren KM 10, Bur Retak Area, Mendale Village, Kebayaan District, Central Aceh Regency, Aceh Province. In further research, it is expected that the installation of coconut coir fibers can be installed at the location of this barren slope as additional cohesion of native soil to increase soil shear strength.

2. Research methods

The method used in data collection was through direct observation conducted by authors at the research location (primary data) and written information or document form (secondary data).

2.1. Primary data

Primary data included the tensile testing of coconut fiber, soil sampling in the field and testing in the laboratory.

2.1.1. Field testing. There were two activities undertaken in the field, examining the tensile strength of coconut fiber and sampling the soil at the location of the barren slope and the bush slope overgrown with plants.

2.1.1.a. Testing the tensile strength of coconut fiber. The tensile strength of coconut fiber stress testing was done by taking into account the diameter size and the number of fibers. The diameter of the coconut fiber was measured using a caliper tool. Before testing, coconut coir fibers were soaked in water until saturated. The tensile stress testing of coconut coir fibers was done by clamping both sides (figure 1), with different diameters and fiber counts. This is measured by a Handifor 100 type recording device (maximum 100 N), as shown in figure 2.

This research had 100 samples with a combination of different diameters and a number of coconut fibers.

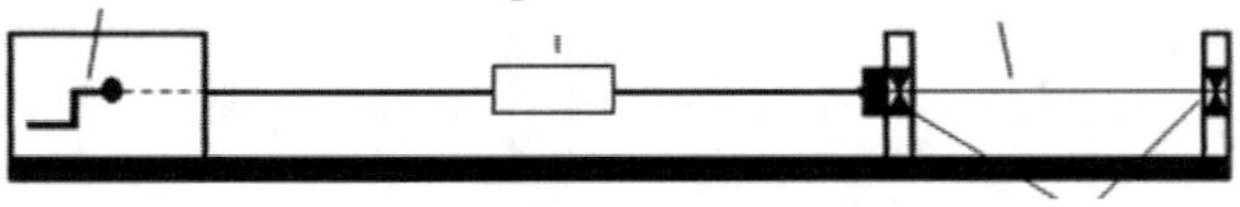

Figure 1. Coconut Fiber tools test.

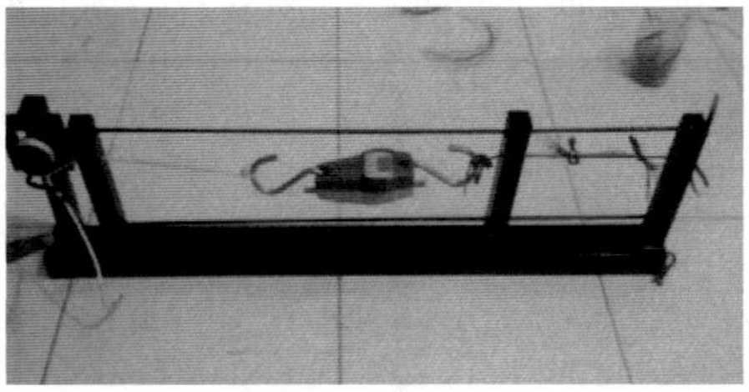

Figure 2. Tensile strength of coconut fiber testing.

2.1.1.b. Field soil sampling. Soil sampling was carried out in two ways: disturbed sample for physical characteristics testing and undisturbed sample for mechanical characteristics. Soil retrieval was done by using a special tube of iron with open both ends. The sample was then brought to the Syiah Kuala University Soil Mechanics Laboratory for direct shear testing.

2.1.2. Laboratory testing
1. Physical Characteristics: soil water content, specific gravity (SG), soil volume weight, liquid and plastic limits, grain size analysis, permeability
2. Mechanical Characteristics: Direct Shear (Direct Shear Test)

2.2. Secondary data
Secondary data includes the map of the research location and daily rainfall data.

2.2.1. Location map. The location map illustrated the location of the study, a barren slope located on Jalan Takengon-Bintang-Blangkejeren KM 10, Bur Retak Area, Kampung Mendale, Kebayakan Sub-District, Central Aceh District, Aceh Province.

2.2.2. Rainfall data. Rainfall data was measured using a rainfall gauge. This rainfall data was obtained from the Geophysical Climatology Meteorology Agency (BMKG), Central Aceh District.

2.3. Data processing and analysis
The data on the tensile strength testing of coconut coir fibers and root pulling vegetation was analyzed using the Analysis of Variance (ANOVA) and the Microsoft Excel (2010) program to display data in the graph.

IOP Conf. Series: Materials Science and Engineering **933** (2020) 012038 doi:10.1088/1757-899X/933/1/012038

3. Results and discussion

3.1. Rainfall

One of the factors causing landslides on the slopes is rainfall. Based on observations from the Geophysical Climatology Meteorology Agency (BMKG) in Central Aceh District, the maximum monthly rainfall intensity in the last four years was in October 2014 (560 mm), while the minimum was in July 2014 (7 mm).

3.2. Physical characteristic of soil

The testing of the soil's physical characteristics was to classify the type of soil used in the study.

3.2.1. Soil characteristics. Soil moisture content was tested in three samples with the same type of soil. The test results revealed the average water content in the soil, and it can be concluded that the soil from the barren slope location contained 10.27% of water. Meanwhile, the bush slope location had 38.12% of water. These results indicate that the soil on the barren slope has a lower water content compared to the bush slope location. The results of specific gravity (SG) tests carried out in laboratories reported the values of the specific gravity were 2.66 and 2.57 for the barren slope location and the bush slope location, respectively. Test results and calculations revealed that the average weights of the soil volume were 1.85 grams/cm^3 and 2.15 grams/cm^3 at the location of the barren slope and the bush slope location, accordingly.

3.2.2. Soil characteristics. The American Association of State Highway and Transportation Officials (AASTHO) classifies soil into eight groups, A-1 to A-7. In classifying the soil with this method three soil parameters are needed, namely the liquid limit value (LL), the plasticity index value (PI) and the percentage of grain filter passing. The liquid limit value, plasticity index and grain analysis, as shown in table 1

Based on the classification of the Unified Soil Classification System (USCS), the classification of soil at the barren slope location indicates that the fine-grained soil types were greater than 50%, based on the percentage of filter passing No.200. Based on the results of the plot in the USCS classification table, the soil at the location of the barren slope was classified in SC type, soil containing silt sand. Whereas the soil at the bush slope location was classified in type CH, soil containing contains inorganic clay.

Table 1. Boundaries of consistency and analysis of grains.

Parameter \ Slope Location	Barren	Bush
Liquid Limit LL (%)	32,79	53,31
Plastic Limit PL (%)	21,63	31,66
Plasticity Index IP (%)	11,16	21,64
Pass #200 (%)	34,11	91,24
Pass #40 (%)	52,04	93,54
Pass #10 (%)	90,91	97,00

Based on table 1 and the American Association of State Highway and Transportation Officials (AASTHO) soil classification system, the soil at the barren slope location was included in group A-2-6, a type of sandy soil. As for the bush slope location, the soil was categorized into group A-7-5, a clay soil, the type of ordinary to bad.

3.2.3. Soil permeability. Permeability is the soil's ability to allow water to pass through the pores or the speed at which it moves the rate of a liquid in a porous media. Soil permeability values were obtained from the results of soil sample testing at the Soil and Environmental Physics Laboratory, Agricultural Product Engineering, Syiah Kuala University.

Table 2. Permeability testing result.

Location	Permeability (cm/hour)	Class
Barren Slope	1,62	Medium
Bush Slope	1,59	Medium

Table 2 presents the permeability test results, and it shows that the permeability values were 1.62 cm/hour and 1.59 cm/hour for the barren slope and bush slope location, respectively, and both were classified in the medium class. The greater the permeability value corresponds to the lower the ability of the soil to absorb water, meaning that more water will pass, and vice versa.

3.3. Soil mechanical characteristics

The mechanical characteristics of the soil, including the soil shear strength, was tested at the Soil Mechanical Laboratory, Civil Engineering, Syiah Kuala University.

3.3.1. Soil shear strength. The direct shear test results in this study generated the parameters of cohesion (c) and the internal friction angle (φ) from the relationship between the normal stress values and soil shear stresses. Before and after the direct shear test, measurements of initial and final water contents were measured. figure 3 is an example of the collapse path at the bush slope location 1.

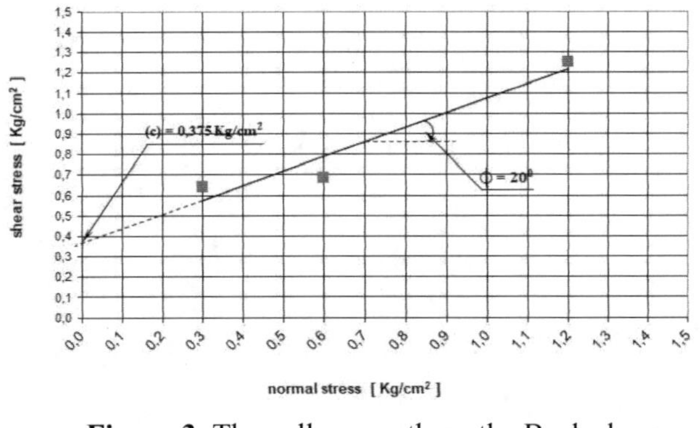

Figure 3. The collapse path on the Bush slope.

Figure 3 shows that the collapse path on the bush slope location 1 as a line diagram connecting the domicile points of the soil sample condition at the time the test took place.

Table 3. Recapitulation of direct shear test results.

Slope Location	Barren Slope			Bush Slope		
Parameter	Point 1	Point 2	Point 3	Point 1	Point 2	Point 3
Cohesion (c)	0.165	0.185	0.180	0.375	0.380	0.370
Internal friction angle (φ)	28	36	27	20	18.5	20.5

IOP Conf. Series: Materials Science and Engineering **933** (2020) 012038 doi:10.1088/1757-899X/933/1/012038

Table 3 shows that the average of the shear angle (ϕ) at the barren slope location was 30.330, and the cohesion value (c) was 0.18 kg/cm2. Meanwhile, for the bush slope location, the average shear angle (ϕ) was 19.670, and the cohesion value (c) was 0.375 kg/cm2.

3.4. Tensile strength of coconut coir fiber

The tensile strength testing of coconut coir fiber revealed the graphical relationship between the tensile strength of the fibers and the diameter and number of fibers.

3.4.1. The relationship of the number of coconut fiber and fiber tensile strength. figure 4 shows that the tensile strength of the fiber significantly increases with the increase in the number of fibers. The fiber tensile strength depended on the number of fibers, even if the influence of the diameter of each fiber was different. Equation (1) displays an equation based on the relationship of the tensile strength (Tr) and the number (n) of coconut coir fibers.

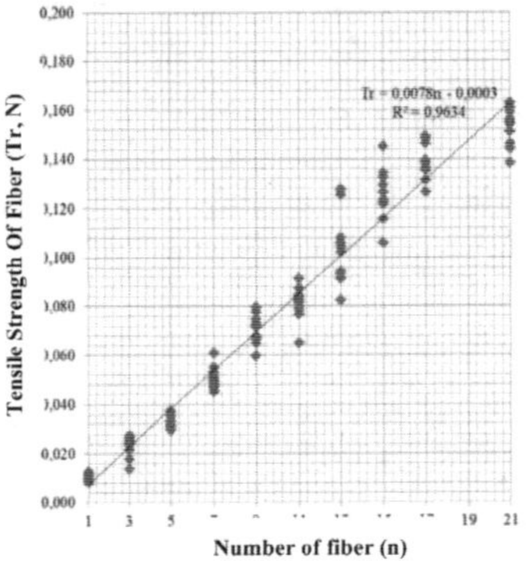

Figure 4. The relationship of number (n) coconut fibers with the tensile strength of fiber (Tr).

$$Tr = 0.0078n + 0.0003 \quad R^2 = 0.9634 \qquad (1)$$

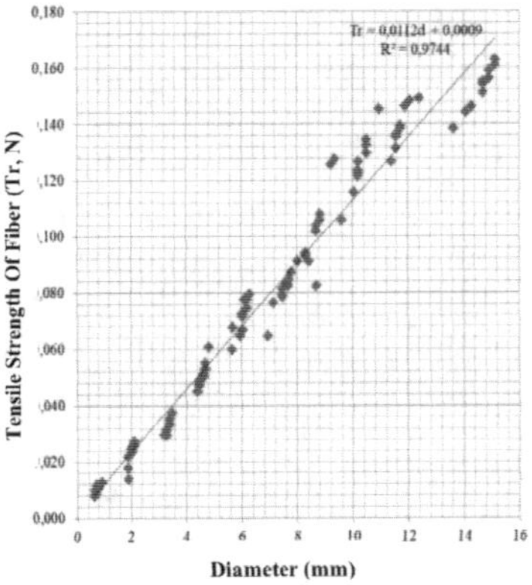

Figure 5. The relationship of diameter (d) of coconut fiber with the tensile strength of fiber (Tr, MPa).

IOP Conf. Series: Materials Science and Engineering **933** (2020) 012038 doi:10.1088/1757-899X/933/1/012038

3.4.2 Relationship of fiber tensile strength with fiber diameter. figure 5 illustrates the relationship between the diameter (d) of coconut fiber and Tensile Strength (Tr). The analysis shows that the fiber tensile strength increases significantly with the increase in the fiber diameter. Based on figure 5, the following equation is created.

$$Tr = 0.0078n + 0.0003 \; R^2 = 0.9634 \quad (2)$$

3.5. Tensile strength of coconut fiber

The tensile strength of fibers is highly influenced by the diameter of each fiber, as shown by the fiber tensile strength equation (3).

$$tf = 4F / (\pi d^2) \quad (3)$$

Where : tf = tensile strength of individual fibers (MPa);
F = maximum load at the breaking point (N); and
d = average fiber diameter at three points (mm).

The results of tensile strength testing of coconut coir fibers generated the graphical relationship between the fiber tensile strength, the number of fibers and the diameter of fibers.

3.5.1. Relationship of the amount of coconut fiber with fiber strength. Figure 6 indicates the relationship between the tensile strength (tr, MPa) and the number (n) of coconut coir fibers and it shows that the strength of coconut coir fibers significantly decreased at the test limit of fewer than five fibers, with different diameters of each fiber. Meanwhile, when more than five fibers were used, with different diameters of each fiber, the strength did not significantly decrease.

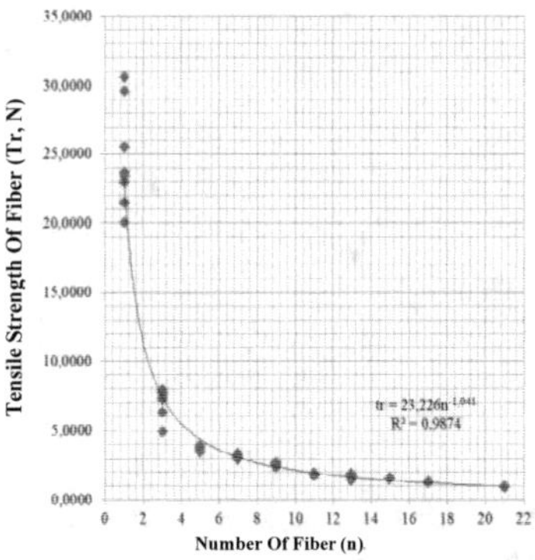

Figure 6. The relationship of number of coconut fiber (n) with tensile strength of fiber (Tr, MPa).

An equation based on the relationship of strength (tr) and the number (n) of coconut fibers is as follows.

$$tr = 23{,}226n\text{-}1{,}041 \; R^2 = 0{,}9874 \quad (4)$$

3.5.2. Relationship of the diameter of coconut fiber with fiber strength. Figure 7 shows that the strength of coconut fiber significantly decreased when the diameter was less than three, and it began to stabilize with a diameter of more than three. Thus, based on the relationship of the fiber strength (tr, MPa) and the diameter (d) of coconut coir fibers in figure 7, the following equation was generated.

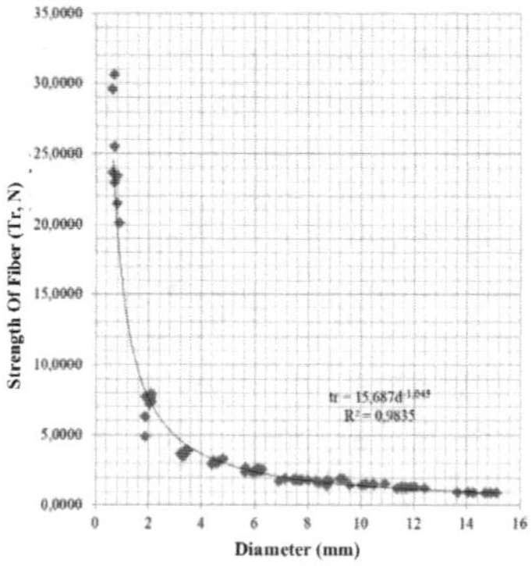

Figure 7. The relationship of the diameter (d) of the coconut fiber and the fiber strength of (tr, MPa).

$$tr = 15{,}687n\text{-}1{,}045 \quad R^2 = 0{,}9835 \tag{5}$$

3.6. Tensile strength of cocomesh anova analysis results

Based on the ANOVA, the F value of the calculated tensile strength of coconut coir fibers is presented in table 4.

Table 4 The Results of ANOVA for the Tensile Strength of Coconut Fiber against the Fiber Diameter.

No	Diameter	Calculate F Value	Table F Value
1	0,63	18,12	18,51
2	0,64	94,79	161,00
3	0,65	6,19	18,51
4	0,66	112,04	161,00
5	0,67	6,61	6,62
6	0,68	5,84	6,62
7	0,69	12,38	18,51
8	0,70	5,67	5,99
9	0,71	16,10	18,51
10	0,72	6,60	161,00

Table 4 shows that the F_{count} is smaller than F_{table}. It means that the fiber tensile strength decreases when the fiber diameter is large, but it is not significant.

3.7. Discussion

Rainfall is one of the triggers for landslides with certain slope conditions. Continuous high rainfall has the potential for landslides, as in this condition, the saturation of the soil by water increases the mass of the soil. The observation results of the Geophysical Climatology Meteorology Agency (BMKG), Central Aceh Regency, showed that the cumulative rainfall intensity in the last four years (from January 2013 to 2016) was 8072 mm. The highest monthly rainfall intensity was in October 2014 (560) mm. This rain intensity is included in the criteria of very heavy rainfall, based on the classification of rainfall of the international standards of the World Meteorological Organization (WMO) (Susilowati, 2010).

Based on the classification results of the AASHTO and USCS systems, the type of soil at the barren slope location is silt sand, while it is clay for the bush slope location. This classification shows that the water content on the barren slopes is smaller than on the bush slopes overgrown with plants. This is in agreement with the test results revealing that the soil water content on the barren slope and the bush slope was 10.27% and 38, 12%, respectively.

The results of the direct shear test generated the value of the deep shear angle (ϕ), with the average at the barren slope location was 30.330, and the cohesion value (c) was 0.18 kg/cm2. On the other hand, for bush slope locations, the average shear angle (ϕ) was 19,670, and the cohesion value (c) was 0.375 kg/cm2. These figures show that the location of the barren slope has a lower shear strength than that of the bush slope, which is at risk of landslides.

The focus of this study is the testing of cocomesh. The test results show that the tensile strength of the fiber increases with the increasing number and diameter of the fiber significantly. While the tensile strength of the fiber (tr) decreases with increasing number and diameter of the fiber. This is evident from the graphic relationship between the tensile strength (tr, Mpa) and diameter (mm) of the fibers in figures 4.8 and 4.9. Anova analysis shows that the tensile strength of the fiber decreases, although not significantly.

In its application, coconut coir fibers will be formed like a net or commonly called Coir Geotextile to strengthen the soil structure. It can also be a medium for growing various types of plants. These coconut coir fibers will be placed on the sloping land or barren rocky slopes. The nature of the coconut coir fibers can store water for a long time, causing the soil surface covered by this material to retain its moisture. This circumstance can affect the structure of the soil so that it is easy to plant seeds. Hence, the combination of coconut coir fibers and Based on the classification of the Unified Soil Classification System (USCS), the classification of soil at the barren slope location indicates that the fine-grained soil types was greater than 50%, based on the percentage of filter passing No.200. Based on the results of the plot in the USCS classification table, the soil at the location of the barren slope was classified in SC type, soil containing silt sand. Whereas, the soil at the bush slope location was classified in CH type, contains inorganic clay. Clay soil on a barren slope has the potential to increase the shear strength of the soil.

4. Conclusions

Based on the results of the discussion on the strength test of coconut coir fibers to prevent landslides on barren slopes, the following conclusions can be drawn:

1. The cohesion value (c) and the average shear angle (ϕ) from the results of the direct shear test show that the barren slope location has a lower shear strength compared to the bush slope location.
2. The tensile strength (tr, MPa) of coconut coir fibers will decrease with the increase in the number and the diameter of fibers.

References

[1] Santosa B, Suprapto H and Suryadi 1998 *Basic Soil Mechanics* (Jakarta: Gunadarma)
[2] Nandi 2007 Avalanche Department of Geography Education (Bandung FPIPS-UPI)
[3] Angkriawan C 2007 *Indonesia still ignores the potential of coconuts* Daily Newspaper Daily Update Voice
[4] Suryana A, Prastowo B, Mahmud Z, Wahyudi S, Hardono G S, Novarianto H, Luntung H T and Rfendi D S 2007 *Prospects and direction of coconut agribusiness development* 2nd Edition Agency for Research and Development, Agriculture Ministry of Agriculture
[5] Mahmud Z dan Ferry Y 2005 Prospects for Processing Coconut By-products. *J. Perspektif* **4**
[6] Ferra Fahriani 2016 Analysis of Soil Stability with Strength of Coconut Fiber *J. of Bangka Belitung University*
[7] Paskawati, Yessica A et al. 2010 Utilization of Coconut Fiber as Raw Material for Making Alternative Composite Paper *Technical J.* **9** 1 12-21 (Surabaya: Widya Mandala Catholic University)
[8] Abe K and Ziemer R R 1991 *Effect of tree roots on shallow-seated land slides* USDA forest Service Gen. Tech. Rep. PSW-GT 130: 11-20
[9] Lastiko L J, Haja Z F and Sulistyorini D 2017 Soil Stability Using Coconut Fibers Through Direct

Shear in the laboratory *J. of Sarjanawiyata* (Yogyakarta: University Taman Siswa)
[10] Ridwan 2014 *Effect of Coconut Fiber Addition to Clay Soil Strength* (Thesis. Department of Civil Engineering, Bangka Belitung University)

Hard bearing layers identification of escape hill design for facing tsunami – earthquake disaster at Deah Baroe rural area for natural disaster risk reduction

M Munirwansyah[1], R P Munirwan[1], M Sungkar[1], H Yunita[1], H Munirwan[2], Saharuddin[1,3], F Fachrurrazi[1] and W Widilestari[4]

[1] Department of Civil Engineering, Universitas Syiah Kuala, Banda Aceh, Indonesia
[2] Urban and Regional Planning Department, Institut Teknologi Sumatera, Bandar Lampung, Indonesia
[3] Department of Civil Engineering, Universitas Negeri Yogyakarta, Yogyakarta, Indonesia
[4] Public Work Office, Banda Aceh, 23243, Indonesia

Corresponding e-mail: nir_geotechnical@unsyiah.ac.id

Abstract. Aceh is surrounded by the Indo-Australia and Euro-Asia subduction plate, which has a profound effect on the occurrence of large tectonic earthquakes above 5 SR which may cause the tsunami. The tsunami which occurred in Aceh on December 26th, 2004 had the large number of the casualties during the 9.3 magnitude earthquake was affected by the lack of temporary evacuation like escape hill. The government's lack of preparedness for escape hill in order to deal with several serious threats such as the earthquake which may cause tsunami disaster in Banda Aceh. To deal with this serious threat, it is necessary to prepare a place of salvation to face the tsunami, in this case is an escape hill (artificial hill). The objective of this research is to determine the carrying capacity of the hard bearing layers identification of escape hill design and the amount of the settlement, due to the load of the embankment height (H_{crit}) and the safety factor value. This research is conducted in Ulee Lheue Banda Aceh which was calculated by the Plaxis 8.2 finite element analysis. The research results found that the settlement value of 17.45 cm still within in the tolerance limits and the bearing stratum is still able to withstand the embankment load. Furthermore, the safety factor value is 1.04 shows that the bearing stratum is still in the safe condition (hard soil) and it does not indicate the failure.

1. Introduction

Aceh Province is located directly on the path of two big earth plates namely the Indo-Australian and Eurasian plates. These plates are the main cause of the strong earthquake M 9.3 which may trigger the tsunami (Munirwansyah, 2018). According to the BMKG InaTEWS (Indonesia Tsunami Early Warning System for the Meteorology, Climatology and Geophysics Agency) (2008), a tsunami that occurred in Aceh on December 26, 2004 was caused by a 9.3 magnitude earthquake centered on UTM (Universal Transverse Mercator) coordinates 3.3 LU - 95.98 BT as shown in figure 1. Given the various earthquake threats that have caused tsunami and the lack of optimal efforts in the construction of evacuation sites in open area [1], it is necessary to conduct studies on temporary evacuation sites, namely escape hills.

Referring to the data taken from BNPB (National Disaster Management Agency), the tsunami disaster which caused the strong vibrations that propagate from the Andaman to the Aceh disaster cause casualities 173,741 people died, 116,368 people were declared missing, it was also causing the thousands of houses and buildings damaged, and almost 500,000 people became refugees [2]. The high number of casualties was influenced by the lack of temporary evacuation sites in Aceh. The government lack of preparedness to provide temporary evacuation sites in Aceh to deal with this serious threat, such as the earthquake that caused the tsunami. However, it is necessary to prepare and build a rescue site, one of them is an escape hill.

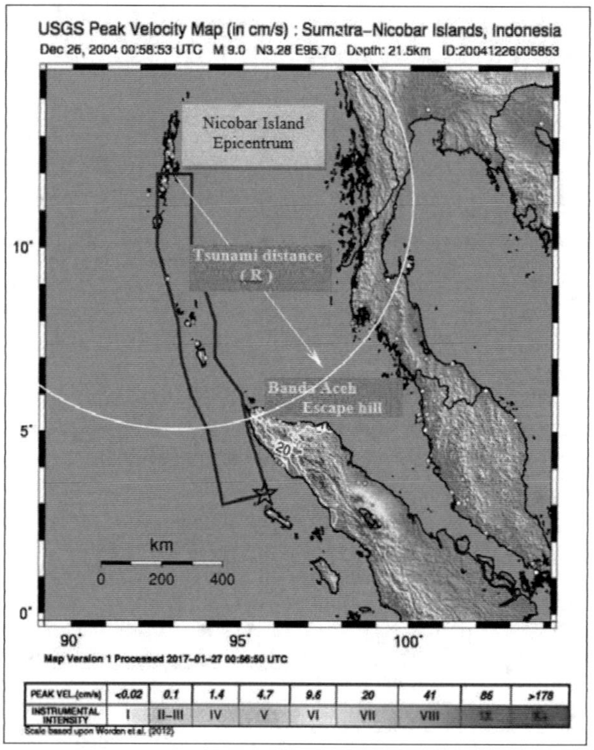

Figure 1. Tsunami distance from epicentre to Tibang escape hill Banda Aceh.

Similar to the other Civil Engineering infrastructure developments, build the escape hill construction required safe and proper planning [3]. Several studies about slope remediation were already conducted by [4] and [5]. Compacted soil-cement fill was conducted by [4] whereas sheet pile with anchor reinforcement was applied by [5]. The temporary evacuation site needs to study the height of critical embankment and the stability of the escape hill slope which is expected to obtain the safety values that meet the stability requirements of the slope so that it will fulfill the appropriate stability value as an evacuation site for the residents to face the tsunami. This study is carried out by using the 8.2 Plaxis program method. The use of the Plaxis program is used to implement the implementation stages in the field into the stages of computational work. The work that describes the conditions of the initial construction which are expected that the implementation in the field can be approached as close and real as possible [6]. The response generated from the program is assumed as a reflection of the actual conditions that will occur in the field. It allows achieving the safe slope design which can be obtained a faster and efficient calculation process.

2. Material and methods

This part shows the method of carrying out the research done. The method implementation included several steps as shown in figure 2. In addition, the detail of each step is explained in the section.

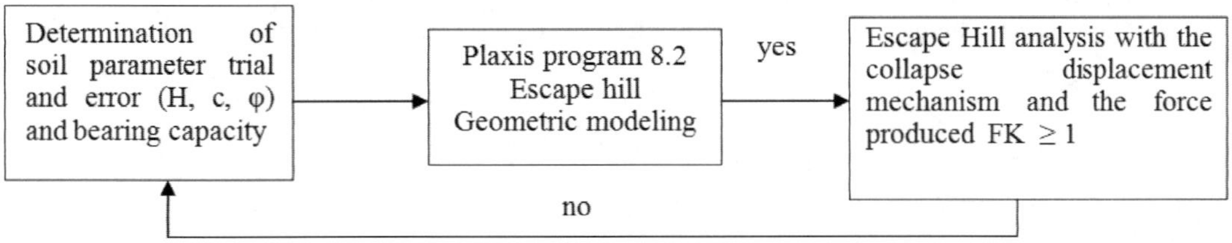

Figure 2. Steps of the research methods implementation.

2.1. The soil parameter determination

This study applied the physical properties testing and the mechanical properties of the soil. In the physical properties testing. There are three classifications of soil used, namely ASTM (American Society for Testing Material), AASHTO (American Association of State Highway and Transportation Officials Classification) and the USCS (Unified Soil Classification System). The ASTM classification is used to determine the criteria for the subgrade material, while the embankment criteria refer to ASTM and AASHTO which is fit into the category from Excellent to Good, those are A-1, A-3, and A-2. The soil mechanical testing is carried out by the direct shear experiments to obtain the cohesion (c) and the friction angles (φ).

2.2. Geometry escape hill on the Plaxis program

The escape hill geometry modeling on plaxis program data entry is made based on the criteria of escape hill design. It is 35 m pile width and the embankment height are 12 m, the slope angle used is the maximum allowable angle which is 26.5°. This related to the Guidelines for Designing a Hill as the Tsunami Evacuation Site (TES) taken from BNPB [2]. Geometry modeling can be seen in figure 3.

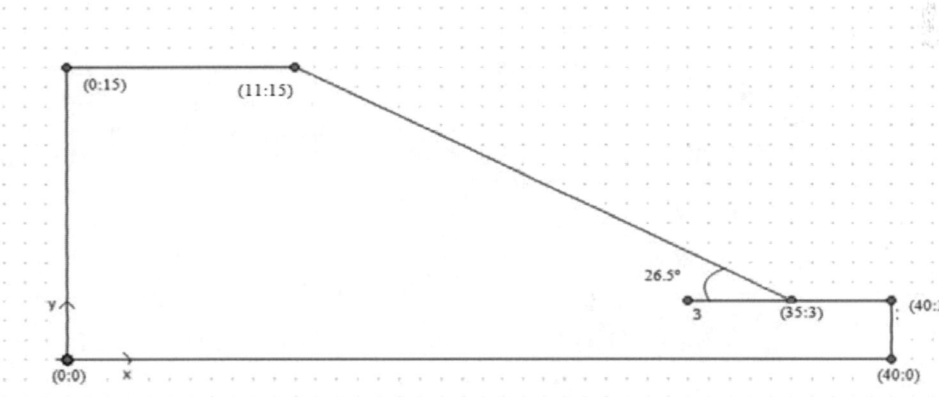

Figure 3. The Geometric escape hill modelling.

2.3. The soil bearing capacity

To construct the escape hill, the carrying capacity of the soil shall be able to bear the pressure without having to collapse because of the shear which is also determined by the ground shear strength. The calculation of soil carrying capacity can be calculated based on Terzaghi's theory in equation (1).

IOP Conf. Series: Materials Science and Engineering **933** (2020) 012039 doi:10.1088/1757-899X/933/1/012039

$$q_{ult} = C.Nc + \gamma.D.Nq + 0{,}5.\gamma.B.N\gamma \quad (1)$$

2.4. The soil shear strength

The soil shear strength is the resistance force carried out by the soil grain against the shear force. In obtaining the value of the ground shear strength proposed by Coulomb can be determined by equation (2).

$$\tau = c + \sigma \tan \varphi \quad (2)$$

2.5. The critical embankment height

In order to prevent the sliding and soil spreading around the embankment where the subgrade is under the saturated condition, it is highly necessary to study the collapse of the bearing capacity of the subgrade due to embankment [6]. And the easiest method to figure out this thing is by taking into account the height of the critical embankment as mentioned in equation (3).

$$H_{crit} = C_u \, . \, N_c \, / \, \gamma_{embankment} \quad (3)$$

2.6. The slope collapse criteria

[6] stated that the slope material has a tendency to occur due to the landslide shear pressure caused by gravity and other forces (water flow, tectonic pressure, earthquake). This tendency is detained by the shear strength of the slope material as it is explained by the Mohr Coulomb theory with this equation:

$$S = c + \sigma_n \tan \varphi \quad (4)$$

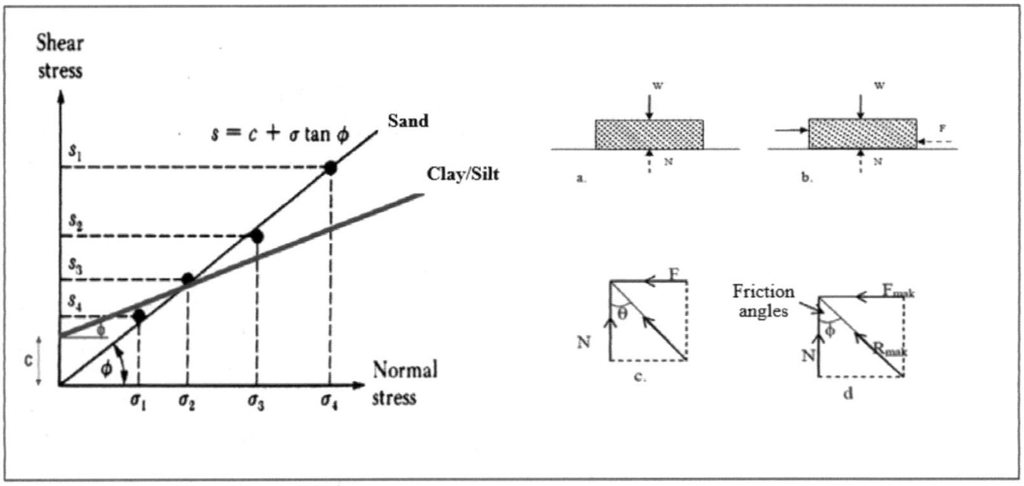

Figure 4. The Mohr-Coulomb collapse criteria slide test principle [6].

The Collapse Criteria explained by Mohr Coulomb on The Shear Test Principle can be seen in figure 4. Based on figure 4, it can be seen that (a) objects achieve a W force, then a reaction arises as much similar as the N, where the direction is opposite to W, The object (b) is given a horizontal force, and the F reaction arises where its magnitude depends on the surface roughness of the shear plane. The (c) Resultant (R) is the combination of the vector N and F, (d) If the P style is given the addition, the the friction force is also increase until it reaches the maximum value (limiting value F_{max}), then the object will move, the combination between F_{max} forms the angle of ϕ, (e). (e) The relation between the normal voltage and the shear stress when collapsing [7].

2.7. The safety factor
Generally, the value of the safety factor is greater than one (FS ≥ 1) to provide the estimation of the safety factor in the slope stability analysis [3]. However, it is important to ensure that slope design is safe to prevent unexpected factors during the analysis and construction is taking place [8]. The unexpected error which may occur is such as the incorrect data, lack of analysis, lack of work skills and the bad field supervision.

2.8. The material modeling
The critical embankment (H_{crit}) and slope stability study are using the plaxis program and calculated by trial and error. The material modeling is proportionally conducted by increasing the value of the soil parameters to obtain a stable and balanced condition.

Table 1. The Soil Parameter Modeling.

c variant model	φ variant model		
	φ	φ + 10%	φ + 20%
c	FS, U_{tot}	FS, U_{tot}	FS, U_{tot}
c + 50%	FS, U_{tot}	FS, U_{tot}	FS, U_{tot}
c + 100%	FS, U_{tot}	FS, U_{tot}	FS, U_{tot}

3. Results and discussion

3.1. The location observation
The location observation was conducted at Deah Baro Village, Meuraxa subdistrict Banda Aceh. The location is chosen based on the data from the previous research and the field review of the Tsunami Height Memorial Poles, with a maximum run-up when the tsunami occurred in 2004 in Desa Deah Baro reaching 7.0 m, larger than the other villages in Meuraxa where the height of the run-up was reaching up < 7.0 m height. The recorded height of the tsunami run-up wave in Meuraxa was 10 m height [3]. So, the escape hill which planned in this study is designed by purpose as shown in figure 3, and as high as > 10 m. For that, the place which can be reached easily and quickly by the people before the tsunami wave reaches was chosen as the location to establish the escape hill.

3.2. The soil parameter
The subgrade parameter and embankment soil used in this study are shown and can be seen in table 2. The parameter values of existing soil for the escape hill subgrade were obtained from the soil sample testing at the Syiah Kuala University Soil Mechanics Laboratory. As a result of physical properties on the existing soil, which obtained from the silty sand types. The land or soil which were used as the embankment material refers to the material classification criteria in building construction guidelines and tsunami testing hills. The soil type which is used as the escape hill heap material in this study is the land/soil with the type of sand covered by categories A-2-4 (AASHTO) and SM (USCS) type.

3.3. The results of escape hill analysis with the Plaxis program
In this study, the analysis of the Escape hill in Plaxis program used nine modelings by reviewing the ability to pxist (bearing capacity) subgrade to withstand the embankment load and the safest slope area. The Modeling is held by varying the value of the cohesion (c) and friction angle (φ) proportionally from the normal conditions of the soil embankment parameters. The result of the cohesion variant modeling

and friction angle modeling can be seen in table 3. As the result from the first modeling with a pile height of 12 m where the phase calculation of the soil loading is calculated every 2.0 m increments then the declining number is 17.45 cm and safety factor = 1.04~1.

Table 2. The soil parameter values in the escape hill planning.

Normal

Parameter	Name	Sub grade		Land Fill	Unit
		Sandy Silt 0-1 m	Sandy silt 2-3 m	Silty sand >2 m	
Material model	Model	MC	MC	MC	-
Type of behavior	Type	*Undrained*	*Undrained*	*Drained*	-
Dry soil weight	γ_{unsat}	9.7	10.3	11.5	kN/m³
Wet soil weight	γ_{sat}	21.74	21.75	21.50	kN/m³
Horizontal permeability	Kx	8.640E-03	8.640 E-03	8.640E-05	m/day
Vertical permeability	Ky	8.640E-03	8.640 E-03	8.640E-05	m/day
Young's modulus	E_{ref}	15000	8000	5000	kN/m²
Poisson's ratio	v	0.33	0.33	0.25	-
Cohession	c	0.4	10.7	1.0	kN/m²
Friction angel	φ	37.6	32.3	25.0	°
Dilatancy angel	ψ	0	0	0	°

Table 3. The cohesion modeling variants and shear angle.

Cohesion (kN/m²)		Friction angle (°)		Total displacement (cm)	Safety factor (FK)
c	1.0	φ	25.0	17.45	1.04
c + 50%	1.5	φ	25.0	17.04	1.10
c + 100%	2.0	φ	25.0	16.71	1.16
c	1.0	φ + 10%	27.5	16.60	1.12
c + 50%	1.5	φ + 10%	27.5	16.31	1.22
c + 100%	2.0	φ + 10%	27.5	16.10	1.27
C	1.0	φ + 20%	30.0	16.03	1.28
c + 50%	1.5	φ + 20%	30.0	15.96	1.33
c +100%	2.0	φ+ 20%	30.0	15.98	1.38

The second modeling is by increasing the cohesion value by 50%, then obtained a reduction rate of 17.04 cm and a safety factor = 1.10~1. The third modeling is conducted by adding 100% cohesion value, then it obtained a reduction rate of 16.71 cm and a safety factor of 1.16. The fourth modeling is conducted by adding a friction angle value of 10%, it obtained a reduction rate of 16.60 cm and a safety factor of 1.12. The fifth modeling is by increasing the cohesion value by 50% and the friction angle value by 10%, then it obtained a reduction rate of 16.31 cm and a safety factor of 1.22. The sixth modeling is by increasing the cohesion value by 100% and the friction angle value by 10%, obtained a decrease of 16.10 cm and a safety factor of 1.27. The seventh modeling is by adding a friction angle value of 20%, obtained a reduction rate of 16.03 cm and a safety factor of 1.28. The eighth modeling is by adding 50% cohesion value and 20% friction angle value, obtained a decrease of 15.96 cm and a safety factor of 1.33. The ninth modeling is by adding 100% cohesion value and 20% friction angle value, obtained a decrease of 15.98 cm and a safety factor of 1.38.

IOP Conf. Series: Materials Science and Engineering **933** (2020) 012039 doi:10.1088/1757-899X/933/1/012039

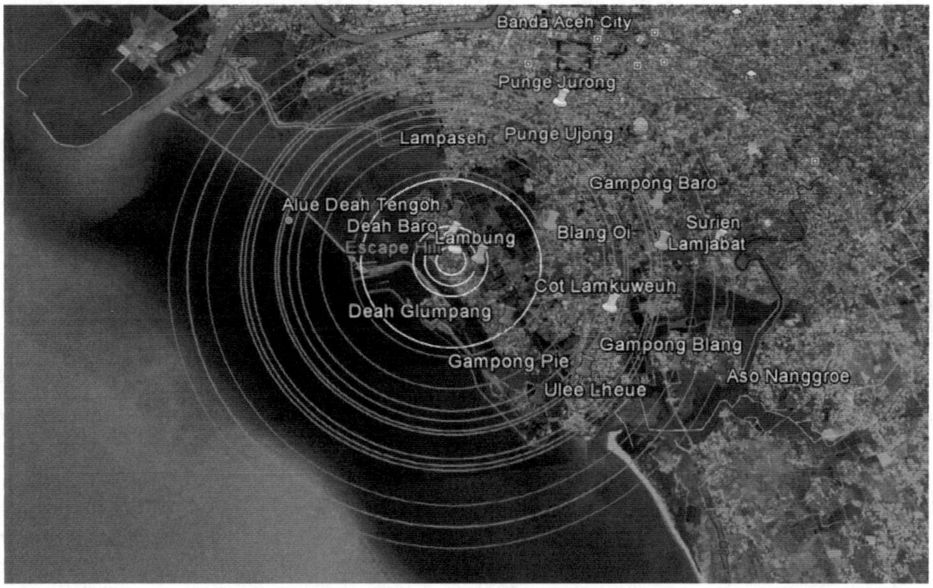

Figure 5. The escape hill radius achievement towards the safe zone for the community.

3.4. The results of the escape hill output with the Plaxis program

The following shows the output of the Plaxis program which is deformed mesh and arrow in the natural conditions and modeling by increasing the cohesion value and the friction angle value which is in the most stable condition.

Figure 6 shows an arrow that shows the direction of deformation motion from the embankment that is passed to the subgrade layer. The shearing force generated by the slope is still smaller than the shearing force of the soil so that the subgrade is still able to hold the embankment as high as 12 m.

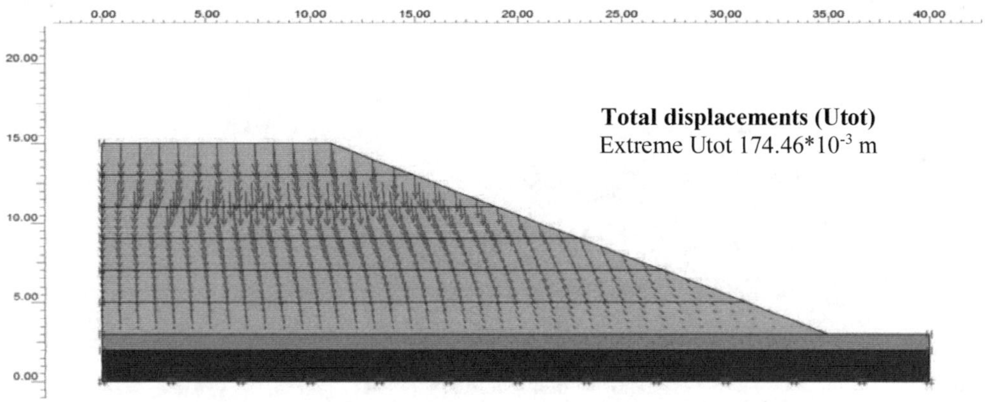

Figure 6. Total displacement with 12 m embankment height.

Figure 7 shows the deformed mesh that occurred. In the calculation of Plaxis by increasing the height of the embankment 12 m from the existing height, it was found that at the stage construction the subgrade was still able to withstand the embankment load with settlement of 17.45 cm and the soil body has not to collapse and the safety factor value shows that the slope in safe condition does not show an indication of landslide with FS = 1.04.

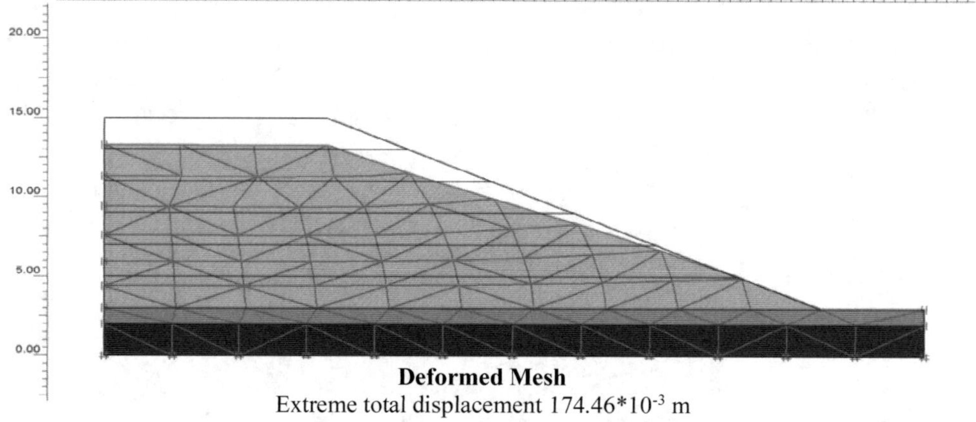

Deformed Mesh
Extreme total displacement $174.46*10^{-3}$ m
(displacements scaled up 10.00 times)

Figure 7. Deformed mesh with 12 m embankment height.

Modeling of soil parameters is carried out by increasing the parameter value of embankment soil, namely cohesion value (c) and friction angle (φ) to obtain a safe condition than the calculated natural conditions [9]. From the various modeling that has been carried out, the most stable condition is obtained by modeling by increasing the value of cohesion (c) by 50% (1.5 kN/m^2) and the value of friction angle (φ) by 20% (30°).

As it is shown in figure 8, shows the arrow that shows the direction of deformation motion from the embankment is passed to the layer of the subgrade. The shear force generated by the slope is smaller than the shear force of the soil. So, the subgrade is able to hold the pile with 12 m height.

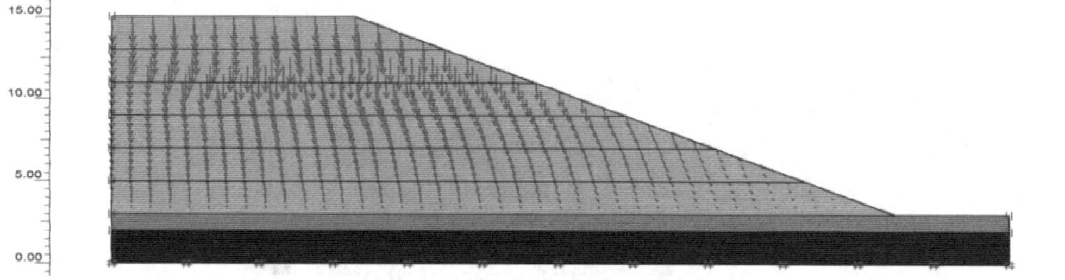

Total displacements (Utot)
Extreme Utot $159.60*10^{-3}$ m

Figure 8. Total displacement with 12 m embankment height (c+50% dan φ+20%).

Figure 9 shows the results of the Plaxis output in the form of deformed mesh at 12 m embankment height from the original soil height where subgrade is still able to withstand the embankment with a decrease of 15.96 cm. So that the soil has not been collapse and the safety factor value shows the slope is in stable condition with Safety Factor (FK) is 1.33 > 1.

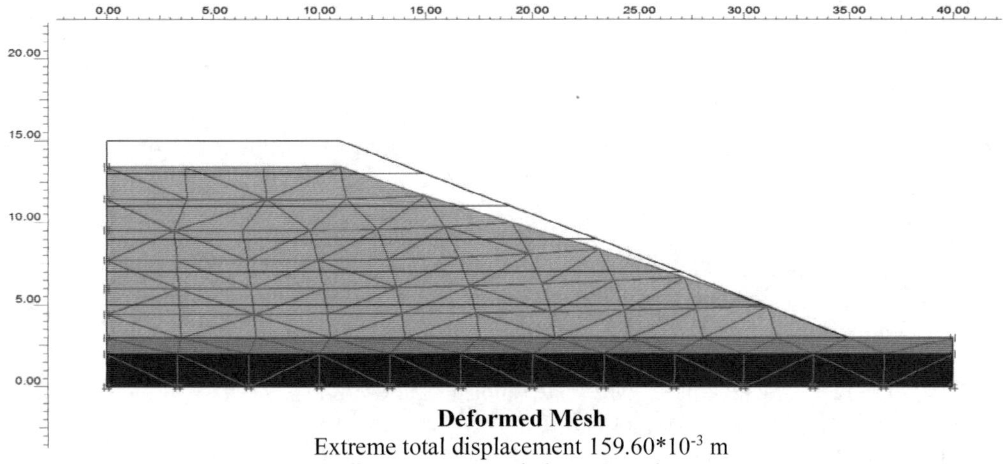

Figure 9. The deformed mesh with 12 m embankment height (c+50% dan φ+20%).

3.5. Discussion

Based on the soil samples testing in the planned location, the researcher found that the subgrade used in this research is the sandy silt. The calculations have been conducted by using the 8.2 Plaxis program with the Mohr-Coulomb model to calculate the safety factor value of the embankment height based on soil the parameters. It is important to set the material that meets and fit to the escape hill planning specifications by the trial and error. The safe characteristic in the embankment is obtained by using the silty sand soil type as the main embankment material.

Based on the theory stated by Terzaghi's, the soil bearing capacity, the forces acting are the forces due to the weight of the embankment load. An arrow shows that the load from the embankment is passed and pointed directly to the layer and can be deployed along the shear plane which does not exceed the subgrade's strength. The soil settlement that occurs due to embankment load is still within the tolerance value so that the subgrade still tends to hold the embankment [9]. In the very natural conditions with the height of embankment is 12 m height (h), the calculations conducted by using the Plaxis computation shows the settlement value condition is in the tolerance limit of 17.45 cm and the safety factor value FK is 1.04~1 which indicates the slope is still safe and do not indicate the slope slides. The modeling is conducted on the landfill parameters from the natural conditions in purpose to obtain the decreasing value and stable safety factor value. The most stable condition is shown in the modeling which increased the value of cohesion (c) by 50% (1.5 kN/m^2) and the value of friction angle is (φ) 20% (30°). So that the result is the value of the settlement at the height of embankment 12 m is 15.96 cm and also the value of the safety factor for the escape hill FK is 1.33 > 1.

4. Conclusion

Based on the analysis results, the researcher found that the subgrade was still able to withstand the load of the embankment. So that the soil had not been about to collapse and the slope condition is safe and did not indicate the landslides. The shear force which generated by the slope is still smaller than the ground shear force. So that the subgrade is able to withstand embankments with 12 m height. Even more, by increasing the cohesion value (c) and the friction angle (φ) increased the safety factor value (FK) by of 0.023% in average and it decreased the average reduction value of 0.097%. This research is expected to be used as a reference to espouse the correct design plan on building an escape hill as an evacuation site from an earthquake which may trigger the tsunami in Aceh. For the last, if the land does not meet the safety factor requirements at the research location, the researcher suppose to import the subgrade material from near borrow area, stabilize the subgrade and land reinforcement with geosynthetics and geogrid material.

Acknowledgment
The author gratefully acknowledges financial support from Ministry of Research, Technology and Higher Education, Universitas Syiah Kuala Banda Aceh Indonesia, for the Profesor Research Funding, Grant No. 520/UN11/SPK/PNBP/2019, February 8th, 2019.

References
[1] Munirwansyah, Irsyam M, Munirwan R P, Yunita H and Usrina M Z 2018 Geotechnical approach for occupational safety risk analysis of critical slope in open pit mining as implication for earthquake hazard *IOP Conf. Series: Materials Sci. and Eng.* vol 352 Conference 1
[2] Munirwansyah M, Munirwan R P and Yunita H 2018 Geotechnical engineering aspect related to Pidie Jaya – Aceh earthquake disaster and mitigation *Intern. J. on Adv. Sci. Eng. Information Techn.* **8** 3 870-5 http://dx.doi.org/10.18517/ijaseit.8.3.4189
[3] Munirwan R P and Munirwansyah 2019 Escape Hill as geotechnical quick response method in facing upcoming tsunami disaster *IOP Conference Series: Earth and Envir. Sci.* vol 273 1
[4] Wu J Y, Huang K and Sungkar M 2017 Remediation of slope failures by compacted soil-cement fill *J. of Performance of Constructed Facilities* **1** (ASCE) 1 1-9
[5] Yoshida M, Mitou M, Kiyomiya O, Tashoro S and Gouda K 2012 Effect of seismic reinforcement of sheet pile quay wall using ground anchor 15 WCEE (Lisboa)
[6] Abramson L W 1996 *Slope stability and stabilization methods* (New York: Wiley and Sons Inc.)
[7] Munirwansyah, Munirwan R P, Sungkar M and Melinda Z 2019 The Effect of soil-structure interaction on multi-storey building resonance and dynamic shear modulus for Pidie Jaya Aceh earthquake, *IOP Conference Series: Materials Science and Engineering* vol 523
[8] Das B M 2000 *Fundamentals of Geotechnical Engineering* (USA: Brooks/Cole Thomson Learning)
[9] Hasyim A 2007 *Slope Stability Analysis in Saturated Slope* (Malaysia: Faculty of Civil Engineering Universiti Teknologi Malaysia, inside.mines.edu)

Numerical simulation of groyne placement in minimising Krueng Aceh river bank erosion

E Fatimah[1], A Fauzi[1], S Rezeki[2] and S Suryati[2]

[1] Department of Civil Engineering, Universitas Syiah Kuala, Banda Aceh, Indonesia
[2] Department of Public Works and Spatial Planning, North Aceh Regency, Lhoksukon, Aceh Province, Indonesia

Corresponding e-mail: eldinafatimah64@gmail.com

Abstract. This paper presents the prediction of the flow velocity in a 575 meters-long reach of the Krueng Aceh river. It focused on the part of the riverbank where erosion occurred due to the pillars and bridge abutments. One alternative to reduce erosion is using groynes for regulation. The simulations were conducted using the Surface Water Modelling System program (SMS 11.2). The effects of the groynes on the flow changes causing erosion were determined by mathematical simulation of two-dimensional flow using RMA2. The calibration of RMA2 was undertaken by determining the Root-Mean-Square (RMS) error between the measured and mathematical model velocities. Nine scenarios of simulations have been done to investigate the effect of the groyne installation in reducing erosion in the river bend, in term of its length (L), and angle (α), according to river condition. From the simulation results, it can be concluded that the distribution of flow velocities towards the outer river bend for the perpendicular groyne of 9.0m long was smaller compared to the 7.0m one. The groyne with the angle of 10° and length of 7.0m was better in reducing the velocities at the outer riverbank than the 9.0m one. Whereas the groyne with 30° orientation and 9.0m long is better in reducing the eddy than that of the 7.0 m.

1. Introduction

Constructing river training structures are necessary for guiding or managing the river water for various beneficial purpose of mankind. The most commonly used river training structures generally are built along the river, placed from the riverbank and extended towards the main channel flow, called a groyne. The construction of structures aims to deflect or repel the velocity away from the bank for reducing river bend erosion and increase the flow depth for smooth navigation. However, local hydrodynamics and the river morphology will change because of these structures, leading to a blockage in the flow.

Krueng Aceh river, located in Aceh, Indonesia, faces many problems related to the bank erosion due to many factors, such as the sand mining at the upstream and downstream of the river, the installation of bridge pillars and abutment along the rivers blocking its cross-section, and the land-use degradation. Hence, it is necessary to eliminate the riverbank erosion due to the bridge pillars using groyne. A case study was selected at the Pango flyover bridge, located in Pango village, Aceh Besar District, Aceh. Before installing the groynes in this reach, it is essential to conduct a model study to evaluate the possible effects of the groynes on the river. These model studies were conducted based on numerical approaches using SMS.

IOP Conf. Series: Materials Science and Engineering **933** (2020) 012040 doi:10.1088/1757-899X/933/1/012040

2. Methodology

Numerical study on groyne placements has been widely studied by previous researchers, such as Suharjoko 1999 [1] and (2001) [2]. He has already analyzed and reviewed the influence of the impermeable groyne on a straight channel, concerning the flow pattern and the best distance between the structures using Computational hydrodynamics simulations. He found that the relationship between the Froude number (Fr.) and the length of embankment protected and the other parameters can be used to show the best distance between the groyne.

Mojtahedi A and Basmenji A B (2017) [3] have investigated the influence of the groynes and the embankment dike placement on Ghezel-Ozan River. After the instalment, they found that the groynes in the given interval, the main channel of the river became significantly deeper while the sediment settled between the groynes and made the dikes more stable. They said that the chosen length, the space between the groynes, and their shapes were optimum, and the correct forms of vortexes were made between them. However, it is better to replace the first and the last groynes in the series with an L-shape groyne. While Dutta D and Kalita H M (2018) [4] compared the performance of straight head and T-head groynes for vulnerable channel bend numerically using MIKE21C modelling tool. They concluded that T-head groynes were better for the stability of the outer bank.

2.1. Groyne characteristics

Groyne construction is a form of indirect protection used to protect river banks indirectly from the danger of the local scour and meandering process due to currents directing the river flow. A series of groyne construction also functions as a way to deepen and narrow the river flow, [6] Sunaryo, Daoed D, and Sari F L (2010) [5].

Sosrodarsono and Tominaga (1985: 174) [6] explained that based on the permeability level, groynes are classified into three types of construction, namely: permeable, impermeable and semi impermeable groynes. Generally, the groyne formation applied on the river is perpendicular to the flow, leaning towards the upstream area, leaning towards the downstream area and a combination of both.

2.2. Groyne placement

Groyne placement is adjusted according to its function, either for the directional and alignment repair or for riverbank protection. To function as a riverbank protection, an integrated river planning must be done, so that the river management can be carried out effectively. The results of the velocity distribution formed greatly influence the angle of the groyne placement.

2.3. Groyne formation

Sosrodarsono and Tominaga (1985: 179) [6] argued that there are three groynes orientation applied, i.e. perpendicular, inclined to the upstream, and inclined to downstream. Groynes orientation refers to the groynes alignment for the direction of the current main flow in a channel. The most compatible angle to various types of groynes is described in Table 1.

Table 1. Flow direction and groyne angle.

	Orientation of the groynes	Flow direction and groyne angle	
1	Straight Section	$10^\circ - 15^\circ$	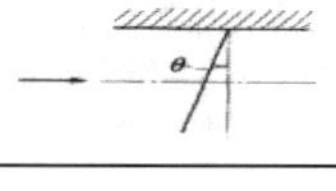
2	Out Section	$5^\circ - 15^\circ$	
3	In Section	$0^\circ - 10^\circ$	

Based on the results of surveys and observations on the groynes that have been built by Sosrodarsono and Tominaga (1985: 179) [6], then, the length of the groynes can be calculated as follows:

$$L_k \leq 10\%\ B \tag{1}$$

In which, L_k=length of the groynes (m), and B=river width (m)

2.4. *Flow pattern around the Groynes*

Mudjiatko (2000) [7] stated that the water flowing through a bend will experience a centrifugal force, causing water to move out of the bend. The centrifugal force will work if there is no transfer of water mass in the transverse direction. Since the velocity distribution affects the depth, leading to the velocity at the surface being higher than that near the base, it will undoubtedly affect the distribution of the centrifugal force. Centrifugal force will be larger on the surface than near the base.

The placement of the groyne to protect a riverbank and direct the flow or improve the alignment of the river will certainly affect the direction of the river flow, resulting in a whirlpool around the groyne. Whirlpools will form flow patterns, scour patterns around the groyne. Thus, a study of flow patterns is important to plan the groyne placement.

2.5. *Surface water modelling system (SMS)*

Surface water Modeling System (SMS) is a program designed to draw flow patterns on the surface, with 1D, 2D and 3D modelling in the form of hydrodynamics. To solve the modelling problem, one RMA2 sub-program is used. RMA2 is a two-dimensional sub-dynamic flow equation settlement program, Boss SMS (1995) [8].

2.6. *Modelling approach*

One of the BOSS SMS (Surface water Modeling System) software modules, RMA2 (Resources Management Association Inc.) version 11.2, is a numerical model for calculating the two-dimensional flow hydrodynamic process at an average depth.

2.6.1. Surface water modelling system (SMS 11.2) governing equations and solution techniques. The SMS software is a post and pre-processing unit, while RMA2 is the running execution program. The water level elevation and flow velocity of each point with the mesh elements as well as water display in digital forms, such as rivers, ports and estuaries are clearly depicted. RMA2 can solve both permanent and temporary flow problems because the boundary conditions (incoming discharge, water level elevation) can be set accordingly. This program is designed to create models with dynamic flow conditions caused by fluctuations in the surface flow or tidal cycles. RMA2 is not used for supercritical flow.

The output of RMA2 is written by the binary solution file. The file contains the completion of one or several time steps depending on whether the flow analysis is specified permanently or temporarily. Solution file can be used as an input for an SMS program to display data in graphically. The general governing equation for shallow water by RMA2 is solved using the following formulas (Boss SMS, 1995) [7].

$$h\frac{\partial u}{\partial t}+hu\frac{\partial u}{\partial x}+hv\frac{\partial u}{\partial y}-\frac{h}{\rho}\left(E_{xx}\frac{\partial^2 u}{\partial x^2}+E_{xy}\frac{\partial^2 u}{\partial y^2}\right)+gh\left(\frac{\partial a}{\partial x}+\frac{\partial h}{\partial x}\right)$$
$$+\frac{gun^2}{\left(1.486h^{1/6}\right)^2}\left(u^2+v^2\right)^{1/2}-\zeta V_a^2\cos\psi-2h\omega v\sin\phi=0 \tag{2}$$

$$h\frac{\partial u}{\partial t}+hu\frac{\partial u}{\partial x}+hv\frac{\partial u}{\partial y}-\frac{h}{\rho}\left(E_{xx}\frac{\partial^2 u}{\partial x^2}+E_{xy}\frac{\partial^2 u}{\partial y^2}\right)+gh\left(\frac{\partial a}{\partial x}+\frac{\partial h}{\partial x}\right)$$
$$+\frac{gun^2}{\left(1.486h^{1/6}\right)^2}\left(u^2+v^2\right)^{1/2}-\zeta V_a^2\cos\psi+2h\omega v\sin\phi=0 \tag{3}$$

$$\frac{\partial h}{\partial t}+h\left(\frac{\partial u}{\partial x}+\frac{\partial v}{\partial y}\right)+u\frac{\partial u}{\partial x}+v\frac{\partial u}{\partial y}=0 \tag{4}$$

Where: h=water depth (m), u,v: velocity on the direction of the x and y axis (m/s), x,y,t=Cartesian coordinate and time, ρ=liquid mass density (kg/m^3), g=gravitational acceleration (m/s^2), E=*Eddy Viscosity* coefficient, xx=for the normal direction on the x axis, yy=the normal direction on the y axis, xy and yx the direction=shear on each surface, a=bed elevation (m), n=Manning roughness value, 1.486=conversion from metric units to English units, ζ=coefficient of wind friction, Va, ψ=wind speed and wind direction, ω, ϕ=Earth's angular rotation rate and local latitude (km/h).

2.6.2. RMA2 hydrodynamic model. The selection of the suitable analytical method to adopt during a river hydraulics study depends on many factors, such as the objectives, the level of detail being called for, the regime of flow expected, the availability of necessary data, and the availability of time and resources to properly address all essential issues (EL-Sammany M S and El-Moustafa A M, 2011) [9]. The RMA2 model under the SMS interface was selected in this study.

The main objective of this study was to develop a hydrodynamic model to study the flow fields of Krueng Aceh river, where various scenarios of groynes were installed. The requirements of input data to run the model were related to spatial and temporal parameters, such as computational mesh, bathymetry, simulation length and overall time step.

2.7. Measuring site and data collection

The measuring site with the length of 575m was located at Pango Village in Ulee Kareng Sub-district of Aceh Besar District. The flood discharge with a five year return period was estimated at 405.456m^3/s. The water elevation data was 6.0m. Krueng Aceh river, as shown in Fig. 1, has an average width of 40.62m for the full bank. Topographic map of the reach considered is given in Fig. 2.

During the mesh generation process, bathymetric values were extracted by the AutoCAD, providing the data compatible with SMS. Mesh independence is essential for obtaining reliable results from the models. In this investigation, Krueng Aceh river was simulated using a grid with flexible meshes, covering the considered reach and a part of the surrounding area (Fig. 3). Simulation time reached its numerical stability after 24 hours running the models.

IOP Conf. Series: Materials Science and Engineering **933** (2020) 012040 doi:10.1088/1757-899X/933/1/012040

Figure 1. Measuring site.

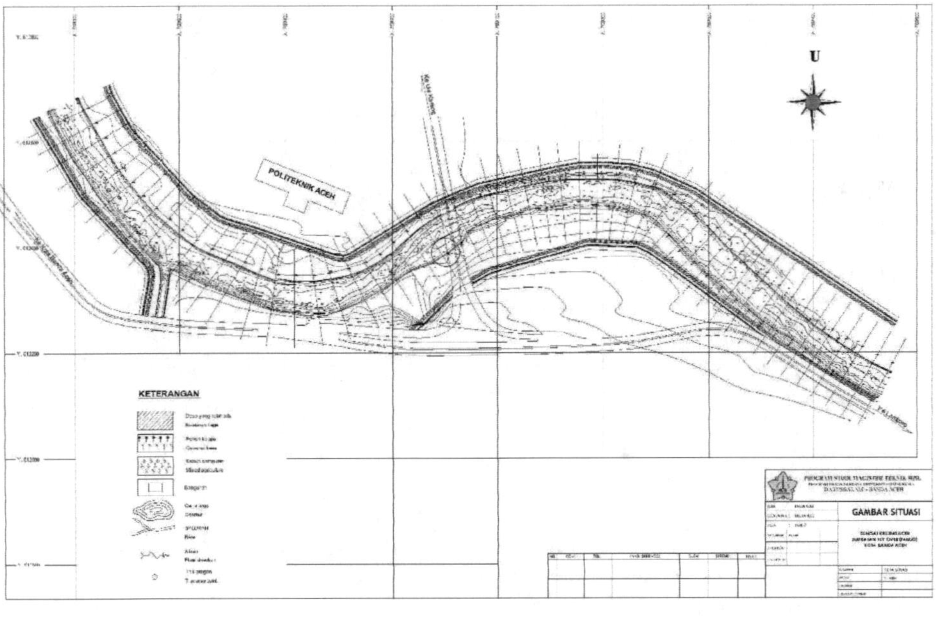

Figure 2. Topographic map.

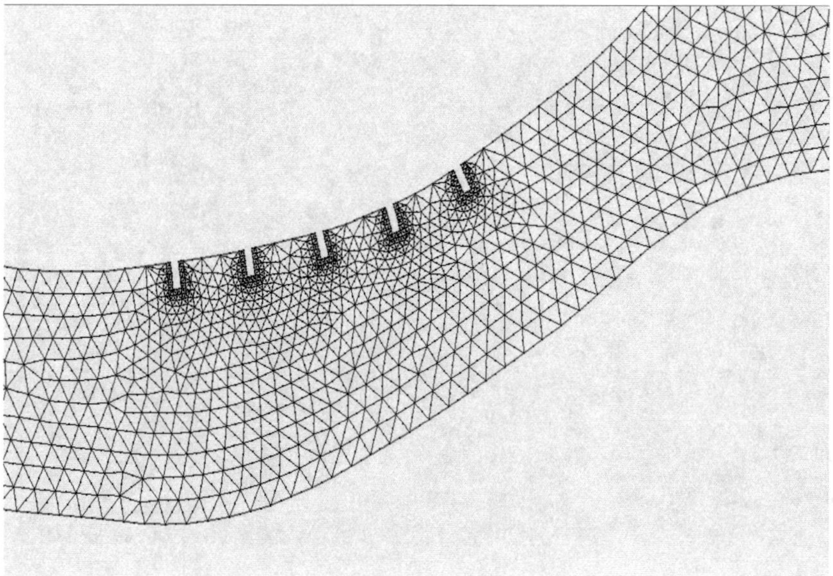

Figure 3. Topographic grid map.

The length of the groyne was 6.16% of the river width (B). The variation of the distance (L) intervals were considered 2.8 and 3.6 times the length of the groyne (L_k), and the variation in the number of groynes (N) was set to be five units. The groyne was placed to flow in three angular variations, namely perpendicular to the riverbank, 10° and 30° facing towards the downstream area. Groyne stability is not considered in this report. The groyne placement scenario is summarized in Table 2.

Table 2. Groyne placement scenario.

Discharge (Q5-m^3/s)	Groyne position		Groyne distance (L-m)	7	9
			Number of groyne (N-Unit)	5	
405,456	Perpendicular		GT	GT5L7	GT5L9
	downstream	10°	GIa	GIa5L7	GIa5L9
		30°	GIb	GIb5L7	GIb5L9

2.8. Validation of numerical model

In this study, the value calibrated is the value of the manning coefficient (n). Based on the results, the velocity measurement in the first (upstream) and the second (downstream) cross-section was 0.659 m/sec and 0.622 m/sec, respectively. While the numerical velocities in the first and second cross were 0.583 m/sec and 0.710 m/sec, respectively. The absolute error values were 0.039 for cross 1 and 0.051 for cross 2. Hence, the data is valid since the magnitude of the absolute error value is less than 1, as shown in Figure 4.

IOP Conf. Series: Materials Science and Engineering **933** (2020) 012040 doi:10.1088/1757-899X/933/1/012040

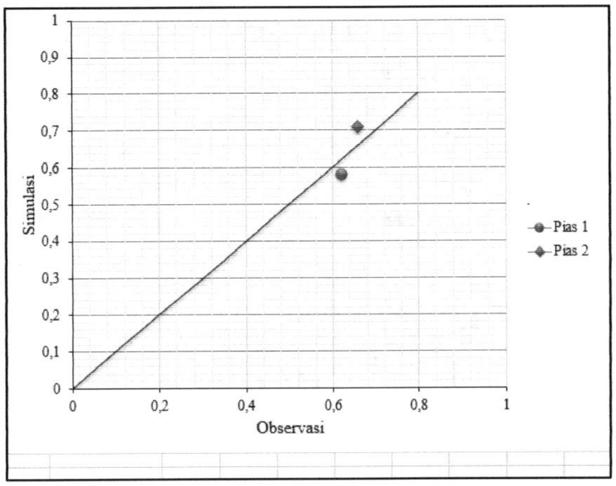

Figure 4. Calibration graph of the measured and simulated velocity.

3. Results and discussion

Fig. 5 shows the flow pattern in the Krueng Aceh river without any groyne instalment. Simulation results indicate that the red covers a very critical area and brick red is a critical area. Whereas, yellow, green, and blue refer to a medium area, a normal area, and no flow area, respectively. It can be seen that there are three bridge pillars transverse the river cross-section. Due to these pillars, the velocity at the downstream of the pillar is high (critical to very critical) compared to the velocity at the pillar vicinity. Besides, the river width at this section is also smaller than that of the pillar vicinity.

Field measurement at the riverbank, in the inner and outer bend, was found eroded. However, velocities at the inner bend were higher than that at the outer bend. Theoretically, river bank at the outer bend is the most vulnerable to the bank erosion than the inner bend. This phenomenon can happen due to the pillar that may alter the thalweg of the river. Thalweg is defined as the middle pattern of river flow with the highest velocity value formed in river flow in the middle of the river, illustrated by a long river crossing line. It occurs because of the changes in river morphology, the changes in flow patterns and the magnitude of flow velocity. The middle flow pattern of the river (thalweg) must be controlled so that it remains in the central position of the flow, to prevent meandering or erosion of the river bank. Brown (1985) [8].

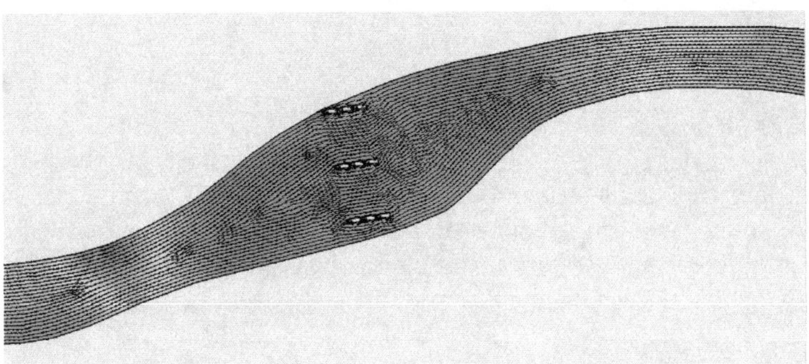

Figure 5. Flow pattern without groyne placement.

IOP Conf. Series: Materials Science and Engineering **933** (2020) 012040 doi:10.1088/1757-899X/933/1/012040

3.1. Flow pattern based on perpendicular angle groyne and two difference length (L)
Fig. 6 depicts the flow pattern around the 5-unit of groynes, 7.0 m and 9.0 m long, installed perpendicular to the river bank. The 9.0m groyne was narrower than the river width than the 7.0m. Flow velocities at these two formations accelerated at the tip of each groyne at the outer bend due to deflection. Velocities at the inner river bend decreased significantly because of the blocked area, but the high velocity was transmitted to the other side (outer river bend) and the downstream area. It is fascinating to note that velocities distribution at the outer river bend for GT5L9 was lower than that of GT5L7. The presence of groyne affected the turbulent flow field, and the separation zone length of GT5L9 was better than that of GT5L7. Eddy distribution occurred at GT5L9 was pushed towards the middle of the river and extended downstream. Whereas at GT5L7 eddy distribution was concentrated at each nose of the groins and was not swept away the inner river bend. Hence, the eddy overlapped each other and was then transmitted away to the outer river bend and downstream as well.

GT5L7 GT5L9

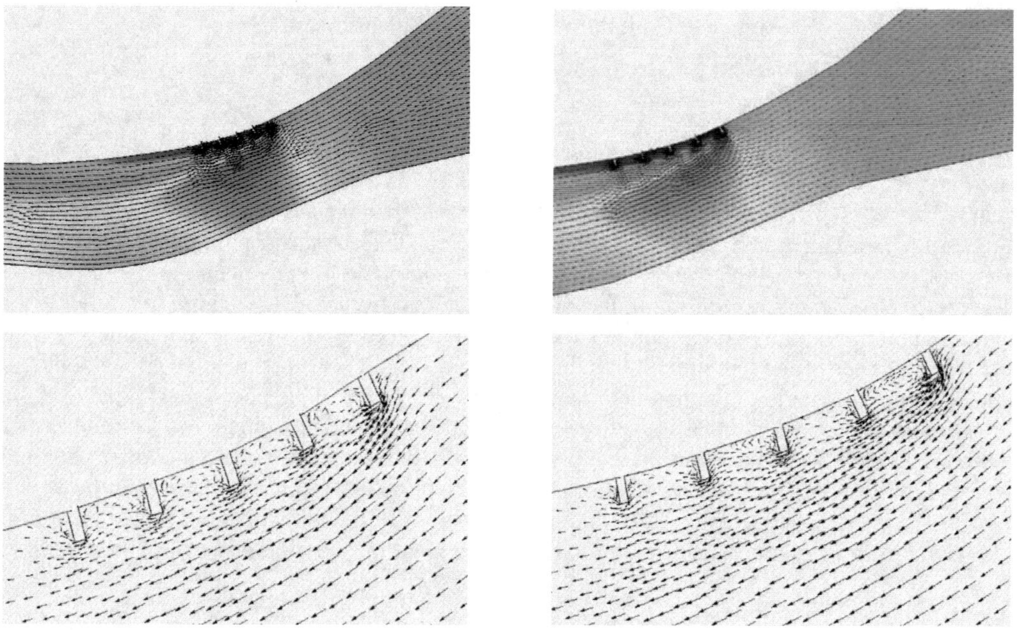

Figure 6. Flow pattern for GT5 groyne placement scenario.

3.2. Flow pattern based on groyne angle and groyne distance
Flow variation for groyne angles of 20° and 30° and the length of 7.0 and 9.0m are shown in Fig. 7. The GIa5L7 formation resulted in smaller velocity vector distribution in the reach. One can see that at the deflection angle of 10°, the groyne spread out eddy to the middle and downstream of the river and the critical velocity was lower compared to that of GIa5L9 groyne type. Whereas for deflection angle of 30° groyne type spreading out eddy occurred at each tip of the groynes and critical velocity extended to the outer river bend.

IOP Conf. Series: Materials Science and Engineering **933** (2020) 012040 doi:10.1088/1757-899X/933/1/012040

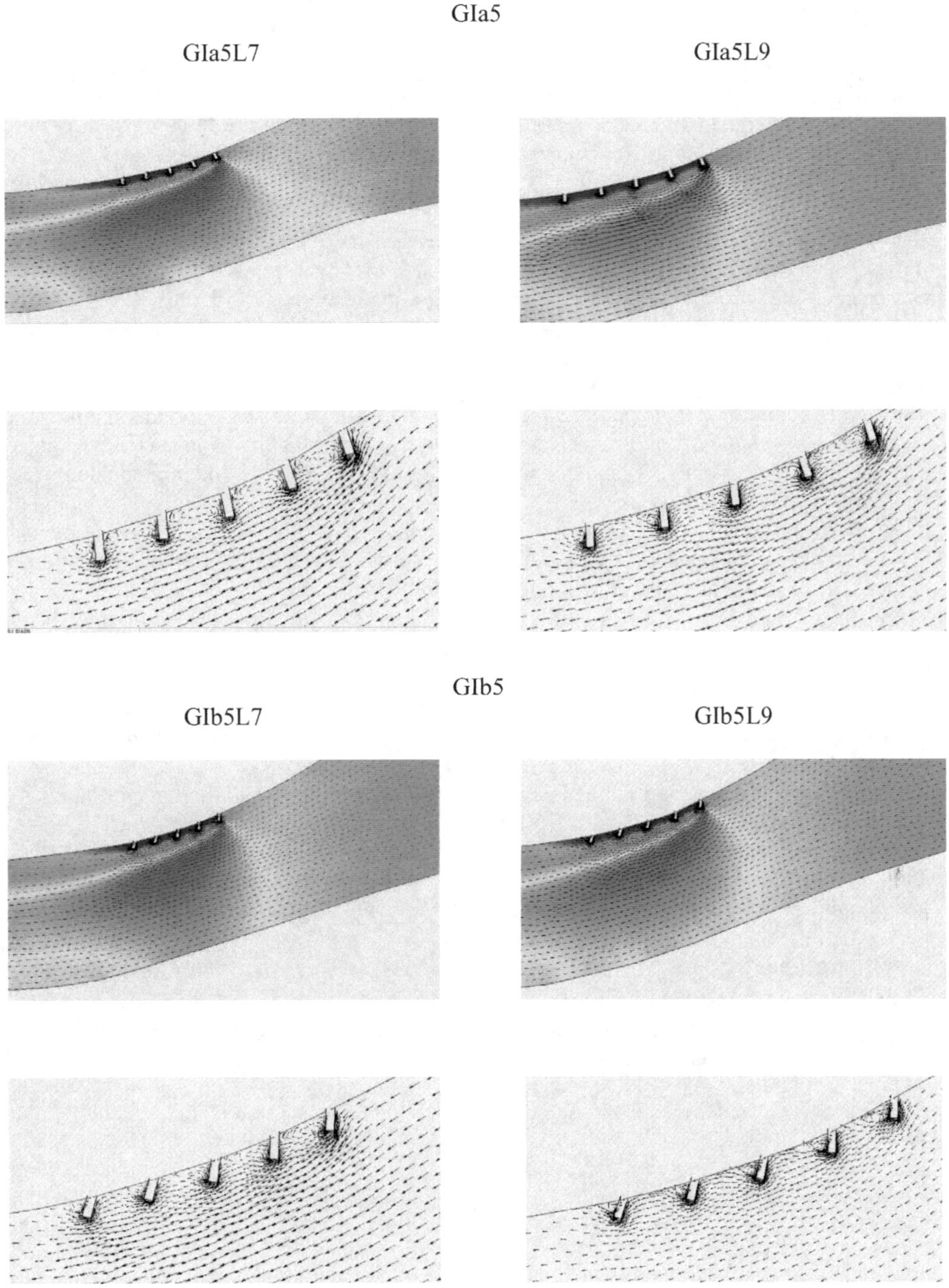

Figure 7. Flow pattern for GIa and GIb groyne placement scenario.

4. Conclusion

Nine scenarios of groyne instalment at the Krueng Aceh river have been simulated using SMS software. It can be concluded that the distribution of flow velocities towards the outer river bend for 9.0m perpendicular groyne was smaller compared to the 7.0m. Groynes with the angle of 10° and the length of 7.0m were better in reducing the velocities at the outer river bank than the 9.0m one. Whereas groyne with 30° orientation and 9.0m long was better in reducing the eddy than that of 7.0 m.

Acknowledgments

This research was supported or partially supported by the Ministry of Research, Technology and Higher Education (Kemenristek Dikti) through LPPM Unsyiah under the research grant for the assistant professor scheme. We would also like to express our gratitude to the LPPM staff administration for helping us during this research, and we also thank the reviewers for their so-called insights.

References

[1] Suharjoko 1998 *Study on Numerical Modeling of Two-Dimentional Horizontal Flow special case Groyne on the River Estuary* Thesis for the degree of Master Science in Civil Engineering Program (Yogyakatra: Departement of Engineering Science, Post-graduate Program, Univercity Of Gajah Mada).

[2] Suharjoko 2001 *Numerical Modeling of Two-Dimentional Horizontal Flow on Groyne Field due to Groyne Placement on the River Straight* Recearches Report (Surabaya: Departement of Recearch and Public Official, Institute of Technologie Sepuluh Nopember).

[3] Mojtahedi A and Basmenji A B *2017 Numerical and Field Investigation of the Impacts of the Bank Protection Projects on the Fluvial Hydrodynamics (Case Study: Ghezel Ozan River) Intern. J. of Eng. and Techn.* **9** 6.

[4] Dutta D and Kalita H M 2018 Performances of straight head and T-head Groynes as river training structures *IOP Conference Series: Materials Science and Engineering* vol 491 012013.

[5] Sunaryo, Daoed D and Sari F L 2010 Pengaruh pemasangan krib pada saluran di tikungan 120° *J. Rekayasa Sipil* **6** 1 45 – 54 (Palembang: Fakultas Teknik – Universitas Andalas).

[6] Sosrodarsono and Tominaga 1985 *Perbaikan dan Pengaturan Sungai*, (Jakarta: Pradnya Paramita).

[7] Mudjiatko 2000 *Pengaruh Meander Sungai Terhadap Perubahan Konfigurasi Dasar dan Seleksi Butiran Sedimen* (Yogyakarta).

[8] Boss SMS 1995 *User's Manual Surface water Modelling System, Version 5.02*, (Madison: Engineering Computer Graphics Laboratory, Brigham Young University).

[9] EL-Sammany M S and El-Moustafa A M 2011 Adaptation of surface water modeling system for sediment transport investigations in Lake Nasser *Nile Basin Water Science & Eng. J* **4** 1.

Community satisfaction levels on the housing reconstruction project after Mentawai tsunami in 2010 – a case study at Sipora Island

T Ophiyandri[1], B Hidayat[1] and C Ghiffari[1]

[1] Civil Engineering Department, Engineering Faculty, Universitas Andalas, Universitas Unand Limau Manis, Padang, 25163, Indonesia

Corresponding E-mail: ophiyandri@eng.unand.ac.id

Abstract. An earthquake accompanied by tsunami that occurred in the Mentawai Islands on 25 October 2010 brought many casualties and damage on housing. As a result, people living on the coastline lost their homes and has to be relocated to safer places. The implementation of the housing reconstruction program was based on the relocation approach. Unsafe coastal locations require the communities to move to higher ground which unfortunately at the forest areas of Mentawai. The success of the reconstruction of the housing sector can be an indicator of success in the implementation of post-disaster reconstruction. The purpose of this research is to identify the community satisfaction levels on the housing reconstruction projects in Sipora island. This research was conducted in South Sipora by conducting structured interview on 20 respondents spread across four villages. Respondents were invited to rate the satisfaction levels from very dissatisfied until very satisfied with points values 1-5. The results show that the level of community satisfaction was 2.91. The main problem was caused by a long delay in the start of the reconstruction program. In addition, there are still problems after the house is completed, such as the lack of availability of clean water and public facilities.

1. Introduction

The tsunami along the coast of the Mentawai Islands triggered by the 7.7 Mw magnitude earthquake centered off the southwestern coast of Pagai Island, at 3,848 ° South latitude and 100,114 ° East Longitude with 20.6 km depth caused a tsunami [1]. Based on the damage and loss assessment, the number of damaged houses was 1,269 housing units, with details of damage including 879 heavily damaged, 116 moderately damaged, and 274 slightly damaged. The impact of the disaster also caused damage to roads, government offices, health facilities, educational facilities, trade facilities, and resorts [2]. Further, based on the assessment of rehabilitation and reconstruction needs and the acceleration of development in the Mentawai Islands region, taking into account the direction of the President of Indonesia and the sectoral recommendations, it is necessary to relocate community settlements in coastal areas to safer locations. Rehabilitation and Reconstruction with the planned Resettlement Relocation approach will be carried out in stages in the 2011 and 2012 fiscal years, in areas affected by the tsunami is North Pagai Island, South Pagai Island and Sipora Island [2]. Basically, any form of reconstruction activities or programs that have been planned and organized should have benefited the community, especially the housing sector. However, if the survivors did not satisfy, sooner or later they clearly want to return to their original lives. Therefore, this paper aims to identify the community satisfaction levels on the housing reconstruction projects at Sipora island of Mentawai.

2. Literature review

2.1. Tsunami impact

The 2010 Mentawai Islands earthquake occurred on October 25th, 2010 with a 7.2 magnitude earthquake at 21:42:20 local time, located 78 km southwest of South Pagai Island. Followed by tsunami waves within only 5-10 minutes [3]. Damage and loss assessment results are based on data as of November 22, 2010, showing that the earthquake and tsunami disasters have caused damage and losses of Rp. 348.92 billion. The greatest damage and loss occurred in the productive economy sector with estimated damage and losses reaching Rp. 117.82 billion (33.9% of the total value of damage and losses), where almost 80% of the damages and losses has occurred in the plantation sub-sector and fisheries sub-sector. Then, followed by damage and losses in the housing sector amounting to Rp. 115.82 billion (33.2%), infrastructure sector Rp. 19.16 billion (5.50%), social sector Rp. 16.03 billion (4.60%) and cross-sector Rp. 79.44 billion (22.81%) [2].

2.2. Principles of mentawai post-disaster recovery

During the post-tsunami emergency response in the Mentawai Islands District, options regarding the rehabilitation and reconstruction strategy were reviewed and consulted with the Vice President of the Republic of Indonesia on November 8th, 2010 [2]. The Mentawai Islands District Government supports the concept of Resettlement of Settlements from tsunami-affected coastal areas to safer areas at an altitude of ± 25 ASL, which is currently a forestry area, which is then followed up with the determination of new locations for housing relocation and the construction of residential environmental infrastructure. To accelerate the development of the Mentawai Islands Regency which is currently an undeveloped region, it is proposed to develop vital infrastructure in the form of interisland shaft roads, construction of air transportation infrastructure (airstrip) and sea transportation (docks) and construction of environmental roads including evacuation routes [2].

2.3. Community satisfaction indicator

Indicators of the community satisfaction toward the reconstruction program used as a reference in this study consisted of 19 indicators, during the reconstruction (14 variables) and to the results of reconstruction (5 variables) referring to Ophiyandri, et al., [4]. The detail of satisfaction level indicator can be seen at table 1.

Table 1. Indicator of community satisfaction level.

No	Satisfaction Indicator
A.	**Before the Reconstruction**
1.	Information and socialization about reconstruction
2.	The time the reconstruction program began
3.	Home damage identification process
4.	Home recipient identification process
5.	Involvement / participation in reconstruction
6.	Right aimed target recipient of aid
7.	Participation in the reconstruction process
8.	Collaboration between local communities in reconstruction
9.	The wishes of the people are fulfilled by the reconstruction
10.	Amount of aid funds received
11.	Easy administration / disbursement process
12.	The role of government in the reconstruction process
B.	**During the Reconstruction**
1.	The Role of Community Assistance Team (TPM)
2.	The Role of the Facilitator in the reconstruction process
3.	Labor availability (craftsman)
4.	Work experience and skills (craftsmen)

No	Satisfaction Indicator
5.	Availability of material for housing reconstruction
6.	Quality material available for housing reconstruction
C.	**After Reconstruction**
1.	With the results of existing assistance
2.	The current state of the house is compared to the past
3.	The house become more earthquake resistant
4.	The comfort of home compared to before
5.	The quality of the house now compared to before
6.	The house was had been as community wish
7.	Satisfaction with the current design
8.	Election of relocation area
9.	Security after relocation
10.	Access road to residence compared to before the relocation
11.	Current availability of street / environment lighting
12.	Availability of clean water
13.	The current condition of the surrounding community
14.	Availability of places of worship (Mosques, Churches, etc.)
15.	Availability of educational facilities

Pribadi et al. [5] state that in some major catastrophic events in Indonesia, the impact of a major disaster felt was damaging to housing owned by residents. Therefore, housing reconstruction is one of the actions taken to restore people's lives back to normal. Remembering the house is one of the basic human needs. The process of restoring conditions after a disaster is an important part of restoring people's lives after a disaster [6]. The principle of "Building Back Better" is important applied in the recovery process after a disaster occurs, by rebuilding people's lives by reducing vulnerability and increasing capacity to avoid the risk of impending disasters [7]. So that it requires post-disaster recovery planning so that recovery can be done better and faster. Coordination between stakeholders including the community, government and non-government institution needs to be considered in order to form a well-integrated reconstruction process [8].

3. Research methodology

In this study, the data collection method used in this study is by conducting a structured interview. The interview guideline was based on table 1 which divided based on the time of implementation: before and after the reconstruction process. The respondents were invited to indicate their satisfaction and provided their reason. The interview targeted the surrounding communities who experienced the earthquake and tsunami disaster who participated in the rehabilitation and reconstruction program after the 2010 earthquake and tsunami disaster in Mentawai Islands, especially in South Sipora. The respondents in this study were the community consisting of village heads and members of the local community. Since the limitations of the researchers, it was not possible for researchers to conduct a survey of all populations, so the sampling method was based on the cluster random sampling type. Interviews were conducted with 20 respondents on June 2019 in South Sipora, Mentawai Islands. The majority of respondents aged 41-50 with a percentage of 45%, followed by aged 31-40 with a percentage of 25%. The jobs of the interviewees are mostly farmers with a percentage of 70%.

Data are collected in Bosua Utara Village six respondents, Gobik Village five respondents, Katiet Village five respondents, and Masokut Village four respondents. In addition, interviews were also conducted with one of the local government parties who had served as Head of Rehabilitation and Reconstruction in the BPBD agency involved in the reconstruction program. The interview guideline consisted of two main parts, the first was the description of the interviewee data and the second was the question of the level of satisfaction. The assessment for the level of satisfaction on the guideline using the Likert Scale method which can be seen in table 2.

Table 2. Satisfaction level.

Code	Level	Weigh
VD	Very Dissatisfied	1
NS	Not Satisfied	2
QS	Quite Satisfied	3
S	Satisfied	4
VS	Very Satisfied	5

With each level of satisfaction converted into points in numbers range 1-5. Then, the community satisfaction level data on the guideline will be averaged per each indicator to obtain the level of community satisfaction for each indicator using Microsoft Excel software.

The conversation, during the interview will be recorded. The result of the recorded interview will be written into a copy of a conversation dialogue. The data analysis method used in this research is content analysis. Content analysis is research that is in-depth discussion of the contents of written or printed information in a media. This analysis is usually used in qualitative research. The content analysis method in this study was carried out using Nvivo software. Recorded data that has been copied will be analyzed using Nvivo software. Statements of problems stated by the respondents will be grouped into several groups. Based on data processing with Nvivo, conclusions can be drawn from the problem factors that exist in the reconstruction.

4. Analysis and discussion

In the pre-reconstruction section, the level of community satisfaction at the location was an average of 2.84 (range dissatisfied - quite satisfied). From these questions, there are a number of questions below the average of three point, including the following : the length of time to start the reconstruction program, transparency in the delivery of aid, the disbursement process, the amount of aid funds, the identification of damage to houses, the process of identifying beneficiaries, and the role of government.

Table 3. Average level of community satisfaction before the reconstruction.

A.	Before the Reconstruction	Average
1.	Information and socialization about reconstruction	2.50
2.	The time the reconstruction program began	1.90
3.	Home damage identification process	2.75
4.	Home recipient identification process	2.85
5.	Involvement / participation in reconstruction	2.80
6.	Right aimed target recipient of aid	3.35
7.	Participation in the reconstruction process	3.75
8.	Collaboration between local communities in reconstruction	3.50
9.	The wishes of the people are fulfilled by the reconstruction	2.95
10.	Amount of aid funds received	2.65
11.	Easy administration / disbursement process	2.45
12.	The role of government in the reconstruction process	2.60

As the causes of dissatisfaction based on community complaints that have been processed using qualitative data processing. Problem about information are the lack of information from the facilitator, material information from suppliers that are lacking, the government that does not directly come down to provide information. Regarding the reconstruction starting time, the reconstruction began in 2013-2015. The problem about the data collections are there is no data collection damage to houses from the government, the absence of data collection of recipients of assistance from the government and there are people who do not get housing assistance. As for the funds, lack of openness of funds, insufficient amount of aid funds, cuts in funds, funds not managed by the community, disbursement of funds a gradual, long-term disbursement process. The other problems are the relocation sites prone to landslides, and the absence of government oversight.

Based on information obtained from the government, the cause of this delay is because the housing land is located in the forest, so it takes time from 2012-2013 for the permit of the forest land, in March 2014 only began to be built. So that caused the start of the reconstruction of housing being delayed.

At the time of reconstruction, the level of community satisfaction at the location gained an average final point of 2.88 (Within the range of dissatisfied-quite satisfied). From these questions, there are four questions that average below 3, including the following: the role of the community facilitating team, the role of the facilitator, the taking and far-reaching materials, the quality of material.

Table 4. Average level of community satisfaction during the reconstruction.

B.	During the Reconstruction	Average
1.	The Role of Community Assistance Team (TPM)	2.75
2.	The Role of the Facilitator in the reconstruction process	2.50
3.	Labor availability (craftsman)	3.15
4.	Work experience and skills (craftsmen)	3.20
5.	Availability of material for housing reconstruction	2.80
6.	Quality material available for housing reconstruction	2.85

Problems found in this section include, for TPM, lack of TPM cooperation with the community, the existence of fund cuts by TPM and the absence of TPM assistance for the community. The role of the facilitator, the deduction of funds by the facilitator and the absence of assistance provided by the facilitator to the community. Regarding the availability of workers, the availability of workers is lacking. For the availability of material materials, among others, the price of expensive wood shaving, material taken from the river, remote location of material taking, inadequate equipment, and the absence of pavement. As for the quality of the material, the quality of the sand material that is lacking is due to the beach. Based on information obtained from the government, it was admitted that there was negligence, such as deduction of money from the facilitator, which of course should not have happened.

In the post-reconstruction section, the community satisfaction level gets an average end point of 2.95 (Being in the dissatisfaction range – quite satisfied). From these questions, there are some questions whose index is below three point, including the following: availability of clean water, availability of religious facilities, availability of economic facilities, homes that are as desired, the distance from house to workplace, the checking from the government, periodic assistance, and the government promises have been fulfilled.

Table 5. Average level of community satisfaction after reconstruction.

C.	After Reconstruction	Average
1.	With the results of existing assistance	3.10
2.	The current state of the house is compared to the past	3.25
3.	The house become more earthquake resistant	3.80
4.	The comfort of home compared to before	3.60
5.	The quality of the house now compared to before	3.65
6.	The house was had been as community wish	2.90
7.	Satisfaction with the current design	3.05
8.	Election of relocation area	3.85
9.	Security after relocation	3.95
10.	Access road to residence compared to before the relocation	3.90
11.	Current availability of street / environment lighting	3.60
12.	Availability of clean water	1.65
13.	The current condition of the surrounding community	3.60
14.	Availability of places of worship (Mosques, Churches, etc.)	2.75
15.	Availability of educational facilities	3.55

C.	After Reconstruction	Average
16.	Availability of economic / market facilities	1.35
17.	Availability of electricity at home	3.40
18.	Distance from home to work / field	2.20
19.	Attention / re-checking by the government after relocation	1.55
20.	Periodic assistance from the government after the house is built	1.70
21.	Fulfill the promise given by the government	1.60

Problems found in the post-reconstruction period include the absence of a kitchen, the absence of a bathroom, the small size of the house, which is the cause of dissatisfaction with the results of existing house. The small size of the house and the design of houses that use wood boards cause dissatisfaction with existing homes compared to the past house. For clean water facilities, most people rely on rainwater and well water. For places of worship are still in the old location and are still promised by the government. For market availability, inadequate roads and remote locations are the cause of dissatisfaction. For the availability of electricity, it still uses PLTS (Solar Power Plants). In case the distance from home access to work, including lack of access to roads, long distances and the absence of transportation advice are causes of dissatisfaction. For checks from the government, the absence of checks is the cause of dissatisfaction. The absence of periodic assistance, assistance that is only given during the emergency response, and providing assistance that is not too frequent. Regarding government promises, there are still government promises that have not been fulfilled such as, kitchen pledges, land certificates, places of worship, clean water, and economic recovery of farmers.

Based on information obtained from the government, the government has never promised to increase kitchen money, and public facilities at the housing site. Lack of information from the government and the circulation of incorrect issues can be a misunderstanding for the community. This problem should be immediately followed up by the government before causing other problems such as lack of public trust toward the government.

5. Conclusion

It can be concluded that the level of satisfaction of people who get house assistance in the reconstruction program are as follows: Before the reconstruction program 2.84, during the reconstruction program 2.88 and after the reconstruction program 2.95. The average level of community satisfaction was 2.91, meaning that it was close to the "quite satisfied" criteria. From 39 indicators used, 21 indicators show that the satisfaction level of beneficiaries are lower than three. The three indicators with the lowest level of satisfaction are the availability of economic facilities, the absence of checking from the government to housing locations, and government promises that still have not been fulfilled completely.

Acknowledgements
Authors wishing to acknowledge the Universitas Andalas for providing fund in conducting the research.

References

[1] Yudhicara et al. 2010 Jejak tsunami 25 Oktober 2010 di Kepulauan Mentawai berdasarkan penelitian kebumian dan wawancara. *J. Lingkungan dan Bencana Geologi* **1** 3 165-81.

[2] BNPB 2010 *Rencana Aksi Rehabilitasi dan Rekonstruksi Pascabencana Serta Percepatan Pembangunan Wilayah Kepulauan Mentawai Provinsi Sumatera Barat Tahun* 2011-2013.

[3] Sinaga L C 2012 Mentawai Pasca Tsunami 2010 (Kompasiana). https://www.kompasiana.com/lidyasinaga/5510c345813311af36bc6f65/mentawai-pasca-tsunami-2010 accessed on 29 Juli 2019.

[4] Ophiyandri et al. 2015 *Tingkat Kepuasan Masyarakat Terhadap Program Rehabilitasi dan Rekonstruksi Perumahan Pasca Gempa 30 September 2009 di Provinsi Sumatera Barat* (Padang: Pusat Studi Bencana Universitas Andalas, Kampus Unand Limau Manis)

[5] Pribadi K S et al. 2013 *Post-Disaster Housing Reconstruction in Indonesia : Review and Lessons from Aceh (2004), Yogyakarta (2006), West Java (2009) and West Sumatra (2009)*

Earthquakes ed. R. Shaw (Kyoto: Disater Recovery :Used or Misused Development Opportunity. Springer)

[6] Olshanky R & Chang S 2009 Planning for Disaster Recovery: Emerging Research Needs and Challenges *Progress and Planning* **72**

[7] Monday J 2002 Building Back Better: Creating A Sustainable Community After Disaster (University of Boulder. Natural Hazards Research and Application Information Center)

IOP Conf. Series: Materials Science and Engineering **933** (2020) 012042 doi:10.1088/1757-899X/933/1/012042

Seismic performance analysis of multi-span masonry infilled RC frame based on analytical strut model

M Maidiawati[1], J Tanjung[2], F Nugroho[1] and Y Hayati[3]

[1] Department of Civil Engineering, Padang Institute of Technology, Padang, Indonesia
[2] Department of Civil Engineering, Universitas Andalas, Padang, Indonesia
[3] Department of Civil Engineering, Universitas Syiah Kuala, Banda Aceh, Indonesia

Corresponding e-mail: maidiawati@itp.ac.id

Abstract. According to previous studies, it was accepted that masonry infills have a contribution to lateral strength, stiffness, and ductility of reinforced concrete (RC) frames structures. In the past studies, an analytical method of diagonal strut model has developed for evaluating the seismic performance of masonry infill. In this model, the masonry infills were presented by compression struts as distributed forces at the frame-infill interfaces in which the frame-infill contact lengths were evaluated by static equilibrium related to compression balance and lateral displacement compatibility. The lateral strength of infill was determined based on frame-infill contact length. In the current study, an experimental study of quasi-static cyclic lateral loading tests was conducted to evaluate the seismic performance of two-bay brick masonry infilled RC frames. This paper discusses the results of seismic performance brick infills based on experimental and analysis results of the multi-span masonry RC frame applying the diagonal strut model. According to the analytical model, antisymmetric strut was acted at the top and bottom of the middle column of the multi-span infilled frame from each infill. As a result, good agreements of lateral strength of structures were obtained between the experimental and analytical values.

1. Introduction

Masonry infills are a non-structure element which generally used as a partition on reinforced concrete (RC) buildings in Indonesia as high seismicity. According to Indonesian code, the masonry infill is ignored in seismic design calculation of RC buildings, only consider for self-weight estimation. A lot of experimental and analytical results have been confirmed that masonry infills contributed to the seismic performance of surrounding frames structures [1-5]. Therefore, neglecting the masonry infills in seismic design calculations may result in unprecise assessments for the lateral stiffness, strength, and ductility of the buildings.

The authors [1-3] have been performed cyclic loading tests on single-span brick masonry infilled frames finding that the lateral strength, stiffness, and ductility performance of overall structures were affected by the presence of the masonry infills. Many analytical approaches have been proposed by many researchers such as Smith and Carter [6, 7], Mainstone [8], Leuchars and Scrivener [9], Paulay and Priestley [10], Maidiawati and Sanada [11, 12] who developed the diagonal strut models for evaluating the seismic performance of masonry infill in framed structures. Most of the developed strut models were concentrated to evaluate the strut width of infill based on the frame-infill contact length. In the case of the diagonal strut model developed by Maidiawati and Sanada [11, 12], the infill contribution was distributed compression transferred diagonally between frame-infill interfaces. The frame-infill contact length can be determined by static equilibriums related to compression balance and lateral displacement compatibility at the frame-infill interfaces. As a result, the equivalent strut width is given as a function of frame-infill contact length. The effectiveness of the proposed model was accepted based on comparisons with the experimental results of a series of structural tests of brick masonry-infilled RC frames representing a typical RC building with non-structural masonry elements.

In the current study, an experimental study of quasi-static cyclic lateral loading tests was conducted to assess the seismic performance of two-bay brick masonry infilled RC frames. This paper presents the comparison of seismic performance between experimental and analytical results based on the diagonal strut model.

2. Methodology

2.1. Specimens

Two 1/4 scale two-bay RC frames specimens, one without infill (BF-2B) and one with brick infills (IF-2B-BM), were prepared. Figure 1 presents the drawing of the configuration and bar arrangements of the two-bay RC frame specimens. In the case of infilled frame specimen, as shown in figure 1(b), bricks with length of 60 mm, width of 30 mm and height of 15 mm were arranged in the interior clear height of one RC frame with mortar beds at a volume ratio of cement: sand: water = 1: 4: 1.4. The infill surfaces were plastered with finishing mortar with a thickness of 10 mm. Consequently, the thickness of the infills was 50 mm. Material properties of specimens included compressive strength of concrete of 28.4 N/mm^2, young modulus of concrete of 25045.0 N/mm^2, compressive strength brick masonry prism of 7.6 N/mm^2, yield strengths of longitudinal dan transversal reinforcements of columns of 449.5 N/mm^2 and 390.2 N/mm^2, respectively.

2.2. Test method

Both specimens, BF-BM and IF-2B-BM, were tested under quasi-static cyclic lateral load. Figure 2 shows the view of the experimental test set up in which the specimen was fixed on rigid-floor by using six post-tensioning rods at the lower beam. Lateral actuator equipment for lateral force action was attached and fastened to the strong wall. Several transducers (LVDTs) were placed at the upper beam and at columns to measure the lateral displacement of the RC frame.

The cyclic lateral load was applied to specimens based on according to the FEMA 461 [13]. Figure 3 was the lateral loading history applied to specimens with initial cycles to R=1/800 followed by two cycles to R=1/400, R=1/200, R=1/100, R=1/50, R=1/25, R=1/12.5 and a final load by a pushover to R=+1/10 rad. The R was drift angle, a ratio of lateral displacement to column height, used for controlling the incremental loading. The incremental lateral load and lateral deformation of RC frame were measured during the tests.

2.3. Test results

Experimental seismic performance of BF-2B and IF-2B-BM specimens were presented by the relationship between lateral force and drift ratio, as shown in figure 4. The maximum lateral strength of bare RC frame of 64.2 kN was observed at a drift ratio of 8.0%, while the maximum lateral strength of brick infilled RC frame of 159.7 kN was detected at a drift ratio of 1.0%, as shown in figure 4. It revealed that the lateral strength of brick infilled frame structure was higher about two and half times than those of bare RC frame. Due to lateral deformation of the structure, a diagonal compression force occurred on infill which acted as a punching shear force at the bottom/top of compressive/tensile column.

IOP Conf. Series: Materials Science and Engineering **933** (2020) 012042 doi:10.1088/1757-899X/933/1/012042

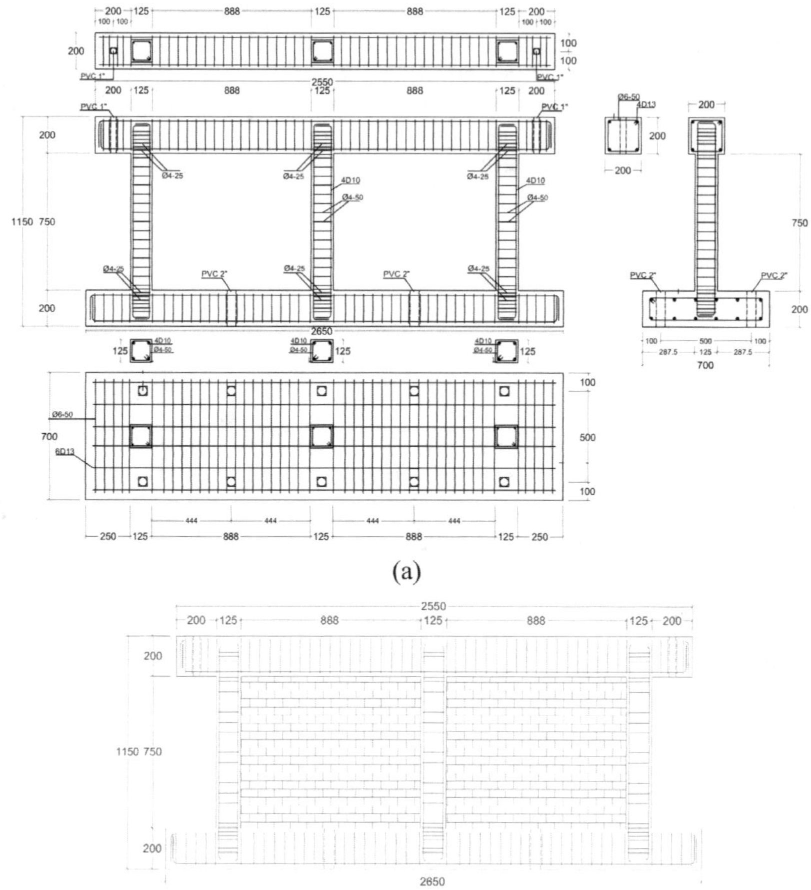

(a)

(b)

Figure 1. a) the configuration and rebars arrangement of two-bay RC frame specimen b) drawing of the two-bay brick infilled specimen (IF-2B-BM).

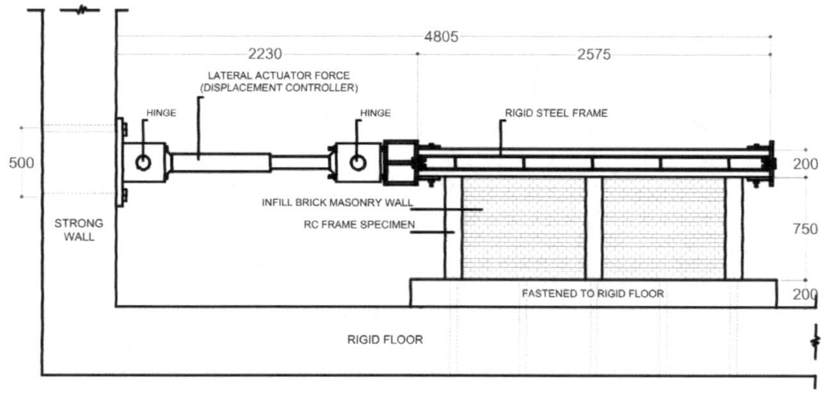

Figure 2. Schematic view of test.

IOP Conf. Series: Materials Science and Engineering **933** (2020) 012042 doi:10.1088/1757-899X/933/1/012042

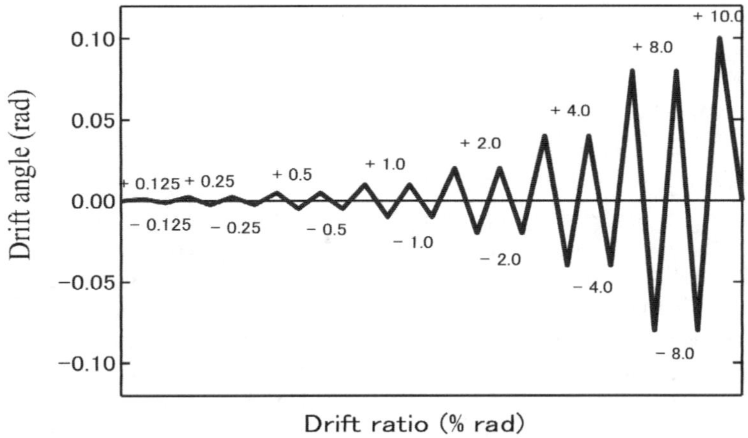

Figure 3. Cyclic loading to specimen.

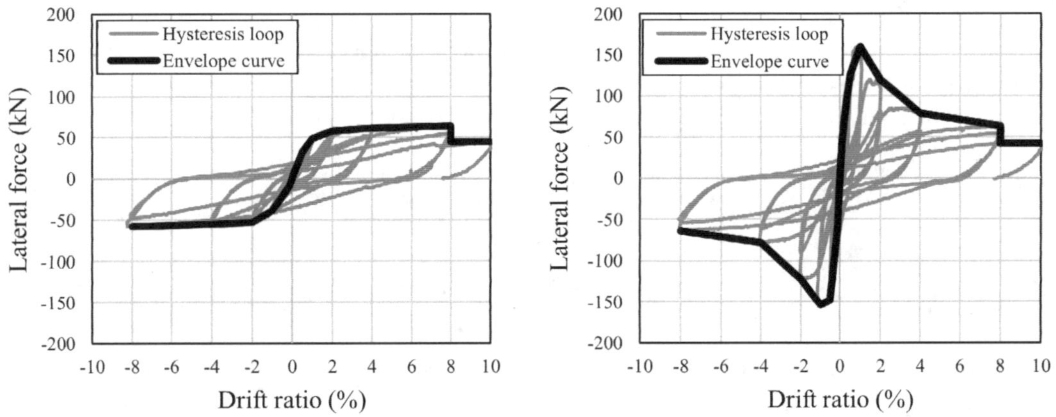

(a) (b)

Figure 4. a) two-bay bare RC frame: b) two-bay brick infilled RC frame.

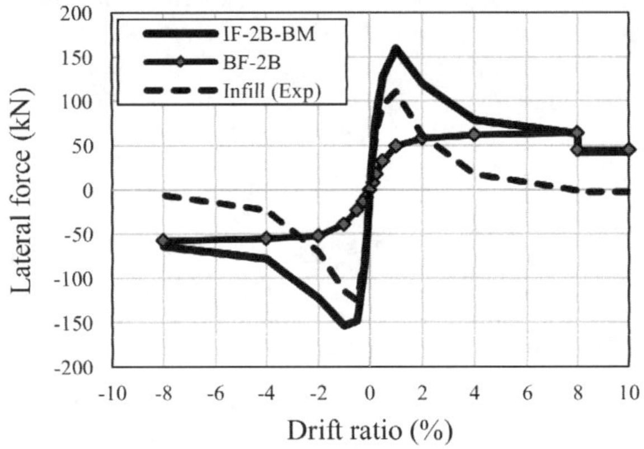

Figure 5. Envelope curves of infilled frame performance.

The infill contribution is presented in envelope curve of the lateral force-drift ratio relationship, as shown in figure 5 which were evaluated by extracting the difference of envelope curve between the lateral forces of the infilled frame and the bare frame at each load step (at the same drift ratio).

3. Results and discussion

3.1 Analytical model

In the previous study, an analytical strut model for evaluating the seismic performance of masonry infilled frames have been developed by the first author. The proposed model has been verified using experimental study results single-bay masonry infilled RC frames, and it has been applied for seismic performance analysis of RC buildings by considering masonry infills as explained in references [11, 12]. The developed analytical model replaces masonry infill with a diagonal compression strut, which represents distributed compression transferred between frame-infill interfaces. The equivalent strut width is presented as a function of the frame-infill contact length h_s, as shown in figure 6(a), which can be evaluated by static equilibriums related to compression balance and lateral displacement compatibility at the frame-infill interfaces. The minimum W' was accepted because two different contact lengths were usually obtained from the two ends of the strut. An equivalent rectangular stress block was applied for replacing the stress distribution at the frame-infill interface, as shown in figure 6(a). Therefore, the averaged compressive strength f_m' can be calculated by multiplying the uniaxial compressive strength of the infill f_m by a reduction factor defined by equation (1a). The α was obtained according to the stress profile at the frame-infill interface, which was assumed to correspond to a strain profile ε(y) by equation (1b) under uniform strains along the strut.

$$\propto = \frac{\varepsilon_{ave}}{\varepsilon_{max}} = \frac{\int_0^{hs} \varepsilon(y)^{dy}/h_s}{\varepsilon_{max}} \tag{1a}$$

$$\varepsilon(y) = \frac{\{\delta_f(y) - \delta_s(y)\}\cos\theta}{l(y)} \tag{1b}$$

where, $\delta_f(y)$ is the flexural deformation of the column and $\delta_s(y)$ is shear deformation of the infill along a height of y, $l(y)$ is the diagonal length of the strut crossing the column at the height of y; and $\varepsilon_{ave.}$, $\varepsilon_{max.}$, and h_s is infill-column contact length, as in figure 6(a), θ is inclination angle of the strut, as shown in figure 6(b).

The diagonal compression C_s', which acts on the bottom/top of the compressive/tensile column, as shown in figure 6(b), is given by equation (2a). As a result, the total diagonal compression C_s was represented by twice Cs', as given by equation (2b). C_s was then resolved into the horizontal and vertical components C_h and C_v, which were represented by uniformly distributed forces along the column height, as shown in figure 6(d). The C_h and C_v were calculated by equations (2c) and (2d), respectively.

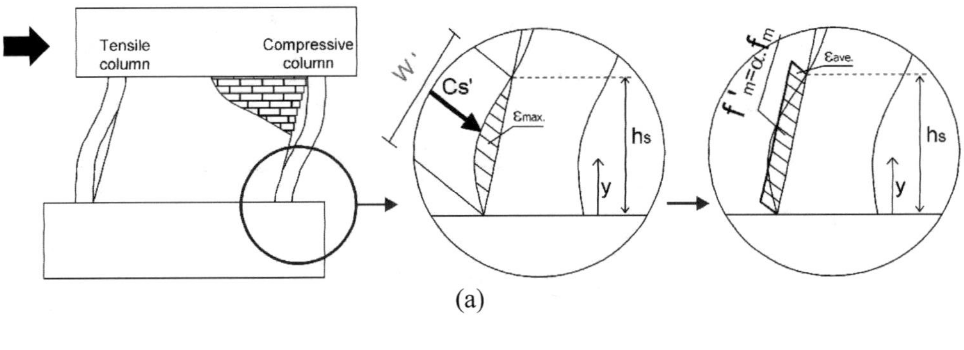

(a)

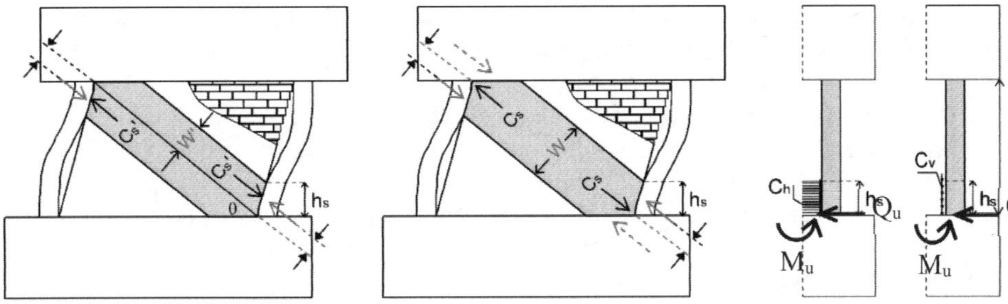

Figure 6. Modelling of masonry-infilled frame; (a) lateral deformation; (b) Diagonal compression force at the frame-infill interfaces, (c) Strut width of infill, (d) Distributed strut force.

$$C_s' = W'.t.f_m' \tag{2a}$$

$$C_s = 2W'.t.f_m' = W.t.f_m' \tag{2b}$$

$$c_h = t.f_m'.cos^2\theta \tag{2c}$$

$$c_v = t.f_m'.\sin\theta\ \cos\theta \tag{2d}$$

where, the total width of strut W is twice of W', and t is the thickness of the infill.
The column was assumed to yield in flexure at the end of the column-infill contact. Therefore, equations (3a) and (3b) present the moment distribution from the end $_cM(y)$ for the cases of $0 \leq y \leq h_s$ and $h_s \leq y \leq H$, respectively.

$$_cM_{(y)} = {}_cM_u - Q_u y + \frac{1}{2}C_h y^2 \tag{3a}$$

$$_cM(y) = {}_cM_u - Q_u.y + C_h h_s y - \frac{1}{2}C_h h_s^2 \tag{3b}$$

where, H is the clear height of column (figure 6d), $_cMu$ is the flexural strength at the column end based on the Japanese standard [14], and Q_u is the shear force at the column end given in equation (5).
Lateral displacement along column height $_c\delta(y)$ was generated by the twice integrals of equation (3)/EI (= elastic flexural rigidity of the column) as follows:

In the case of $0 \leq y \leq h_s$

$$c\delta_{(y)} = \frac{1}{EI}\left(\frac{1}{24}C_h y^4 - \frac{1}{6}Q_u y^3 + \frac{1}{2}M_u y^2\right) \tag{4a}$$

In the case of $h_s \leq y \leq H$

$$c\delta(y) = \frac{1}{EI}\left[\left(\frac{1}{6}C_h h_s - \frac{1}{6}Q_u\right)y^3 + \left(\frac{1}{2}M_u - \frac{1}{4}C_h h_s^2\right)y^2 + \frac{1}{6}C_h h_s^3 y - \frac{1}{24}C_h h_s^4\right] \tag{4b}$$

Q_u for equations (3) and (4) is given by equation (5) with assuming that rotation of zero at a column height of H $(d_c\delta(y = H)/dy = 0$ from equation (4b) assuming a column-sway mechanism.

$$Q_u = \frac{2M_u}{H} + C_h h_s - \frac{C_h h_s^2}{H} + \frac{C_h h_s^3}{3H^2} \tag{5}$$

IOP Conf. Series: Materials Science and Engineering **933** (2020) 012042 doi:10.1088/1757-899X/933/1/012042

Even though the above equations were derived based on the elastic rigidity, only the drift profile along the column height was applied to obtain the frame-infill contact length in the following calculations. The contact length h_s was calculated based on a lateral shear drift profile along the infill height ${}_i\delta(y)$ which was obtained by equation (6), considering the infill shear deformation assumed above and the compatibility between the column and infill drifts at the ends. Ultimately, an intersection height y_i between the column and the infill can be evaluated by solving equation (7). The process for determining of h_s are presented in figure 2, where the intersection height should equal h_s. The unknown h_s was found iteratively after satisfying equation 6. The minimum h_s between both ends of the strut was used to determine the width of the compression strut W, as given in equation (8).

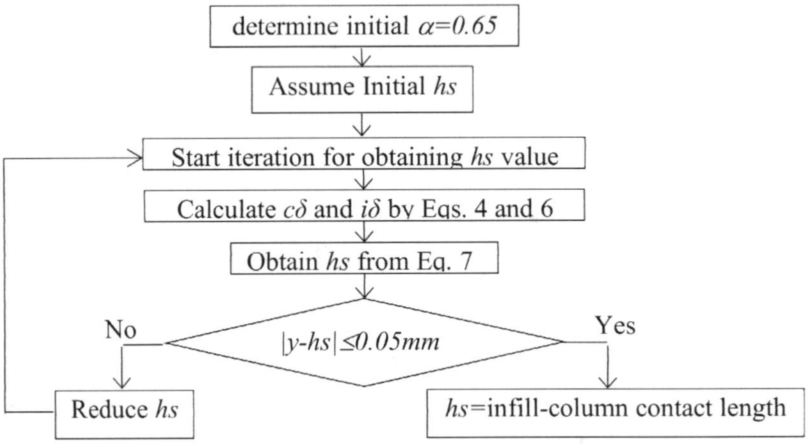

Figure 7. Flowchart for evaluating the contact length of the infill/column.

$$ {}_i\delta(y) = \frac{{}_c\delta(y=H)}{H}\, y \tag{6}$$

$$ {}_c\delta(y_i) = {}_i\delta(y_i) \tag{7}$$

$$ W = 2h_s \cos\theta \tag{8}$$

In the case of modeling the multi-bay infilled frames, each column was classified into end columns on the tensile and compressive sides and a middle column, as shown in figure 8(a). In particular of the middle column, antisymmetric strut actions were considered for the top and bottom of the column, as shown in figure 8(b), assuming the formation of a single diagonal strut in each infill. Similar to analytical strut model of single-infilled frame mentioned above, C_s was resolved into the horizontal and vertical components, which represented the distributed forces along column height at the bottom and top of the column, as shown in figure 8(c). As a result, the shear force at the end of the middle column is given in equation (9), where h_s is the minimum contact length from both column-infill interfaces.

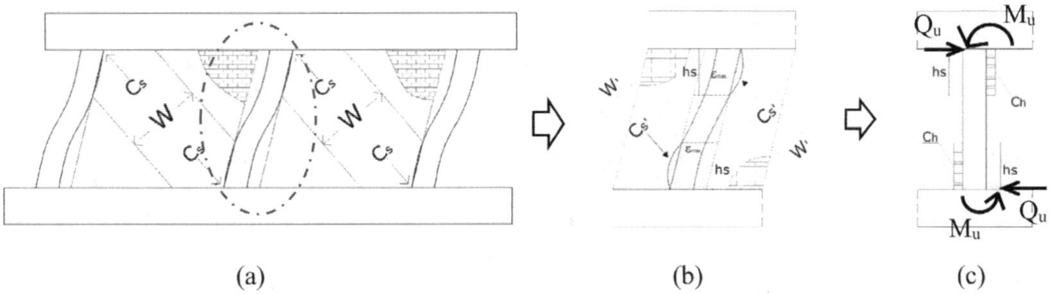

Figure 8. Modelling of multi-span masonry-infilled frame; (a) lateral deformation; (b) strut model on middle column, (c) distributed strut model.

$$Q_u = \frac{2M_u}{H} + c_h.h_s - \frac{c_h.h_s^2}{H} \qquad (9)$$

3.2 Analytical analysis of the seismic performance of masonry infills

The seismic performance of brick infills in the two-bay infilled frame was evaluated by applying the analytical method mentioned above. Consequently, the frame-infill contact lengths h_s were evaluated at 234.3 mm. Thus, the strut width of infill was 357.9 mm. Moreover, the infill performance was replaced by a bilinear curve with a yield point of $({}_iV_y, {}_i\delta_y)$, where the yield strength ${}_iV_y$ and drift ${}_i\delta_y$ are given in equations (10) and (11), respectively.

$$V_y = C_s \cos\theta = Wtf'_m \cos\theta \qquad (10)$$

$$\delta_y = \frac{V_y}{K_y} = \frac{V_y}{E_m Wt \cos^2\theta / d} \qquad (11)$$

where, K is secant stiffness, and d is the diagonal length of infill.

As a result, the maximum lateral strength of brick infills was 135.1 kN at a drift of 0.37%. The evaluated lateral strength and drift of infill are compared to the experimental results in figure 9. Good agreements were obtained between experimental and analytical results. It was verified that the strut model was applicable for estimating the seismic performance of masonry infill of multi-span infilled frame.

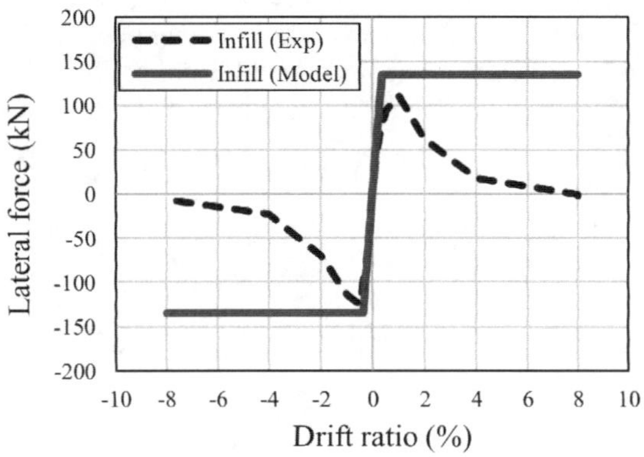

Figure 9. Comparisons of infill performance.

4. Conclusion

Experimental study of static cyclic lateral loading was conducted on specimens of two-bay RC frames, without infill and with infills. Test results revealed that brick infills in two-bay RC frame increase lateral strength of overall structure about two and half time. A developed diagonal struct model was implemented to evaluate the infill-frame contact length and struct width of infills as parameters for assessing the lateral strength and drift of brick infills at a yield point. As a result, good agreements of lateral strength were obtained between experimental and analytical results. It indicates that the diagonal strut model is compatible with the multi-span brick infilled RC frame.

Acknowledgments

This research was financially supported by the Ministry of Research, Technology, and Higher Education, Indonesia (326/27.O10.5/PN/II/2019). The authors also sincerely acknowledge to head and staffs of Structure and Construction Material Laboratory, Civil Engineering Department, Syiah Kuala University, for supporting during prepared the specimens, provided the structural testing facilities and conducted the structural tests.

References

[1] Maidiawati, Sanada Y, Konishi D and Tanjung J 2011 *JAABE* **10** 1
[2] Tanjung J, Maidiawati and Nugroho F 2017 *Proc. Int. Conf. on AIP* vol 1892 020009, doi: 10.1063/1.5005640
[3] Maidiawati, Sanada Y, and Tanjung J 2018 *IJASEIT* **8** 6
[4] Suzuki T, Choi H, Sanada Y, Nakano Y, Matsukawa K and Paul D 2017 *Bull Earthquake Eng*, doi:10.1007/s10518-017-0139-1
[5] Dautaj A D, Kadiri Q and Kabashi N 2018 *J. of Eng. Structures* **165**
[6] Smith B S 1967 *Building Science* **2** 3 doi: 10.1016/0007-3628(67)90027-8
[7] Smith B S and Carter CA 1969 *Proc. Int. Conf. on the Institution of Civil Engineers* vol 44
[8] Mainstone R J 1971 *Proc. Int. Conf. on the Institution of Civil Engineers* vol 49 No 2
[9] Leuchars J M and Scrivener J C 1976 *Bulletin of the New Zealand National Society for Earthquake Engineering* **9** 2
[10] Paulay T and Priestley M J N 1992 *Seismic Design of Reinforced Concrete and Masonry Buildings*, John Wiley & Sons: New York
[11] Maidiawati and Sanada Y 2013 *Proc. Int. Conf. on the 13th East Asia-Pacific Conference on Structural Engineering and Construction*
[12] Maidiawati and Sanada Y 2017 *Earthquake Engineering Structural Dynamics* **46** 2
[13] FEMA 461 2007 *Interim Testing Protocols for Determining the Seismic Performance Characteristic of Structural and Non-structural Component* p. 21-25
[14] The Japan Building Disaster Prevention Association English Version 1st 2005 *Standard for Seismic Evaluation of Existing Reinforced Concrete Buildings 2001*

IOP Conf. Series: Materials Science and Engineering **933** (2020) 012043 doi:10.1088/1757-899X/933/1/012043

Public perception of the transit bus plan in the city of Lhokseumawe

H Haryadi[1], Y Darma[2] and M Isya[2]

[1] Civil Engineering postgraduate program, Department of Civil Engineering, Universitas Syiah Kuala, Banda Aceh, Indonesia
[2] Department of Civil Engineering, Universitas Syiah Kuala, Banda Aceh, Indonesia

Corresponding e-mail: hendra_h@mhs.unsyiah.ac.id

Abstract. Community dependence on private transportation is a major factor in traffic congestion. Community perception is one of many important aspects in policy-making because by involving the community, the decision-makers will be able to capture the views, needs, and expectations of the community. The purpose of this study is to examine people's perceptions of the transit bus plan. The data collection method used is Stated Preference (SP) with a total of 100 respondents in stratified random sampling. The SP questionnaire contains information about socio-economic status and respondents' perceptions. The results of the perception survey show that the overall psychological questions have values above 2.5 from the 4 scale Likert, which means that respondents' perceptions of the transit bus plan are very positive. The respondents agree that the transit bus can be received with positive values.

1. Introduction

Transportation is a very important means of supporting the success of development [1]. In urban areas, poor public transportation services result in the increasing of the ownership of private vehicles every year. Problems then arise due to traffic congestion, air pollution, traffic accidents, and the cost of repairing road infrastructure. Transportation problems need special attention from the government. There are several problems related to the transportation system in Lhokseumawe City, which is one of the Centers of National Activities (CNA), and also the place of Arun Lhokseumawe Special Economic Zones (SEZ). Lhokseumawe city serves the flow of people, goods, and services from outside into the city area and vice versa for both domestic and international purposes. As a policy measure to alleviate these problems, the Government of Lhokseumawe City plans to create transit mass bus transport, which aims to advance the system and management of public transportation and improve the quality of transportation services to be comfortable, fast, timely, and affordable for the community. The transit bus is planned to provide a new bus service that is different and better than other public transportation services in Lhokseumawe City.

The perception of community acceptance is one of the keys to the successful implementation of transportation policies [2]. Understanding community needs is one of the ways to achieve the best policy because, by involving the community, the decision-makers will be able to capture the views, needs, and expectations of the community, following the logic that the views and preferences of the community become valuable input that then enables the realization of responsive decisions. The use of public transportation is determined by the satisfaction of its users [3]. Therefore, this information is very important and useful to develop the transit bus plan into acceptable policies because of valuable input from the community. This study analyzes the public perceptions of the transit bus plan by using

IOP Conf. Series: Materials Science and Engineering **933** (2020) 012043 doi:10.1088/1757-899X/933/1/012043

descriptive statistical analysis with a Likert scale. The discussion and conclusions of the study are presented at the end of the paper.

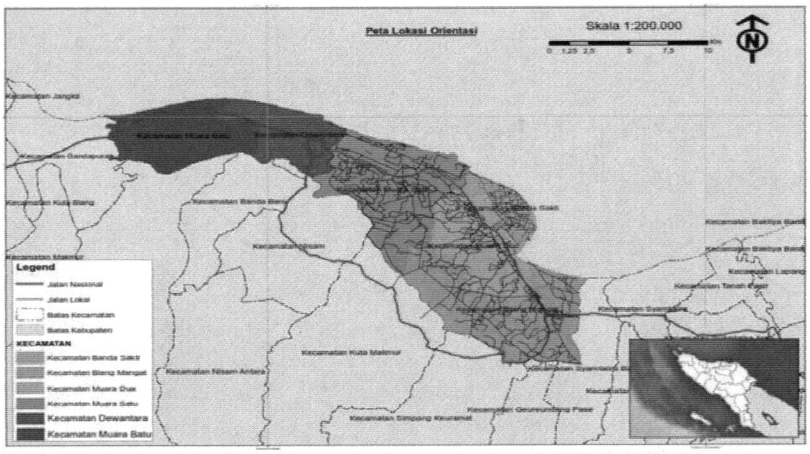

Figure 1. Lhokseumawe city (map data © 2019 Google).

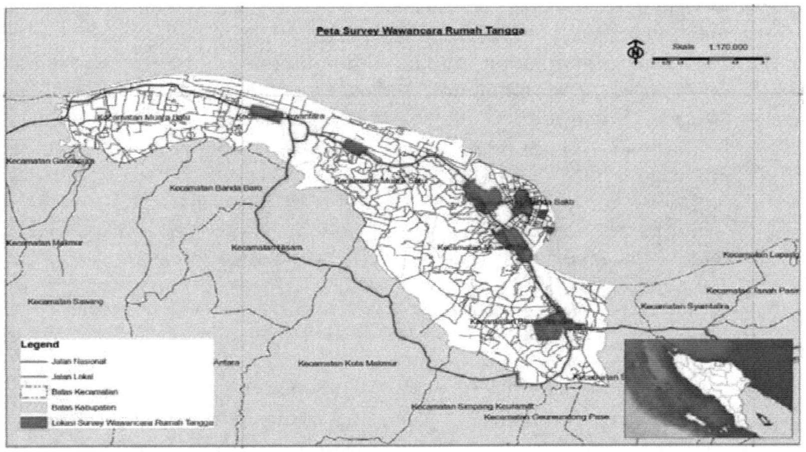

Figure 2. Area of survey (map data © 2019 Google).

2. Methodology

Community perception is the response or environmental knowledge from a group of individuals who interact with each other because they have values, norms, ways, and procedures. The degree of user satisfaction is one of the determinants of the model of selection [4]. The stated preference (SP) technique is an approach to have the best option of respondents' choice by making an alternative hypothetical situation. The hypotheses of the situation are based on an experimental design that guides the questionnaire to be given to respondents. The revealed preference data is collected using a questionnaire survey based on the Likert scale choices [5]. The questionnaire inquired about what choices respondents prefer or what ranking/rating they would give for a series of variables or certain choices in one or several possible situations. The study was conducted in 2019. Lhokseumawe city was the area of the study and the survey, which explanation can be seen in table 1 and 2.

Table 1. Summary of the SP survey.

Description	Detail
The year of survey	2019
Target location	Lhokseumawe city
Distribution sampling method	Direct interviews and collected by the enumerators
Sampling aggregation	100 samples
Distribution	Weekdays (75%); Weekends (25%)

Table 1 describes the SP survey which consists of the time of data collection, survey location, sampling number, and distribution of the sample's information. The target respondents consisted of commuters, commercial visitors, students, employees within the targeted area of study. More than 100 respondents were asked about their level of satisfaction and how important different aspects were to them; the respondents' respond was in terms of a 4-point Likert scale from 1 to 4 (1 representing strongly unsatisfied/strongly unimportant; 4 representing strongly satisfied/strongly important). Furthermore, the measured variables are translated into the indicator variables. This questionnaire included questions regarding the socio-demographic characteristics of respondents and travel behavior included during the survey. The questionnaire was designed to attain several bus system service elements, as shown in table 2.

Table 2. Itemed questions related to the bus service satisfaction and importance measures.

Element ID	Description	*Content of Element Question
Q1	Acceptability	Planned operation of the transit bus
Q2	Security1	Security aspect from crime while in the bus
Q3	Security2	Security aspect from crime while at the bus stop
Q4	Safety	Feeling safe aspect
Q5	Convenience1	The convenience of bus cleanliness
Q6	Convenience2	Convenience while boarding the bus
Q7	Convenience3	Bus capacity comfortable
Q8	Convenience4	The convenience of bus facilities
Q9	Convenience5	The convenience of bus stop facilities
Q10	Accessibility	Easy for passengers to reach the bus stop
Q11	Punctuality1	Easy to obtain the route and the timetable of the bus
Q12	Punctuality2	The accuracy of bus schedules
Q13	Equality	Facilities for disabilities and elderly people
Q14	Time Regularity1	Travel time within the bus stops
Q15	Time Regularity2	The headway between consecutive bus

* *using a 4-point Likert scale such as 1 to 4 (strongly unsatisfied/unimportant to strongly satisfied/important)*

Researchers used descriptive statistics to analyze data by describing the data that had been collected. Descriptive statistics present the results of the survey through a form of frequency distribution and percentage of recapitulation results of respondents' SP questionnaires. Frequency distribution provides both statistical and graphic figures. This frequency will be used to see the overall picture of the parameters, which were gender, age, the highest level of education, type of work, transportation costs, frequency of usage, income per household, and travel behavior.

According to [6], the Likert scale is a scale used to measure the attitudes, opinions, and perceptions of a person or group about social events or symptoms. Likert scale measurement is used for the purposes of quantitative analysis. Then, the answer can be given as a score, for example, strongly agree with a score of 4, agree with a score of 3, disagree with a score of 2, and strongly disagree with a score of 1.

A Structural Equation Modeling (SEM) framework is applied to model and analyze the collected data. The regression parameters were calibrated by using the applied maximum likelihood estimator (MLE) and were performed by using Amose 24. This method aimed to test a model hypothesis. This method also aimed to find out the vector of indicator variables (y) connected by a latent variable (y), with a covariate (x) included in the analysis calculation. The SEM model tested the consistency of structural theory and measurement through data and had the purpose of estimating parameters (coefficients, variances) and assessing the fit of the model.

Table 3. Respondent's Socio-demographic distribution.

Item	Category	Respondent's (Share)
Gender	Male	59 (59%)
	Female	41 (41%)
Age	16 - 25 years	29 (29%)
	26 - 35 years	13 (13%)
	36 - 45 years	34 (34%)
	46 - 55 years	17 (17%)
	56 years or more	7 (7%)
Highest level of Education	Primary School	38 (38%)
	College	16 (16%)
	University/Bachelor	42 (42%)
	University/postgraduate	4 (4%)
Monthly Income	1 million IDR or less	12 (12%)
	1- 1.9 million IDR	23 (23%)
	2 - 2.9 million IDR	26 (26%)
	3 - 3.9 million IDR	33 (33%)
	4 million IDR or more	6 (6%)
Vehicle Ownership	Has no vehicle	0
	Has motorcycle	76 (76%)
	Has car	3 (3%)
	Has car and motorcycle	21 (21%)
Occupation	College/University Student	33 (33%)
	Non-Government	38 (38%)
	Employee	17 (17%)
	Government Employee	12 (12%)
	Housewife	
Driver's License	Has the driver's license	87 (87%)
	Has no driver's license	13 (13%)
Purpose of traveling on the day of the questionnaire survey	Studying and lessons	37(37%)
	Work	49 (49%)
	Shopping/Entertainment	7 (7%)
	Other reasons for traveling	7 (7%)

Table 3 shows the distribution of respondent's socio-demographic and mobility attributes. Of the respondents, 59% were male and 41% were female. It can be seen that the questionnaire has been distributed equally, so it is expected that there are various different travel characteristics. More than 76% of respondents were young people below the age of 45 years. Concerning education and occupation, both datasets show that respondents having education of high school and college are predominant; and many of the respondents are employed. As a result, it can be seen that the age ratio of respondents interviewed can represent the number of respondents having a productive age.

Based on the respondents' monthly income variable, the largest plurality of respondents, at 33%, had an income of 3-3,9 million IDR per month per household; 26% of respondents had a middle monthly income of 2-2,9 million IDR per month per household; 23% of respondents had a low monthly income of 1-1,9 million IDR per month per household [6]. It can be seen that the respondents' family income was taken as a sample variable. However, 76% of the respondents have a motorcycle, and the remaining, 13%, have at least a car. Regarding owning a driver's license, nearly 87% of the respondents have a driver's license. Finally, for the purpose of traveling on the day of the questionnaire survey performed, about 86% of the respondents were traveling on a mandatory trip, meaning for the purpose of studying for university/college students and for the purpose of working. Thus, it can be seen that the travel destination of the respondents varies greatly.

3. Results and discussion

3.1. Likert scale

This part describes the research results obtained after the process of data collection, data processing, and data analysis to determine the frequency associated with the characteristics of respondents. The results obtained from field observations were then recapitulated. One hundred samples of respondents were obtained. This analysis is intended to determine the frequency of perceptions of the plan that the transit bus plan has proposed to the public. Perception is a related psychological concept used to observe respondents' intentions related to psychological determinants and scenarios. Table 4 shows that the respondents agree that the transit bus plan is acceptable and give an average value of 3.55, meaning that the transit bus plan is an agreeable policy. For the safety factor, the transit bus plan was rated as 3.23 and 3.40. In terms of safety, the majority of respondents agreed that mitigating the possibility of accidents caused by human, vehicle, road, and environmental factors, is very important or valued as 3.48.

Another part of the respondent's desire for transit bus is convenience, namely for cleanliness, which earned a value of 3.41. This means that the respondents expect the availability of garbage bins on the bus, so that bus cleanliness can be maintained. Another part of convenience is the convenience in boarding the bus, which is rated as 3.46; the respondents recommend the distance between the bus door and the bus stop must not be so far. Other indicators, related to convenience, is bus capacity, which rated as 3.2, and bus facilities, which rated as 3.5. Finally, the convenience at the bus stop facilities was rated as 2.94. From the five indicators of convenience, it can be seen that convenience is an important thing desired by respondents for transit bus services.

In terms of accessibility, ease of reaching the bus stop had a value of 3.17. Punctuality of route and timetable was rated as 2.97, and the accuracy of the timetable was rated as 2.79. Concerning equality, the majority of respondents rated the availability of infrastructure facilities for people with disabilities, pregnant women, elderly as 3.41. In terms of time regularity, e.g. travel time and headway, the respondents rated as 3.2 and 2.84.

Descriptive analysis of community perception indicators shows that all indicators have values above 2.5 from the 1-4 Likert scale, which means that respondents' perceptions of the transit bus plan are very positive at 3.23 with an average percentage value of 80.86 of the reference Likert scale 1-4. The greatest indicator of acceptance level was 3.55, which means the respondents agree that the transit bus has many positive aspects. The punctuality, travel time, and waiting time (headway) should be the primary attribute to improve in order to attract transit bus ridership. It is likely that the transit bus is the new bus system to be introduced by the government and it will replace existing public transport because the existing public transport lacks reliability in terms of punctuality, travel time, waiting time, and bus facilities (convenience).

Table 4.The mean value and percentage of the factors that the respondents desire for the transit bus service.

ID	Description	Mean Likert scale	Percentage (%)
P1	Acceptability (transit bus plan)	3.55	88.75

ID	Description	Mean Likert scale	Percentage (%)
P2	Security1 (crime in the bus)	3.23	80.75
P3	Security2 (crime in bus stop)	3.40	84.88
P4	Safety	3.48	87.00
P5	Convenience1 (cleanliness)	3.41	85.25
P6	Convenience2 (boarding the bus)	3.46	86.50
P7	Convenience3 (bus capacity)	3.20	79.88
P8	Convenience4 (bus facilities)	3.50	87.38
P9	Convenience5 (bus stop facilities)	2.94	73.50
P10	Accessibility (ease reach the bus stop)	3.17	79.13
P11	Punctuality1 (route and timetable)	2.97	74.13
P12	Punctuality2 (accuracy of timetable)	2.79	69.63
P13	Equality (disabilities and elderly)	3.41	85.13
P14	Time regularity1 (travel time)	3.20	80.00
P15	Time regularity2 (headway)	2.84	71.00
	Mean	3.23	80.86

3.2. Measurement model (confirmatory factor analysis/CFA part)

The results of the measurement model calibration are presented using the Structural Equation Modeling (SEM) approach. The latent variables such as security, convenience, accessibility and time regularity were measured by using several indicators. As expected, most of the indicators in this latent construction group have a significant factor loading (t-value > 1.96; at a significant error of 5%) to the measured latent construction. There are four latent variables built. The variables represent 14 indicators. In table 5, there are twelve observed variables with loading factors ≥ 0.5 and t-value ≥ 1.96. The twelve measurement models show that the calibrated parameters are standardized (standardized loading factor) by adding a value of 1 for the first indicator within a group of latent construction.

Table 5.Calibrated parameters of the measurement model (CFA part).

Latent Variable		Indicators	Coefficient	t-value
Latent1	Security	Crime in bus stop	1.00	1.13
		Safety	0.88	
Latent2	Convenience	Cleanliness	1.00	6.56
		Boarding the bus	0.86	
		Bus capacity	1.04	
		Bus facilities	1.46	6.69
		Disabilities and elder people	1.05	
Latent3	Accessibility	Ease reach the bus stop	1.00	10.35
		Route and timetable	0.90	
		Accuracy of timetable	0.50	
Latent4	Time Regularity	Travel time	1.00	6.35
		Headway	1.43	

Table 5 illustrates the correlation between individual perception as indicators and latent construction such as 'security (latent1)', 'convenience (latent2)', 'accessibility (latent3)' and 'time regularity (latent4)'. Looking to the latent variables of latent1 to latent4, the latent3 variable of 'accessibility' has been significantly influenced by three indicators such as ease to reach the bus stop, route and timetable, and accuracy of the timetable.

4. Conclusion

Based on the results of the discussion and analysis of the data conducted, several conclusions can be drawn. Because the Likert scale measurement is used for the purposes of quantitative analysis, then the answer can be given as a score, for example, strongly agree with a score of 4, agree with a score of 3, disagree with a score of 2 and strongly disagree with a score of 1. The answer to each variable that uses the Likert scale has a gradation from very negative to very positive. Each answer is linked to a question or an attitude of support. More than 59% of respondents were male and 41% of respondents were female. It can be seen that the questionnaire has been distributed equally, so it is expected to provide different travel characteristics. More than 76% of respondents are young people aged below 45 years. Based on their monthly income, the majority respondent, 33%, have an income of 3-3,9 million IDR per month per household. Regarding owning a driver's license, nearly 87% of the respondents have the driver's license. Finally, as the purpose of traveling on the day of the questionnaire survey performed, about 86% of the respondents have conducted the mandatory trip.

Regarding the descriptive analysis of community perception indicators, all indicators have values above 2.5 of the 4 scale Likert, which means that respondents' perceptions of the transit bus plan are very positive at 3.23 with an average percentage value of 80.86 of the reference Likert scale 1-4 with the greatest indicator of acceptance level of 3.55. This fact means that the respondents agree that the transit bus can be received with positive values. The punctuality, travel time, and waiting time (headway) were the primary attribute to attract transit bus ridership. The fact might be because the transit bus is the new bus system introduced by the government to replace existing public transport that lacks reliability in terms of punctuality, travel time, waiting time, and bus facilities (convenience). This quality of service measured from the system encounter remains a future work of this study.

Acknowledgments

The authors would like to express their honest gratitude to all those who have helped in this research, the respondents who were willing to take the time and thought to answer and fill out the questionnaire in this study, as well as all survey teams who worked hard in conducting interviews in the survey location of Lhokseumawe city.

References

[1] Sugiarto S, Saleh S M, Anggraini R and Merfazi M 2019 Investigating public perceptions and its implication toward Trans Koetaradja policy considering latent motivation *IOP Conf. Series: Materials Science and Engineering* vol 523 012036

[2] Schade J and Schlag B 2000 *Acceptability of Urban Transport Pricing* VATT Research Report. 72

[3] Winarno B and Okto Risdianto Manullang 2018 Parameters For Determining The Use Of General Transportation In Urban Pati *TATALOKA* **20** 1 75-86

[4] Gebeyehu M and Takano S 2008 Modeling the Relationship between Travelers' Level of Satisfaction and Their Mode Choice Behavior using Ordinal Models. *J. of the Transportation Research Forum 2008* **47** 2 103-118

[5] Saleh S M, Sugiarto S and Salmannur A 2019 Attitudinal Dataset for Mediating the Effects of Public Acceptance on Bus Reform Scheme in a Developing Country Context *Data in Brief.* **25** 104035 1-10.

[6] Aprilianto A 2012 Statistics h*ttp://hayat08.files.wordpress.com/2012/03/stat1.pdf* **online first**

IOP Conf. Series: Materials Science and Engineering **933** (2020) 012044 doi:10.1088/1757-899X/933/1/012044

Mechanical properties improvement of clay soils using dry dust collector and silica sand wastes

A S S Gunarti[1], S Nuryati[1], F Primatama[1] and I Raharja[2]

[1] Department of Civil Engineering, Universitas Islam 45, Bekasi, Indonesia
[2] Department of Accounting Information Systems, Universitas Bina Sarana Informatika, Karawang Barat, Indonesia

Corresponding e-mail: anita_ft@unismabekasi.ac.id

Abstract. Wastes from industrial parks increase rapidly; they are also difficult to manage. They should be carefully managed and used for other purposes, e.g., improving the infrastructure. The study did some laboratory experiments, e.g., California bearing ratio test, direct shear test, and triaxial (unconsolidated and undrained condition) test. Soil stabilization through chemical addition was done by mixing the original soil with some compositions of dry dust collector and silica sand wastes. California bearing ratio value of the 5mm penetration of soil with an additive composition of 7.5 % silica sand and 1.5 % dry dust collector was increased by 86 %. Internal friction angle value based on triaxial test showed that soil with an additive composition of 7.5 % silica sand (without dry dust collector) was increased by 2.5 %. The cohesion of soil having the additive composition of 7.5 % silica sand and 1.5 % dry dust collector increased by 60 % from the original soils.

1. Introduction

Industrial parks always create wastes from their factories in a production process, for example a coloring and compounding plastic resin in EJIP industrial park, located in Cikarang, West Java, Indonesia, which throws the dry dust collector (DDC) for about 3 tons monthly. Another example is a mineral factory in JABABEKA industrial park that creates silica sand wastes (SS) for about 50 tons monthly. DDC and SS wastes compound sand contents that actually can be utilized productively for soils stabilizer. Such kind of waste has not been managed optimally; the waste collection also depends on some collectors at a high price. This situation happens because of the lack of innovation in waste management. In addition, the factory owners strongly depend on the collector. A research in [1], entitled "direct shear strength improvement through soil stabilization using DDC and SS from industrial waste", indicated the best friction angle for the mixed soil containing SS (5 %) and DDC and the increase of shear strength for about 32.26 % when 1.5 % of DDC and 5 % of SS were mixed in the soil. This study concluded that mechanical characteristic was improved when clay was mixed with DDC and SS. Therefore, new materials should be studied from waste processing to get cheaper additive materials for an optimum clay additive composition.

This study aims to know the mechanical properties change of the clay mixed with DDC and SS and to find an optimal composition for getting maximum shear strength. The studies on the uses of DDC and SS for additive are still lacking.

Another study [2] showed a different character based on the aggregate's contribution. A sample with 10 % of diabase aggregate had higher compressive strength than a reference sample. A study [3] on the "utilization of drinking water treatment sludge for the manufacturing of ceramic products" showed that

IOP Conf. Series: Materials Science and Engineering **933** (2020) 012044 doi:10.1088/1757-899X/933/1/012044

the property of ceramic bodies—both physical and mechanical—varied depend on the amount of sludge added. Moreover, the additive was proved to be an effective coloring agent which makes the ceramic have a darker red color. A previous study [4] on clay soil stabilization using powdered glass concluded that the highest California bearing ratio (CBR) values of 14.9 % and 112.91 % were obtained at the content of 5 % glass powder and penetration of 5 mm for both unsoaked and soaked treated samples, respectively. In addition, the maximum cohesion and angle of internal friction values of 17.0 and 15.0 were obtained at 10 % of glass powder content.

Another study [5] of the experimental study showed that the addition of SDA resulted in a significant increase in CBR and unconfined compressive strength. The CBR values obtained are also within limits recommended by the Asphalt Institute for highway sub-base and sub-grade. Thus, the inexpensive industrial wastes may be used for the satisfactory stabilizing agents for the sub-base and base course in clayey fills, although their performances are often improved by combining waste to reduce the construction cost of road, particularly in the rural areas.

2. Methodology

This study used several materials which were clay soils [6]-[8], from Bekasi – Indonesia, gathered from 20 to 50 cm deep from original soil (disturbed condition), water, SS, and DDC. SS and DDC were used as the additive. Figure 1 shows the research flowchart. Sample preparation was done by mixing the soil with DDC and SS wastes at a predetermined composition. While the curing was carried out for seven days up to be a homogeneous mixture and ready to test. For the composition of additives, SS is 7.5% of the dry weight of soil and DDC is 1.5% of the dry weight of soil. The variations of the soil mixture, which uses waste as the additive, are given in table 1. A compaction test was applied to the original soil and the stabilized soil to get the optimum moisture content (OMC). Then, the mechanical properties of the original soil and the stabilized soil (OMC conditions) were tested using the CBR test and the triaxial unconsolidated undrained test based on the ASTM standard.

Table 1. Mix Variation.

No	Additive composition
1	Original soil
2	Original soil + 7.5 % SS
3	Original soil + 7.5 % SS + 1.5 % DDC

3. Results and discussion

3.1. Compaction

Compaction test results are given in table 2. The maximum dry density (γd_{max}) of stabilized soil, which has 7.5% of SS additive % and has no DDC, decreased to about 3.71 % of γd_{max} of original soil. The γd_{max} of stabilized soil, having the additive composition of 1.5% DDC and 7.5% SS, increased for about 7.43%, compared to the original soil.

Table 2. Compaction Test Result.

Mix Variation	Optimum Moisture Content (%)	Maximum Dry Density (γd_{max}) (gram/cm^3)
Original soil	33.00	1.345
Original soil + 7.5% SS	32.00	1.340
Original soil + 7.5% SS + 1.5 % DDC	31.55	1.355

Stabilized soil with an additive composition of 7.5 % SS (without DDC) showed that OMC was decreased by 3.03 %, compared to the original soil. The OMC value of the soil with an additive composition of 1.5 % DDC and 7.5 % SS showed that the OMC was decreased by 4.39 % compared to the original soil. Figure 2 shows the change in the OMC value.

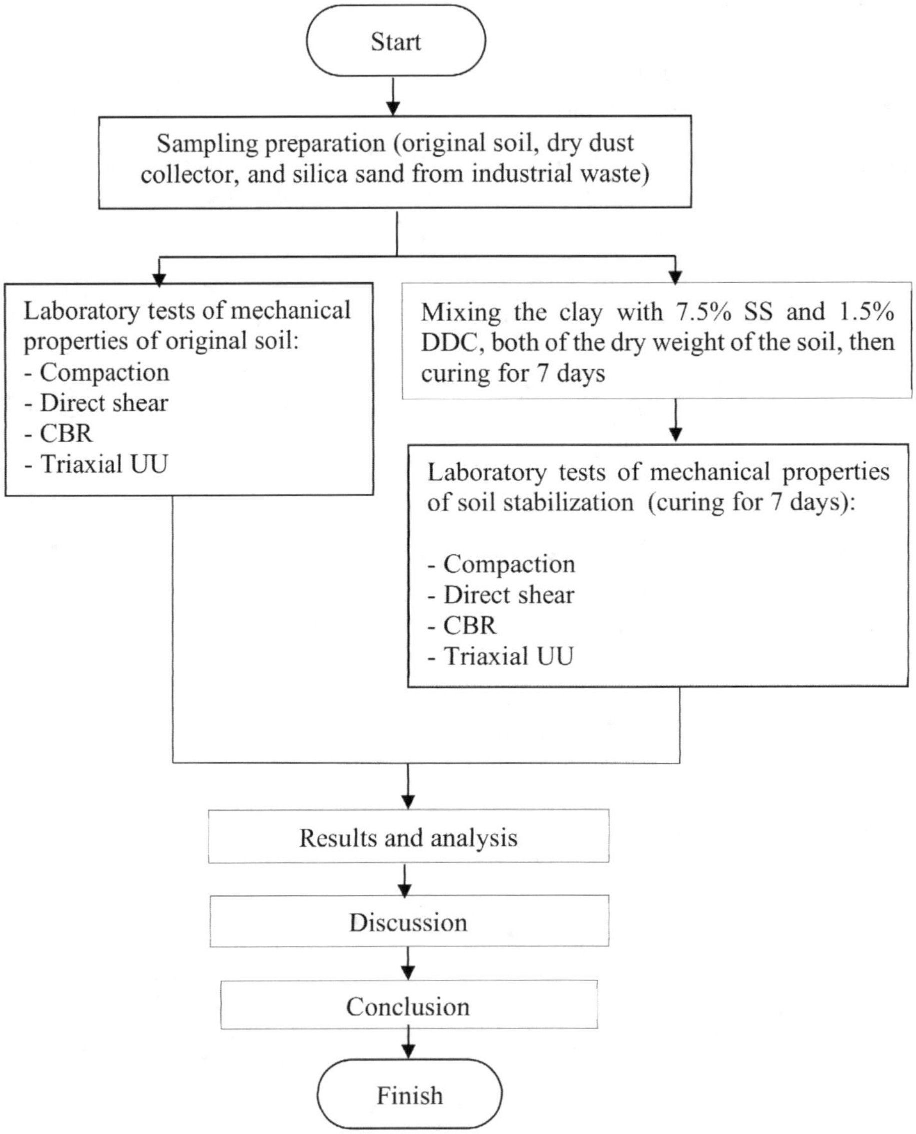

Figure 1. Research flowchart.

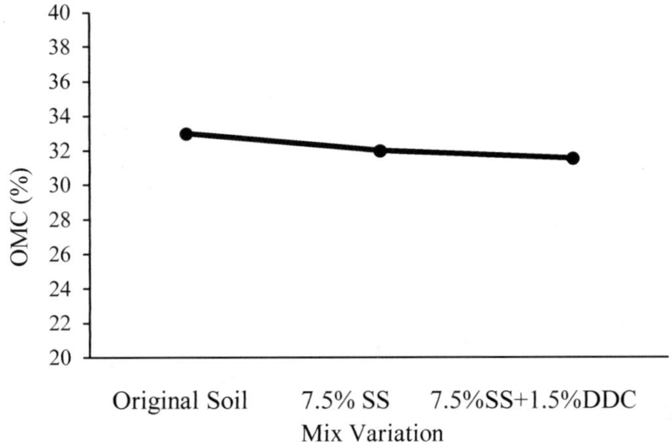

Figure 2. OMC value chart.

3.2. *Mechanical properties*

The test results of the mechanical properties of the soil are given in table 3.

Table 3. Mechanical properties test result.

Type of Test		Original Soil	Soil + 7.5% SS	Soil + 1.5% DDC + 7.5% SS
CBR	2.5 mm	3.20	3.93	5.09
	5 mm	2.77	3.91	5.15
Triaxial	C (kPa)	69.56	61.57	69.28
	ϕ (^{0})	26.67	27.34	10.58
Direct Shear	ϕ (^{0})	26.33	26.33	24
	C (kg/cm^2)	0.025	0.028	0.040

3.2.1. California bearing ratio. The CBR value (2.5 mm penetration) in stabilized soils increases compared to the CBR value (2.5mm penetration) in the original soil. The CBR value (5mm penetration) increased by 41 %, compared to the CBR of original soil. For the stabilized soil with the additive composition of 1.5% DDC and 7.5% SS, the CBR increases significantly by 86%, compared to the CBR of original soil. The CBR value of the soil, which was stabilized using SS and DDC, improves. Figure 3 shows the change in 5 mm penetration - CBR value.

3.2.2. Triaxial UU. The triaxial test result (figure 4) showed a cohesion value of 61.57 kPa for stabilized soil with SS composition of 7.5% (without DDC). This cohesion value decreased by 11 % compared to the original soil. Stabilized soil with the additive composition of 1.5 % DDC and 7.5 % SS had a cohesion value of 69.28 kPa or decreased 4.02 %, compared to the cohesion value of original soil.

Internal friction angle value (ϕ) from the triaxial test of a stabilized clay with the additive composition of 7.5 % SS (without DDC) was 27.34°. This value showed the increment of ϕ by 2.5 %, compared to the ϕ value of original soil. The ϕ value of soil using additive DDC decreases against the value of the original clay. Figure 5 shows the change of the internal friction angle based on the triaxial test result.

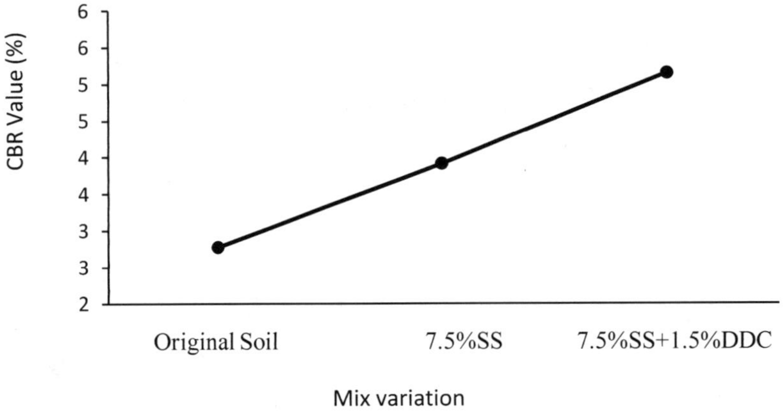

Figure 3. CBR's (penetration 5 mm) value chart.

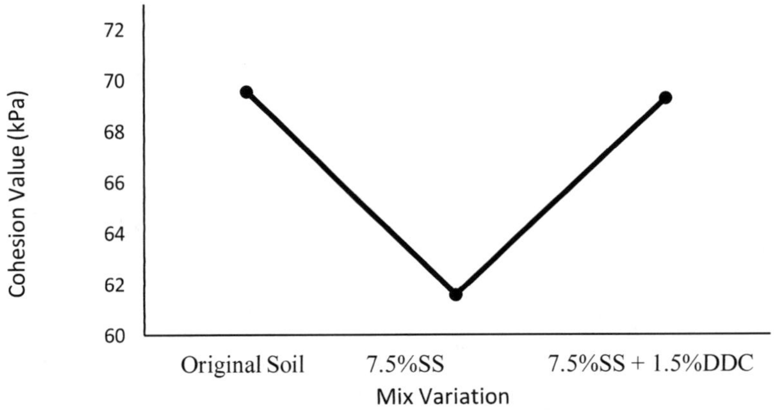

Figure 4. Cohesion value chart.

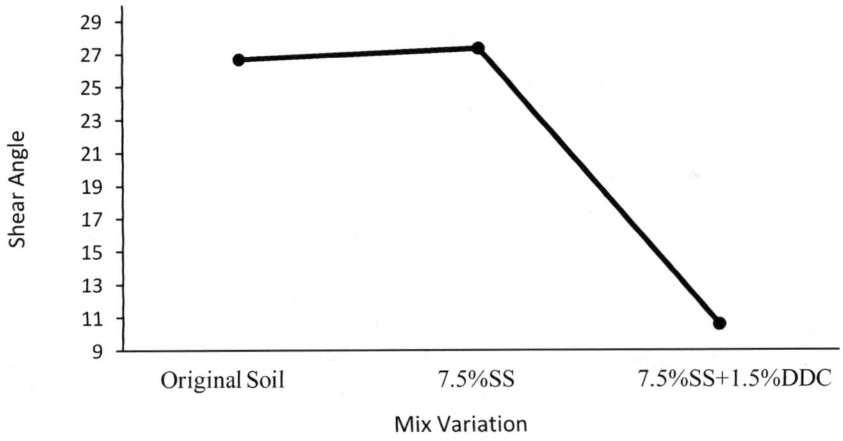

Figure 5. Shear angle value chart.

3.2.3. Direct shear. There was no significant change in internal friction angle (ϕ) based on a direct shear test for stabilized soil with the additive composition of 7.5% SS (without DDC). Figure 6 shows the change in the internal friction angle. On the other hand, for stabilized soil with the additive composition of 1.5% DDC and 7.5 % SS, the value of ϕ obtained from a direct shear test decreased by 8.84%, compared to the original.

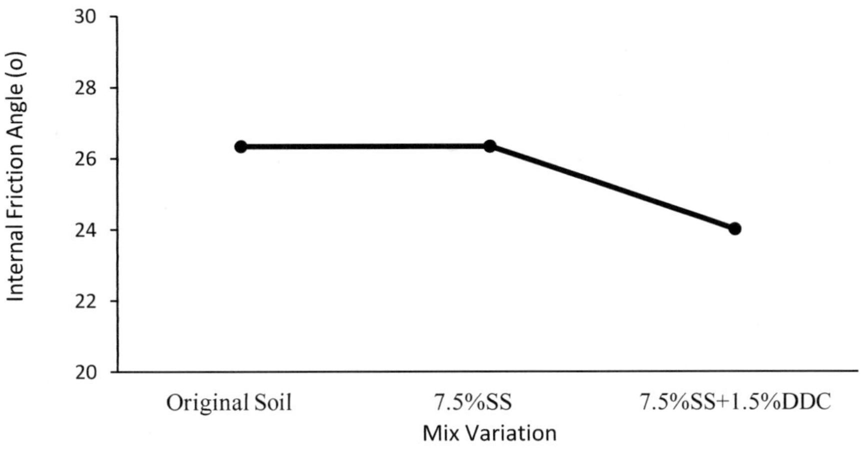

Figure 6. Internal friction angle (ϕ) chart.

For stabilized soil with an additive composition of 7.5% SS (without DDC), from the direct shear test, the cohesion value was 0.028 kg/cm² or increased by 12 %, compared to original soil. For stabilized soil with the additive composition of 1.5% DDC and 7.5% SS, the cohesion value increased by 60%, as shown in figure 7.

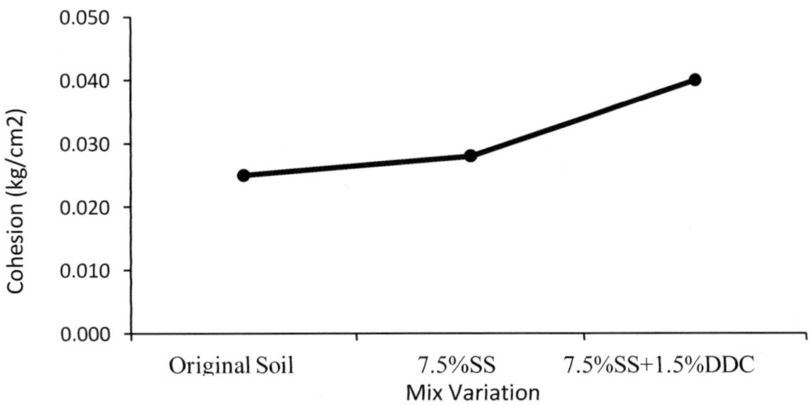

Figure 7. Cohesion value chart.

4. Conclusion

The result indicated that mechanical properties were improved based on the soil test. CBR value of stabilized soil with the additive composition of 7.5 % SS and 1.5 % DDC was increased by 86 %, compared to the CBR value of original soil. Triaxial test in this study, which was in unconsolidated and undrained condition, indicated that for stabilized soil with the additive composition of 7.5 % SS (without DDC), the internal friction angle value increased by 2.5 %, compared to original soil's internal friction angle. Based on the direct shear test of stabilized soil with the additive composition of 1.5 % DDC and 7.5 % SS, there was an increment of cohesion value up to 60 %, compared to original soil's cohesion.

Acknowledgments
Thanks to DRPM RISTEKDIKTI for 2019 applied research grant funding. Thanks to Soil Mechanics Laboratory staff of Universitas Indonesia.

References

[1] Gunarti A S S 2018 Direct shear strength improvement through soil stabilization using dry dust collector and silica sand from industrial waste *Proceeding ICRIIS*
[2] Komisarczyk K and Stępień A 2017 Optimization the composition of sand-lime products modified of diabase aggregate *IOP Conf. Series: Materials Sci. and Eng.* vol 251 1 012019 (IOP Publishing)
[3] Kizinievič O and Kizinievič V 2017 Utilisation of drinking water treatment sludge for the manufacturing of ceramic products *IOP Conf. Series: Materials Sci. and Eng.* vol 251 1 p 012018 (IOP Publishing)
[4] Olufowobi J, Ogundoju A, Michael B and Aderinlewo O 2014 Clay soil stabilisation using powdered glass *J. of Eng. Science and Techn.* **9** 5 541
[5] Butt W A, Gupta K, and Jha J N 2016 Strength behavior of clayey soil stabilized with saw dust ash *Intern. J. of Geo-Eng.* **7** 1 18
[6] Kholis N, Gunarti A S S and Sylviana R 2018 Stabilisasi tanah lempung menggunakan semen dan renolith *BENTANG: J. Teoritis dan Terapan Bidang Rekayasa Sipil* **6** 1 62-77
[7] Gunarti A S S 2014 Daya dukung tanah lempung yang distabilisasi dengan SpentCatalyst RCC 15 dan kapur *BENTANG: Jurnal Teoritis dan Terapan Bidang Rekayasa Sipil* **2** 1 38-52
[8] Hardiyatmo H C 2012 *Mekanika Tanah 1* Edisi 6

IOP Conf. Series: Materials Science and Engineering **933** (2020) 012045 doi:10.1088/1757-899X/933/1/012045

Passenger's choice in responding to inter-city railway operation for Makassar – Parepare line: a conditional logit model based on stated preference method

M I Ramli[1], H Yatmar[1], D B E Dharmowijoyo[2], S A Adisasmita[1], M Pasra[1], A Arif[1], H Dwiatmoko[3] and M Z Irawan[4]

[1] Department of Civil Engineering, Universitas Hasanuddin, Indonesia
[2] Department of Civil and Environment Engineering, Universiti Teknologi Petronas, Malaysia
[3] Department of Civil Engineering, Universitas Mercu Buana, Indonesia
[4] Department of Civil and Environment Engineering, Universitas Gadjah Mada, Indonesia

Corresponding e-mail: isranramli@unhas.ac.id

Abstract. The present paper aims to model the passenger's choice in responding to the inter-city railway operation for Makassar-Parepare line. The railway line is the first railway track in Eastern Part of Indonesia, and it is still under construction. In this regard, the study attempted to construct the passenger's preferences through a conditional logit model approach based on the stated preference method. The study considered some main attributes of transport mode services, i.e., travel cost, travel time, and frequency. In further, the study also taking account into the model two additional attributes of the mode, i.e., transfer travel time and transfer travel cost. The data collection of the study carried out an interview method using questionnaire sheet based the stated preference method. The study experimented with varying the level of service for the main attributes of the two available travel modes, i.e., railway and private car modes. The study results showed that passengers prefer to choose one of both travel modes when the mode has travel time faster than the other one, as well as for the lower travel cost. In addition, the inter-city travelers more considered the transfer travel cost attribute than the transfer travel time attribute, when they face both travel mode choice, the new railway mode, and the existing private car mode.

1. Introduction

In recent years, Indonesia Ministry of Transportation has been constructing a new railway track in South Sulawesi Province, as the first railway infrastructure in Eastern Part of Indonesia [1]. For the first stage, the central government constructs the railway track in Makassar City to Parepare City, namely Makassar – Parepare line [1]. The infrastructure project is significant to improve travel demand management in the region. In his regard, the transportation demand management has travel choices and links as the derived problem [2]. For example, in Malaysia, many people still use their cars at the beginning [3], as well as at Ohio State, USA [2].

To explore the problem, at least there two are factors which play an essential role, i.e., the characteristics of the inter-city travel, and the characteristics of the travel using railway mode. In this view, some previous studies have elaborated the characteristic of the inter-city travel such as in Libya

IOP Conf. Series: Materials Science and Engineering **933** (2020) 012045 doi:10.1088/1757-899X/933/1/012045

[4, 5], in China [6, 7], and Great Britain [8]. Related to characteristics of the inter-city travel which utilize a railway mode, there are previous studies which focus on London-Paris line [9], and on Beijing – Guangzhou line [10]. Besides, psychological factors and specific attitudes also influence on the inter-city travel behavior [11-14], as well as the socioeconomic characteristics, land use, and travel time [15].

Furthermore, there were some previous studies which focused on the model development on estimating the inter-city travel demand. Majority of the previous studies applied multidimensional system and modeling such as some variation of the logit models [16-20]. Further, structural equation modeling was also applied [21]. Focused on inter-city travel by using railway mode, some previous studies have a concern about the travel mode choice [22-24]. In a case study for the choice problem of the traveler or passengers of the inter-city travel on Makassar – Parepare line, the travelers have two choice modes, i.e., the new railway mode, and the private car mode. This case is similar to traveler's problem in Kuala Lumpur [25].

Regarding the inter-city problem for operating the new railway mode in Makassar – Parepare line, the present study aims to model the passenger's choice in responding to the inter-city railway operation for the line. The study attempted to construct the passenger's preferences through a conditional logit model approach based on the stated preference method. In particular, this study provides the methodology in developing and calculating a stated preference method for determining the significant variables influencing the mode choice behavior model.

2. Methodology

2.1. The study location

The study located in Parepare City, the second biggest city in South Sulawesi Province, after Makassar City. The main reason to select the citizen of the city to become the respondent target of this study is due to the city is the final terminal for the first stage construction of the new railway mode in the province. Indonesia government has planned to operate a new railway mode in South Sulawesi Province as the first railway operation in the eastern part of Indonesia. In the first stage, the government has been constructing railway track for Makassar – Parepare line. The railway line has track length of about 142 kilometers, and through three regency area, i.e., Maros, Pangkep, and Barru regencies [1]. Figure 1 shows the map of the railway line.

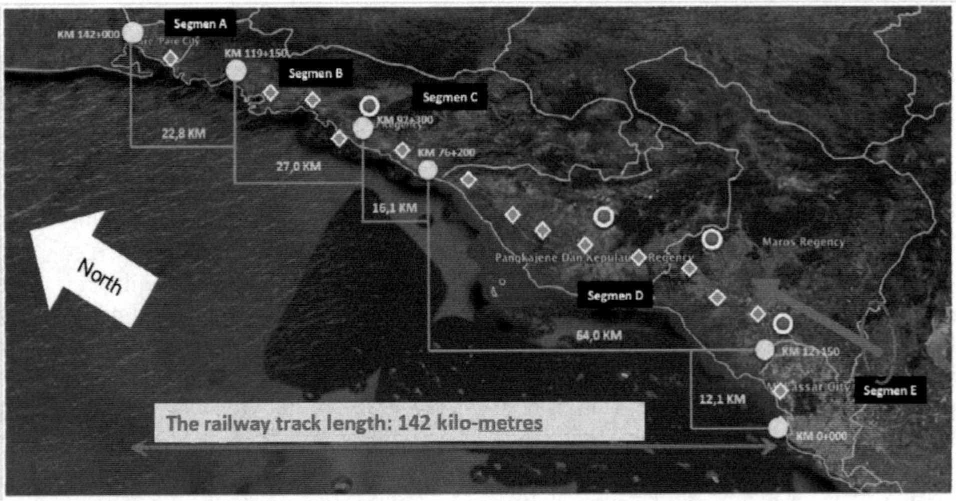

Figure 1. The route of the railway track for Makassar – Parepare line [1].

2.2. Data collection

The study carried out the data collection through an interview survey using a questionnaire sheet. We randomly selected respondent participants of Parepare City citizens who are already have traveled for Makassar – Parepare line using private cars. Two hundred fifty respondents participated in the survey. The questionnaire in this study involves two categories of questions, i.e., socio-demography characteristics, and the travel attributes of the respondents. The socio-demography characteristics of the respondents include age, family size, occupation field, education, and income. Table 1 presents the categories of each character.

Table 1. The traveler's characteristics and their categories.

Socio-Demography Characteristics	Categories of the characteristics		
Age (Years old)	a. 18-23	b. 24-29	c. 30-35
	d. 36-41	e. 42-47	f. 48-53
	g. 54-59	h. 60-65	i. 66-71
Family Size (Person)	a. 1	b. 2	c. 3
	d. 4	e. 5	f. 6
	g. > 6		
Occupation Field	a. Industry/Manufacture	b. Construction	
	c. Communication/Transportation	d. Bank/Finance/Insurance	
	e. Business	f. Electrical/Gas/Water	
	g. Central Government	h. Local Government	
	i. Service Industry	j. Military/Police	
	k. Teacher/Lecturer	l. Others	
Education	a. Elementary School	b. Junior High School	
	c. Senior High School	d. Diploma (D3)	
	e. Bachelor	f. Master/Doctor	
Income (IDR)	a. < 2,000,000	b. 2,000,001- 3,000,000	
	c. 3,000,001 – 5,000,000	d. 5,000,001- 7,500,000	
	e. 7,500,001- 10,000,000	f. 10,000,001 – 15,000,000	
	g. > 15,000,000		

In the second category, the study considers three main attributes of inter-city travel commonly, i.e., travel time, travel cost, and trip frequency. Due to the study focused on the traveler's preference when a new railway mode operated, we added two additional attributes, i.e., transfer cost, and transfer travel time from origin place to railway station, or on the contrary. All attributes will become the main variables considered in the experiment choice situation based on the stated preference method. The next sub-section will present this experiment choice situation.

2.3. The model's variables, and choice experiment situation

Regarding the primary purpose of this study, we attempted to generate some choice situation or scenarios based on the stated preference method [26]. By applying the five attributes as the variables in a conditional logit model approach, the study developed eight choice scenarios through varied the values of the variables. Table 2 provides eight experimental choice situations.

2.4. The model construction and the calculation method

The study adopted the random utility maximization approach in constructing and calculating the parameters of the individual preference [27-30]. In this regard, a conditional logit model of choice

IOP Conf. Series: Materials Science and Engineering **933** (2020) 012045 doi:10.1088/1757-899X/933/1/012045

probability was applied. The concept assumes that a decision-maker which has a set alternative to be chosen, obtain a certain level of utility from each alternative. Then, the individual decides to choose the alternative that has the highest utility. In the context of mode choice in travel behavior, the utility of each alternative is decomposed as follows [29, 30]:

$$U_{nj} = V_{nj} + \varepsilon_{nj} \quad (1)$$

Where:

- V_{nj} is observed aspect as the deterministic portion, and
- ε_{nj} is unobserved aspect as a stochastic portion.

Table 2. The experimental choice situation of the mode travel choice model.

Choice Condition Scenarios	Travel Cost/Fare (IDR)		Travel time (Hours)		Trip Frequency (Times/Hours)		Transfer cost (IDR)		Travel time from origin place to the railway station (minutes)	
	Railway Mode	Private Car	Railway Mode	Private Car	Railway Mode	Private Car	Railway Mode	Private Car	Railway Mode	Private Car
A	100,000	150,000	2	3	4	Every time	0	0	45	0
B	100,000	150,000	2	3	2	Every time	50,000	0	45	0
C	100,000	150,000	3.5	3	4	Every time	50,000	0	20	0
D	100,000	150,000	3.5	3	2	Every time	0	0	20	0
E	175,000	150,000	2	3	4	Every time	50,000	0	20	0
F	175,000	150,000	2	3	2	Every time	0	0	20	0
G	175,000	150,000	3.5	3	4	Every time	0	0	45	0
H	175,000	150,000	3.5	3	2	Every time	50,000	0	45	0

Therefore, the probability that decision-maker n chooses alternative i is stated below [29, 30]:

$$P_{ni} = \frac{e^{-V_{ni}}}{\sum_j e^{-V_{nj}}} \quad (2)$$

The logit probabilities become:

$$P_{ni} = \frac{e^{-\beta_1 x_{ni} + \beta_0}}{\sum_j e^{-\beta_1 x_{nj} + \beta_0}} \quad (3)$$

Where:

- x_{nj} is a vector of observed variables that represent attributes relevant to the jth travel mode.
- β_1 is the parameter of x_{nj} that should be estimated, and
- β_0 is a specific constant of the model for mode j.

Due to the experimental of the choice condition of the inter-city travel passengers by using stated preference method, this study only considered the attributes of the public transport mode in constructing the travel mode choice model of the passenger's preference. Then, the study applied a conditional logit model for the choice model construct in case travelers have two choice modes, the new railway mode, and auto cars as the existing mode. In this regard, we generated two choice models, model-1 and model-2, which considered the three main attributes, and the five main attributes, respectively, from table 1. The three main attributes for Model-1 involves travel time (x_1), travel cost (x_2), and service frequency (x_3) of the travel mode. We extended model-1 to generate Model-2 through added both additional

IOP Conf. Series: Materials Science and Engineering **933** (2020) 012045 doi:10.1088/1757-899X/933/1/012045

attributes, additional travel time (x_4), and additional travel cost (x_5). Therefore, the hypothetical model of Model-1 and Model-2 by applying equation (3) is expressed in equation (4) and equation (5) as follows, respectively.

Table 3. The respondent's characteristics.

Variables	Categories	Frequency	Percentage (%)
Age (Years old)	18-23	29	11.6
	24-29	24	9.6
	30-35	48	19.2
	36-41	56	22.4
	42-47	44	17.6
	48-53	28	11.2
	54-59	11	4.4
	60-65	6	2.4
	66-71	4	1.6
Family Size (Person)	1	4	1.6
	2	23	9.2
	3	50	20
	4	67	26.8
	5	69	27.6
	6	21	8.4
	> 6	16	6.4
Occupation Field	Industry/Manufacture	10	4
	Construction	10	4
	Communication/Transportation	3	1.2
	Bank/Finance/Insurance	13	5.2
	Business	42	16.8
	Electrical/Gas/Water	2	0.8
	Central Government	4	1.6
	Local Government	114	45.6
	Service Industry	7	2.8
	Military/Police	5	2
	Teacher/Lecturer	14	5.6
	Others	26	10.4
Education	Elementary School	1	0.4
	Junior High School	2	0.8
	Senior High School	59	23.6
	Diploma (D3)	9	3.6
	Bachelor (S1)	154	61.6
	Master/Doctor (S2/S3)	25	10
Income (IDR)	< 2,000,000	22	8.8
	2,000,001- 3,000,000	39	15.6
	3,000,001 – 5,000,000	93	37.2
	5,000,001- 7,500,000	62	24.8
	7,500,001- 10,000,000	24	9.6
	10,000,001 – 15,000,000	7	2.8
	> 15,000,000	3	1.2

$$P_i = \frac{e^{-\beta_1 x_{1i} - \beta_2 x_{2i} - \beta_3 x_{3i} + \beta_0}}{e^{-\beta_1 x_{1i} - \beta_2 x_{2i} - \beta_3 x_{3i} + \beta_0} + e^{-\beta_1 x_{1j} - \beta_2 x_{2j} - \beta_3 x_{3j} + \beta_0}} \tag{4}$$

$$P_i = \frac{e^{-\beta_1 x_{1i} - \beta_2 x_{2i} - \beta_3 x_{3i} - \beta_4 x_{4i} - \beta_5 x_{5i} + \beta_0}}{e^{-\beta_1 x_{1i} - \beta_2 x_{2i} - \beta_3 x_{3i} - \beta_4 x_{4i} - \beta_5 x_{5i} + \beta_0} + e^{-\beta_1 x_{1j} - \beta_2 x_{2j} - \beta_3 x_{3j} - \beta_4 x_{4j} - \beta_5 x_{5j} + \beta_0}} \tag{5}$$

In calculating the values of the parameters of both conditional logit models, we utilized *STATA 12.0*, an econometric package tool which suitable for calculating parameters of a discrete choice model.

3. Results and discussion

3.1. The characteristics of the respondents

Table 3 presents the result of the data compilation from the interview survey. The table shows that the respondents have age on the work age level, from twenty-four until fifty-three years old. However, there is a significant portion of the respondents who have age around eighteen until twenty-three years old. In further, table 3 shows that the respondents are dominant, having three until five persons for their family size. For occupation filed, a large portion of the respondents has an occupation as staff of the local government and a businessman. As well as, the respondents have dominant education from senior high school until bachelor degree. Also, majority of the respondents have monthly income around three million rupiahs until seven point five million rupiahs.

3.2. The calculation results of the model's parameters

Regarding the calculation results of the model's parameters using STATA, an econometric tool which suitable for a discrete choice model, we got the results as shown in table 2, and table 3 for the result of the first scenario model (Model-1), and the second scenario model (Model-2), respectively.

Table 4. The calculation results of Model-1.

Variables	Coef.	P(Value)
Travel cost	-0.0304	0.000
Travel time	-0.7886	0.000
Frequency	0.2395	0.000
Number of obs		3,086
Pseudo R^2		0.2606

Table 5. The calculation results of Model-2.

Variables	Coef.	P(Value)
Travel cost	-0.0320	0.000
Travel time	-0.9064	0.000
Frequency	0.1236	0.000
Transfer Travel time	-0.0060	0.186
Transfer travel cost	-0.0170	0.000
Number of obs		3,086
Pseudo R^2		0.2865

Table 2 shows that generally, the calculation of model-1 has excellent performance, as shown by the likelihood ratio around 0.2606. This value is available in the range 0.2 until 0.4 which indicated that the model calculation result has accepted. In further, the values of the parameters of model-1 has achieved a significant level, which indicated by P(value) of all parameters more than 0.05. Also, the sign of the three attributes of the mode is inline with the expected sign.

Table 3 shows that the likelihood ratio of the model-2 is around 0.2865, which indicated the model calculation generally has excellent performance. As similar to model-1 results, the ratio also is in the range 0.2 until 0.4 which mean the model has accepted. Furthermore, the model's parameters have values smaller than 0.05, which indicated a significant level of achievement. Briefly, the parameter's sign of the five attributes are in line with the expected sign.

Model-1 which only consider three main attributes taking account into the model calculation indicated that travel time, travel cost, and the service frequency had played the critical role for the preference of the inter-city travel in choosing the new railway mode and the passenger car mode. These results are suitable for all previous study results in the field research area [13, 14]. The phenomena are also correspondence to our habit commonly when facing the inter-city travel situation.

Furthermore, the choice experiment results of model-2 showed that the inter-city travel passengers have a little bit different in responding to the additional travel cost and the additional travel time for travel mode choice. In this regard, the passenger more calculates the additional travel cost than the additional travel time. The phenomenon is in line with the collective behavior of the passengers for inter-city travel condition.

4. Conclusion

The present study has elaborated the inter-city travel passengers' preference when the passenger will face a new railway mode implementation shortly. Through a case study for railway operation plan in Makassar – Parepare line, the study has analyzed some main attributes of both available modes, railway mode as a new mode, and the passenger car as the existing mode in the line. By applying the stated preference survey and conditional logit model, we have found that the passengers have significantly considered travel time, travel cost, and frequency of the travel modes. However, the passengers have more prefer the additional travel cost than the additional travel time for travel from their origin place to the station of the travel mode. These results provide a look forward for the development of seamless connectivity for multi-mode transport in the operation plan of the new railway mode in Makassar – Parepare line.

Acknowledgments

The authors would like to many thanks for JICA C-BEST Project in Faculty of Engineering, Universitas Hasanuddin, which provide research grant for this study. Also, we address to Prof. Junyi Zhang from Hiroshima University, many thanks for his advisor as to the research advisor of the research grant.

References

[1] https://kppip.go.id/en/priority-projects/railway/makassar-parepare-railway/

[2] Akar G, Flynn C and Namgung M 2012 Travel choices and links to transportation demand management: case study at Ohio State University *Transport. Research Record: J. of the Transp. Research Board* **2319** 77-85

[3] Kamba A N, Rahmat, R. A and Ismail I 2007 Why do people use their cars: a case study in Malaysia *J. of Social Sciences* **3** 3 117-122

[4] Miskeen M A A B, Alhodairi A M and Rahmat R A A B O 2013 Modeling a multinomial logit model of intercity travel mode choice behavior for all trips in Libya. World Academy of Science, Engineering and Technology, *Intern. J. of Civil, Environm., Struct., Constr. and Architec. Eng.* **7** 9 636-45

[5] Manssour A, Ahmed M and Riza A 2013 Modeling of intercity transport mode choice behaviour in Libya: a binary logit model for business trips by private car and intercity bus, Australian *J. of Basic and Applied Sciences* **7** 1 302–11

[6] Wang Y, Wu B, Dong Z and Ye X 2016 Joint modeling analysis of passengers' intercity travel destination and mode choices in Yangtze river delta Megaregion of China, *Mathematical Problems in Eng.* **2016** 1-10
[7] Dong Z, Li L, Wu B and Zhao X 2015 Investigating mode preference of intercity business travellers in the urban agglomeration of China: application of fractional multinomial logit model *Proc. of the 94th Annual Meeting of the Transp. Research Board* (Washington DC: USA)
[8] Dargay J M and Clark S 2012 The determinants of long distance travel in Great Britain *Transp. Research Part A: Policy and Practice* **46** 3 576–87
[9] Behrens C and Pels E 2012 Intermodal competition in the London–Paris passenger market: high-speed rail and air transport *J. of Urban Economics* **71** 3 278-88
[10] Li Z C and Sheng D 2016 Forecasting passenger travel demand for air and high-speed rail integration service: a case study of Beijing-Guangzhou corridor, China. *Transp. Research Part A: Policy and Practice* **94** 397-410
[11] Kaewkluengklom R, Satiennam W, Jaensirisak S and Satiennam T 2017 Influence of psychological factors on mode choice behaviour: sase study of BRT in Khon Kaen city, Thailand *Transport. Research Procedia* **25** 5072-82
[12] Kroesen M and Chorus C 2018 The role of general and specific attitudes in predicting travel behavior–a fatal dilemma? *Travel Behaviour and Society* **10** 33-41
[13] Kroesen M, Handy S and Chorus C 2017 Do attitudes cause behavior or vice versa? An alternative conceptualization of the attitude-behavior relationship in travel behavior modeling *Transport. Research Part A: Policy and Practice* **101** 190-202
[14] Madhuwanthi R, Marasinghe A, Rpc J, Dharmawansa A D and Nomura S 2016 Factors influencing to travel behavior on transport mode choice *Intern. J. of Affective Eng.* **15** 2 63-72
[15] Limtanakool N, Dijst M and Schwanen T 2006 The influence of socioeconomic characteristics, land use and travel time considerations on mode choice for medium- and longer-distance trips *J. of Transport Geography* **14** 5 327–41
[16] Koppelman F S 1989 Multi-dimensional model system for intercity travel choice behaviour *J. of the Transport. Research Board* **1241** 1–8
[17] Yao E and Morikawa T 2005 A study of an integrated intercity travel demand model *Transport. Research Part A: Policy and Practice* **39** 4 367–81
[18] Mattson J, Peterson D, Thomas W, Ripplinger D and Hough J 2010 Assessing Demand for Rural Intercity Transportation in a Changing Environment *Transport. Research Record* **2145**, pp 108-112.
[19] Moeckel R, Fussell R and Donnelly R 2013 Mode choice modeling for long distance travel Transportation Letters **7** 1 35–46
[20] Miller E 2004 The trouble with intercity travel demand models." *Transport. Research Record: J. of the Transport. Research Board* **1895** 94-101
[21] Wang Y, Yan X, Zhou Y and Xue Q 2017 Influencing mechanism of potential factors on passengers' long-distance travel mode choices based on structural equation modeling. *Sustainability* **9** 11 1943
[22] Wong B and Habib K M N 2015 Effects of accessibility to the transit stations on intercity travel mode choices in contexts of high-speed rail in the Windsor–Quebec corridor in Canada *Canadian J. of Civil Eng.* **42** 11 930-9
[23] Wang Y, Li L, Wang L, Moore A, Staley S and Li Z 2014 Modeling traveler mode choice behavior of a new high-speed rail corridor in China *Transport. Planning and Technology* **37** 5 466-83
[24] Sperry B, Ball K and Morgan C 2011 Cluster analysis of intercity rail passengers in emerging high-speed rail corridor. *Transport. Research Record: J. of the Transport. Research Board* **2261** 31-38

[25] Kwan S C, Sutan R and Hashim J H 2018 Trip characteristics as the determinants of intention to shift to rail transport among private motor vehicle users in Kuala Lumpur, Malaysia *Sustainable Cities and Society* **36** 319-26

[26] Yang L, Choudhury C F, Ben-Akiva M, e Silva J A and Carvalho D 2009 Stated preference survey for new smart transport modes and services: Design, pilot study and new revision *MIT Transport. Systems* (Lisbon: Portugal)

[27] Train K E 2009 *Discrete Choice Methods with Simulation* 2nd Edition (Cambridge University Press)

[28] Koppelman F S and Bhat C 2006 *A self-instructing course in mode choice modeling: multinomial and nested logit models* (U.S. Department of Transportation Federal Transit Administration)

[30] Ben-Akiva M and Lerman S R 1985 *Discrete Choice Analysis: Theory and Application to Travel Demand* (Cambridge, Mass, USA MIT Press)

IOP Conf. Series: Materials Science and Engineering **933** (2020) 012046 doi:10.1088/1757-899X/933/1/012046

Effect of vertical distance of geogrid layers on slope stability

B Chairullah[1], H Yunita[1], M Munirwansyah[2] and R Ambiya[3]

[1] Department of Civil Engineering, Universitas Syiah Kuala, Banda Aceh, Indonesia
[2] Professor and Lecturer of Civil Engineering, Universitas Syiah Kuala, Banda Aceh, Indonesia
[3] Graduated from Civil Engineering, Universitas Syiah Kuala, Banda Aceh, Indonesia

Corresponding e-mail: halida_yunita@unsyiah.ac.id

Abstract. The landslide that occurred on the Meulaboh - Geumpang road section on May 10, 2017, resulted in the interruption of land transportation connecting West Aceh Regency and Pidie Regency, Aceh Province. The landslide on this road had an impact on the obstruction of logistics distribution and other needs from the direction of Meulaboh to Geumpang and vice versa. Based on these problems, the appropriate solution is needed in the prevention and planning of slopes to prevent landslides from happening again. This research was carried out by analyzing the slope stability under natural slope conditions and analyzing the slope stability of the conditions given slope reinforcement in the form of geogrids. Slope stability analysis is performed using the finite element method. The slopes are given alternative geogrid reinforcement with varying vertical distances. The variation of vertical geogrid distance is adjusted to the condition of the curve of the natural slope collapse before reinforcement. Based on the analysis of slope stability for natural conditions without reinforcement (existing condition), the safety factor values (SF) obtained at STA 84 + 280 and STA 84 + 910 were respectively 1.14 and 1.16. The results of the analysis of the slope stability that has been given reinforcement show that the vertical geogrid distance influences the slope safety factor.

1. Introduction

Indonesia has diverse geological and geomorphological structure conditions. One of the various structural conditions is the existence of a slope. Slopes can occur naturally or be formed by humans with a specific purpose. If the surface forms a slope, then the soil mass component above the slip plane tends to move downward due to gravity. If the gravity component is large enough, then it can cause landslides on the slope [1].

Landslides in Aceh have been occurring very often, especially on provincial roads. Like a landslide (figure 1) that occurred on Wednesday (10/5/2017) night around 04:00 GMT. It caused Meulaboh-Geumpang transportation links on Thursday (11/5/2017) to be totally paralyzed. Meulaboh-Geumpang Street is a road that connects West Aceh Regency with Pidie Regency. This road is used by residents to transport staples or vegetables from Pidie Regency to West Aceh Regency, and vice versa.

Based on these problems, it is necessary to find the right solution in handling slides due to unstable slope conditions. The research location is on Meulaboh - Geumpang street, namely STA 84 + 280 and STA 84 + 910, which are located in Lancong Village, Sungai Mas District, West Aceh Regency. The handling of landslides is done by strengthening the slopes using geogrid.

Figure 1. The condition of the Meulaboh-Geumpang KM 84 road segment due to a landslide (2017).

In line with the increasing use of slopes for various human interests, it is necessary to develop the concept of slope stability which aims to overcome the problem of slope collapse. An understanding of the factors associated with slope collapse, slope analysis, and selection of effective slope strengthening methods is needed to achieve this goal.

The stability of slopes should be very thoroughly analyzed since their failure may lead to loss of human life as well as colossal economic loss. The failure of a mass of soil located beneath a slope is called a slide [2]. The aims of slope stability [3] are :

- To understand the development and form of natural slopes and processes that occur in different natural conditions;
- To assess slope stability in short-term and long-term conditions;
- To assess the possibility of natural slopes and artificial slopes;
- To analyze and to understand the mechanism of collapse and the factors that cause it.

The value of safety factor ≥ 1.25 is a normal design to provide an estimate of the safety factor in the slope stability analysis. This is important to ensure that the slope design is safe and to prevent unexpected factors during analysis and construction such as incorrect data, analysis errors, work skills, and the lack of field supervision [4].

The slope stability analysis at this location was analyzed by using Finite Element Method [5], which the calculation process is faster and more efficient. The slope at this location is the artificial slope. The slope stability analyses give the output of safety factor at STA 84 + 280 and STA 84 + 910, which respectively is 1.14 and 1.16. These safety factors are less than 1.25; it means the slope unstable. So, the geogrid is used as the reinforcement.

Geogrids [1] can be used to strengthen slopes with sharp or steep slopes. The use of geogrids as a retaining wall is a development of the concept of reinforced earth discovered by Vidal in France in the mid-1960s. Reinforced earth walls with reinforced soil systems are a soil reinforcement system using metal strips with enclosure elements from panels - concrete or steel panels. Geogrid stress transfer mechanism involves two components, namely horizontal ground friction on the surface of the grid and passive resistance or called horizontal ground support to the transverse grid. An example of a geogrid is shown in figure 2.

Holes between the ribs in geogrids allow soil and rock aggregates to enter and create high friction resistance. Polymer geogrids have been used in the construction of walls with sharp, and even vertical slopes such as abutments.

The variation of vertical geogrid distance is adjusted to the condition of the curve of the natural slope collapse before reinforcement. The vertical distances of geogrid layers are 0.2m, 0.4m, and 0.6m. The results of the analysis of the slope stability that has been given reinforcement show that the vertical geogrid distance influences the slope safety factor.

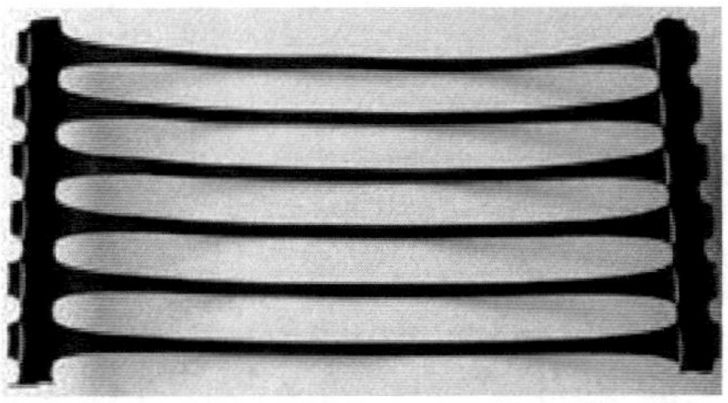

Figure 2. Uniaxial Geogrid [1].

2. Methodology

There are two locations in this study, Km 84 + 280 and Km 84 + 910. The coordinates of the study location are shown in the following table 1.

Table 1. Coordinates of the study location.

Point	STA	Coordinate
A	84 +280	N 4 30 39.6828 - E 96 7 46.5852
B	84+910	N 4 30 37.9692 - E 96 7 48.6052

2.1. Preparation

The preparation stages are a series of activities, before the starting data collection and the processing, which include:

- Literature study of existing problems;
- Determining data requirements;
- Direct survey to the research location to find out the conditions in the field and the initial data collection needed.

2.2. Collecting data

The data in this study consisted of primary data and secondary data. Primary data includes measurements of slope geometry, location coordinates, documentation, data from soil sample testing in the laboratory, and modeling of slope conditions. Whereas the secondary data needed is the map of West Aceh Regency, map of research location points, parameter of Young modulus value, permeability, and Poisson ratio.

2.3. Soil sampling

Soil sampling for this study was carried out by the Pit Test Method. Pit Test Method is a soil investigation by manually digging the soil to the desired depth, then taking samples using a steel tube as an undisturbed sample. The samples were three samples at each research location. Samples consisted of disturbed and undisturbed soils on the Meulaboh-Geumpang Road STA 84 + 280 and STA 84 + 910.

2.4. Slope stability analysis

Slope stability analysis has three stages: the data input, the calculation, and the data output. Slope stability was analyzed using Finite Element Method which requires the slope modeling that suitable with the existing data to obtain accurate results. The existing slope geometry in the Meulaboh-Geumpang Road Section Km 84, at the point of view, can be seen in figure 3 below.

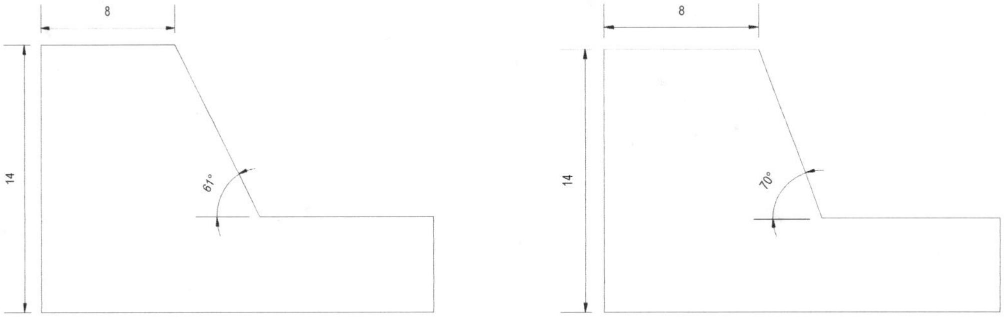

(a) (b)

Figure 3. (a) Geometric of the existing slope at STA 84+280 and (b) Geometric of the existing slope at STA 84+280.

The soil parameters used as input in Finite Element Method were obtained from the laboratory testing and the literature. The parameters obtained from laboratory testing are:

- Wet volume weight (γ_b);
- Dry volume weight (γ_d);
- Cohesion (c); and
- Friction angle (ϕ).

Parameters obtained from literature sources are:

- Dilatancy angle (Ψ);
- Young Modulus (E);
- Poisson ratio (υ);
- Permeability (k).

The research method, which is carried out in the study of the use of geogrids as slope reinforcement, can be seen in the research flowchart in figure 4.

3. Results and discussion

Slope stability analysis is carried out for existing- and reinforced slopes. Table 2 shows the results of the analysis conducted for the existing slope. Table 3 shows the results of the reinforcement analysis with variations in the thickness of the geogrid layer.

Table 2. The result of slope stability analysis for existing condition.

No.	STA (Km)	Safety Factors
1	84+280	1.14
2	84+910	1.16

Table 2 shows that the safety factor is less than 2.5. It means that the existing slope for both points of study location is unstable and even landslides occurred at that location. Therefore, reinforcement for these slopes is needed. The geogrid with variations of vertical distance of geogrid layers was chosen in this study.

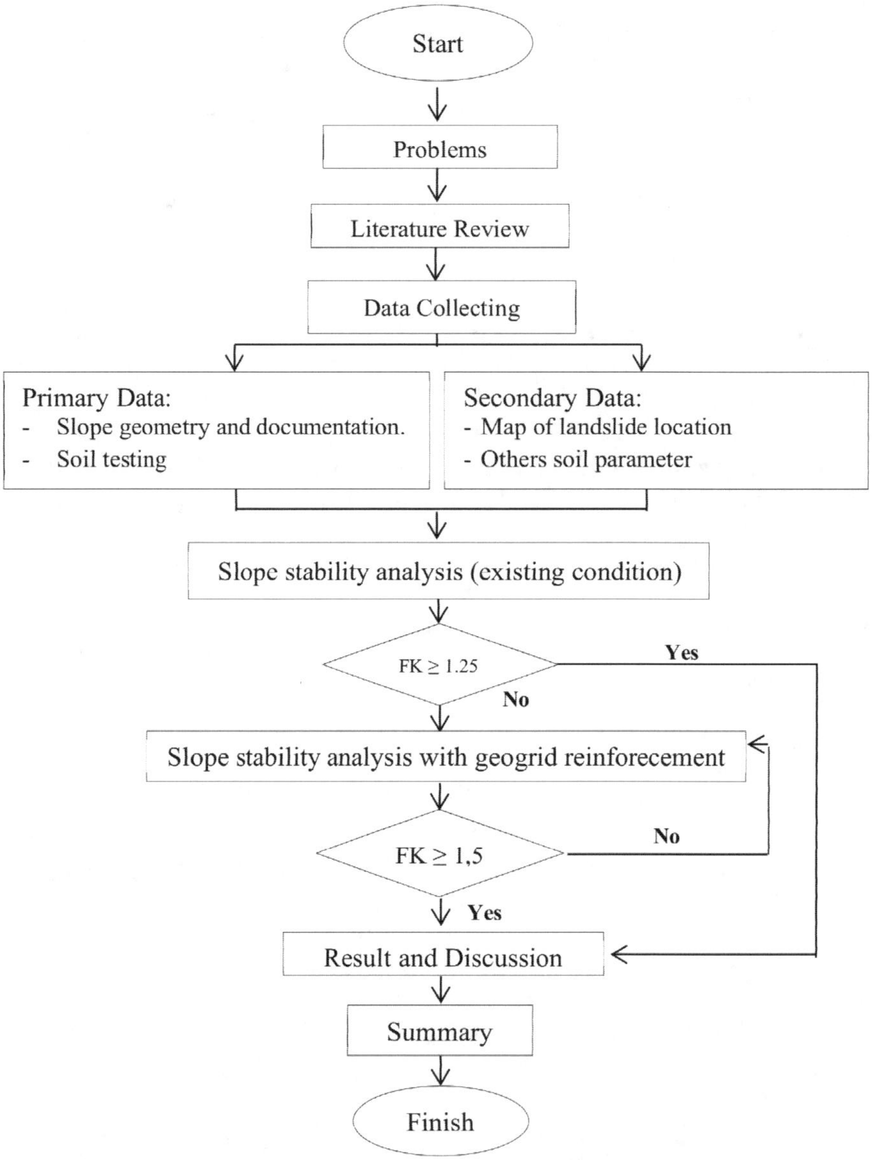

Figure 4. Research method.

Table 3 The result of slope stability analysis for the reinforced slope.

No	STA	Vertical Distance of Geogrid Layers		
		0.2 m	0.4 m	0.6 m
1	84+280	1.71	1.54	1.37
2	84+910	2.27	2.02	1.7

IOP Conf. Series: Materials Science and Engineering **933** (2020) 012046 doi:10.1088/1757-899X/933/1/012046

The slope stability analyses, shown in table 3, explain that the thickness of the geogrid layer significantly affects the stability of the slope. The closer the geogrid vertical distance on the slope, the slope safety factor is increasing.

4. Conclusion

There are several conclusions that can be drawn from this study. The safety factor of the existing slope obtained at STA 84 + 280 and 84 + 910 from the slope stability analysis using Finite element method are 1.14 and 1.16. It means that the landslides occur. The use of geogrids can increase the slope safety factor. The safety factor will increase when the vertical distance of the geogrid gets closer.

Acknowledgments

We would like to thank the Research Team for valuable contributions in data analysis.

References

[1] Hardiyatmo H C 2007 *Mekanika Tanah II* 4th Edition (Yogyakarta: Gadjah Mada University Press)

[2] Burman A, Acharaya S P, Sahay R R and Maity D 2015 Comparative study of slope stability analysis using traditional limit equilibrium method and finite element method *Asian J. of Civil Engineering* BHRC **16** 4 467

[3] Abramson L W 2001 *Slope Stability and Stabilization* (New York: John Wiley & Sons Inc) p 2

[4] Bowles J E 1993 *Physical and Geotechnical Properties of Soils* (New York: McGraw-Hill Inc USA)

[5] Griffiths D V and Lane P A 1999 Slope stability analysis by finite elements *Géotechnique* **49** 3 387-403

Effect of utilizing geopolymer fly ash on potential and corrosion rate of reinforcement in high-strength concrete

T B Aulia[1], M Muttaqin[1], M Afifuddin[1], M Zaki[1] and Firdaus[1]

[1]Civil Engineering Department, Universitas Syiah Kuala, Banda Aceh, Indonesia

Corresponding e-mail: aulia@unsyiah.ac.id

Abstract. With the increasing use of high-strength reinforced concretes as a structural element of construction, their resistance to corrosion in environmental conditions needs to be studied. It is well known that cement manufacturing at plants is un-environmental friendly due to plenty of fuel consumption and CO_2 emission to the atmosphere leading to greenhouse effect. Therefore, substituting cement with environmentally friendly natural materials and natural wastes become a smart solution. This study aims to measure the potential level and corrosion rate of reinforcing steel in high-strength concrete with variations in cement replacement by geopolymer fly ash, environmental conditions, and concrete covering thickness. Corrosion potential measurement was carried out with the half-cell potential mapping method, while the corrosion rate used the linear polarization resistance method. Data collection was executed every two weeks during eight weeks of immersion. High-strength concrete beams with compressive strength target 70 MPa having dimension 15cm x 20cm x 80cm were substituted with 15% coal fly ash (CFA), 15% palm oil blast furnace slag fly ash (POSFA), and 10% pozzolanic fly ash (PFA) from cement weight. High-strength concrete specimens without fly ash substitution were made as a comparison (HSCWFA). It was installed 4 longitudinal reinforcement Ø10 mm and 7 stirrups Ø6 mm with a distance of 10 cm. The concrete covering thickness was 2 cm and 4 cm. Immersion medium uses artificial seawater (3.5% NaCl) and shallow well water. The results showed that specimens resulting in the lowest corrosion rate were PFA, CFA, POSFA, and HSCWFA respectively. Corrosion potential was obtained at low to medium levels, while corrosion rates were at low corrosion level criteria. It can be clearly concluded that the utilizing of a geopolymer fly ash could increase the permeability of high-strength concretes through their higher resistance to the penetration of chloride ions.

1. Introduction

High-strength concrete is made with a low cement water ratio requiring a large amount of cement besides needing additives and admixtures with the best quality aggregates. As a result, a dense structure with low porosity and permeability as well as high durability is physically obtained. Mechanically, a structure having high carrying capacity and serviceability is gained which is very suitable for use in advanced and complex infrastructures. It is well known that the cement production process in the factory is not environmentally friendly because it requires many raw natural materials in the form of limestone and clay containing a lot of silica oxide, alumina oxide, and iron oxide. These raw materials must be refined and heated to a temperature of 1500°C until a new mixture is formed called clinker, so it will need high fuel. To produce 1 ton of cement, 1.7 tons of raw material are needed and 1 ton of CO and CO2 gas is released into the atmosphere which can cause the greenhouse effect as a trigger for global warming [1].

Based on the fact, efforts to reduce the high amount of cement in high-strength concrete production are necessary. One of them is using geopolymer fly ash from volcanic eruption material such as pozzolanic ash and industrial waste materials like coal fly ash and palm oil blast furnace slag ash. Such a geopolymer can be used as a cement replacement because it also contains cementitious materials such as SiO_2, Al_2O_3, Fe_2O_3, and CaO.

Aulia et al. [1] and Aulia et al. [2] state that the substitution of coal fly ash 15%, palm oil blast furnace slag fly ash 15%, and pozzolanic fly ash 10% of cement weight can still produce good shear capacity and bending capacity comparing with high-strength reinforced concrete beams without fly ash substitution. Further, the use of a combination of natural nano fly ash and nano lime with strength target f'_c 35 MPa and w/c ratio 0.49 generated higher compressive strength at 7 days compared to plain specimens without nano fly ash and nano lime [3].

Nowadays, investigation of the corrosion resistance and mechanism of reinforcement in reinforced concrete subjected to corrosive environments has been widely investigated. Several factors are affecting the corrosion potential of reinforcement in reinforced concrete, namely concrete moisture content, water to cement ratio, concrete cover thickness, as well as chloride content [4]. The effect of water to cement ratio variations and the type of cement on the electrical resistance of unsaturated normal concrete has been reported in [5] and the results showed that the water to cement ratio did not affect the concrete electrical resistivity value when using the same type of cement, but, on the other hand, the cement type had a significant effect, where the type of cement having additional blast furnace slag and pozzolan produced higher electrical resistance.

The measurement of corrosion potential of reinforcement and its possibility in reinforced concrete using half-cell potential mapping method has also been widely carried out as reported in [6], and it was concluded that the method can be used as an indicator to detect rebar corrosion initiation [7]. Based on the measurement of corrosion potential using half-cell potential method in concrete incorporating geopolymer fly ash, it was concluded that geopolymer concrete generated higher electrical resistance and yielded, therefore, better corrosion performance than corresponding concrete without fly ash [8]. Furthermore, measuring the corrosion rate using the linear polarization resistance (LPR) method has been conducted in [9], whose results showed that LPR could identify the initial phases of concrete structures degradation subjected to aggressive environments containing chloride ions. The reinforcement corrosion rate of geopolymer concrete was reported lower than that of cement-based concrete. High alkalinity in geopolymer concrete and strong silicate membranes found in geopolymer concrete could provide a passive layer on reinforcement [10]. According to Zainal et al. [11], better corrosion resistance and low corrosion rate were produced by geopolymer fly ash concrete added with slag (fly ash-slag geopolymer).

High-strength concretes are characterized by a dense structure leading to low porosity and high durability. These characteristics contribute obviously to better reinforcement corrosion resistance. However, the behavior of high-strength concretes against the corrosion resistance with the use of geopolymer fly ash as cement replacement needs to be thoroughly investigated. Therefore, this research aims to obtain the potential level and corrosion rate of reinforcement in high-strength concretes containing geopolymer fly ash, i.e., 15% coal fly ash, 15% palm oil blast furnace slag fly ash, and 10% pozzolanic fly ash from cement weight with variations in environmental conditions and concrete covering thickness. The corrosion potential level was measured with the half-cell potential mapping method, while the corrosion rate was measured with the linear polarization resistance method. Data collection was taken by soaking high-strength concrete beams with compressive strength target 70 MPa having dimension 15cm x 20cm x 80cm every two weeks during eight weeks of immersion using artificial seawater (3.5% NaCl) and shallow well water. High-strength concrete specimens without fly ash substitution were made as a comparison. Four longitudinal reinforcement Ø10 mm and seven stirrups Ø6 mm with a distance of 10 cm was embedded in the beams, with concrete covering thickness 2 cm and 4 cm.

2. Methodology

2.1. Materials

The used coal flyash emanated from electric steam power plant Nagan Raya, Aceh Province, with the chemical composition content as follows: SiO_2 = 26.65%, Al_2O_3 = 9.6%, Fe_2O_3 = 17.56%, SO_3 = 2.51% and MgO = 1.74% [1-2]. The amount of SiO_2, Al_2O_3, and Fe_2O_3 of coal flyash is 53.81%, which according to ASTM C618 [12], it is classified into Class C flyash with a minimum content about 50% [12]. The used palm oil blast furnace slag possessed the composition of chemical content as follows: SiO_2 = 34.11%, Al_2O_3 = 3.57%, Fe_2O_3 = 2.06%, SO_2 = 0.2%, and MgO = 1.95% [1]. Furthermore, the pozzolanic flyash used in this research originated from Beurandeh Village, Regency of Aceh Besar which have composition of chemical content as follows: SiO_2 = 42.96%, Fe_2O_3 = 1.92%, CaO = 0.42%, MgO = 0.28%, Na_2O = 0.13%, K_2O = 0.36%, and TiO_2 = 0.12% [1].

These fly ashes were utilized as cement replacement in the amount of 15% for coal fly ash, 15% for palm oil blast furnace slag ash, as well as 10% for pozzolanic fly ash, all from cement weight, to produce high-strength reinforced concrete (HSRC) beams. The replacement percentage was taken based on the optimum value from previous research [2]. Plain beams without fly ash were cast and tested as a comparison.

HSRC beams having a dimension of 15 cm x 20 cm x 80 cm were cast with a w/c-ratio 0.30 consisting of 600 kg/m^3 Portland cement type I using polycarboxylate ether-based superplasticizer ViscoCrete-10 with an amount of 1.5% from cement weight. Crushing stones with a maximum diameter of 4.76 mm (2-5 mm) were used as fine aggregates and with a maximum diameter of 12 mm (5-8 and 8-12mm) were used as coarse aggregates. Fine aggregate with a maximum diameter of 2 mm (0-2 mm) used was river sand. The concrete mixture was designed to have a compressive strength of 70 MPa.

Testing of potential and corrosion rate of reinforcement was carried out with a variation of the concrete covering thickness, i.e., 2 cm and 4 cm, and immersion medium, i.e., artificial seawater (using 3.5% NaCl) and shallow well water. The total amount of specimens was 16 with each treatment had one specimen. Reinforcements were embedded in the beams comprising of 4 D 10 mm as longitudinal reinforcement as well as 7 pieces Ø6 mm – 100 mm for stirrups. The specimen dimension and reinforcement arrangement were shown in figure 1. Prior to testing of potential and corrosion rate of reinforcement embedded in HSRC beams, the specimens were beforehand cured with immersion in water for 28 days to achieve the designed compressive strength.

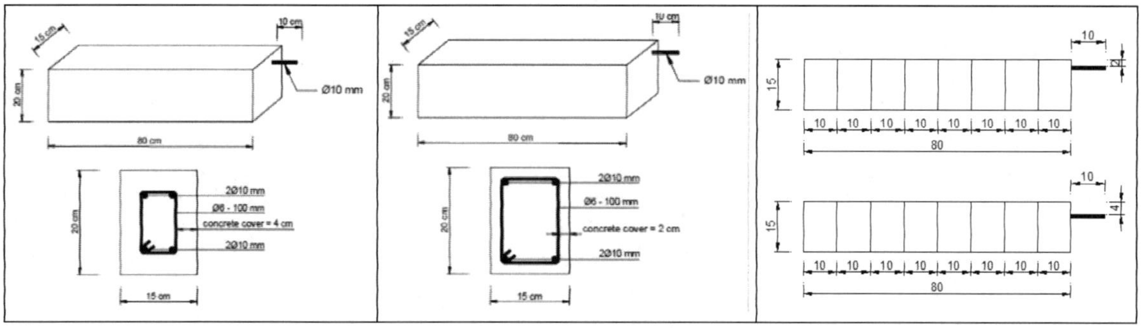

Figure 1. Specimen dimension and setup.

2.2. Instrument and testing

The main equipments used in the testing of potential and corrosion rate of reinforcement in high-strength reinforced concrete beams include: profometer to map the location of reinforcement and concrete covering thickness; digital half-cell meter to measure potential corrosion value of reinforcement in concrete with a non-destructive test method; potentiostat/galvanostat; digital multimeter; reference electrode; as well as counter electrode.

Testing of resistance and corrosion behavior of the reinforcement was preceded by mapping the reinforcement in concrete beams with drawing grids on the specimen surface, which was carried out on two sides, namely side A (top) and side B (bottom). Each meeting point of the longitudinal reinforcement and stirrups of the mapping was numbered, point 1 to 14 on side A, and point 15 to 28 on side B, as shown in figure 2.

Data of reinforcement corrosion potential values were taken on the concrete surface using Digital Half-Cell Potential Meter (PC1018). The method used was according to ASTM C 876 standards. Firstly, the excessed reinforcement tip was cleaned with sandpaper. Then the connecting cable from the reference electrode was linked to the reinforcement. After the number on the Digital Half-Cell device was stabilized (close to zero), this instrument was ready for use by attaching the tip of the reference electrode to the concrete surface vertically. The number that came out from the instrument was recorded as corrosion potential value data up to three times, to then be estimated so that the data obtained was more accurate. The set-up of the corrosion potential value measurement is shown in figure 2. The criteria for reinforcement corrosion potential values using the half-cell potential mapping technique for several half-cell standards can be seen in table 1 [13].

Table 1 The criteria for reinforcement corrosion in several half-cell standards.

No	$Cu/CuSO_4$ (CSE)	Ag/AgCl	Hydrogen Standard (SHE)	Calomel (SCE)	Corrosion Condition
1	*>-200 mV*	*>-106 mV*	*>+ 116 mV*	*>-126 mV*	Low (10% Corrosion Risk)
2	*-200 to -350 mV*	*-106 to -256 mV*	*+116 to -34 mV*	*-126 to -276 mV*	Medium Corrosion Risk
3	*<-35 mV*	*<-256 mV*	*<-34 mV*	*<-276 mV*	High (<90% Corrosion Risk)
4	*<-500 mV*	*<-406 mV*	*<-184 mV*	*<-426 mV*	Very High Corrosion Risk

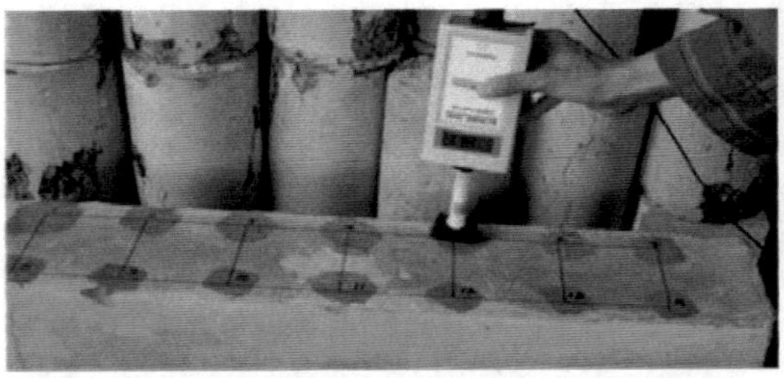

Figure 2. Setup and Measurement of Corrosion Potential Value with Half-Cell Potential Mapping.

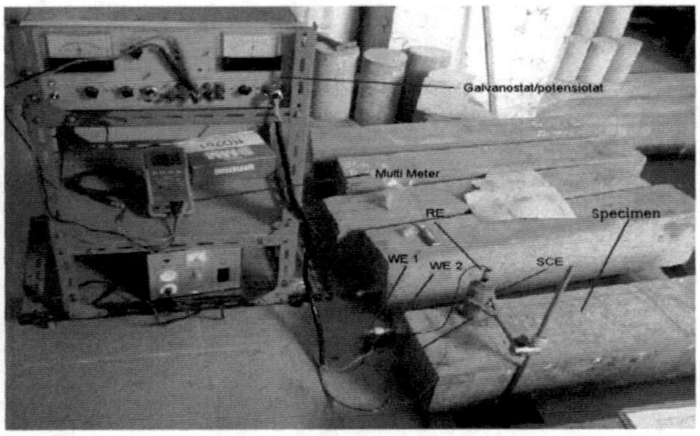

Figure 3. Setup of Corrosion Rate Measurement.

The measurement of the corrosion rate by LPR (linear polarization resistance) method was executed by placing the measuring instrument on the outer surface of the reinforced concrete beam specimen. The required data were the potential value (*E*) and current (*I*) which was controlled using potentiostat/galvanostat, as well as the value of the anodic and cathodic corrosion current density (*i*) as the result of the calculation between the current and the cross-sectional area of measurement. During measuring, all sets of equipment must be in a state of quiet and balance. Measurement with a swaying circuit will produce incorrect measurement data.

The measurement mechanism using potentiostat/galvanostat was initiated by the entry of the current (*I*) given by the potentiostat/galvanostat to the concrete through the counter electrode. The current that entered into the concrete will then be polarized. The potential of resulting polarization of reinforcement will be measured by the reference electrode to produce potential data (*E*) in the form of anodic potential and cathodic potential. The device used to measure the corrosion rate consisted of a series of instruments, which can be seen in figure 3.

After reinforcement mapping on the concrete surface, the potential value and its corrosion rate were measured at initial conditions (Week 0). Measurements were made at each numbering point. The next step was to immerse the specimens in the medium of artificial seawater (NaCl 3.5%) and shallow well water. The subsequent measurement of the potential value and corrosion rate was carried out every two weeks for eight weeks of immersion. In the second week onwards, the test specimens were removed from the immersion tub and aerated for 24 hours before measurement. This was carried out so that the reading of potential values was more stable. The concentration of NaCl solution in the immersion media of artificial seawater was examined every 2 weeks using boumemeter. If the concentration changes due to the effect of evaporation, the water or NaCl was added until the concentration returned 3.5%.

The potential data of the resulting measurement, the supplied current, and density of the resulting calculation were written in tabular form for each column point of the resulting measurement. Then a polarization curve was created based on data from the table of resulting measurement. This curve was plotted with the x-axis as the current density value (i) and the y-axis as the measured potential value. Further, the polarization curve was lined up to get a slope pattern and determine the straight line. By using the straight-line equation of the curve, it can be determined the corrosion potential value in the anodic region by entering the desired current density value, as well as the corrosion potential value and the current density in the cathodic region. For the accuracy of the corrosion rate calculation, the measured reinforcement corrosion potential value was in the range of 10 to 20 mV in the polarization curve.

The amount of polarization resistance (Rp) was calculated using the following equation [14]:

$$R_p = \frac{\Delta E}{\Delta i_{app}} \tag{1}$$

Rp = polarization resistance ($\Omega.m^2$),
ΔE = potential change (Volt), and
Δi_{app} = difference in measured current (A/m^2).

Calculating the corrosion current density (i_{corr}) can use the following equation:

$$i_{corr} = \frac{B}{R_p} \tag{2}$$

where B is a constant value determined from anodic and cathodic table slope with a value of 0.074 V.

The value of the corrosion rate (CR) can be determined based on Faraday's law.

$$CR = i_{corr} \text{ x } K.\frac{a}{nD} \text{ (mpy)} \tag{3}$$

where CR is corrosion rate (mpy), i_{corr} is corrosion current density ($\mu A/cm^2$), K is a constant (0.129 mpy), a is reinforcing steel atomic weight (55.8), i is density of corroded iron current (1 $\mu A/cm^2$), n is number of electrons lost during the corrosion process for iron (2), and D is density (7.86 g/cm^3).

3. Results and discussion

3.1. Measurement of corrosion potential

Test results on deep well and shallow well water sources at the Structural and Material Laboratory of Syiah Kuala University, Banda Aceh, showed that both water sources contained negligible corrosive materials, as shown in table 2. The pH value of water was also within the limit that could not damage the passive layer of reinforcement, which was around 7.00 so that the two well water could be used as water for making NaCl solution by adding a 3.5% chloride compound.

Table 2. The examination results of the water chemical content.

No.	Testing Parameters	Testing Method	Standard Value mg/L	Deep Well mg/L	Shallow Well mg/L
1	Dissolved Oxygen	Probe In-situ	-	8.5	8.3
2	Electrical Conductivity	Probe	-	1017	661
3	Sulfate (SO4)	Gravimetric	400	1.922	3.517
4	Chloride (Cl)	Titrimetric	500	99.4	49.98
5	pH	Probe	4.5-8.5	7	7.03

Based on the measurement results of corrosion potential, locations with the most negative potential values on the specimen surface are obtained. This location is the area with the highest corrosion possibility (top and bottom) as shown in table 3. The corrosion potential values for both beam surfaces are given in table 4.

According to table 3, it can be seen that the highest (negative) corrosion potential value was found in HSCWFA, followed then by POSFA, CFA, and PFA. This value was obtained for corrosion potential at the end of immersion, i.e., week 8 for all variations and treatments. In accordance with table 1, the corrosion risk criteria for all high-strength concrete beams using geopolymer fly ash with the variation of concrete covering thickness and immersion medium are at a medium corrosion risk (-126 to -276 mV).

Table 3. The results of mapping the highest potential location (negative) on each surface of the specimen.

Concrete Covering Thickness	HSCWFA		CFA		PFA		POSFA	
	Well Water	NaCl 3.5%	Well Water	NaCl 3.5%	Well Water	NaCl 3.5%	Well Water	NaCl 3.5%
	Point	Point	Point	Point	Point	Point	Point	Point
2 Cm	A.3 B.15	A.1 B.15	A.2 B.17	A.4 B.22	A.13 B.28	A.14 B.24	A.5 B.28	A.1 B.15
4 Cm	A.1 B.20	A.3 B.16	A.8 B.21	A.2 B.19	A.2 B.21	A.8 B.28	A.9 B.16	A.8 B.22

Table 4. The results of the measurement of the highest potential value (negative) on each surface of the specimen.

Concrete Covering Thickness	HSCWFA		CFA		PFA		POSFA	
	Well Water	NaCl 3.5%	Well Water	NaCl 3.5%	Well Water	NaCl 3.5%	Well Water	NaCl 3.5%
	mV	mV	mV	mV	mV	mV	mV	mV
2 Cm	-110 to -230 -108 to -257	-106 to -279 -79 to -295	-111 to -183 -85 to -181	-109 to -230 -88 to -234	-89 to -147 -77 to -146	-105 to -220 -82 to -248	-106 to -208 -95 to -203	-102 to -245 -64 to -245
4 Cm	-111 to -176 -92 to -151	-111 to -190 -83 to 192	-89 to -166 -77 to -161	-99 to -196 -79 to -195	-94 to -121 -75 to -121	-97 to -163 -81 to -168	-115 to -151 -76 to -156	-108 to -172 -84 to -177

The difference in concrete covering thickness of 2 cm and 4 cm affects the measurement results of the reinforcement potential values at all test specimens. From table 3 it can be seen that the corrosion potential values of each beam tend to be more positive (lower) on the concrete covering thickness of 4 cm. This shows the ability of concrete as a conductor is becoming less good for thicker layers, where the resistance or barriers to the electric current will be even greater. In addition, the diffusion of chloride ions in the immersion medium using 3.5% NaCl became smaller on larger concrete covering thickness.

Corrosion potential values for immersion treatment in well water and artificial seawater (3.5% NaCl) in general showed the same trend, but the high-strength concrete beams immersed in water with 3.5% NaCl produced a more negative value (high) as shown in figure 4. High-strength concrete substituted with pozzolanic fly ash with 2 cm concrete cover immersed in water with 3.5% NaCl resulted in a fluctuating corrosion potential values at each measurement point, but with 4 cm concrete cover it was returned almost uniform. Cement replacement with coal fly ash, pozzolanic fly ash, as well as palm oil blast furnace slag fly ash could reduce the corrosion risk at high-strength concrete. This was proved by the corrosion potential values which were increasingly positive (decrease) compared with high-strength concretes without using additives.

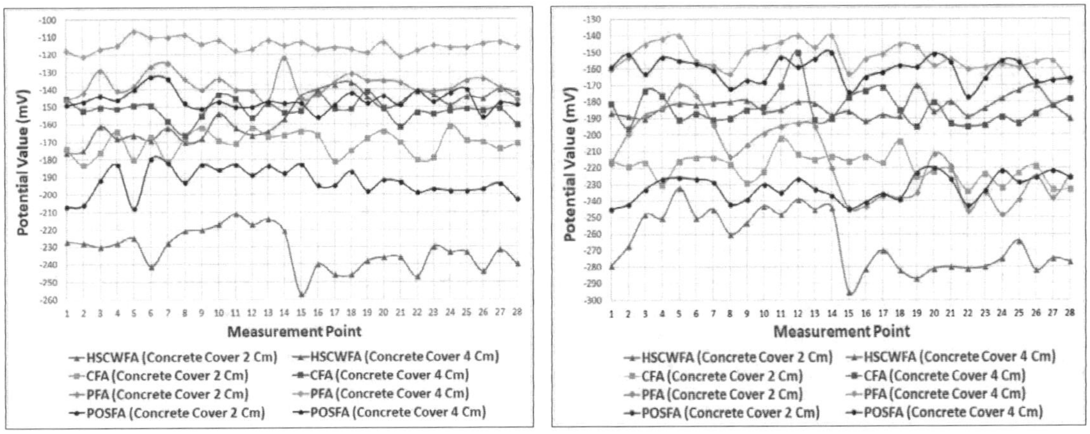

Figure 4. Comparison of corrosion potential values at week 8 immersed in well water (left); and using artificial sea water (NaCl 3.5%) (right).

3.2. Measurement of corrosion rate

As the results of corrosion potential measurements on high-strength concrete beams, corrosion rate measurement results also showed the same tendency, as shown in table 5. The same high-strength concrete with the same concrete covering thickness, but using different immersion media, produced different effects on the corrosion rate of reinforcement. Immersion in artificial seawater (3.5% NaCl) resulted in higher corrosion rate values when compared to the immersion in well water. Concrete covering thickness of 2 cm also yielded a greater corrosion rate than a 4 cm concrete cover. The thicker the concrete covering thickness, the smaller the permeability of the concrete, so that penetration of corroding elements such as oxygen, carbon dioxide, water, and chloride ions entering the concrete through the pores contained in the concrete cover will be less.

Table 5. Calculation of the corrosion rate for all specimens with all variations in week 0 to week 8.

Corrosion Rate Values of Concrete Having Covering Thickness 2 cm with Immersion in Well Water (μm/y)						
No.	Specimen	Week-0	Week-2	Week-4	Week-6	Week-8
1	HSCWFA	2.20935	2.44881	2.64384	2.66942	2.76232

2	CFA	1.91983	2.08370	2.21537	2.43351	2.63557
3	PFA	1.86577	2.00005	2.13534	2.32378	2.51364
4	POSFA	1.99345	2.10845	2.23507	2.43153	2.65295

Corrosion Rate Values of Concrete Having Covering Thickness 4 cm with Immersion in Well Water (μm/y)						
No.	Specimen	Week-0	Week-2	Week-4	Week-6	Week-8
1	HSCWFA	1.79292	1.89450	2.05039	2.16835	2.35646
2	CFA	1.67645	1.75859	1.86003	1.96104	2.10109
3	PFA	1.59697	1.67692	1.77880	1.90648	2.04413
4	POSFA	1.71654	1.79684	1.87330	2.05598	2.26531

Corrosion Rate Values of Concrete Having Covering Thickness 2 cm with Immersion in artificial seawater (3.5% NaCl) (μm/y)						
No.	Specimen	Week-0	Week-2	Week-4	Week-6	Week-8
1	HSCWFA	2.30074	2.61047	2.72610	2.90190	3.05636
2	CFA	1.92473	2.19280	2.37315	2.59478	2.79566
3	PFA	1.85830	2.16138	2.38820	2.60596	2.86332
4	POSFA	1.99608	2.30953	2.44529	2.59586	2.86470

Corrosion Rate Values of Concrete Having Covering Thickness 4 cm with Immersion in artificial seawater (3.5% NaCl) (μm/y)						
No.	Specimen	Week-0	Week-2	Week-4	Week-6	Week-8
1	HSCWFA	1.77828	2.10037	2.34547	2.65181	2.81780
2	CFA	1.67552	1.88149	2.01341	2.20481	2.38727
3	PFA	1.59824	1.76737	1.88560	2.01881	2.19117
4	POSFA	1.71312	2.04830	2.15672	2.27983	2.40629

This is consistent with the findings by Law et al. [14] who found that the electrical resistivity of concrete improved with increasing concrete strength and concrete covering thickness. Concrete electrical resistivity served to reduce the corrosion rate by inhibiting the electric current in the concrete due to electrochemical corrosion.

According to Neville [15], one of the causes of corrosion in concrete is caused by the bond between Mg and SO_4, forming Magnesium Sulfate ($MgSO_4$) which reacts with Calcium Hydroxide $Ca(OH)_2$ in concrete. These sulfate compounds attack free $Ca(OH)_2$ resulting from the hydration process in cement. According to the results of the chemical examination of the pozzolanic fly ash, coal fly ash and palm oil blast furnace slag fly ash, as explained in Section 2.1, it appeared that Mg compound was more contained in the palm oil blast furnace slag fly ash than in coal fly ash and pozzolanic fly ash, this is assumed to cause high-strength concretes with coal fly ash and especially with pozzolanic fly ash having lower corrosion rate than high-strength concrete using palm oil blast furnace slag fly ash.

4. Conclusion

Substituting cement with environmentally friendly geopolymer materials and natural wastes, i.e., coal fly ash, palm oil blast furnace slag fly ash, and pozzolanic fly ash, could increase the corrosion resistance of high-strength reinforced concrete beams in the form of corrosion potential level and corrosion rate. The lowest corrosion potential values and corrosion rate were given by PFA, CFA, POSFA, and HSCWFA respectively. Corrosion potential level was obtained at low to medium levels, while corrosion rates were at low corrosion level criteria. In terms of concrete covering thickness, the thicker the covering thickness, the smaller the permeability of the concrete, so that penetration of corroding elements in the form of oxygen, carbon dioxide, water, and chloride ions entering the concrete through the pores contained in the concrete cover will be less. The electrical resistivity of concrete improved with applying high-strength concretes and increasing concrete covering thickness. Concrete electrical resistivity served to reduce the corrosion rate by inhibiting the electric current in the concrete due to electrochemical corrosion.

Acknowledgments
Authors wishing to acknowledge the Ministry of Research, Technology, and Higher Education of Indonesia through the Directorate of Research and Community Services for providing funds (Contract No. 103 /UN11.2/PP/SP3/2019). The authors sincerely express appreciation for the generous supports for this research while the findings, opinions, and recommendations resulted are those of the authors.

References

[1] Aulia T B, Muttaqin, Afifuddin M, Zaki M and Merriza S 2018 Shear capacity analysis of high-strength reinforced concrete beams using geopolymer fly ash and palm oil blast furnace slag as additives and aggregate substitution *IOP Conference Series: Materials Science and Engineering* **434** 012199
[2] Aulia T B, Afifuddin M and Opirina L 2016 Bending capacity analysis of high-strength reinforced concrete beams using geopolymer fly ash and environmentally friendly slag as fine aggregate substitution *Proc. of The 7th Civil Eng. Conf. in the Asian Region (CECAR7)* (Waikiki: USA)
[3] Tudjono S, Purwanto and Apsari K T 2014 Study the effect of adding nano fly ash and nano lime to compressive strength of mortar *Procedia Engineering* **95** 426 – 32
[4] Medeiros M H F, Rocha F C, Medeiros-Junior R A and Helene P 2017 Corrosion potential: influence of moisture, water-cement ratio, chloride content and concrete cover, *Rev. IBRACON Estrut. Mater.* (São Paulo) **10** 4
[5] Medeiros-Junior R A and De Lima M G 2016 Electrical resistivity of unsaturated concrete using different types of cement *Construction and Building Materials* **107** 11–6
[6] Pour-Ghaz M, Isgor O B and Ghods P 2009 Quantitative interpretation of half-cell potential measurements in concrete structures, *J. of Mater. in Civ. Eng* **21** 9 467-75
[7] Pradhan B and Bhattacharjee B 2009 Half-cell potential as an indicator of chloride-induced rebar corrosion initiation in RC, *J. of Mater. In Civ. Eng.* **21** 10 543-52
[8] Olivia M and Nikraz H R 2011 Corrosion Performance of Embedded Steel in Fly Ash Geopolymer Concrete by Impressed Voltage Method *Incorporating Sustainable Practice in Mechanics of Structures and Materials* (London: Taylor & Francis Group) 781-6
[9] Romano P, Brito P S D and Rodrigues L 2013 Monitoring of the degradation of concrete structures in environments containing chloride ions, *Construction and Building Materials* 47 827–32
[10] Farhana F, H. Kamarudin H, Rahmat A, Abdullah M M A B, Abdullah A and Saud N 2013 Corrosion performance of reinforcement bar in geopolymer concrete compare with its performance in ordinary portland cement concrete: a short review *Advanced Materials Research* **795** 509-12
[11] Zainal F F, Amli S F M, Hussin K, Rahmat A and Abdullah M M A B, 2017 Corrosion studies of fly ash and fly ash-slag based geopolymer *IOP Conf. Series: Materials Science and Engineering* 209 012026
[12] ASTM C618 2003 Standard specification for coal fly ash and raw or calcined natural pozzolan for use in concrete *American society for testing and materials* (West Conshohocken: USA ASTM International)
[13] Broomfield J P 2007 *Corrosion of Steel in Concrete, Understanding, Investigation and Repair* 2nd Edition (Oxon: UK Taylor and Francis)
[14] Law D W, Cairns J J, Millard S G and Bungey J H 2003, Evaluation of corrosion loss of steel reinforcing bars in concrete using linear polarization resistance measurements *Intern. Symposium (NDT-CE 2003) Non-Destructive Testing in Civil Engineering* vol 8 no 10
[15] Neville A M 1995 *Properties of concrete* (London: Longman)

IOP Conf. Series: Materials Science and Engineering **933** (2020) 012048 doi:10.1088/1757-899X/933/1/012048

The feasibility construction of Alas-Pedesi bridge in Kuta Cane Southeast Aceh district based on the analysis of the transportation economic value

J Harja[1], M Isya[2] and S Sugiarto[2]

1 Civil Engineering Postgraduate Program, Department of Civil Engineering, Universitas Syiah Kuala, Banda Aceh, Indonesia
2 Department of Civil Engineering, Universitas Syiah Kuala, Banda Aceh, Indonesia

Corresponding e-mail: jetnoharja834@gmail.com

Abstract. Kutacane is a city located in Southeast Aceh district, the majority of the community's economic income comes from agricultural and plantation products, the production of corn and cocoa is distributed to the city center from several sub-districts located across the Alas river which is a river that separates some of these districts from the city center. Therefore, the Lawe Alas-Pedesi bridge was built, to facilitate the distribution of production results by reducing distance and travel time. The purpose of this study was to determine the effect of the Alas-Pedesi Lawe bridge, on the operational costs of vehicles, agricultural production, plantations based on producer surpluses, and calculate the economic viability of development including Benefit Cost Ratio (BCR), Net Present Value (NPV), Economic Internal Rate of Return (EIRR). The results of this study get a total time difference of IDR 68,583, - / vehicle, a total difference in vehicle operating costs (BOK) of IDR.4,441, / vehicle, the profit of farmers (corn) for 10 years can be as much as IDR200.567.937.375, - / planting season, the total difference for the benefit of farmers (cocoa) for 10 years can be as much as IDR 3.919.484.400 / planting season. The economic feasibility of the construction of the Lawe Alas-Pedesi bridge obtained in 2030 has met the eligibility standards for all discount rates. In the BCR obtained a 10% discount rate is 1.74, a 12% discount rate is 1.46 and at a 15% discount rate is 1.12, and at the Net Present Value 10% discount rate is obtained NPV IDR 23,422,534,229, - discount 12% rate is obtained NPV IDR. 13,491,749,019, and at a 15% discount rate is obtained NPV IDR. 3,243,265,869. While the value of EIRR obtained at a discount rate of 16, 33 %, this shows the bridge Lawe Alas-Pedesi feasible to build the bank lending rate at 16.33%.

1. Preface

Kutacane is a district capital located in Aceh province, namely Southeast Aceh district, which has administrative boundaries west of GayoLues district, north of East Aceh district, south of Subulussalam District, and east borders directly with Tanah Karo, North Sumatra Province. This city is often called as Lembah Alas because the position of the city is located in the valley of the Bukit Barisan mountain range and crossed by several rivers, one of which is the largest river Alas, a river that divides from the north of the city extends to the south of the city so that several districts those found across the river must cross the bridge to move towards the city center to distribute their natural resource production or for other purposes.

The purpose of this study is to determine the time value and Vehicle Operating Costs (VOC) between the Mbarung bridge *(existing)* and the Lawe Alas - Pedesi bridge (*alternative*), to determine the effect of the construction project of the Lawe Alas - Pedesi bridge on agricultural production (corn) and plantation (cocoa)) based on the *producer surplus* method, determining of economic feasibility is based on *Benefit Cost Ratio* (BCR), *Net Present Value* (NPV), *Economic Internal Rate of Return* (EIRR), on the construction of the Lawe Alas - Pedesi bridge (*alternative*).

2. Literature review

2.1. Basic theoretical
The Operation of public roads is done by giving priority to the development of road networks centered-production center and the roads that can be connecting production centers with the market area, the implementation of public road directed to the construction of the road network in order to strengthen the unity of the national territory in order to reach remote areas, the implementation of a public road directed to realize harmonious people's lives with the same level of progress that is evenly balanced as well as the efficacy and results of state defense and security efforts, the road technical requirements include speed of plan, width of the road body, capacity, intersection entrance, a plot of complementary buildings, road equipment, use road in accordance with its function and uninterrupted [1].

2.2. Traffic volume
The Traffic Volume is the number of vehicles passing through one point or the cross-section in a time unit expressed in vehicles/ hour/ lane. The number consists of various types of vehicles such as passenger cars, buses of all sizes, light or heavy trucks, two-wheeled vehicles, as well as vehicles that do not use engines such as bicycles, paddle rickshaws, carts, and pedestrians. Each vehicle is counted per unit in traffic flow [2]. The volume of traffic is also the number of vehicles passing, expressed in units of passenger cars (*smp*) multiplied with by using the following equivalence of passenger cars (*PCE*) for each type of vehicle, depending on the type of road and the current total traffic expressed in veh/h. Traffic volume calculation is done by using the following equation.

$$Q = Q_{LV} + Q_{MHV} x emp_{MHV} + Q_{LB} x emp_{LB} + Q_{LT} x emp_{LT} + Q_{MC} \text{ x } emp_{MC} \quad (1)$$

2.3. Speed and travel time
Speed is defined as the average for the average speed of vehicle space along the road segment. Calculation of speed and travel time is done by using the following equation.

$$V = L / TT \quad (2)$$

Where: V = average speed (km / h);
L = Path length (km); and
TT = Average travel time (hours).

2.4. Time value
The savings of travel time value obtained from the difference calculation of travel time for the condition to the project and without the project (without project). The travel time in the base year for various types of vehicles obtained through field surveys. The time value used can be determined from the results of time value studies using productivity methods, stated preferences, or revealed preferences calculation of the time value using the equation.

$$NW \text{ (people)} = GRDP / (40\% \times 2100 \times JP) \quad (3)$$

Where : NW = time value of people;

GRDP = GRDP (without oil and gas);
JP = total population;
40% = proportion of working population; and
2100 = assumption as the number of hours worked in a year.

2.5. Vehicle operation cost
Vehicle Operational cost vehicle is the total cost required to operate the vehicle on road conditions for the types of vehicles per kilometer of distance traveled, the units of rupiah per kilometer. The cost of vehicle operation consists of two main components of variable cost and fixed costs. Equation conducted on multiple components of VOC, which includes Gasoline; Use of lubricating oil; Use of tires; Maintenance cost; Mechanical costs; Depreciation; Interest rate; Travel time of passengers; and unexpected costs (*overhead*); VOC can be calculated by using the following equation.

1. The equation of fuel consumption on the old road based on PCI are:
   Light Vehicles :$Y=0{,}05693.S^2 - 6{,}42593.S+269{,}18576$;
   Bus :$Y=0{,}21692.S^2 - 24{,}15490.S+954{,}78824$;
   Truck :$Y=0{,}21557.S^2 -24{,}17699.S+947{,}80882$;
2. The equation of fuel consumption on the new road based on PCI are:
   Light Vehicles :$Y= 0{,}04376.\ S^2 - 4{,}94078.S + 207{,}0484$;
   Bus :$Y =0{,}14461.\ S^2 - 16{,}10285.S + 636{,}503431$;
   Truck :$Y= 0{,}13485.\ S^2 - 15{,}12463.S + 592{,}60931$;
3. The equation of lubricating oil consumption on the old road based on PCI are:
   Light Vehicles :$Y=0{,}00037.S^2 - 0{,}04070.S+2{,}20403$;
   Bus :$Y = 0{,}00209.S^2 - 0{,}24413.S+13{,}29445$;
   Truck :$Y=0{,}00188.S^2 - 0{,}22035.S+12{,}06488$;
4. The equation of lubricating oil consumption on the new road based on PCI are:
   Light Vehicles :$Y=0{,}00029.S^2 - 0{,}03134.5 + 1{,}69613$;
   Bus :$Y=0{,}00131.S^2 - 0{,}15257.S + 8{,}30869$;
   Truck :$Y=0{,}00188.S^2 - 0{,}13770.S + 7{,}54073$;
5. The equation of tire used on both roads based on PCI are:
   Light Vehicles :$Y=0{,}0008848.S - 0{,}0045333$;
   Bus :$Y=0{,}0012356.S - 0{,}00064667$;
   Truck :$Y=0{,}0015553.S - 0{,}0059333$;
6. Equation of maintenance costs on both roads based on PCI are:
   Light Vehicles :$Y=0{,}0000064.S + 0{,}0005567$;
   Bus :$Y=0{,}0000332.S + 0{,}00020891$;
   Truck :$Y=0{,}0000191.S + 0{,}00015400$;
7. The mechanical cost equation on both roads based on PCI are:
   Light Vehicles :$Y=0{,}00362.S + 0{,}36267$;
   Bus :$Y=0{,}02311.S + 1{,}97733$;
   Truck :$Y=0{,}01511.S + 1{,}21200$;
8. The depreciation cost equation for both roads based on PCI are:
   Light Vehicles :$Y=\frac{1}{2{,}5.S+100}$
   Bus :$Y=\frac{150}{9.S+315}$
   Truck :$Y=\frac{1}{6.S+210}$
9. The interest rate equation on both roads based on PCI are:
   Light Vehicles :$Y=\frac{150}{500.S}$
   Bus :$Y=\frac{150}{2571{,}42857.S}$
   Truck :$Y=\frac{150}{1714{,}28571.S}$

10. The equation vehicle crew travel time on both roads are:
    Bus :$Y=\frac{1000}{S}$
    Truck :$Y=\frac{1000}{S}$
11. The equation of unexpected costs on both roads are:
    Bus :10% from total sub;
    Truck :10% from total sub.

2.6. Producer surplus method
Producer surplus analysis is an evaluation parameter of project feasibility. In this case, the benefits used are all surpluses enjoyed by producers of goods and services sold in the area of Project's influence. The surplus producer approach is carried out by paying attention to certain surrounding areas (for example agriculture and certain products). For these products, it is necessary to pay attention to various production data, production factors and selling prices from year to year, some assumptions used in the surplus producer method are as follows [3]:

a. With the existence of new road construction projects, resulting in an increase in planted area or increase in production;
b. Transportation costs are reduced because of good accessibility from the investments;
c. Products are sold at the same prices at the same specified distance (on average) throughout the reviewed area;
d. Additional production will not result in a collapse in market prices, in other words, fixed selling prices, so farmers can sell as much production as possible;
e. Household consumption remains unaffected by production; and
f. Transportation and production costs are the same for all farmers.

Surplus producer calculation is done by using the following equation:

$$FS_2 - FS_1 \tag{4}$$

Where: FS1 = Farmer Surplus Before the Bridge, and
FS2 = Farmer Surplus After The Bridge.

2.7. Analysis of the economic value of transportation

2.7.1. Benefit cost ratio The benefit cost ratio is the ratio between the Present Value Benefit divided by the present value cost. The BCR results of a project are declared financially feasible if the BCR value is greater than 1 (> 1). This value is done based on the present value, namely by comparing the difference in benefits and costs smaller than zero [3]. This method is used to analyze the economics of project development by comparing the total profits against the total costs that have been equivalent to the base year using the applicable discount rate. Calculation of benefit cost ratio is using the following equation:

$$BCR = \frac{\text{Present value benefit}}{\text{present value cost}} \tag{5}$$

2.7.2. Net present value. The Net present value (NPV) is the difference between the present value benefit reduced by the present value cost. The NPV results from a project is stated to be financially feasible are those that produce a positive NPV value [3]. The Calculation of net present value using the following equation:

$$NPV = \sum_{t=0}^{n} \frac{Bt-Ct}{(1+r)^t} \tag{6}$$

Where: NPV = net present value;

IOP Conf. Series: Materials Science and Engineering **933** (2020) 012048 doi:10.1088/1757-899X/933/1/012048

Bt	=	total amount of project benefit components in year t;
Ct	=	total amount of cost component in year t;
n	=	the economic age of the project being studied;
r	=	interest rate (% / year); and
t	=	economic life of the project, starting from the planning stage to the end of the life of the road plan.

2.7.3. Economic internal rate of return. Economic Internal Rate of Return (EIRR) is a rate of return based on the determined value of the discount rate, where all future benefits that are valued now at a certain discount rate are the same as the cost of capital or the present value of the total cost. In order to calculate the EIRR value the way is to try some of the interest rates. The calculation of the EIRR value is calculated based on the interest rate that produces the smallest positive NPV and the interest rate that produces the smallest negative NPV[3]. Then interpolation is carried out using the equation:

$$EIRR = i_1 + (i_2 - i_1)\frac{NPV_1}{NPV_2 - NPV_1} \qquad (7)$$

Where:

EIRR	=	economic internal rate of return;
i_1	=	discount rate that produces the smallest negative NPV;
i_2	=	discount rate that produces the smallest positive NPV;
NPV1	=	current value and use i_1; and
NPV2	=	present value and using i_2.

3. Research methodology

3.1. Research object

In this research object, the traffic volume is calculated covers all types of vehicles includes the motorcycle (MC); Light Vehicle (LV); Medium Heavy Vehicle (MHV); Large Bus (LB); and Large Truck (LT), which passes through the location of the object of research as basic data to get the movement of people from the area to the city center.

3.2. Data type and sources

3.2.1. Primary data

a. Traffic Volume Survey ;
b. Vehicle Speed Survey ;
c. Vehicle Travel Time Survey.

3.2.2. Secondary data

a. The population number is obtained from the Badan Pusat Statistik of Southeast Aceh District.
b. Gross Regional Domestic Product (GRDP) from the Badan Pusat Statistik of Southeast Aceh District;
c. Percentage of vehicle growth from the DinasPerhubungan of Southeast Aceh District;
d. Regional Spatial Planning from BAPPEDA, Southeast Aceh District;
e. Long Term Development Plan (RPJP) from BAPPEDA, Southeast Aceh District ;
f. Annual data of agricultural and livestock production from the Badan Pusat Statistik of Southeast Aceh;
g. Road Network Map of the Southeast Aceh from Bina Marga; and
h. Planning, physical work and supervision costs from the Bina Marga of Southeast Aceh.

3.3. Data collection technique

a. Field Survey;

b. Documentation;
c. Data Collection from related agencies.

3.4. Analysis technique
After data from the field were collected in the form of traffic volumes, documentation and data from relevant agencies, an analysis is carried out on that data using the help of *Microsoft Office Excel Software.* The results of the field survey were analyzed using the productivity method, *stated preference* or *revealed preference*, to obtain the time value, and with MKJI 1997 and the *PCI* method (2000) in order to obtain the VOC value, while the data from the related agencies mentioned above were analyzed using the producer surplus method, after obtaining the results of the three methods, then it is entered into an analysis of the economic value of transportation to get the value of *Benefit Cost Ratio* (BCR), *Net Present Value* (NPV), at a *discount rate of* 10%, 12%, 15%, so that it will get an *Economic Internal Rate value of return* (EIRR) later, as a reference to the construction of the Lawe Alas - Pedesi bridge (*alternative*) to find out the percentage of loan interest at the bank.

4. Results and discussion

4.1. Time value
From the research on Mbarung Bridge (existing), total n use values while passenger cars, buses, and trucks on the bridge Mbarung (existing) is IDR 102 850, - /vehicle, while assuming total time value of passenger cars, buses and trucks on the bridge Lawe Alas -Pedesi (*alternative*) is IDR .34.267,-/ vehicle, the difference in the time value of the vehicle is the time value of the vehicle on the *existing* Mbarung bridge minus the time value of the vehicle on the Alas-Pedesi bridge (*alternative*) obtained at IDR 68.583,-/vehicle. The results of calculating the time value can be seen in the following table.

Table 1. Time value on the existing and alternative bridge.

No	Road	Travel Time (hour)			Time Value			Total of Time Value
					PC	BUS	TRUCK	
		LV	BUS	TRUK	(IDR/Veh)	(IDR/Veh)	(IDR/Veh)	(IDR/Veh)
1	Existing	0,285	0,354	0,385	21.225	61.667	19.957	102.850
2	Alternative	0,114	0,114	0,114	8.483	19.872	5.912	34.267
Difference in Value		0,171	0,240	0,271	12.743	41.795	14.045	68.583

4.2. Vehicle operational costs
From the results of the research on the existing (bridge Mbarung) for passengers cars amounted to IDR1.875, -veh/km , the bus is IDR5.631, -veh/km and for trucks IDR.4.480, -veh/km, while for alternative roads (the Lawe Alas-Pedesi bridge) for passenger cars is IDR1.439, -veh/km , buses IDR.3.382, -veh/km and trucks obtained IDR2.724, -veh/km. The total savings of VOC or difference in VOC obtained between existing roads and alternative roads for passenger cars is IDR435, -veh/km, for buses obtained IDR2.250, -veh/km and for trucks is IDR1.756, -veh/km . the calculation of VOC can be seen in the following table.

Table 2. Calculation of vehicle operational costs (VOC).

No	Type of Vehicle	VOC of existing bridge	VOC of alternative bridge	Difference in VOC
		(IDR/veh./km)	(IDR/veh./km)	(IDR/veh./km)
1	Passenger Car	1.875	1.439	435

2	Bus	5.631	3.382	2.250
3	Truck	4.480	2.724	1.756

4.3. Surplus producer

4.3.1. Surplus of corn farming. Corn production in the sub-districts of Lawe Alas and Tanoh Alas is 58,215,600 kg or (58,215.6 tons), with an average production of 12.56 tons/ha. While the planting area of these two districts is 5,571 Ha, with a harvesting area of 4,635 ha in 2021. The farmers 'profit (without project) is IDR979.469.446.500 / season for 10 years, while the farmers' profit (with project) is obtained for IDR 1.180.037.383.875/ planting season, the difference in profit is IDR200.567.937.375. The results of the calculation of producer surplus on corn production can be seen in the following table.

Table 3. Corn production *(without project).*

No	Year	Sub-district	Planted area (Ha)	Harvest area (Ha)	Corn Production (ton/ Ha)	Corn Production (kg)	Production Results (@IDR. 3800/kg)	Intensification cost /Operational (IDR)	Farmer's Benefits (IDR)
1	**2**	**3**	**4**	**5**	**6**	**7**	**8**	**9**	**10**
1	2021	Lawe Alas + Tanoh Alas	5.571	4.635	12,6	58.216	2.212.193	1.880.364	331.829
2	2022	Lawe Alas + Tanoh Alas	5.571	4.635	12,6	58.216	2.212.193	1.880.364	331.829
3	2023	Lawe Alas + Tanoh Alas	5.571	4.635	12,6	58.216	2.212.193	1.880.364	331.829
4	2024	Lawe Alas + Tanoh Alas	9.170	7.864	16,7	131.486	4.996.471	4.247.000	749.471
5	2025	Lawe Alas + Tanoh Alas	9.170	7.864	16,7	131.486	4.996.471	4.247.000	749.471
6	2026	Lawe Alas + Tanoh Alas	9.170	7.864	16,7	131.486	4.996.471	4.247.000	749.471
7	2027	Lawe Alas + Tanoh Alas	13.542	12.694	20,1	255.022	9.690.853	8.237.225	1.453.628
8	2028	Lawe Alas + Tanoh Alas	13.542	12.694	20,1	255.022	9.690.853	8.237.225	1.453.628
9	2029	Lawe Alas + Tanoh Alas	15.073	14.119	21,3	301.158	11.444.014	9.727.412	1.716.602
10	2030	Lawe Alas + Tanoh Alas	16.221	15.187	22,3	338.063	12.846.380	10.919.423	1.926.957
Amount			**10.260**	**9.219**	**17,2**	**171.837**	**6.529.809**	**5.550.338**	**9.794.714**

Note:
- Assuming seed price of IDR. 92.000/kg

Table 4. Corn production *(with project).*

No	Year	Sub-district	Planted Area (Ha)	Harvest Area (Ha)	Corn production (ton/ Ha)	Corn Production (kg)	Production Results (@IDR. 4500/ kg)	Intensification costs/ Operational (IDR.)	Farmer's Benefits (IDR)
1	**2**	**3**	**4**	**5**	**6**	**7**	**8**	**9**	**10**
1	2021	Lawe Alas + Tanoh Alas	5.571	4.635	12,6	58.216	2.619.702	2.226.747	392.955

2	2022	Lawe Alas + Tanoh Alas	5.571	4.635	12,6	58.216	2.619.702	2.226.747	392.955
3	2023	Lawe Alas + Tanoh Alas	5.571	4.635	12,6	58.216	2.619.702	2.226.747	392.955
4	2024	Lawe Alas + Tanoh Alas	9.170	7.864	16,7	131.486	5.916.874	5.029.343	887.531
5	2025	Lawe Alas + Tanoh Alas	9.170	7.864	16,7	131.486	5.916.874	5.029.343	887.531
6	2026	Lawe Alas + Tanoh Alas	9.170	7.864	16,7	131.486	5.916.874	5.029.343	887.531
7	2027	Lawe Alas + Tanoh Alas	13.542	12.694	20,1	255.022	11.476.011	9.754.609	1.721.402
8	2028	Lawe Alas + Tanoh Alas	14.307	13.406	20,1	269.327	12.119.694	10.301.740	1.817.954
9	2029	Lawe Alas + Tanoh Alas	15.455	14.475	21,3	308.752	13.893.829	11.809.754	2.084.074
10	2030	Lawe Alas + Tanoh Alas	16.603	15.543	22,3	345.987	15.569.423	13.234.010	2.335.413
Amount			**10.413**	**9.362**	**17,2**	**174.819**	**7.866.868**	**6.686.838**	**11.800.303**
Difference in profits of farmers = corn production (*with project*) – corn production (*without project*)									**2.005.589**

Note:
- Assuming seed price IDR. 92.000/kg

4.3.2. Surplus of cocoa plantation

Cocoa production in the sub-districts of Lawe Alas and Tanoh Alas is 2.134.800 kg or (2.134,8 tons), with an average production of 1.80 tons/ha. While the planting area of these two districts is 2.475 Ha, with a harvesting area of 1.186 ha in 2021. Farmer profits *(without project) is*IDR29.047.502.100/ planting season for 10 years, while farmers profits *(with project) is*IDR32.966.986.500/planting season, the difference in profit is IDR3.919.484.400. The results of the calculation of producer surplus on cocoa production can be seen in the following table.

Table 5. Cocoa production (without project).

No	**Year**	**Sub-districts**	**Planted Area (Ha)**	**Harvest Area (Ha)**	**Cocoa Pro-duct-ion (ton/ Ha)**	**Cocoa Pro-duct-ion (kg)**	**Produc-tion Results (@IDR. 22000/ kg)**	**Intensi-fication /Operat-ional cost (IDR.)**	**Farm-er's Benefits (IDR)**
1	**2**	**3**	**4**	**5**	**6**	**7**	**8**	**9**	**10**
1	2021	Lawe Alas + Tanoh Alas	2.475	1.186	1,8	2.135	46.966	39.921	7.045
2	2022	Lawe Alas + Tanoh Alas	2.475	1.186	1,8	2.135	46.966	39.921	7.045
3	2023	Lawe Alas + Tanoh Alas	2.475	1.186	1,8	2.135	46.966	39.921	7.045
4	2024	Lawe Alas + Tanoh Alas	4.316	1.321	3,0	3.950	86.895	73.861	13.034
5	2025	Lawe Alas + Tanoh Alas	4.316	1.321	3,0	3.950	86.895	73.861	13.034
6	2026	Lawe Alas + Tanoh Alas	4.316	1.321	3,0	3.950	86.895	73.861	13.034
7	2027	Lawe Alas + Tanoh Alas	6.550	1.490	4,5	6.690	147.182	125.105	22.077

8	2028	Lawe Alas + Tanoh Alas	6.550	1.490	4,5	6.690	147.182	125.105	22.077
9	2029	Lawe Alas + Tanoh Alas	8.039	1.603	5,5	8.800	193.610	164.569	29.042
10	2030	Lawe Alas + Tanoh Alas	9.156	1.687	6,2	10.527	231.591	196.853	34.739
Amount			**5.067**	**1.379**	**3,5**	**5.096**	**112.115**	**95.298**	**168.172**

Note:
- Assuming seeds price ofIDR. 3.000/stick

Table 6. Cocoa production *(with project).*

No	**Year**	**Sub-districts**	**Planted Area (Ha)**	**Harvest Area (Ha)**	**Cocoa Pro-duct-ion (ton/ Ha)**	**Cocoa Pro-duct-ion (kg)**	**Produc-tion Results (@IDR. 31000/ kg)**	**Intensi-fication /Operat-ional** (IDR.)	**Farm-er's Benefits** (IDR)
1	**2**	**3**	**4**	**5**	**6**	**7**	**8**	**9**	**10**
1	2021	Lawe Alas + Tanoh Alas	2.475	1.186	1,8	2.135	66.179	56.252	9.927
2	2022	Lawe Alas + Tanoh Alas	2.475	1.186	1,8	2.135	66.179	56.252	9.927
3	2023	Lawe Alas + Tanoh Alas	2.475	1.186	1,8	2.135	66.179	56.252	9.927
4	2024	Lawe Alas + Tanoh Alas	4.316	1.321	3,0	3.950	122.443	104.077	18.367
5	2025	Lawe Alas + Tanoh Alas	4.316	1.321	3,0	3.950	122.443	104.077	18.367
6	2026	Lawe Alas + Tanoh Alas	4.316	1.321	3,0	3.950	122.443	104.077	18.367
7	2027	Lawe Alas + Tanoh Alas	5.401	1.423	4,5	6.389	198.067	168.357	29.710
8	2028	Lawe Alas + Tanoh Alas	5.401	1.423	4,5	6.389	198.067	168.357	29.710
9	2029	Lawe Alas + Tanoh Alas	6.124	1.491	5,5	8.186	253.753	215.690	38.063
10	2030	Lawe Alas + Tanoh Alas	6.667	1.542	6,2	9.622	298.284	253.542	44.743
Amount			**4.397**	**1.340**	**3,5**	**4.884**	**151.404**	**128.693**	**227.106**
Difference in profits of farmers = chocoalte production (with project) – chocolate production (without project)									**58.934**

Note:
- Assuming seeds price IDR. 5.000/stick

4.4. Feasibility analysis

4.4.1. Benefit cost ratio. BCR is a comparison of the value of benefits with costs. BCR value obtained in 2030 is an analysis since the road was opened or the 10th year at a 10% discount rate is 1,74, the 12% discount rate is 1,46 and at a 15% discount rate is 1,12. Based on the value of the BCR it has been in accordance with the eligibility requirements (BCR requirements > 1), showing that the construction of the Lawe Alas-Pedesi bridge (alternative) began to provide benefits in the 10th year for the three discount rates.

IOP Conf. Series: Materials Science and Engineering **933** (2020) 012048 doi:10.1088/1757-899X/933/1/012048

4.4.2. Net present value. Net present value is the value of net profit or value the benefits of the project after deducting the cost of the project. NPV value obtained in 2030, since the road opened or year to 10 at a discount rate of 10% was obtained NPV IDR23.422.534,229,- discount rate of 12% was obtained NPV IDR 13.491.749.019,- and at a discount rate of 15% was obtained NPV IDR3.243.265.869,-. Based on the three-discount rate of the n use values NPV at year 10 analysis for a third discount rate at the above all are positive (+).

4.4.3. Economic internal rate of return. Economic Internal Rate of Return is a sizes that shows the rate of return on investment due to the implementation of the construction of the Lawe Alas-Pedesi bridge (alternative) based on the interest rate that produces BCR equal to 1 (one), the result is at a value of NPV = 0, - obtained at a discount rate 16.33 %. So, the construction of the Lawe Alas-PedesiBridge (alternative) is feasible to be built with a bank loan interest rate at 16.33%. The results of calculating the economic feasibility of transportation can be seen in the following table.

Table 7. Economicf feasibility analysis with interest rate of 10%, 12%, 15% using the BKBOK equation calculation (Part 1 of 4).

Year	To	Cost estimation and benefits			
		Cost	Benefits		
		Project	BKBOK and Time value	Producer Surplus (corn)	Producer Surplus (cocoa)
2013	0	1.238.103.806			
2014	1	6.721.134.950			
2015	2	5.895.732.412			
2016	3	5.895.732.412	Project Implementation Period		
2017	4	5.895.732.412			
2018	5	5.895.732.412			
2019	6	5.895.732.412			
2020	7	5.895.732.412			
2021	8	206.350.634	219.483.684	6.112.638.000	224.154.000
2022	9	206.350.634	230.200.800	6.112.638.000	224.154.000
2023	10	206.350.634	241.513.787	6.112.638.000	224.154.000
2024	11	206.350.634	253.455.776	13.806.038.400	414.727.950
2025	12	412.701.268	266.061.740	13.806.038.400	414.727.950
2026	13	206.350.634	279.368.595	13.806.038.400	414.727.950
2027	14	206.350.634	293.415.312	26.777.358.300	499.400.250
2028	15	206.350.634	308.243.026	36.437.132.550	499.400.250
2029	16	206.350.634	323.895.161	36.748.870.425	510.048.450
2030	17	412.701.268	340.417.555	40.848.546.900	493.989.600
Total		45.809.840.835	2.756.055.436	200.567.937.375	3.919.484.400

Table 7. Economic feasibility analysis with interest rate of 10%, 12%, 15% using the BKBOK equation calculation (Part 2 of 4).

Year	To	Present Value			
		Discount Rate 10,00%		Discount Rate 12,00%	
		Cost	Benefits	Cost	Benefits
2013	0	1.238.103.806		1.238.103.806	
2014	1	6.110.122.682		6.001.013.348	

2015	2	4.872.506.126	Project Implementation Period	4.700.041.782	Project Implementation Period
2016	3	4.429.551.023		4.196.465.877	
2017	4	4.026.864.567		3.746.844.533	
2018	5	3.660.785.970		3.345.396.905	
2019	6	3.327.987.245		2.986.961.522	
2020	7	3.025.442.950		2.666.929.930	
2021	8	96.264.094	3.058.550.993	83.341.560	2.647.969.787
2022	9	87.512.812	2.785.046.006	74.412.107	2.368.123.438
2023	10	79.557.102	2.536.221.652	66.439.382	2.118.038.405
2024	11	72.324.638	5.073.126.555	59.320.876	4.160.992.987
2025	12	131.499.342	4.615.949.880	105.930.136	3.718.407.946
2026	13	59.772.428	4.200.172.595	47.290.239	3.323.056.683
2027	14	54.338.571	7.260.088.465	42.223.428	5.641.403.809
2028	15	49.398.701	8.916.103.213	37.699.489	6.804.481.295
2029	16	44.907.910	8.179.115.339	33.660.258	6.130.571.082
2030	17	81.650.746	8.246.750.244	60.107.604	6.070.886.371
Total		31.448.590.713	54.871.124.943	29.492.182.785	42.983.931.804

Table 7. Economic feasibility analysis with interest rate of 10%, 12%, 15% using the BKBOK equation calculation (Part 3 of 4).

Year	To	Present Value Discount Rate 15,00%		IRR = 16,33%	
		Cost	Benefits	Cost	Benefits
2013	0	1.238.103.806	Project Implementation Period	1.238.103.806,00	
2014	1	5.844.465.174		5.777.782.926,56	
2015	2	4.458.020.727		4.356.873.751,42	
2016	3	3.876.539.763		3.745.360.115,81	
2017	4	3.370.904.142		3.219.676.125,00	
2018	5	2.931.220.993		2.767.775.068,18	
2019	6	2.548.887.820		2.379.301.063,41	
2020	7	2.216.424.191		2.045.351.739,53	
2021	8	67.456.388	2.143.258.151	61.539.597,83	1.955.266.921
2022	9	58.657.729	1.866.749.213	52.902.142,31	1.683.580.908
2023	10	51.006.721	1.626.056.583	45.477.006,03	1.449.773.360
2024	11	44.353.670	3.111.135.950	39.094.032,62	2.742.204.863
2025	12	77.136.818	2.707.691.754	67.213.896,43	2.359.372.840
2026	13	33.537.747	2.356.677.304	28.890.009,36	2.030.083.582
2027	14	29.163.258	3.896.455.672	24.835.121,46	3.318.180.338
2028	15	25.359.355	4.577.177.552	21.349.361,65	3.853.403.178
2029	16	22.051.613	4.016.278.768	18.352.849,36	3.342.619.846
2030	17	38.350.631	3.873.425.467	31.553.831,45	3.186.946.609
Total		26.931.640.546	30.174.906.414	25.921.432.444	25.921.432.445

Table 7. Economic feasibility analysis with interest rate of 10%, 12%, 15% using the BKBOK equation calculation (Part 4 of 4).

Criteria Feasibility of Economy		Discount Rate			
		10,00%	12,00%	15,00%	16,33%
Benefit-Cost Ratio	BCR	1,74	1,46	1,12	1,00
Net Present Value (Rp.)	NPV	23.422.534.229	13.491.749.019	3.243.265.869	0
Economic Internal Rate of Return	EIRR		**16,33%**		

IOP Conf. Series: Materials Science and Engineering **933** (2020) 012048 doi:10.1088/1757-899X/933/1/012048

5. Conclusions and recommendations

5.1. Conclusions

1. The total time difference obtained in the research or at the Alas-Pedesibridge (alternative) is IDR 68.583/vehicle, and the total difference in the operational costs of the vehicle (VOC) is IDR 4.441/vehicle.
2. The total difference in profits of corn farmers for 10 years in Lawe Alas and Tanoh Alas district, in the amount of IDR 200.567.937.375,-/planting season, and the total difference in profits for cocoa farmers for 10 years can be as much as IDR3.919.484.400,-/growing season.
3. The economic feasibility of the construction of the Lawe Alas-Pedesi bridge (*alternative*), Southeast Aceh district in 2030 or the 10th year, has met the economic feasibility standard for all *discount rate*. That's proved by BCR at *a discount rate* of 10% is 1,74, *a discount rate* of 12% is 1,46 and the *discount rate of* 15% is 1,12, then to *Net Present Value* at *a discount rate* of 10% was obtained *NPV of* IDR 23.422.534.229,- *discount rate* of 12% was obtained *NPV* IDR 13.491.749.019,- and at a *discount rate* of 15% was obtained *NPV* IDR3.243.265.869,-. Then obtained value of values *EIRRs* at the *discount rate of* 16.33, it indicates that the construction of the Lawe Alas-Pedesi bridge is very feasible to be carried out, with bank loan rates up to 16,33%.

5.2. Recommendations

1. The need to accelerate the completion of the construction of the Lawe Alas-Pedesi bridge (*alternative*), in the Southeast Aceh, because it can save time and operational costs of the community in distributing agricultural and plantation products in the two sub-districts, so as to improve the economy and standard of living of the people in the region; and
2. The need for more studies on this research by increasing the variable or using methods such as modeling, APV, and others.

References

[1] Peraturan Pemerintah Republik Indonesia No. 34 Tahun 2006 *Tentang Jalan (Aspek Teknis Jalan dan Jembatan).*

[2] Bukhari R A and Saleh S M 2002 *Rekayasa Lalu lintas I* (Banda Aceh: Syiah Kuala University)

[3] Ofyar O Z 2008 *Perencanaan, Pemodelan dan Rekayasa Transportasi* (Bandung: Institut Teknologi Bandung)

[4] Pedoman Konstruksi dan Bangunan 2005 *Tentang Pra Studi Kelayakan Proyek Jalan dan Jembatan*

[5] Pacific Consultant International (PCI) 2005 *Metode Perhitungan Biaya Operasional Kendaraan*

[6] Departemen Pekerjaan Umum 2005 *Pedoman Perhitungan Biaya Operasi Kendaraan* (Jakarta: Badan Penelitian dan Pengembangan Pekerjaan Umum)

[7] Giatman 2005 *Ekonomi Teknik; Evaluasi Investasi* (Jakarta: PT. Raja Grafindo Persada)

[8] Raharjo 2007 *Ekonomi Teknik; Analisis Pengambilan Keputusan* (Yogyakarta: Andi)

[9] Putranto and Suryo L 2008 *Rekayasa Lalu lintas* Cetakan Pertama (Jakarta: PT Mancana Jaya Cemerlang)

IOP Conf. Series: Materials Science and Engineering **933** (2020) 012049 doi:10.1088/1757-899X/933/1/012049

Oversights in road planning leading to unsafe road features: a safety auditor's perspective

Y Darma[1], M R Karim[2] and S Abdullah[3]

[1] Transportation Science Society of Malaysia.
[2] Department of Civil Engineering, Universitas Syiah Kuala, Banda Aceh, Indonesia
[3] Accredited Road Safety Auditor, Malaysia.

Correspondence e-mail: yusriadarma@unsyiah.ac.id

Abstract
Road safety auditors are required to identify safety deficiencies that may lead to potential hazard on a road project. The available audit tools may only provide the prompt lists that should be followed. Auditors often become a design checker rather than an auditor. They may overlook the critical safety issues. The common oversights during road design have been identified. The experience in auditing a total of 24 road projects from 1998 to 2008 was referred and summarised. As a result, lack of or poor provision for vulnerable road users is the most common oversight done during the road designs.

1. Introduction
Road engineers often pay more attention to have their proposed design to comply with the adopted geometry standard. It is believed that if road geometry requirements are complied with, the road safety features have also been ensured. The design for other road provisions such as the facilities for pedestrian, road lighting and road furniture can be provided at a later stage if necessary or requested. This practice may not be always right as a hazardous situation on a road does not necessarily relate to or depends on geometric parameters. As such, lack of provisions for vulnerable road users and unsafe roadside features shall also increase the probability of road accident and casualty.

Road safety audit (RSA) in road design has been imposed by road authority in many countries. This is to address and solve or minimize the safety issues of road projects from the earliest stage and considered as a proactive approach. RSA guidelines to streamline and ease the audit process have been developed and published. However, the auditors may still get difficulties when conducting the RSA as road projects contain various conditions and many features are needed to be checked. Thus, experience in RSA is required as the more experienced auditors would have better knowledge to avoid items that may be overlooked when doing the RSA.

This paper examines the RSA outcomes of selected road projects. Common oversights experienced by the road engineers during the planning and feasibility stage have been identified. Future road projects may use the findings to avoid the need to redesign due to the unfulfilled safety requirements. The road safety auditors and road engineers could also minimise their oversights leading to unsafe road features when carrying out the road design.

1.1. RSA definition and implementation in road project
In the literature, there are various definitions of RSA [1-4]. Although the RSA guidelines use different words when describing the definition of RSA, however, some similarities are observed as follows:
- RSA is a formal examination;
- It is conducted for existing and future road;

- Conducted by an independent team; and
- To improve safety performance on the road.

The first procedure for conducting a RSA was developed in the United Kingdom in the 1980's. In 1990, the Institution of Highways and Transportation published the first *Guidelines for the Safety Audit of Highways* [5]. Then, in 1993 and 1994, guidelines for the audit were established in New Zealand and Australia respectively. Later in the 1990's, several countries such as Denmark, Finland, France, the Netherland, the USA, Norway and Germany began to explore and then published RSA procedures. In Malaysia, the guideline for safety audit of roads and road projects was published by the Public Works Department, Ministry of Works in 1997. At the initial stage, the RSA was imposed for federal road projects.

Generally, the RSA is conducted for both the existing road and new road project. The stages of auditing for a new road project may comprise of audit on planning and feasibility, preliminary design, detailed design, construction and pre-opening, and post construction stage respectively. Audit for existing road is usually similar to audit during post construction of a new road project.

RSA Guidelines [4] also specify the procedure to carry out RSA in Malaysia. The key points in the procedures involve [1] preparing the audit brief, [2] assembling background information, [3] initial meeting with auditors, [4] doing the audit, [5] preparing the audit report, [6] completion meeting involves the presentation on audit findings, and [7] deciding the action.

1.2. The use of RSA tools

At the moment, there may not be any special tool or software that can detect all the safety features of a road. The auditors will audit most of the road design elements manually. The available software may be used to check the plan and profile, sight distance and vehicle tracking. The earliest tool that has been introduced is the use of RSA prompt list or checklists. Later, several RSA toolkits exist and in recent years, they are available on web-based applications.

The checklists are given within the guidelines and aimed to help the auditors when conducting the safety auditing. The checklists are used as a prompt to help auditors identify potential safety issues in a design. They are particularly useful for less-experienced auditors.[5] Basically, checklists contain a sequence of questions to remind the auditors on items needed to be checked. They may provide general questions and should be associated with the project road. Thus, the auditors should keep the checklists in mind when vetting through the design drawings or while driving throughout the project road. This is to identify the deficiency in road safety aspects of the project road.

The availability of RSA checklists is found very helpful by the parties involved in planning, design and construction of a project road with RSA requirement. This could minimize the overlooks during the road design and construction, in particular for those related to safety features. However, the available checklists may not specify their relation to accident severity. While knowing how severe accident occurs if the particular road element not properly designed is essential. This will also assist the auditors when advicing the road authority on whether the oversight is accepted or should be improved later at operational stage or must be rectified during the design stage.

1.3. More comprehensive rsa tool is required

There are reports on evidence of crash reduction due to the improvement or treatment on the safety feature of a road [6-8]. The crash reduction effectiveness of any treatment on the existing road is very useful, particularly for road safety auditors. The data could be the basis for them to define whether the risk level of the deficiency is intolerable, high, medium or low. Austroads and iRAP have developed the web-based RSA toolkit that deals with crash reduction effectiveness. Although the toolkit does not specify the common oversights during a RSA, it can be utilised as an advanced tool by a road safety auditor. How serious the unsafe road features are after they have been implemented can be identified based on the information provided.

Studies on RSA, however, have focused on how to conduct an auditing on road elements, attempt to improve the safety audit procedures or conduct the before and after treatment measurements. For example, the safety audits conducted for intersection/interchange, road bends, pedestrian zone and school area. Nevertheless, an adequate RSA tool is needed especially by the less-experienced auditors. Moreover, improvement on the RSA has been outlined since the beginning and being one of the communiqué points during the Austroads International RSA Forum in 1998. The forum also stated that research into the processes, methods and effectiveness of the RSA must be encouraged and supported [9].

1.4. RSA checklists and crash reduction

Several researchers have investigated and published the crash reduction effectiveness of the improved road design and road equipment. This involves the crash reduction due to the road or traffic countermeasures developed [6-8]. Based on those data, the correlation between audit items (if they complied with the safety requirements) and the crash reduction effectiveness was obtained.

RSA Guidelines [4] provide the main audit items within its checklists for Stage 1 audit (planning and feasibility stage). The checklists may assist the auditors to identify any potential safety issues and not overlook something important. The auditors often include the audit items as the sub chapter titles of their RSA report. At the moment, similar to other RSA guidelines, the checklists are not associated with the crash reduction data if the items complied with the safety requirements. Thus, auditors may not know how important an audit item to be fulfilled.

- Road Network Effects.
- General Geometric Standards.
- Outline Provision for Users with Special Needs.
- Access Control Details.
- Environmental Considerations.
- Consideration of Alternatives.

Correlating the audit items in RSA Guidelines [4] and the available information provided by Ogden [6], Austroads [7] and Elvik [8], it was found that the treatment on facility for vulnerable road users provides up to 60% crash reduction. The treatment on road networks, geometry, intersection and design standard adopted achieved the crash reduction up to 40%. While the treatment on lane arrangement at intersection and the traffic management provide up to 25% crash reduction. This shows that a treatment for vulnerable road users' facility will result in the highest crash reduction. Overall, at least about 25% of crash reduction achieved if the RSA properly carried out.

1.5. Exploring oversights on safety features in road design

Oversights in RSA may result in hazardous location on a road being left unidentified. Property damaged only, serious injury or even casualty may occur frequently due to the hazardous road section. An experienced safety auditor may have good knowledge to identify the potentially unsafe road feature in the design drawings as well as on the existing road. The less-experienced safety auditor, however, may need to give more attention during the audit to avoid the oversights. Although the checklists for auditing a road are available in the guidelines, knowledge on the common oversights leading to unsafe road feature, particularly for those causing severe accidents is needed.

The experience in auditing a total of 24 road projects from 1998 to 2008 was referred and summarized. All the sample projects are road construction projects and not the roadside development projects. As shown in figure 1, 67% of the reviewed road projects are the upgrading projects while the rests (33%) are new road projects. Moreover, 79% of the projects are situated in rural area while 21% in urban area. The audit summary of each sample road project was analysed. The analysis, however, is limited to audit conducted for the planning and feasibility stage, the first stage conducted for a road project. Nevertheless, poor auditing during this stage may cause the requirement for major redesign or rectification works during or after road construction. The unsafe road feature may even exist when the road is in operation due to the oversights during the first audit stage.

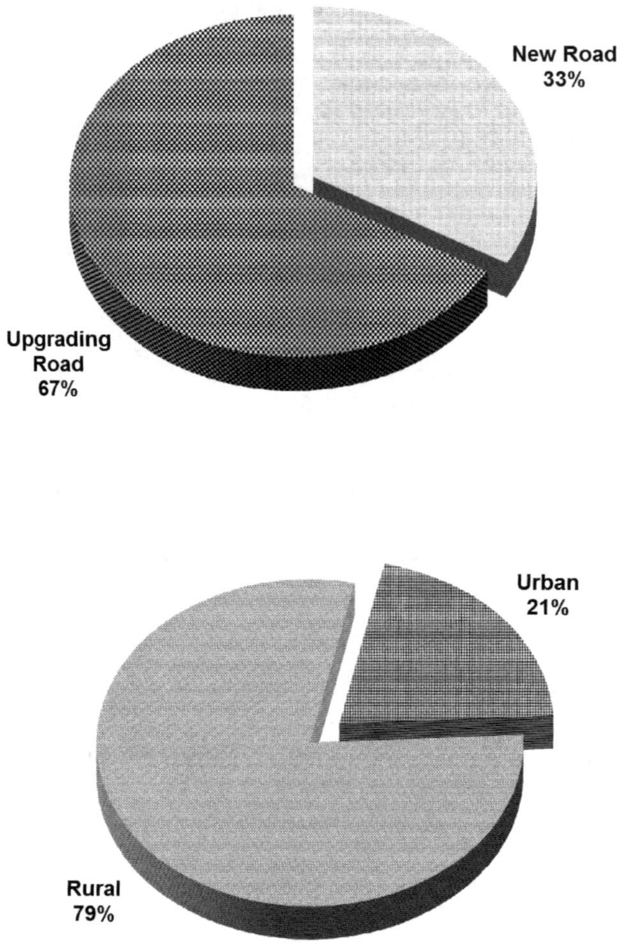

Figure 1. Types and areas of the 24 road projects reviewed in this study.

The RSA conducted for the sample road projects is based upon the audit items and checklists in RSA Guidelines [4]. Every audit item in all project reports was checked. This was to determine whether the design, in particular the audit item complied with the standard and/or accepted by the auditors. Then the number of the audit items accepted or refused by the auditors was counted. Having completed the descriptive statistics, the order from the most to the least unacceptable item was obtained.

1.6. Common oversights leading to unsafe road features

Figure 2 shows the percentage of road projects with unsafe features identified in this study. Most of the road projects were unacceptable by the road safety auditors due to lack of provisions for vulnerable road users. It can be seen in figure 2 that up to 92% of the road projects did not provide adequate provision for vulnerable road users. Thus, the road design was unacceptable to the auditors and recommended to be improved. The second most common audit item that was not acceptable in the road projects is the appropriateness of cross-section and traffic planning strategy. Out of 24 road projects, 16 projects (63%) were identified to be potentially hazardous due to inappropriate cross-section and traffic planning strategy.

Moreover, the location and type of control of the intersection at 10 road projects (42%) had received comments from the auditor as they did not fulfill the safety requirements. Surprisingly, about 33% of road projects received comments from the auditor in terms of road geometry and standard adopted, whereas road geometric standards should be ensured by road engineers. The other audit items such as suitability of design criteria, access control, location of terminal ends of the project, lane and carriageway arrangement, and road capacity were not fulfilled by less than 30% of the project.

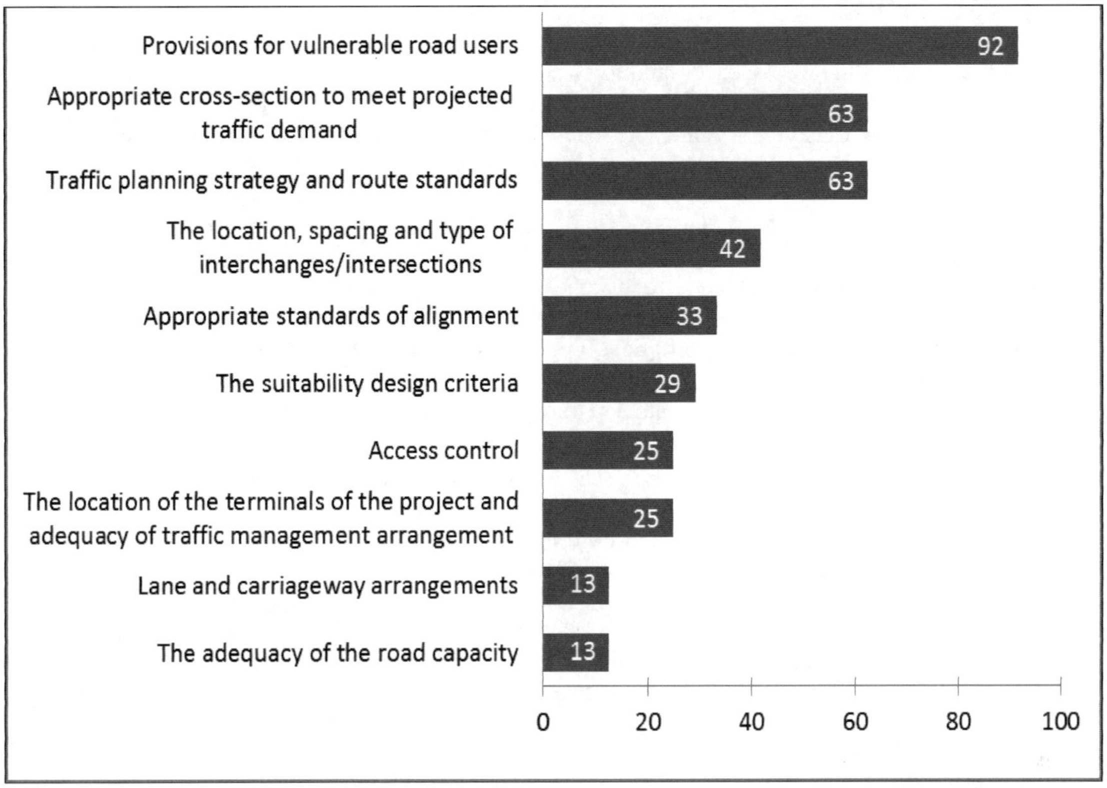

Figure 2. Percentage of road projects with unsafe feature due to unacceptable design.

As mentioned earlier, the treatment in the provision for vulnerable road users has the highest crash reduction effectiveness i.e. up to 60%. However, this item received the most comments by the auditors. The lack of information for the location where the vulnerable road users' activities exist is the main possible explanation to this finding. Of the 24 road projects, it was noticed that six projects (21%) do not have a comprehensive traffic study report. The other projects only quoted traffic data at the nearest traffic census stations. Unfortunately, the available traffic data at traffic census stations was only for mid-block segments of the road network. There were no turning volumes or pedestrian data available.

The inadequate paved shoulder, inadequate clear zone, and median width received the most comments from the auditors for the cross-section design. In practice, provision of adequate paved shoulder will help the drivers of the run-off-road vehicle to redirect their vehicles to the travel lanes. Adequate paved shoulder can also be used by the motorcyclists and breakdown vehicle whenever necessary. Proper physical median shall be used by pedestrians as their refuge when crossing the road.

Traffic planning strategy and the adopted route standard that may cause an increase traffic conflicts often receive comments or refusal from the auditors. For example, many road projects upgraded the trunk roads that traverse through the residential and commercial areas to a higher standard. The activities of vulnerable road users were relatively high within the area. After the construction, the road provides higher operational speed due to better alignment and formation. Through traffic would have the priority rather than local traffic as well as the vulnerable road users. Road accident or the severity of accidents may increase as more high-speed traffic passes through the community areas. This concept is not desirable and thus, bypassing the community areas is recommended by the auditors.

Designing suitable intersection location and spacing may be a problem for road engineers, particularly for a dual carriageway road. Junctions that are too close to one another will result in inadequate weaving section, while junctions that too far between one another may cause difficulty for local residents' movements. In many cases, the provision of a service road is 11.

2. Conclusion

Common oversights in road design that lead to unsafe road features have been identified and shown as follows. These oversights are based on auditors' comments and findings in the 24 road projects in Malaysia from 1998 to 2008.

1. Lack of or poor provision for vulnerable road users, i.e. provisions for pedestrian and cyclist. The road engineers often focus on the provisions for motorists and overlook to facilitate the non-motorized;
2. Inadequate paved shoulder, clear zone, and median width. This may be due to site or budget constraints;
3. Preferring to upgrade the existing road rather than construct a new road even though the road traverse through the community areas;
4. Design for intersections is inappropriate. Often this is due to the absence of a comprehensive traffic study report; and
5. Road standard chosen is inappropriate. Upgrading the existing road is often meant to increase the vehicles' speed.

References

[1] Austroads 2002 *Road Safety Audit* 2nd Edition (Sydney: NSW)
[2] FHWA 2006 *Road Safety Audit Guidelines* (Washington DC: USA)
[3] IHT 2008 *Road Safety Audit* (London)
[4] Public Works Department, Ministry of Works (JKR) 1997 *Road Safety Audit - Guidelines for the Safety Audit of Roads and Road Projects in Malaysia. Kuala Lumpur*
[5] Morgan R 2005 Road safety audits – practice in Australia and New Zealand *ITE J.* 22 – 5
[6] Ogden K W 1996 *Safer Roads - A Guide to Road Safety Engineering. Avebury Technical* (Aldershot: Ashgate Publishing Limited, England)
[7] Austroads *Road Safety Risk Assessment Project* (Sydney: NSW) 2002-7
[8] Elvik R, Hoye A, Vaa T and Sorensen M 2009 *The Handbook of Road Safety Measures* 2nd Edition (Emerald Group, United Kingdom)
[9] Jordan P 1999 ITE and road safety audit – a partnership for traffic safety *ITE J.* 24 – 7

IOP Conf. Series: Materials Science and Engineering **933** (2020) 012050 doi:10.1088/1757-899X/933/1/012050

The effect of water cement ratio (WCR) on compressive strength of interlocking bricks with mix design variations

N Malahayati[1], Y Hayati[1], Mahlil[1], D Sundary[1] and F Maulina[1]

[1] Department of Civil Engineering, Universitas Syiah Kuala, Banda Aceh, Indonesia

Corresponding e-mail: malahayatijj@unsyiah.ac.id

Abstract. Interlocking brick is a material used for wall work in home construction. Called Interlocking Brick or Brick Interlock because the method of arranging the bricks is interlocked with one another. This interlocking brick itself began to be known in Aceh Province after the 2004 tsunami disaster. Interlocking brick is a block of compacted soil, made of cement, silt soil, and dried sand, measuring 300 × 150 × 100 cm^3. Interlocking brick can function as a wall that can withstand structural and non-structural components. The purpose of this research is to get the compressive strength value of interlocking bricks designed with variations in the water cement ratio (WCR) so that the optimum compressive strength of interlocking bricks is obtained. The WCR used was 0.3; 0.4; and 0.5, while for the composition of cement, silt, and sand the ratio of 1:2:2 is used. The making of test specimens for testing the compressive strength of interlocking bricks is by cutting the interlocking bricks into cubes measuring 50 × 50 × 50 mm by 5 specimens for each variation so that the total of the total specimens are 45 specimens. Testing is done at the age of 7, 14, and 28 days. The results of the compressive strength of interlocking bricks using WCR 0.3 average compressive strength at the age of 7 days reached 8.10 MPa, 14 days reached 9.49 MPa and 28 days reached 10.94 MPa. For WCR 0.4, compressive strength on average at 7 days reaches 5.69 MPa, 14 days reaches 7.61 MPa and 28 days reaches 9.77 MPa. And for WCR 0.5 the average compressive strength at the age of 7 days reaches 5.36 MPa, 14 days reaches 8.25 MPa and 28 days reaches 9.93 MPa. From the compressive strength test results the interlocking brick with a WCR variation of 0.3; 0.4; and 0.5 with a mixture of 1:2:2 can be used as wall work material in house construction for structural construction and the optimum compressive strength value is obtained at WCR 0.3.

1. Introduction

The wall is one part of a nonstructural building which is generally categorized as a burden on a building. Determination of the wall as a non-structural part of the Indonesian National Standard (SNI) makes the wall not counted as a structural component in planning. However, technological developments in the field of wall materials continue to increase so that the wall material can also function as a structural substitute for sloof, column, and ringbalk. The wall material used is interlocking brick. Interlocking brick is a pair of walls that are installed by the method of interlock, locked, or interlocked. This brick is made of cement, silt, and sand added with water, then printed using a manual press and maintenance by splashing water until it reaches the concrete age of 28 days. In the last two decades, the development of interlocking wall material continues to increase because of this the method of implementing the installation can save time and does not require special labor [1]. Using interlocking bricks can save time and costs when compared to using red bricks because these interlocking bricks do not need to be plastered and painted because of their attractive appearance [2]. For this reason, several innovations in construction materials with interlocking masonry continue to be developed so that these materials become environmentally friendly, sustainable, and affordable costs [3-7]. Problems associated with interlocking bricks are low strength, higher water absorption, low fire resistance, and high porosity [8].

To be able to make interlocking brick products according to the planned quality, it is suggested that the concrete composition should be determined through an analysis which in concrete technology is called concrete mix design. One step is to determine the WCR value which is the ratio between the amount of free water and the amount of cement used in a concrete mixture. This WCR value can be predicted by using a curve that connects the compressive strength of concrete and WCR. The smaller the value of the WCR is used, the better the concrete strength will be. Concrete mixes that use a large WCR value, will require less cement paste, conversely, concrete mixes that use a small WCR value, will require more cement paste.

Interlocking brick is included in the type of concrete brick because it is made from the main material of portland cement, water, aggregate which is used for wall pairs. Concrete brick is divided into solid concrete brick and hollow concrete brick. Interlocking bricks are included in hollow concrete bricks because bricks have a cross-sectional area of more than 25% of the cross-sectional area of the brick and hole volume of more than 25% of the total boundary volume. Solid or hollow concrete bricks are distinguished according to their quality level, which is the quality level I for construction which carries loads and is unprotected (outside the roof) and has an average gross compressive strength of at least 7 MPa, quality level II for construction which bears the burden but its use is protected from external weather and has an average gross compressive strength of 5 MPa, quality level III and IV only for constructions that do not carry loads and are protected from rain and sun which have an average gross compressive strength of 3.5 MPa and 2 MPa [9].

In general, industries that produce wall-filling materials do not distinguish their production from wall functions as structural or non-structural and tend classified to be only one quality. This interlocking brick is produced using molding through the compaction process. The quality of interrelated bricks products is influenced by many factors including the ratio of the mixture between the forming materials. The purpose of this study is to determine the compressive strength of interlocking bricks that are designed with WCR type and material composition so that the optimum and minimum compressive strength of interlocking bricks is obtained.

2. Methodology

This research was conducted in several stages of work. Starting from the preparation of materials, examination of materials, the mixture plan is continued with the manufacture of test specimens, maintenance of test specimens, and test specimens. All work is carried out based on the applicable rules/ standards. Material inspection is limited to certain materials that are important in mixture calculations. In this study, the testing and examination of the physical properties of the aggregates refer to the Indonesian National Standard (SNI). The mixture planning carried out in this study uses a mixture used in interlocking brick production industries, namely by a mixture composition of 1:2:2 consisting of cement, silt and sand and WCR 0.3, 0.4, and 0.5. The interlocking brick forming material comes from Aceh Besar District which is used by the industry in producing it.

Preparation of test specimens was carried out in the interlocking brick production industry located in Mieruk Lamreudeup Village, Baitussalam District, Aceh Besar Regency, Indonesia. This industry produces interlocking bricks using manual brick machines, called Soeng Thai BP-6 compactor models that produce interlocking bricks with sizes of 100 mm x 150 mm x 300 mm. After interlocking bricks reach the age of 7, 14, and 28 days, they will be cut using concrete cutting tools which are carried out at the Structure and Building Materials Laboratory at the Faculty of Engineering, Syiah Kuala University as a cube-shaped test object with a size of 50 mm x 50 mm x 50 mm as per the compressive strength testing standards for cement paste. However, in this study not all of these measurements were achieved due to difficulties in the process of cutting test specimens, so in this study, 45 specimens were obtained consisting of 5 specimens for each WCR and with interlocked brick ages reaching 7, 14, and 28 days.

Compressive strength testing in this study used a compressive strength universal compression testing machine with the following steps: placing the test object in a centric position in the compressive test instrument; run the pressure test until the test object is destroyed; record the maximum amount of crushed load acting on the test specimen printed on the monitor of the pressure test device.

Compressive strength is the ability of concrete to accept the compressive force of broad unity. Whereas based on SNI the compressive strength is defined as the magnitude of the wide unity load, which causes the concrete specimens to be destroyed when burdened with certain compressive forces produced by the press machine. Compressive strength (kg/cm^2) is defined as the Force P (kg) divided by the cross-sectional area of test specimen A (cm^2). Many factors affect the strength of concrete including the effect of the concrete constituent material itself consisting of cement, silt soil, aggregates, water, conditions at work, conditions during treatment, caping, and testing processes that also greatly affect the strength of concrete samples made. For concrete compressive strength, measurements refer to the ASTM C-133-97 standard.

3. Results and discussion

The compressive strength test of interlocking brick specimens with cubes measuring 50 × 50 × 50 mm is carried out at the age of 7 days, 14 days, and 28 days. The results of the average compressive strength test of mortar with variations of WCR 0.3, 0.4, and 0.5 with a mixture of 1:2:2 are shown in tables 1, 2, and 3 below.

Table 1. Compressive strength of interlocking brick at 7 days old with a mixture of 1:2:2.

WCR	Sample	Dimension			Weight	Load	Compressive Strength	Average Compressive Strength
		Length	Width	Height				
		(mm)	(mm)	(mm)	(g)	(kgf)	(MPa)	(MPa)
	1	57.75	54.11	54.09	327.50	2420	7.60	
	2	54.90	50.03	53.46	273.90	2500	8.35	
0.3	3	54.84	55.11	52.14	308.10	2440	8.37	8.10
	4	53.89	52.86	51.84	283.50	2360	8.28	
	5	51.73	53.54	50.37	275.30	2100	7.90	
	1	54.60	51.66	52.49	273.40	1640	5.61	
	2	56.35	52.80	51.06	283.00	1740	5.93	
0.4	3	53.98	50.89	52.23	265.20	1460	5.08	5.69
	4	50.82	54.41	51.48	266.00	1620	6.07	
	5	54.91	52.92	50.87	267.70	1640	5.76	
	1	52.82	51.35	54.22	259.20	1640	5.62	
	2	51.84	55.23	51.06	260.60	1600	5.93	
0.5	3	52.53	51.42	52.41	266.90	1480	5.27	5.36
	4	50.62	52.68	52.19	265.30	1280	4.75	
	5	53.44	53.51	51.36	261.60	1460	5.22	

Table 2. Compressive strength of interlocking brick at 14 days old with a mixture of 1:2:2.

WCR	Sample	Dimension			Weight	Load	Compressive Strength	Average Compressive Strength
		Length	Width	Height				
		(mm)	(mm)	(mm)	(g)	(kgf)	(MPa)	(MPa)
	1	54.14	52.30	51.42	273.30	2860	10.07	
	2	54.59	52.75	49.92	271.80	2900	10.44	
0.3	3	53.54	50.60	52.25	269.40	2440	8.55	9.49
	4	52.49	54.23	52.00	271.10	2440	8.77	
	5	52.39	54.21	51.65	264.10	2660	9.64	

	1	52.68	50.61	52.27	245.00	1880	6.70	
	2	52.24	53.90	49.67	238.20	2360	8.92	
0.4	3	51.77	53.54	52.87	262.30	1960	7.02	7.61
	4	52.41	50.96	53.07	244.10	2140	7.55	
	5	52.39	51.99	55.74	278.40	2340	7.86	
	1	53.42	51.40	53.79	269.10	2400	8.19	
	2	53.13	54.30	51.35	265.60	2020	7.26	
0.5	3	51.35	52.50	55.09	270.30	2340	8.11	8.25
	4	50.95	52.49	55.21	269.30	2660	9.27	
	5	52.76	52.76	50.42	254.70	2280	8.41	

Table 3. Compressive strength of interlocking brick at 28 days old with a mixture of 1:2:2.

WCR	Sample	Dimension			Weight	Load	Compressive Strength	Average Compressive Strength
		Length	Width	Height				
		(mm)	(mm)	(mm)	(g)	(kgf)	(MPa)	(MPa)
	1	51.47	54.13	51.48	267.70	3080	11.40	
	2	54.87	54.14	51.11	281.50	2880	10.07	
0.3	3	51.24	51.22	52.69	262.80	3180	11.55	10.94
	4	52.28	52.60	52.89	269.10	3060	10.85	
	5	52.05	51.92	53.98	275.40	3100	10.82	
	1	53.58	50.29	54.86	259.50	2620	8.74	
	2	54.51	54.36	52.13	271.30	2920	10.08	
0.4	3	51.78	52.54	59.93	272.70	3480	11.00	9.77
	4	52.51	55.55	52.13	272.00	2780	9.96	
	5	52.95	51.07	51.80	248.90	2540	9.08	
	1	54.05	53.70	50.17	266.40	3080	11.14	
	2	50.50	52.85	53.17	262.30	2920	10.66	
0.5	3	53.54	51.50	53.26	265.10	2640	9.08	9.93
	4	50.81	52.59	53.89	265.50	2800	10.03	
	5	54.08	49.97	53.97	259.40	2600	8.74	

From the results of the compressive strength test of the interlocking brick, it can be seen the decrease in compressive strength along with the magnitude of the WCR value, and the longer the life of the concrete, there will be an increase in the compressive strength of the interlocking brick. For WCR 0.3, compressive strength averages at 7 days 8.10 MPa, 14 days 9.49 MPa, and 28 days 10.94 MPa. For WCR 0.4, compressive strength averages 7 days 5.69 MPa, 14 days 7.61 MPa, and 28 days 9.77 MPa. And for WCR 0.5 the average compressive strength at the age of 7 days is 5.36 MPa, 14 days 8.25 MPa, and 28 days 9.93 MPa.

In the mixture of WCR values of 0.4 and 0.5, the mixture of interlocking brick material is difficult to do (workability) so the compaction is not perfect. As a result, the specimen becomes hollow which causes the compressive strength of the concrete to decrease. In general, the higher the WCR value, the lower the concrete quality will be, but the lower WCR value does not always mean the concrete strength will be higher.

The highest compressive strength of interlocking bricks is 10.94 MPa at WCR 0.3 at 28 days and the lowest compressive strength is 5.36 MPa at WCR 0.5 at 7 days. The results of the compressive strength test of the interlocking brick with a WCR variation of 0.3; 0.4; and 0.5 with a mixture of 1:2:2, all specimens meet the mechanical requirements of SNI 03-0349-1989 with the quality level I can be used as construction material in wall work for construction that is burdened and unprotected and the compressive strength planned is fulfilled as hollow concrete brick material with a compressive strength of not less than 7 MPa.

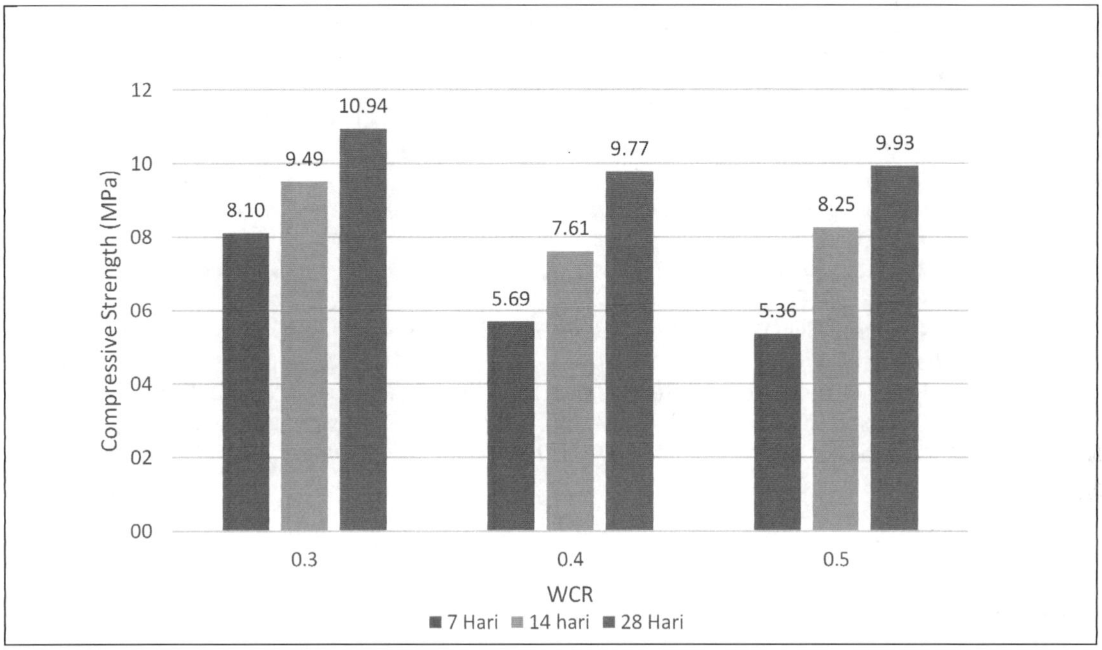

Figure 1. The graph of compressive strength relationship at 7 days, 14 days, and 28 days with a mixture of 1:2:2.

Interlocking brick can be used as a nonstructural and structural wall material such as replacing sloof, column, and ringbalk. The results showed that interlocking bricks produced included in the level of quality I so that when used for non-structural walls is very cost-saving. Interlocking bricks need to be re-tested using different mix compositions so that we get interlocking bricks with different functions, namely that there are constructions that can carry loads and do not carry loads.

4. Conclusion

Based on the results of the study it can be concluded that the highest compressive strength of interlocking bricks at WCR 0.3 is the concrete age reaching 28 days. Interlocking brick compressive strength test results found that all the test specimens with a mixture of 1:2:2 and with variations of WCR 0.3; 0.4; 0.5 meets the mechanical requirements of SNI 03-0349-1989 with a quality level I.

This research needs to be developed further by using other types of materials with different mixture designs so that an effective and efficient mixture is obtained.

Acknowledgments

The research reported in this paper was supported in part by the Institute for Research and Community Services (LPPM) of Syiah Kuala University based on Contract No. 95/UN11.2/PP/SP3/2019. The author wishes to acknowledge that any opinions, findings, conclusions, and recommendations presented in this paper are the property of the author and do not reflect the views of the sponsors.

References

[1] Al-Fakih A, Mohammed B S, Nuruddin F and Nikbakht E 2018 Development of interlocking masonry bricks and its' structural behaviour: a review paper *IOP Conf. Series: Earth and Environ. Sci.* vol 140 no 1 p 012127 (London: IOP Publishing)

[2] Malahayati N, Hayati Y, Nursaniah C and Firsha T 2017 Comparative study on the cost of building public house construction using red brick and interlock brick building material in the city of Banda Aceh *The 7th AIC-ICMR on Sciences & Eng.: Annual Intern. Conf.* 18-20

[3] Wheeler G 2005 Interlocking compressed earth blocks volume II *Manual of Construction. Center for Vocational Building Technology* (Thailand) 110

[4] Adedeji Y M D 2008 Interlocking masonry: Panacea for sustainable low-cost housing in Nigeria *Pakistan J. of Social Sciences* **5** 8 744-50

[5] Calkins M 2008 *Materials for sustainable sites: a complete guide to the evaluation, selection, and use of sustainable construction materials*

[6] Raheem A A, Bello O A and Makinde O A 2010 A comparative study of cement and lime stabilized lateritic interlocking blocks *The Pacific J. of Science and Techn.* **11** 2 27-34

[7] Assiamah S, Abeka H and Agyeman S 2016 Comparative Study of Interlocking and Sandcrete Blocks for Building Walling Systems *International Journal of Research in Engineering and Technology* **05** 1

[8] Irwan J, Zamer M and Othman N 2016 A Review on interlocking compressed earth blocks (ICEB) with addition of bacteria *MATEC Web of Conf.: EDP Sciences* pp 0-5

[9] Badan Standarisasi Nasional 1989 *SNI 03-0349-1989 Bata Beton Untuk Pasangan Dinding.* (Jakarta: BSN)

IOP Conf. Series: Materials Science and Engineering **933** (2020) 012051 doi:10.1088/1757-899X/933/1/012051

The effect of diameter and distance between vegetation in the riverbank to flood discharge by Eco-hydraulic in the Lawe Alas River, Southeast Aceh, Indonesia

Ziana[1], M Rizalihadi[1], Azmeri[1] and A Nugraha[1]

[1] Department of Civil Engineering, Universitas Syiah Kuala, Banda Aceh, Indonesia

Corresponding author: ziana@unsyiah.ac.id

Abstract. One of the eco-hydraulic methods used to reduce flooding and erosion in rivers is by vegetation along the riverbanks. The main problem in the Lawe Alas River is repeated flooding in the downstream areas that need to be addressed upstream. The purpose of this study is to design the proper vegetation diameter and the distance between plants that can reduce the velocity and flow of the river. The method of study in this research consists of conducting field surveys, measuring river discharge, and calculating vegetation diameter on riverbanks to reduce velocity and discharge based on the equation proposed by Merten. The planned diameter of vegetation with diameter is 0,1 m – 1,1 m, and the distance between vegetation 1 m - 11 m. The results showed that the maximum discharge capacity of the Lawe Alas River was 514,360 m^3/s. For the return period of 50 years, 50 m from the riverbank design, vegetation with a diameter of 10 cm to 110 cm can reduce flooding in the resulting discharge 222,95 m^3/s – 241,68 m^3/s. Discharge is greater than 50 yearly flood discharge, which is 528,670 m^3/s. In conditions without vegetation (diameter = 0) velocity = 1,725 m/s, the velocity produced by the design is 0,727 m/s – 0,789 m/s. Discharge and velocity are reduced by 54% - 58% so that the vegetation diameter of 10 cm is suitable for environmentally friendly designs on the riverbanks 50 m. In conclusion, all planned vegetation diameters can reduce discharge and velocity. Recommended vegetation is local vegetation, which is a tree that produces fruit that can be benefited to the community so that it can be planted and managed by local residents.

1. Introduction

Kutacane City is one of the cities in Aceh Province which when the rainy season is often hitting by flooding. Flood disasters often occur due to the overflow of river water. One of the rivers in Kutacane City is the Lawe Alas river. The Lawe Alas River is a large river that crosses the Gunung Leuser National Park, and this river is located along the Gayo Lues District, Southeast Aceh District, and South Aceh District so that the Lawe Alas River is in the Alas-Singkil River Area Unit (SWS). The Lawe Alas River is locating at 03^0 29' LU - 03^0 59' LU and 97^{0}10 'BT - 97^{0}45' BT. The Lawe Alas River is often hit by floods, with the most affected areas being the Bambel District and Babul Makmur District. Based on information obtained from [1], land belonging to Bambel residents who were damaged and failed to harvest reached 20 hectares, due to flash floods, 15 houses were damaged, and hundreds of houses were submerged. To overcome the flooding, the local government built structural buildings, namely gabions on the banks of the river. The function of making the gabion is to maintain the banks of the river against the flow of water flow and business to keep the flow of water flow from the banks of rivers that damage cliffs (erosion) and as flood control.

IOP Conf. Series: Materials Science and Engineering **933** (2020) 012051 doi:10.1088/1757-899X/933/1/012051

The purpose of this study is to design an eco-hydraulic riverbank area to cope with flooding in the Lawe Alas River by optimizing the riverbank width and the selection of vegetation diameters that can reduce flood discharge. The scope of this research is to determine the eco-hydraulic river design that can accommodate 50-year flood discharge.

The methods study in this research consist of field survey, literary study, measurement cross hydrometric river, and eco- hydraulic study. The benefit of this research is to be able to know the condition of river discharge and the maximum capacity of river basins based on water level elevation. We can find out the eco-hydraulic river design that is right on the Lawe Alas River so the river design can reduce flood energy towards the downstream. The research location can see in figure 1.

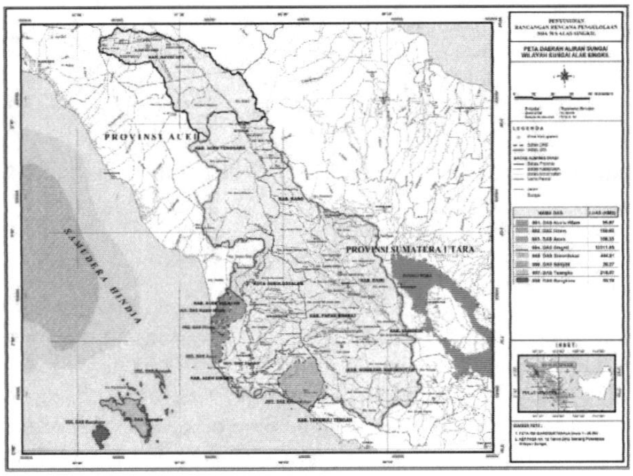

Figure 1. Map of the Lawe Alas-Singkil watershed.

In this study, the Melchior method was used to calculate flood discharge because it has a watershed area of more than 100 km^2. Data needed are river hydrometry data, daily rainfall data, and topographic maps. River hydraulic data is obtaining from measurements for seven days in the field, such as measurements of river depth, river width, flow velocity, and river water level. Rainfall data were analyzed using statistical parameters which can then be determined the appropriate type of distribution between Normal distribution, Log-Normal, Gumbel, and Pearson Log.

2. Literature review

The flood is affected by the geomorphometric of watershed [2]. According to [3], flood control is generally carried out with pure hydraulic concepts, namely increasing physical or structural development such as embankment making, river diversion, dredging of river basins, making riverbanks, and others. Be aware that this is an improper step in applying the eco-hydraulics concept. Flood control with physical development causes a decrease in river retention, an increase in river slope, increasing flow, peak flow discharge and decreasing time to reach peak discharge resulting in an increase in flood retention in the downstream area of the river.

According to [4], the negative effect of structural development is the condition of the watershed will lose the habitat of flora and fauna around the river environment and change the characteristics of the shape of the river. What is including in pure eco-hydraulic? According to [5], pure hydraulics means that the problem-solving in the area of water is only essential on the hydrology function of the water area, without considering the negative impact and its relation to the existing ecological components [6]. According to the pure hydraulic concept, the river is only seeing as a hydraulic channel that discharges excess water towards the sea. This concept concludes that all rivers should be straightened or grounded so that water can quickly flow downstream.

With the rate of development of environmental awareness and the enormous negative impact of pure hydraulic engineering as mentioned above, a new interdisciplinary engineering pattern has been developed by including ecological considerations in solving water problems. This interdisciplinary

IOP Conf. Series: Materials Science and Engineering **933** (2020) 012051 doi:10.1088/1757-899X/933/1/012051

economic approach is seeing as a new approach that can be accepted from both ecological and hydro-civil engineering people and has high sustainability.

According to Pertiwi [4], vegetation planted on the banks of the river will reduce the velocity of water to the ground. By reducing the velocity of water in the river, the problem of flooding in the downstream area can reduce, and the natural conditions of the river can be maintained.

The development of this research has been entirely publishing in [4, 7-10]. According to [11], in natural rivers shaped close to the trapezoid, where on the banks of the vegetation are dense vegetation, there will be a wide interaction area and the process of losing energy due to the friction of the velocity between the faces. With this interaction area, a lower velocity reduction will occur. As a consequence, the water level will rise, and the flowrate capacity will decrease (for the same water level). The method of calculating the influence of vegetation in the riverbank area on water velocity and discharge is basing on the equation proposed by Merten in 1989 [7-11].

1. Coefficient of resistance

$$\frac{1}{\sqrt{\lambda}} = -2{,}03\, log\left(12{,}27\frac{R}{K_s}\right) \tag{1}$$

2. Roughness Value

$$Ks = Cb + 1{,}5dp \tag{2}$$

with:
C = vegetation composition coefficient,
b = riverbank width (m), and
dp = vegetation diameter (m).

3. Vegetation composition coefficient

$$C = 1{,}2 - 0{,}3(B/1000) + 0{,}06(B/1000)^{1{,}5} \tag{3}$$

$$B = \left[\frac{\alpha_x}{d_p} - 1\right]^2 \frac{\alpha_y}{d_p} \tag{4}$$

with:
B = vegetation parameter,
αx = transverse vegetation distance (m), and
αy = distance between vegetation in longitudinal direction (m).

4. Water velocity according to Darcy – Weisbach

$$V = (\frac{1}{\lambda}8gRI)^{0{,}5} \tag{5}$$

with:
λ = resistance coefficient,
g = gravity speed (m/dt^2), and
R = hydraulic radius (m).
I = river slope

3. Methodology

3.1. Field measurements

The research location is in the Badar Subdistrict area of Kutacane City, Aceh Province, with the primary location of the research being the upstream area of the Lawe Alas River. The study conducted in September 2018 to November 2018 which included the results of direct measurements in the field,

IOP Conf. Series: Materials Science and Engineering **933** (2020) 012051 doi:10.1088/1757-899X/933/1/012051

namely river depth measurements aimed at obtaining a cross-sectional shape of the river which can then calculate the cross-sectional area of the river. The width of the river is divided into several pieces and measured the depth of each stake using a weighted rope. Measurements of water level measurements measured by using wood placed at the edge of the river every day for seven days and the flow rate calculated for each different water level elevation. Flow velocity is measured using a float from the bridge. On the bridge the width of the river is measured:

1) measured the height of the bridge to the surface of the river;
2) tie afloat on a 50 m long rope;
3) the buoy place in the cross I (above the bridge) and is allowed to drift along the 50 m (cross II) rope;
4) measured t (float time) using a stopwatch; and
5) this work is done repeatedly until three times, and the time and results recorded.

River hydraulic data is obtained from measurements for 7 days in the field such as measurements of river depth, river width, flow velocity, and river water level, as presented in figure 2.

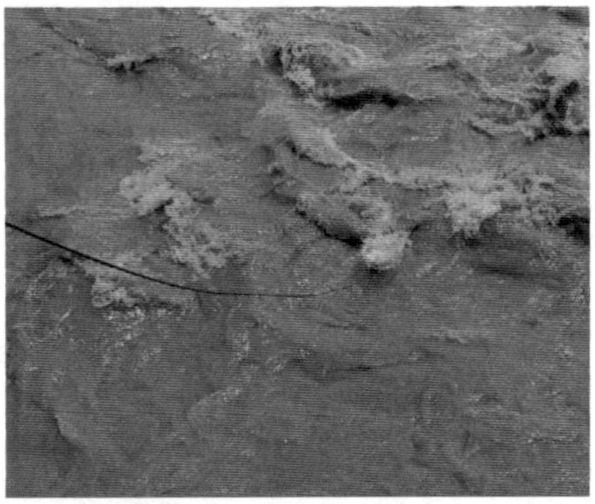

a. River Depth Measurement.

b. Measurement of River Width.

c. River Velocity Measurement

d. Water Level Measurement

Figure 2. River Cross Section Measurement.

Secondary data in the form of topographic maps of the Alas-Singkil River Area, Southeast Aceh Topographic Map, maximum daily rainfall data from 1980-2006 BPP Blang Keujereun station obtained from the Sumatra I-BWS, 2011 governance map obtained from the Sumatra-1 Aceh BWS, serves to determine the value of the coefficient runoff (C) based on the land cover of the Alas River Basin.

River discharge is calculated for several different river water level elevations so that a curve rating obtained. From the rating curve, the amount of flood discharge can be calculated based on the flood water level that occurs. The planned flood discharge analysis is to find out the magnitude of the discharge that occurs in the return periods of 2, 5, 10.25, and 50 years — calculation of the planned flood discharge using the Melchior Method.

IOP Conf. Series: Materials Science and Engineering **933** (2020) 012051 doi:10.1088/1757-899X/933/1/012051

3.2. Eco-hydraulics analysis

The hydraulic analysis aims to determine eco-hydraulics planning, namely in the form of riverbank designs. The steps to be carried out in the hydraulic analysis in this study are as follows:

1) Plan the width of the riverbank 50 m;
2) Vegetation planned with a diameter of 0,1 m, 0,2 m, 0,3 m, 0,4 m, 0,5 m, 0,6 m, 0,7 m, 0,8 m, 0,9 m, 1 m and 1,1 m with longitudinal and transverse spacing between 1 m, 2 m, 3 m , 4m, 5m, 6m, 7m, 8m, 9m, 10m and 11m;
3) Calculate the roughness value of the channel using equation (2);
4) Calculation based on various vegetation diameters consisting of diameters of 10 cm, 20 cm, 30 cm, 40 cm, 50 cm, 60 cm, 70 cm, 80 cm, 90 cm, 100 cm and 110 cm;
5) Calculating the water velocity values for various vegetation diameters using equation (5);
6) Calculate the flood discharge after the influence of vegetation planting; and
7) Comparing flood discharge before and after vegetation planting.

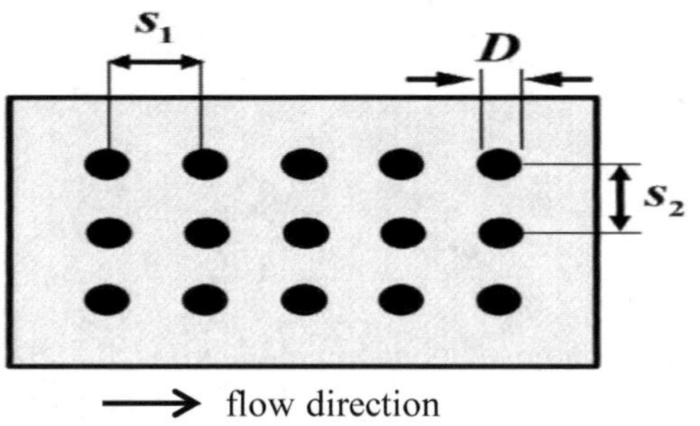

with: D= diameter vegetation

S_1= the distance between vegetation in the transverse direction (ax)

S_2= the distance between vegetation is elongated (ay)

Figure 3. Configuration of vegetation planting models on river banks.

4. Results and discussion

Based on measurements in the field, the total watershed area was obtained = 312,05 km^2 river length = 89,7 km and Slope = 0,0245. The following is a table of results of velocity and river discharge measurements for a week.

Table 1. Results of measurement of velocity and discharge (flow) for a week.

Tanggal	Velocity(V) (m/d)	Discharge (Q) (m^3/d)
09-Nov 2018	0,76	514,36
10-Nov 2018	0,67	441,55
11-Nov 2018	0,61	388,13
12-Nov 2018	0,56	346,02
13-Nov 2018	0,50	297,70
14-Nov 2018	0,46	260,48
15-Nov 2018	0,43	236,75

The relationship figure of vegetation diameter to roughness; figure of vegetation diameter to velocity; and figure of vegetation diameter to steam flow (discharge) are presented in figure 4, 5, and 6. Based on figure 4, it can be following that when the vegetation diameter is 0, the roughness = 0 after the vegetation is planting, the higher the vegetation diameter, the greater the roughness. Based on figure 5 it can be seen that in conditions without vegetation (diameter = 0), velocity = 1.725 m/s while after improvement of the banks with vegetation planting 0.1 to 1.1 m, the velocity decreases to 0.727 - 0.789 m/s, which indicates the occurrence decrease in velocity by 53 - 57%. Based on figure 6, at the time before the river bank structuring river discharge was 514.36 m^3/s, after the arrangement of the banks with vegetation river discharge decreased to 222,95 – 241,68 m^3/s, resulting in a decrease in discharge of 54 - 58% from before the riverbank was structured.

Table 2. Results of calculation of roughness parameters, obstacle coefficient, velocity and discharge several variations diameter distance between vegetation for return period 50 years and riverbank 50 m.

Vegetation Diameter	Space Horizontal	Space Vertical	Vegetation Parameter	Vegetation Coefficient	Rouhgness	Cross Section Area	Hydraulic radius (R)	Roughness Coefficient	Velocity (V)	$Q_{ekohidrauik}$
(m)	(m)	(m)	(B)	(C)	(Ks)	(m^2)	(m)	(λ)	(m/d)	(m^3/d)
0	0	0	0,00	0,000	0,000	306,500	2,881	0,000	1,725	528,67
0,1	1	1	810,00	1,001	50,187	306,500	2,881	10,468	0,727	222,95
0,2	2	2	810,00	1,001	50,337	306,500	2,881	10,292	0,734	224,85
0,3	3	3	810,00	1,001	50,487	306,500	2,881	10,121	0,740	226,74
0,4	4	4	810,00	1,001	50,637	306,500	2,881	9,954	0,746	228,63
0,5	5	5	810,00	1,001	50,787	306,500	2,881	9,792	0,752	230,51
0,6	6	6	810,00	1,001	50,937	306,500	2,881	9,635	0,758	232,39
0,7	7	7	810,00	1,001	51,087	306,500	2,881	9,482	0,764	234,26
0,8	8	8	810,00	1,001	51,237	306,500	2,881	9,333	0,770	236,12
0,9	9	9	810,00	1,001	51,387	306,500	2,881	9,187	0,776	237,98
1	10	10	810,00	1,001	51,537	306,500	2,881	9,046	0,782	239,83
1,1	11	11	810,00	1,001	51,687	306,500	2,881	8,908	0,789	241,68

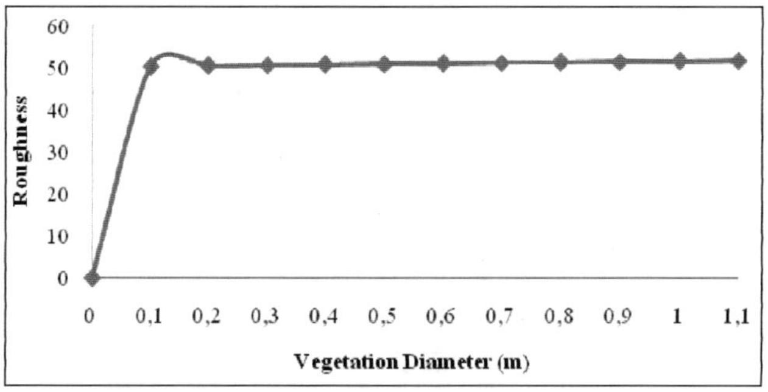

Figure 4. Relationship of Vegetation Diameter to Roughness.

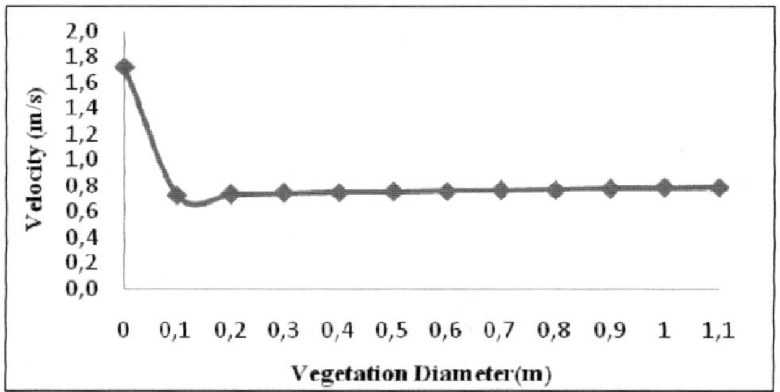

Figure 5. Relationship of Vegetation Diameter to Velocity.

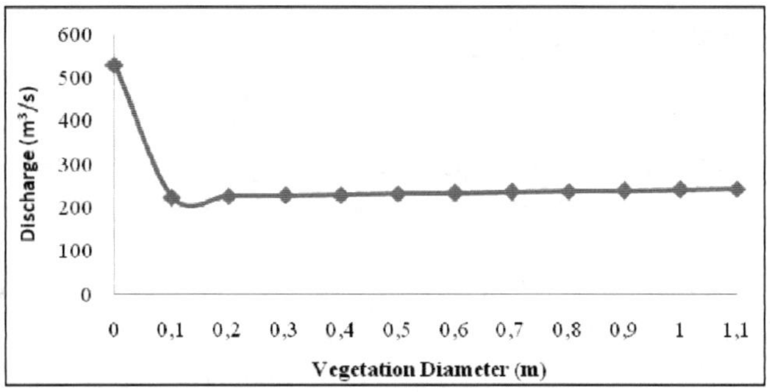

Figure 6. Relationship of vegetation diameter to streamflow (discharge).

5. Conclusion

Sungai Lawe Alas until the outlet has a length of 89.7 km with an area of the watershed is 312.05 km².The width of the Lawe Alas River is 90 m, and the cross-sectional area is 306,5 m^2.Rating curve shows the maximum discharge that can be accommodated by the river is 514,360 m^3/s. Vegetation diameter is 0, the roughness = 0, after the vegetation is planting, the higher the vegetation diameter, the greater the roughness. For return period of 50 years, 50 m of riverbank design, vegetation with a diameter of 1 m to 1,1 m can reduce flooding in the discharge produced by this design 222,95 m^3/s – 241,68 m^3/s. The discharge is higher than the annual 50 flood discharge, which is 528,670 m^3/s. The discharge is reducing 54% - 58% from before there is an eco-hydraulic backing arrangement. The velocity produced by design is 0,727 m/s – 0,789 m/s. The velocity reduces by 53% - 57% from before there is an eco-hydraulic bank arrangement so that the vegetation diameter of 10 cm is feasible for eco-hydraulic design on the riverbank of 50 m. In conclusion, all planned vegetation diameters can reduce discharge and velocity, so vegetation planting on the riverbank should be recommended. Recommended vegetation is local vegetation, which is a tree that produces fruit that can be benefited by the community so that it can be planted and managed by local residents.

Recommendation

To get a more accurate value of velocity, we should use better equipment such as current meter. In planning eco-hydraulics, planning should not be applied only to one area or region that experiences flooding. Instead, the river must be considered one, in the sense that planning must be carried out thoroughly starting from upstream to downstream. In implementing real-time eco-hydraulic planning and implementation in the field, there is a need for public and government awareness, understanding, and participation regarding these eco-hydraulic functions and objectives.

References

[1] https://www.goaceh.co/berita/baca/2017/11/12/sungai-alas-meluap-puluhan-hektare-sawah-dan-kebun-warga-terendam
[2] Azmeri A and Isa A I 2018 An analysis of physical vulnerability to flash floods in the small mountainous Watershed of Aceh Besar Regency, Aceh province, Indonesia *Jàmbá – J. of Disaster Risk Studies* **10** 1 1-6
[3] Maryono A 2014 *Menangani Banjir, Kekeringan, dan Lingkungan* (Yogyakarta: Gadjah Mada University Press, Indonesia)
[4] Pertiwi N 2011 *Pengembangan Model Pengelolaan Sungai Berbasis Pada Konsep Ekohidrolik* (Bogor: Institut Pertanian Bogor, Indonesia)
[5] http://worldfromeyes.blogspot.com/2010/01/kenali-eko-hidraulik.html
[6] Kodoatie R J 2013 *Rekayasa dan Manajemen Banjir Kota* (Yogyakarta: Andi Offset, Indonesia)
[7] Fauzi I N 2016 *Perancangan Ekohidrolika Untuk Pengendalian Banjir Pada Morfologi Sungai Non Simetris di Desa Manggasang, Sungai Barabai Kalimantan Selatan* (Bogor: Institut Pertanian Bogor)

[8] Sari J 2015 *Perancangan Eko-hidrolik Untuk Pengendalian Banjir Pada Morfologi Sungai Simetris* (Bogor: Institut Pertanian Bogor, Indonesia)
[9] Ulmi E I 2015 *Kajian Eko-hidraulik Sungai Martapura* (Banjar Baru: Universitas Lambung Mangkurat, Indonesia)
[10] Ziana, Azmeri and Lidya F 2018 Perbaikan bantaran sungai secara eko-hidraulik untuk menanggulangi banjir di sungai Lae Soraya Kota Subulussalam *Seminar PIT HATHI Ke XXXV* (Medan: Indonesia)
[11] Maryono A 2017 *Eko-Hidraulik Pengelolaan Sungai Ramah Lingkungan*, (Yogyakarta: Gadjah Mada University Press, Indonesia)

IOP Conf. Series: Materials Science and Engineering **933** (2020) 012052 doi:10.1088/1757-899X/933/1/012052

Study on the impact of canal blocking on groundwater fluctuation for tropical peatland restoration

S Sutikno[1], R Rinaldi[1], R A Putri[1] and G K Khotimah[1]

[1] Civil Engineering Department, Universitas Riau, Pekanbaru, Indonesia

Corresponding e-mail: ssutiknoyk@yahoo.com

Abstract. Peatland fires occur almost every year in Indonesia that might be caused by over-drained peatland due to canalization for a certain purpose. Canalization in the peatland leads to a significant decreasing change of groundwater level that might cause peatland very vulnerable against fire. The rewetting approach by the Peatland Restorations Agency of Indonesia using canal blocking is expected to keep groundwater level high so that the peatland is maintained in an always wet condition. Therefore, it is important to study the impact of canal blocking on the fluctuation of groundwater level in peatland ecosystems. This study was carried out at a research site which is located in Pulau Tebing Tinggi peatland hydrological unit, Kepulauan Meranti Regency, Riau, Indonesia. Groundwater level fluctuation as the impact of canal blocking was monitored using eight dipwells that were set with a distance of 1 m, 51 m, 101 m, and 201 m from the canal for each transect of the three transects. The result of this study showed that the canal blocking has a good impact for keeping groundwater level and keeping peatland on an always wet condition for peatland restoration and peat fire prevention until 400 m distance to the upstream area from canal block and about 1 m distance perpendicular to the canal. The canal blockings will have a good impact on rewetting at the peatland area up to 201 m distance perpendicular to the canal if the rise of water level at the canal is more than 0.6 m for peatland restoration.

1. Introduction

Peat is an organic material that has been formed from the weathering process of plant growth and has unusual characteristics. Because about half of the peat is carbon, peat is also a form of carbon sinks on the surface of the earth. They contain nearly 30% of all carbon on the land although covering only 3% of the land area [1]. It means that peatlands are the most efficient terrestrial ecosystems in storing carbon in the world. Most of the tropical peatland (approximately 65-75%) is found in Southeast Asia, particularly in Indonesia. Indonesia has the largest area of tropical peatland in the world with a total area of 20-21 Mha [2]. Because of the unappropriated management of those tropical peatlands in relation to massive utilization for various purposes, the environmental problems become a serious problem including peat fires, decreasing massive ecosystem functions, and increasing carbon emission [3].

In the last decade, most of the peatlands have been used as dried plantations where so many canals have been made in order to manage the land to dry. This process caused over-drained in the peatland that reduced the groundwater table elevation. Therefore, the peatlands ecosystems became dry and very easy to be burnt during the dry season. Peatland fire becomes the most popular issue in Indonesia since this case often occurs in almost every year. In 2014 and 2015, there were severe peatland fires in Riau, that caused severe haze disaster not only in Riau area but also most of the area in Sumatera and

Kalimantan. Forest and land fires in Indonesia in 2015 are estimated at about 1.7 million hectares, which about 35.9% and, 45.5% of them are in Kalimantan and Sumatra respectively. The peatlands fire in Indonesia has caused many problems, including the damage of ecosystems and the degradation of environmental conditions, human health, and socio-economic aspects of society [4]. Therefore, an understanding of the factors that can trigger the fire in peatland is very important for such worst disaster prevention. Severe tropical peatland fires are very difficult to extinguish because the incidence of peatland fires occurs below the surface that can only be extinguished in the presence of rainfall or artificial rainfall [5]. Therefore, the prevention of peatland fire has been established by the Indonesian government through the 3R approach, such as rewetting, revegetation, and revitalizations of livelihood that aims to restore the hydrological functions of the peatland areas.

The rewetting approach attempts to keep groundwater elevation high so that the peatlands are always in a water-saturated condition or high soil moisture. The approach that aims to rehabilitate the peatlands hydrologically to a near-natural state and to decrease the effect of greenhouse gas emissions is carried out by canal blocking, canal backfilling, and construction of deep wells [6], [7]. To understand the effectiveness of the canal blocking for the rewetting approach, therefore, it is important to study the impact of canal blocking on the fluctuation of groundwater level in peatland ecosystems. The initial research to investigate the effectiveness of the canal blocking has been studied [8]. This research was still limited to the one transects in the upstream of the canal block, so that the impact was not investigated along the upstream area of the canal. The fluctuation of groundwater lever becomes one of the essential parameters to understand the effectiveness of canal blocking on rewetting of the tropical peatland to prevent the fire.

The purpose of this study is to investigate the impact of canal blocking on groundwater fluctuation for tropical peatland restoration. The study was carried out at a research site which is located in Pulau Tebing Tinggi peatland hydrological unit, Kepulauan Meranti Regency, Riau Province, Indonesia. Groundwater level fluctuation as the impact of canal blocking was monitored using eight dipwells that were set with a distance of 1 m, 51 m, 101 m, and 201 m from the canal for each transect of the three transects. The groundwater level fluctuations were recorded automatically using the water loggers instrument with a one-hour time interval for every eight dipwells.

2. Methodology

2.1 *Study area*

This research was investigated in Pulau Tebing Tinggi Peatland Hydrological Unit (PHU), which is located in Kepulauan Meranti Regency, Riau Province (See Figure 1a). Pulau Tebing Tinggi PHU is one of the most priority areas for the peatland restoration program by Peat Restoration Agency, Republic of Indonesian (BRG-RI) due to the previous peatland fire and degradation (Anonim, 2016). The peatland area of Tebing Tinggi island is about 94% of the total area of the island. This condition makes Tebing Tinggi island very vulnerable to a peat fire. The investigation was carried out at a series of canal blocks of about 6 km canal at Lukun Village (See Figure 1b). Historically, the village experienced severe peat fires in the last few years.

2.2 *Experiment set up*

In order to analyze the fluctuations of groundwater level as the impact of canal blocking, eight dipwells were set around the canal with a distance of 1 m, 101 m, and 201 m from the canal for each transect of the three transects as presented at Figure 1b. The dipwells in the first transect are D 1-1, D 1-101, D 1-201, in the second transect are D 2-1, D 2-101, D 2-201, and in the third transect are D 3-1, D 3-101, D 3-201 with the distance of 1 m, 101 m, and 201 m from the canal respectively. The first transect is located in the most upstream of the canal, the second transect is in the middle, and the third transect is in the most downstream of the canal. There are two canal blocks in the study site that were constructed in November 2018. The first canal block is located about 100 m in the downstream side of the first transect and about 200 m upstream side of the second transect. The second canal block is located about 100 m in the downstream side of the second transect, and about 100 m in the upstream side of the third transect. Additionally, the rainfall data, air temperature and water depth were also considered as a parameter of the analysis that was taken from a real-time telemetry technology, namely SESAME (Sensory Data Transmission Service Assisted by Midori Engineering). The data were recorded every 10 minutes by the SESAME system are rainfall, air temperature, and water depth.

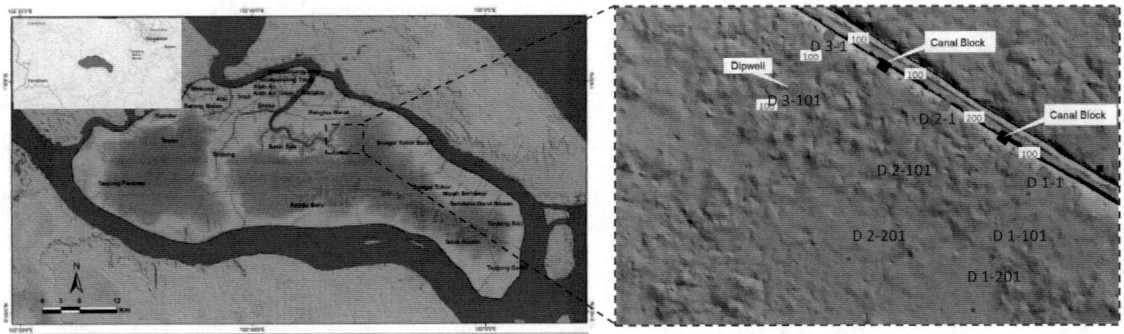

(a) (b)

Figure 1. Study area at Pulau Tebing Tinggi peatland hydrological unit (a) and the configuration of dipwells and canal blocks (b), Kepulauan Meranti Regency, Riau Province, Indonesia

3. Results and discussion

3.1 Impact along the canal

The recorded data of the groundwater level fluctuation along the canal at D 1-1, D 2-1, and D 3-1 dipwells in which the distance from the canal is about 1 m, together with the rainfall data, are presented in Figure 2. The figure represents the response of the water table fluctuation at the dipwells because of the water level change before and after canal blocking. The first canal block which is on the upstream side was built in the middle of November 2018, and the second one was built at the beginning of November 2018. When the first canal block was built, the fluctuation of groundwater level at D 1-1 dipwell which is about 100 m distance on the upstream side of canal block, increased significantly, but two other dipwells were not. It means that the canal blocking only has an impact in the upstream area. The groundwater fluctuation at the D 1-1 dipwell was stable at the elevation about of 8.5 m. This means that the canal block has a good function for keeping groundwater elevation and keeping peatland on always wet conditions.

When the second canal block was built, there was groundwater fluctuation impact at the D 2-1 dipwell, which is the distance about 100 m to the upstream side. However, this blocking has no impact at the D 1-1 dipwell which is the distance about 400 m to the upstream side. It means that in this case, the canal blocking has the rewetting impact no more than 400 m distance to the upstream area. The groundwater level fluctuation at the downstream side of the canal block was not impacted by canal blocking in the upstream. It was impacted only by rainfall as presented on D 3-1 line in Figure-2.

The groundwater depth which is the subtraction between groundwater level and the ground level, is an important parameter in correlation with peatland fire risk. Based on the Indonesian regulation, the minimum groundwater depth in the peatland is 0.4 m in order to keep the peatland area from fire (Anonim, 2016). Considering from the ground level elevation (G 1-1, G 2-1 and G 3-1) and groundwater elevation (D 1-1, D 2-1 and D 3-1), it could be investigated the groundwater depth before and after the canal blockings. The groundwater depth before canal blocking was about 1.5, 1.4, and 1.0 m at upstream area (Transect-1), middle area (Transect-2), and downstream area (Transect-3) respectively. This situation makes peatland was very vulnerable to the fire because the groundwater level was very low. After canal blocking, the groundwater level increased significantly so that the groundwater depth became about 1.0, 0.8, and 1.0 m in the upstream area, middle area, and downstream area respectively. Although the groundwater level increased significantly because of canal blocking, the groundwater depth has not fulfilled the minimum requirement by Indonesian regulation. However, the canal blocking has reduced the risk of peatland fire because of increasing groundwater level. The crest of canal blocks was not able to be built more raised in order to fulfill the Indonesian regulation because it would be caused flooding in other areas.

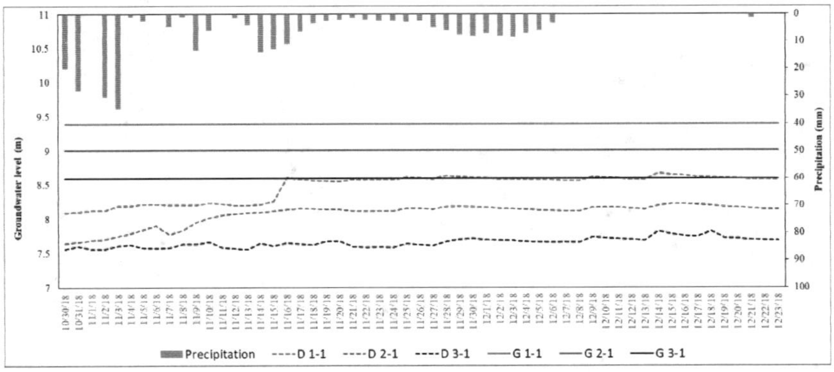

Figure 2. Daily water table fluctuation because of canal blocking and precipitation of three dipwells along the canal with 1 m distance from the canal from October to December 2018

3.2 Impact to the peatland

To investigate the impact of canal blocking for rewetting effort in the peatland area, the groundwater level fluctuation of the three dipwells of each transect with the 1 m, 101 m, and 201 m distance perpendicular to the canal were analyzed. Figure 3(a) presents groundwater fluctuation at Transect-1 of three dipwells with the 1 m, 101 m, and 201 m distance perpendicular to the canal before and after canal blocking. It shows that the groundwater level at D 1-1 dipwells (1 m distance to the canal) increased significantly and quickly after the canal blocking on November 15th. In this area, the groundwater level was stable until the end of monitoring. It means that the rewetting effort for peatland restoration was in a good performance in this area. However, the groundwater level at D 1-101 and D 1-201 (101 m and 201 m distance to the canal respectively) increased slightly in the same time that might be because of canal blocking and rainfall. It means that the canal blocking was not so significantly impacted for keeping groundwater high in this area.

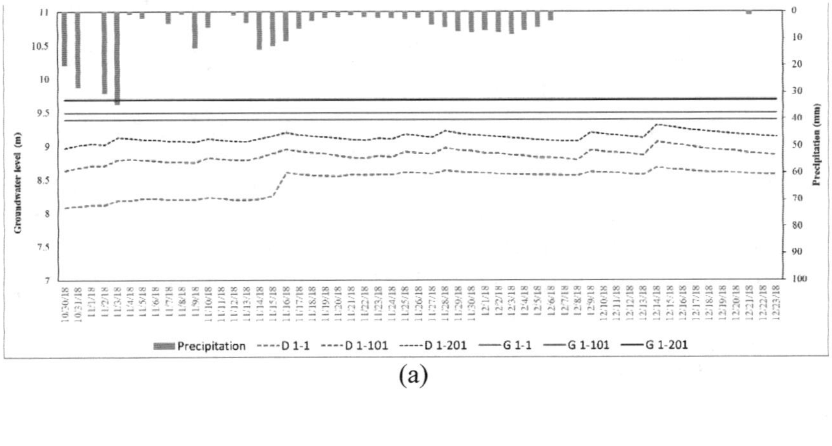

(a)

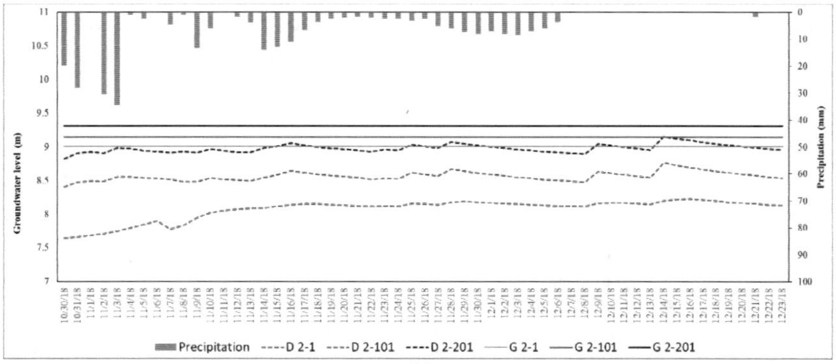

(b)

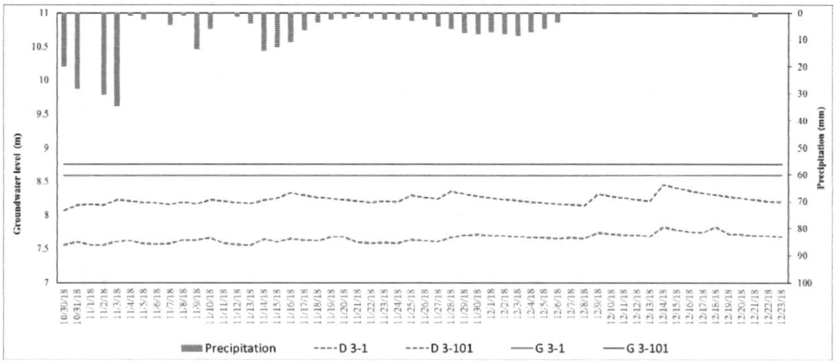

(c)

Figure 3. Daily water table fluctuation because of canal blocking at the Transect-1 (a), Transect-2 (b), and Transect-3 (c) from October to December 2018

Similar situation at the Transect-2 that the canal blocking has a significant impact on groundwater level at 1 m distance to the canal but has a low impact at 101 and 201 m distance to the canal respectively. After the canal block was built on November 7th, 2018, the water table in D 2-1 increased slowly and was stable after reach the maximum level. It means that the rewetting effort for peatland restoration was in a good performance in this area. The groundwater level at 101 m and 201 m distance to the canal respectively increased slightly at the same time that might be because of canal blocking and rainfall. It means that the canal blocking was not so significantly impacted for keeping groundwater high in this area. There was no groundwater level change at the dipwells which is located in the most downstream transect before and after canal blocking, as presented in Figure 3(c). This area has no rewetting impact for peatland restoration due to canal blockings which is located in the upstream area. This means that the upstream areas would be in the same condition with this area if the canal blockings were not applied.

To investigated how big the impact of the canal blockings in the peatland area on increasing groundwater level, the raised of GWL (groundwater level) at 1 m distance to the canal was compared to the raised of GWL at the 101 m and 201 m distance to the canal respectively on each transect (Figure 4). The figures imply that the canal blockings will no impact on rewetting at 101 m and 201 m distance to the canal if the raise of GWL at 1 m distance to the canal is less than 0.45 m. The GWL fluctuation at the canal less than 0.45 m was caused by the rainfall that was not impacted by canal blocking. The raise of GWL because of canal blocking varied at each location, depending on the distance to the canal. The raise of GWL at the Transect-1 for 101 m and 201 m distance from the canal because of canal blocking varied from 0.19 m to 0.51 m and from 0.1 m to 0.41 m that caused groundwater depth from 0.67 m to 0.34 m and from 0.62 m to 0.31 m respectively (Figure 4.a and Table-1). This means that the rewetting efforts could fulfill the groundwater depth based on the Indonesian regulation (0.4 m) if the raise of GWL at the canal (from monitoring data at 1 m distance from the canal dipwells) minimum of 0.6 m. The raise of GWL at the Transect-2 because of canal blocking has similar phenomena with the situation in the Transect-1. However, The transect-3 has no impact on rewetting because of canal blockings in this research area.

Table 1. The impact of canal blocking on the raise of GWL at each transect

Raise of GWL D 1-1 (m)	Raise of GWL D 1-101 (m)	Groundwater Depth (m)	Raise of GWL D 1-201 (m)	Groundwater Depth (m)
0.45	0.19	0.67	0.10	0.62
0.50	0.25	0.60	0.16	0.56
0.55	0.36	0.50	0.26	0.46
0.60	0.51	0.34	0.41	0.31
Raise of GWL D 2-1 (m)	Raise of GWL D 2-101 (m)	Groundwater Depth (m)	Raise of GWL D 2-201 (m)	Groundwater Depth (m)
0.45	0.11	0.63	0.12	0.57
0.50	0.17	0.58	0.18	0.51
0.55	0.25	0.49	0.28	0.41
0.60	0.38	0.36	0.43	0.26

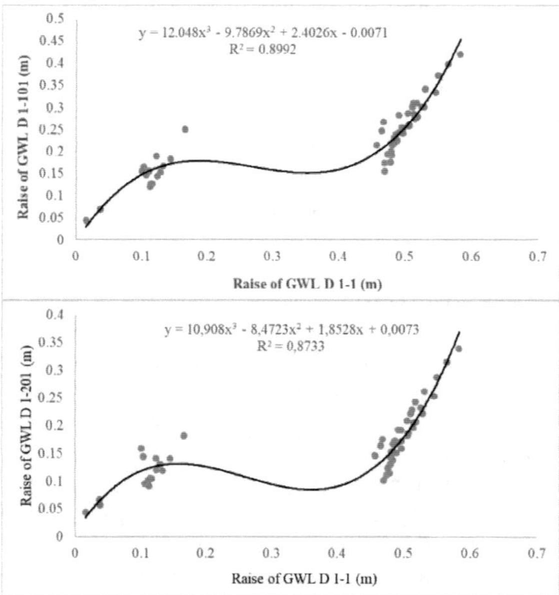

(a)

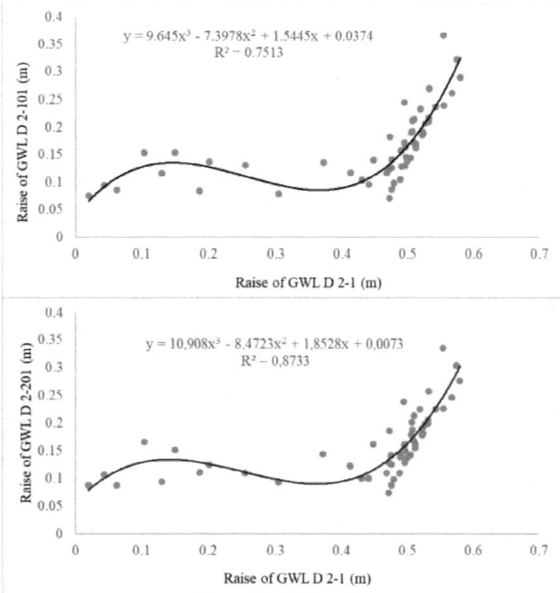

(b)

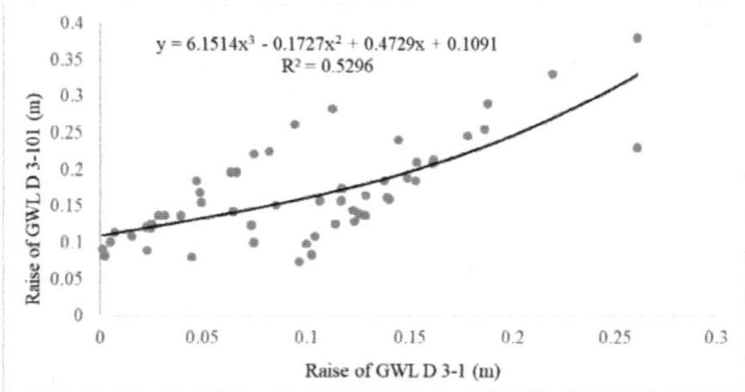

(c)

Figure 4. Daily water table fluctuation because of canal blocking at Transects-1 (a), Transect-2 (b) and Transect-3 (c) from October to December 2018

4. Conclusion

This research investigated the impact of canal blocking on the fluctuation of groundwater level in peatland ecosystems. Groundwater level fluctuation as the impact of canal blocking was monitored using eight dipwells that were set with a distance of 1 m, 51 m, 101 m, and 201 m from the canal for each transect of the three transects. The result of this study showed that the canal blocking has a good impact for keeping groundwater elevation and keeping peatland on always wet condition until 400 m distance to the upstream area from canal block and about 1m distance perpendicular to the canal. The canal blockings will have a good impact on rewetting at the peatland area up to 201 m distance perpendicular to the canal if the rise of water level at the canal is more than 0.45 m. However, at least it needs about 0.6 m of raise water level at the canal to fulfill the groundwater depth in the peatland area to satisfy the Indonesian regulation (0.4 m) for peatland restoration.

Acknowledgments
The authors would like to thank the Directorate General of Higher Education (DIKTI) - Ministry of Research, Technology, and Higher Education for its research grants program.

References

[1] Parish F et al. 2008 *Assessment on Peatlands, Biodiversity and Climate Change: Main Report* (Wageningen: Global Environment Centre Kuala Lumpur & Wetlands International, Wageningen)
[2] Giesen W and Sari E N N 2018 *Tropical Peatland Restoration Report : the Indonesian case*
[3] Sabiham S, Winarna, Pulunggono H B, Novarina D and Sahari B 2016 What is the way forward on Indonesian Peatland? *15th International Peat Congress* pp. 39–45.
[4] Wasis B, Putra E I and Saharjo B H 2019 Impacts of peat fire on soil flora and fauna, soil properties and environmental damage in Riau Province, Indonesia *Biodiversitas* **20** 1770–5
[5] Sandhyavitri A, Perdana M A, Sutikno S and Widodo F H 2018 The roles of weather modification technology in mitigation of the peat fires during a period of dry season in Bengkalis, Indonesia *Talenta-Cest IOP Conf. Series: Materials Science and Eng.* pp 0–9
[6] Dohong A 2017 Bolstering Peatlands Restoration in Indonesia through 3Rs Approach *Developing International Collaborations to Address Fire and Other Conservation Issues in Central Kalimantan, Indonesia*
[7] Dohong A et al. 2017 *Modul pelatihan Pembangunan Infrastruktur Pembasahan Gambut Sekat Kanal Berbasis Masyarakat* (Jakarta: Kedeputian Bidang Konstruksi, Operasi dan Pemeliharaan Badan Restorasi Gambut (BRG), Indonesia)
[8] Sutikno S et al. 2019 The effectiveness of canal blocking for hydrological restoration in tropical peatland *MATEC Web Conf.* vol 6003 no. 276 pp. 1–7
[9] Peraturan Presiden Republik Indonesia (PPRI) 2016 *Nomor 1 Tahun 2016 Tentang Badan Restorasi Gambut* 1–16

Water management for hydrological restoration and fire prevention in tropical peatland

S Sutikno[1], R Rinaldi[1], E Saputra[1], M Kusairi[1], B H Saharjo[2] and E I Putra[2]

[1] Civil Engineering Department, Universitas Riau, Pekanbaru, Indonesia
[2] Faculty of Forestry, Institut Pertanian Bogor, Bogor, Indonesia

Corresponding e-mail: ssutiknoyk@yahoo.com

Abstract. Water management is an important aspect of hydrological restoration and fire prevention in tropical peatland because the availability of water is not well distributed in the dry and rainy seasons. The degraded peatlands with massive canalization are very vulnerable against peat fire especially in the dry season. A pilot project for water management was conducted at Tanjung Leban, Bengkalis Regency, Riau Province with the purpose not only for hydrological restoration but also for fire prevention. In 2013, 2014, and 2015, the big peat fires occurred at Tanjung Leban and many peatland areas in Riau which caused a haze disaster. Peatland restoration and peat fire prevention efforts have been continuously undertaken by Peatland Restoration Agency and stakeholders at Tanjung Leban and many peatland areas in Riau. The rewetting, revegetation, and revitalization of livelihood were the integrated approaches that were carried out to support the peatland restoration. The aim of this research is to monitor the hydrological condition as the impact of rewetting activities using canal blocking at Tanjung Leban. Forty dip wells in four transects and an automatic rainfall record equipment were set up in the research area to monitor the groundwater level (GWL) and the rainfall event as the impact of rewetting activities. The result shows that the rewetting activity using canal blocking has a significant impact on the rewetting effort so that the risk of peat fire can be minimized. The historical evidence shows that the peatland restoration has been improving after the peatland areas were kept in always wet conditions.

1. Introduction

Peatlands are wetland ecosystems that occur as a result of the accumulation of organic material called "peat," which comes from the dead plant material and decays under water-saturated conditions [1]. Peatland ecosystems are the most efficient carbon sinks on the earth because peatland plants capture CO2 that is released naturally from peat to maintain the balance of nature. The Peatlands only cover about 3% of the earth's surface, but they are able to store up to 30% of carbon on earth in the form of peat or organic matter [2]. Peatlands are one type of wetland which is the most valuable ecosystem on earth. They are very important for conserving global biodiversity, providing safe drinking water, minimizing the risk of flooding, and helping to tackle climate change. Basic characteristics of peatlands are in extreme conditions high water and low oxygen content, toxic elements, and low nutrient availability [3]. Peatland now becomes the global topic concerning massive utilization for various purposes that can cause environmental problems including land fires, declining massive ecosystem functions, and increasing carbon emission [4]. Severe peatland fires are very difficult to be extinguished because the incidence of peatland fires occurs below the surface that can only be extinguished in the presence of rain or artificial rain [5].

Indonesia holds the largest tropical peatland area, comprising approximately 50% of the world's total tropical peatlands. The area of peatland in Indonesia is estimated to range from 14.91 million hectares spread over Sumatra Island 6.44 million hectares (43%), on Kalimantan Island 4.78 million hectares (32%), and 3.69 million hectares in Papua Island (25%) [6]. Recently, large areas of those peatlands have already been cleared and drained for food crops and cash crops such as palm oil and other plantations. However, large-scale drainage of peatlands for those purposes has often generated mayor problems of subsidence, fire, flooding, and deterioration in soil quality [7-9]. Indonesian Government trough Peatland Restoration Agency (BRG) and stakeholders have been continuously undertaking peatlands restoration and peat fire prevention efforts on those degraded peatlands using 3R approach, such as Rewetting, Revegetation, and Revitalization of livelihood [10-11].

The key important factor for the restoration of tropical peatlands is how to keep the peat always wet [12]. Water management is an important aspect of hydrological restoration in tropical peatland due to the availability of water is not well distributed in the dry and rainy seasons [13]. The degraded peatlands with massive canalization are very vulnerable against peat fire especially in the dry season due to over-drained. A pilot project for water management was conducted at a research area of Tanjung Leban Village, Bengkalis Regency, Riau Province, Indonesia with the purpose not only for hydrological restoration but also for fire prevention. Tanjung Leban Village is a very vulnerable area against peat fire in Riau. In 2013, 2014 and 2015, the big peat fires occurred at Tanjung Leban Village and many peatland areas in Riau which caused haze disaster for about two months.

This disaster produces the sickening and deadly cloud of smoky pollution, which threatens not only the nation but also neighboring countries. Peatland restoration and peat fire prevention efforts have been continuously undertaken by Peatland Restoration Agency (BRG), Republic of Indonesia and stakeholders at Tanjung Leban Village and many peatland areas in Riau. The rewetting, revegetation, and revitalization of livelihood were the integrated approaches that were carried out to support the peatland restoration. The aim of this research is to present the hydrological condition as the impact of rewetting activities at Tanjung Leban village for hydrological restoration and fire prevention. Forty dip wells in the four transects and an automatic rainfall record equipment were set up in the research area to monitor the groundwater level fluctuation and the rainfall event as the impact of rewetting activities.

2. Methodology

This study is located at Tanjung Leban Village, Bengkalis Regency, Riau Province, Indonesia, as shown in Fig. 1. The village and many peatland areas in Riau suffered big peat fires in 2013, 2014 and 2015 that caused a haze disaster as presented in Fig. 1b. As shown in the figure, a lot of hotspots were detected at those villages in 2013, 2014 and 2015 which the worst fire was in 2014. This disaster produces the sickening and deadly cloud of smoky pollution which threatens not only the nation but also neighboring countries. The research about water management for hydrological restoration and fire prevention was carried out at a series of canal blocks of about 5 km canal at Tanjung Leban Village (See Fig. 1c). There are two canals at the study area which are located at the right and left side of the road, as shown in Fig. 1e. There are two canal blocks at the downstream area of the right canal and a canal block at the upstream area of the left canal. The topographic map (Fig. 1d) shows that the ground surface along the canal is lower than that of in peatland area.

Forty dip wells were set around the canal with four transects, as presented in figure 1c for monitoring the fluctuations of GWL as the impact of water management. Six dip wells were set up with the distance of 2 m, 52 m, 102 m, 202 m, 302 m, and 502 m from the canal for each Transect-1, Transect-2 and Transect-3 and four dip wells for the Transect-4 with the distance of 2 m, 52 m, 102 m and 202 m. The Transect-1is located in the most upstream of the canal and the Transect-4 is in the most downstream of the canal. There are three canal blocks in the study site that were constructed in 2009, 2018, and 2019, respectively. The first and the second canal block are located in the right-side canal which is about 10 m in the downstream side and 10 m in the upstream side of Transect-4 respectively. The third one is

IOP Conf. Series: Materials Science and Engineering **933** (2020) 012053 doi:10.1088/1757-899X/933/1/012053

located about 10 m upstream side of Transect-2. The rainfall data, air temperature and water depth were also considered as a parameter of the analysis that was taken from a real-time telemetry every 10 minutes recorded by the SESAME (Sensory Data Transmission Service Assisted by Midori Engineering) system.

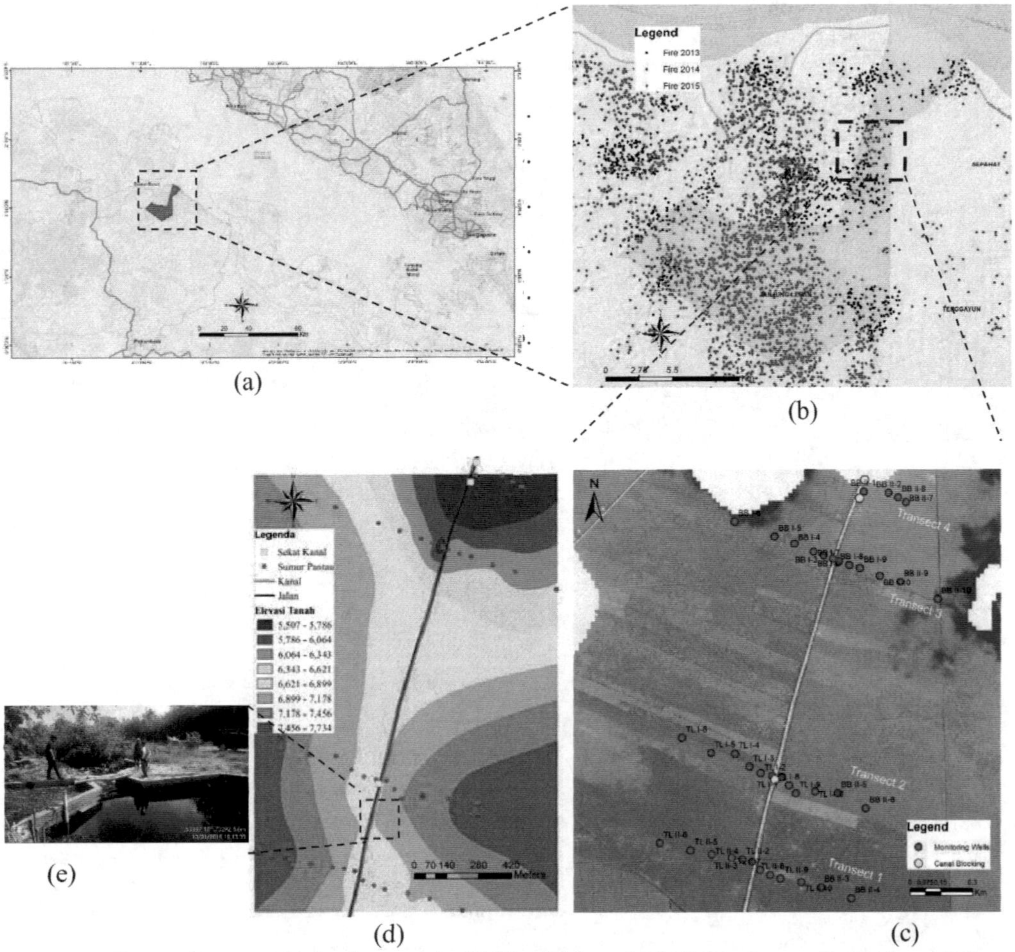

Figure 1. The study area (a), fire hotspot in 2013, 2014, and 2015 (b), the configuration of dip wells and canal blocks (b), the topographic map (c) and a photo of canal block (e) at Tanjung Leban village Bengkalis Regency, Riau Province, Indonesia.

3. Results and discussion

3.1. Impact of rewetting using canal blocking

To understand the impact of rewetting using canal block, the recorded data of the GWL fluctuation before and after canal blocking are mapped as presented in Fig. 2. The third canal block which was the last construction, was constructed on August 26th, 2018. Fig. 2 presents the maps of GWL before (July 20th, 2018) and after (January 10th, February 5th and April 23rd, 2019) canal block construction. The maps show that the GWL in the peatland around the Transect-1 and Transect-2 was about 5.5-6.0 m and 6.0-6.5 respectively on July 20th, 2018 (before canal blocking), as presented in Fig. 2(a). It raised after canal blocking on August 26th, 2018 about 6.0-6.5 m and 6.5-7.0 m in Transect-1 and Transect-2 respectively as presented in GWL map on January 10th, 2019 (Fig. 2b), February 5th, 2019 (Fig. 2c) and April 23rd, 2019 (Fig. 2d). The raised of GWL as the impact of canal blocking for peatland rewetting was also

IOP Conf. Series: Materials Science and Engineering **933** (2020) 012053 doi:10.1088/1757-899X/933/1/012053

presented in Fig. 3. The figure shows the cross-section of GWL at Transect-1 (10 m downstream side of canal block (Fig. 3a)) and Transect-2 (400 m upstream side of canal block (Fig. 3b) as the impact of canal blocking on at Tanjung Leban, Bengkalis Regency, Riau Province. It clearly shows the raised of GWL as the impact of canal blocking which was about 0.5 m for the left side of the canal. However, the peatland in the right-side canal was not raised because it was not blocked.

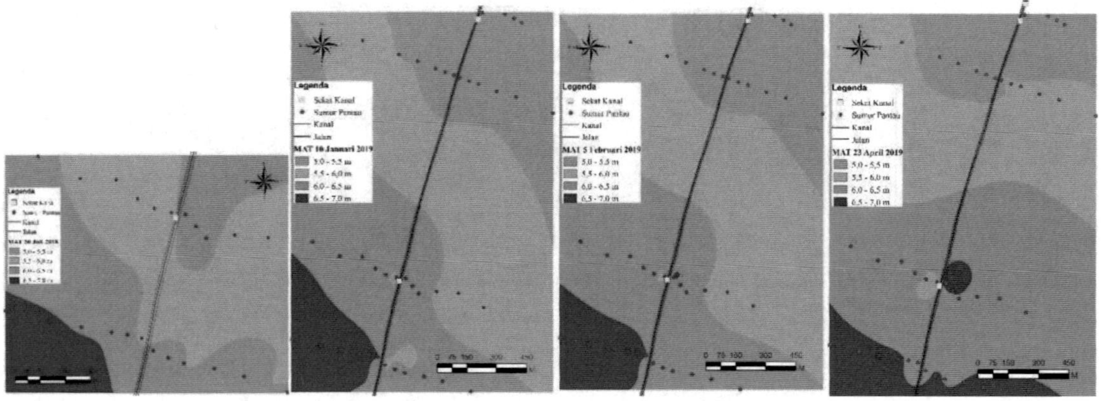

(a) (b) (c) (d)

Figure 2. Map of GWL before canal blocking on July 20th, 2018 (a) and after canal blocking on January 10th, 2019 (b), February 5th, 2019 (c), April 23rd, 2019 (d) at Tanjung Leban, Bengkalis Regency, Riau Province, Indonesia.

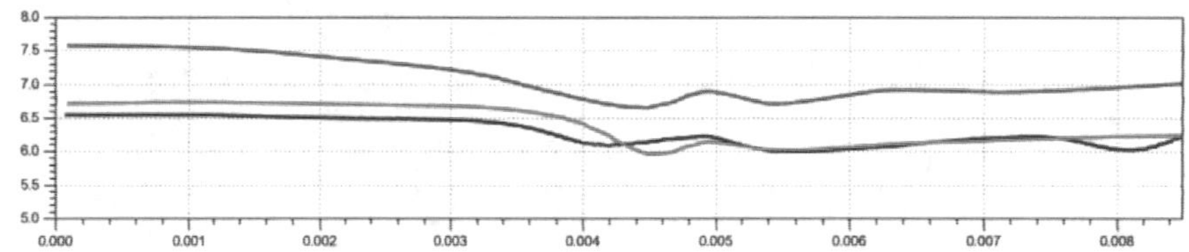

(a)

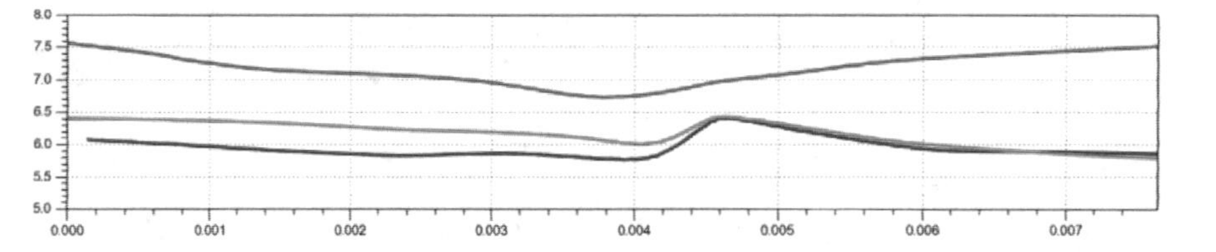

(b)

Figure 3. Cross-section of GWL at Transect-1, 10 m downstream side of canal block (a) and Transect-2, 400 m upstream side of canal block (b) as the impact of canal blocking on at Tanjung Leban, Bengkalis Regency, Riau Province, Indonesia.

In order to investigate the impact of canal blocking, continuous monitoring of GWL in the Transect-2 has also been recorded daily using water loggers from August 1st (before blocking) to October 22nd (after blocking). Fig. 4 presents the daily water level fluctuation at the canal and GWL fluctuation of two dip wells with a distance of 60 m and 210 m perpendicular to the canal because of canal blocking and precipitation from August to October 2018. Minimum water level at canal before canal blocking was about 5.8 m. It raised significantly after canal blocking on August 26th, 2018, and the minimum water level after canal blocking was 6.5 m. It means that the raised water level at the canal after canal blocking was 0.7 m. The 0.7 m raised the water level at the canal caused raised of GWL about 0.3 m and 0.2 m at the peatland with a distance of 60 m and 210 m to the canal respectively. However, the heavy rainfall that was occurred on October 11th, 2018, has a big impact on raising water level not only at the canal but also at the peatland. The existence of the canal block can keep the water level for a long time because the drainage flow can be slowed down.

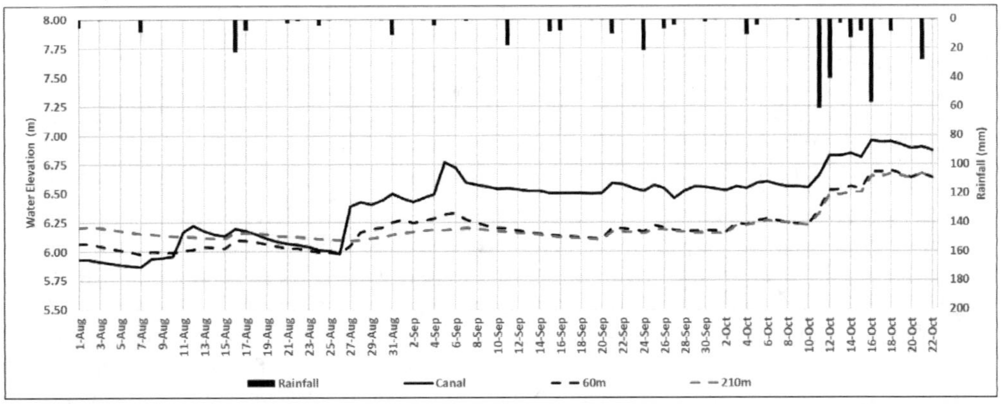

Figure 4. Daily water level fluctuation at the canal and GWL fluctuation of two dipwells with the distance of 60 m and 210 m perpendicular to the canal because of canal blocking and precipitation from August to October 2018.

3.2. Impact of water management on peat fire risk

To analyze the impact of water management on peat fire risk, the maps of groundwater depth were developed. To develop the map of peat fire risk, we classified the level of peat fire risk base on the groundwater depth, such as 0-0.25 m depth is safe, 0.26-0.40 is fire alert, and more than 0.4 is in danger status [14]. The maps of peat fire risk before and after canal blocking is presented in Fig. 5. Before canal blocking, all of the area in the dry season was in danger status (Fig. 5.a), but after canal blocking is reduced about 6% for construction of a canal block (Fig. 5b). For the larger area of reducing the peat fire risk area, it should be added more canal blocking in this area with the shorter distance between each canal block.

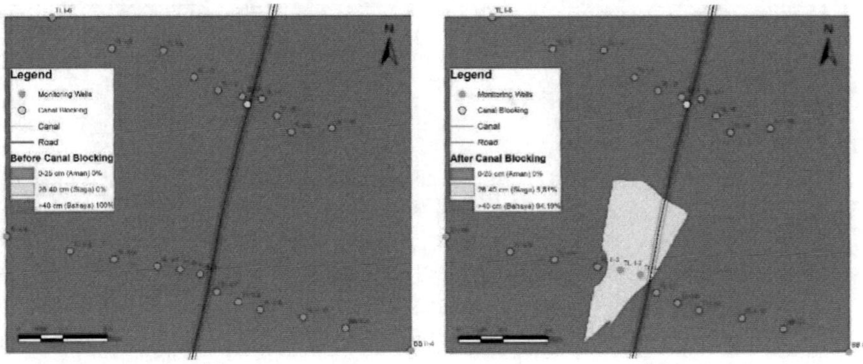

(a) (b)

Figure 5. Map of peat fire risk before canal blocking on July 20th, 2018 (a) and after canal blocking on January 15th, 2019 (b), at Tanjung Leban, Bengkalis Regency, Riau Province, Indonesia.

3.3. Impact of water sharing

Water management in the peatland should be considered in terms of the hole water management in the one peatland hydrological unit (PHU). In a PHU consist of many ownership and management, which usually in the upstream areas are managed by plantation companies and in the downstream areas are managed by local communities. The basic concept for water management is how to save the water as long as possible in the peatland in order to keep peatland always wet. The peat domes that are usually in the upstream area and under the company concession should be designed for the water reservoir to save the water from the rainy season and still available in the dry season. The water flows from peat dome trough canals are blocked to reduce the velocity and to save them in the canals. The water in the canals is supposed to infiltrate to the peatland, and the exceeded water flow is drained to the sea to prevent flooding.

The Tanjung Leban area which is located in downstream of the PHU, is supposed to get more additional water from the upstream reservoir managed by the company concession especially in the dry season as the application of water sharing. Fig. 6 presented the GWL fluctuation at each transect before and after water-sharing application at Tanjung Leban, Bengkalis Regency, Riau Province, Indonesia. The figure shows that the application of water sharing in the right-side canal could raise the GWL about 0.5 m, 0.4 m, 0.3 m and 0.55 m in Transect-1, Transect-2, Transect-3 and Transect-4 respectively. It also has an impact on reducing the level of peat fire risk up to 10% for setting a canal blocking, as presented in Fig. 7. The more canal blocking, the wider area of peatland will reduce against peat fire risk in case of water sharing application.

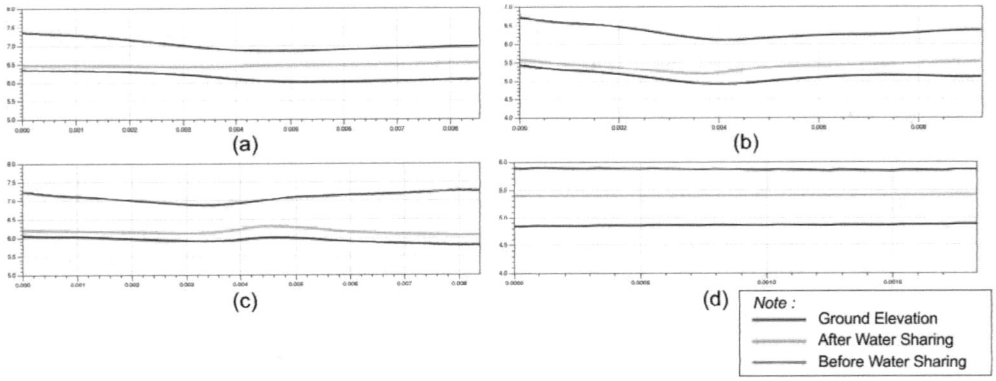

Figure 6. GWL fluctuation at Transect-1 (a), Transect-2 (b), Transect-3 (c) and Transect-4 before and after water-sharing applied at Tanjung Leban, Bengkalis Regency, Riau Province, Indonesia.

3.4. Impact of water management on peatland restoration

The impact of rewetting using canal blocking in the peatland restoration for a long-term period is presented historically in Fig. 8. Those areas are located in the Transect-4, which are in high GWL as presented in Fig 7b. Previously, those areas were burnt severely but after canal block construction in 2009, there was no more peat fire. After canal blocking, the peatland areas are kept always wet that make those areas free from fire and the plantation growing very well. In 2017, those areas were assessed as one of the successful areas for tropical restoration exercises according to many scientists who visited here.

4. Conclusion

This research monitors the hydrological condition as the impact of rewetting activities using canal blocking at Tanjung Leban village. Forty dip wells in the four transects and an automatic rainfall record equipment were set up in the research area to monitor the GWL and the rainfall event as the impact of rewetting activities. The result shows that the rewetting activity using canal blocking has a significant impact on the rewetting effort so that the risk of peat fire can be minimized. The canal blocking could raise the water level at the canal about 0.7 m that could raise GWL about 0.3 m and 0.2 m at the peatland with a distance of 60 m and 210 m to the canal respectively. The canal blocking also could reduce the level of peat fire risk area about 6% for the construction of a canal block. For the larger area of reducing peat fire risk, it should be added more canal block in this area with the shorter distance between them. The application of the water sharing system could raise the GWL about 0.3-0.55 m in the peatland area, which can reduce the level of peat fire risk up to 10% for setting a canal blocking. The more canal blocking, the wider area of peatland will reduce against peat fire risk in case of water sharing application. The historical evidence shows that the peatland restoration has been improving after the peatland areas were kept in always wet conditions.

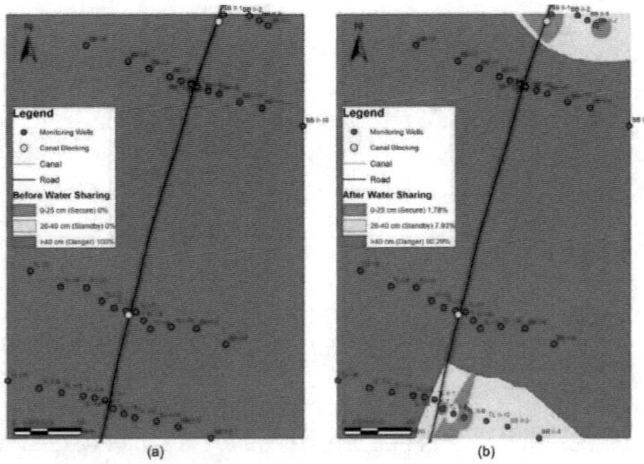

Figure 7. Map of peat fire risk before and after water-sharing application at Tanjung Leban, Bengkalis Regency, Riau Province, Indonesia.

Figure 8. The timeline history of tropical peatland restoration as the impact of rewetting using canal blocking at Tanjung Leban, Bengkalis Regency, Riau Province, Indonesia.

Acknowledgments
The authors would like to thank the NASA for research funds under UMCES/IPB-NASA Tropical Peat Fire Research Project.

References

[1] Parish F et al. 2008 *Assessment on Peatlands, Biodiversity and Climate Change: Main Report* (Wageningen: Global Environment Centre, Kuala Lumpur & Wetlands International, Wageningen)
[2] Hugron S, Bussières J and Rochefort L 2013 *Tree Plantations within the Context of Ecological Restoration of Peatlands: Practical Guide*
[3] Joosten H and Clarke D 2002 *Wise Use of Mires and Peatlands - and Including Framework for Decision-Making* (Finland: International Mire Conservation Group and International Peat Society)
[4] Sabiham S, Winarna, Pulunggono H B, Novarina D and Sahari B 2016 What is the way forward on Indonesian Peatland? 15th *Intern. Peat Congress* pp 39–45
[5] Sandhyavitri A, Perdana M A, Sutikno S and Widodo F H 2018 The roles of weather modification technology in mitigation of the peat fires during a period of dry season in Bengkalis, Indonesia in TALENTA-CEST *IOP Conf. Series: Materials Science and Eng.* pp 0–9
[6] Osaki M and Tsuji N 2015 *Tropical Peatland Ecosystems* (Sapporo: Springer Tokyo Heidelberg New york Dordrecht London)
[7] Adesiji A, Mohammed T, Daud N N, Saari M, Gbadebo A and Jacdonmi I 2015 Impacts of land use change on peatland degradation: a review *Ethiop. J. Environ. Stud. Manag.* **8** 2 225
[8] Putra E I, Graham L L B and Vetrita Y 2016 *Degraded peatlands, Ground Water Level and severe peat fire occurrences*
[9] Ritzema H P 2008 The Role of Drainage in The Wise Use of Tropical Peatlands *Restoration of Tropical Peatlands* (S. E. Wösten, J.H.M., Rieley, J.O. and Page, Ed)
[10] Dohong A 2017 Bolstering Peatlands Restoration in Indonesia through 3Rs Approach *Developing International Collaborations To Address Fire And Other Conservation Issues In Central Kalimantan, Indonesia*
[11] Dohong A et al. 2017 *Modul pelatihan Pembangunan Infrastruktur Pembasahan Gambut Sekat Kanal Berbasis Masyarakat* (Jakarta: Kedeputian Bidang Konstruksi, Operasi dan Pemeliharaan Badan Restorasi Gambut (BRG) Republik Indonesia)
[12] Giesen W and Nirmala E 2018 *Tropical Peatland Restoration Report : the Indonesian case*
[13] Sutikno S et al. 2019 The effectiveness of canal blocking for hydrological restoration in tropical peatland *MATEC Web Conf.* vol 6003 276 pp 1–7
[14] Susilo G E B, Yamamoto K, Imai T, Ishii Y, Fukami H and Sekine M 2013 The effect of ENSO on rainfall characteristics in the tropical peatland areas of Central Kalimantan, Indonesia *Hydrol. Sci. J.* **58** 3 539–548

IOP Conf. Series: Materials Science and Engineering **933** (2020) 012054 doi:10.1088/1757-899X/933/1/012054

Spatial analysis of land subsidence potential due to lowland conversion (case study at Tanjung Api-Api)

Andriani[1], E Ibrahim[2], D D A Putranto[2], A K Affandy[2] and H G Putra[1]

[1] Civil Engineering Departement, Andalas University, Padang, West Sumatera
[2] Sriwijaya University, Palembang, South Sumatera

Corresponding E-mail: andriani@eng.unand.ac.id

Abstract. The conversion of lowland area into special economic zones at Tanjung Api-Api will trigger environmental degradations, one of which is land subsidence. Land subsidence potential is caused by several natural environmental factors, which are soil type, soft soil thickness, topography, geological conditions, and rainfall, and non-natural factors, which are caused by land use or land cover, groundwater extraction, building density and others. Multicriteria analysis was conducted to obtain the weight of each factor, then the factor was overlayed to determine the level of land subsidence based on land use or land cover. Spatial analysis has been done to many factors causing land subsidence. The results showed that industrial estates and housing areas have a high potential for land subsidence. Mitigation efforts must be carried out to reduce severe land subsidence at the built-up area.

1. Introduction

Lowland area is a vulnerable area, which is unstable to environmental changes. The conversion of lowland areas on a large scale can trigger environmental problems. The Problems are environmental degradation such as land subsidences, floods, forest fires, carbon emissions and biodiversity reductions in lowlands, the decrease in the quality of life, and even economic losses. Indications of land subsidence at Tanjung Api-Api have been seen in residential areas such as damage to house floors and the roads [1].

Tanjung Api-Api is a very strategic area because it functions as a seaport area and industrial area. Tanjung Api-Api is expected to encourage economic growth in South Sumatra. The determination of this area as an exclusive economic zone will trigger the growth of high residential and changes in lowland ecosystems. The hoarding of lowland is done to get land areas that are used as residential, industrial, agricultural and plantation. The massive conversion of lowlands will cause positive and negative impacts on the environment and humans. The construction of infrastructure and residential areas in Tanjung Api-Api will require special handling due to unstable soil conditions. One of the negative impacts caused by the conversion of lowlands is land subsidence.

Land subsidence is caused by excessive groundwater extraction, building loads, soil types, and tectonic influences [2,3,4,5]. The high potential of soil subsidence will occur in the Tanjung Api-Api area because the types of soil in the Tanjung Api-Api region are in the form of organic soil, organic clay, and clay [6,7]. The condition of groundwater fluctuations and the types of constituent soil found in lowland areas will trigger land subsidence due to the landfill process [8].

Changes in land use or land cover into industrial estates and residential areas have caused land subsidence in Indonesia and the world. In general, the land subsidence in Jakarta, Semarang, Bandung, and Surabaya is caused by the increasing growth of industrial estates and housing areas. The changes in land use result in excessive groundwater extraction and soil consolidation due to building loads [9, 10,11,12,13]. According to the study of land use or land cover change at Tanjung Api-Api, there was a conversion of land use or land cover of the mangrove forest area into a plantation and aquaculture area in the Tanjung Api-Api area (Air Telang, Banyuasin) from 1989 to 2013 [14]. Changes in land use or land cover due to the increase in built-up areas, which are housing areas, infrastructure buildings and buildings, will cause an increase in surface temperature and environmental degradation [15].

Spatial analysis is one of the models developed to link various spatial environmental information. By using multicriteria analysis, it can be seen that the area has the potential to experience land subsidence so that mitigation efforts can be carried out in the area in order to create sustainable development in lowlands [16-26].

This paper aims to present a spatial model of land subsidence potential that will occur in lowland areas, especially in the Tanjung Api-Api area due to land conversion because of the growth of industrial estates and residential areas.

2. Material and method

2.1. Study area

Tanjung Api-Api is one of the lowland areas on the east coast of Sumatra island in the Banyuasin district, South Sumatra. Geographically, the Tanjung Api-Api area is located at 104° 45' to 104° 55' E and 2° 17' to 2° 24' S (figure 1).

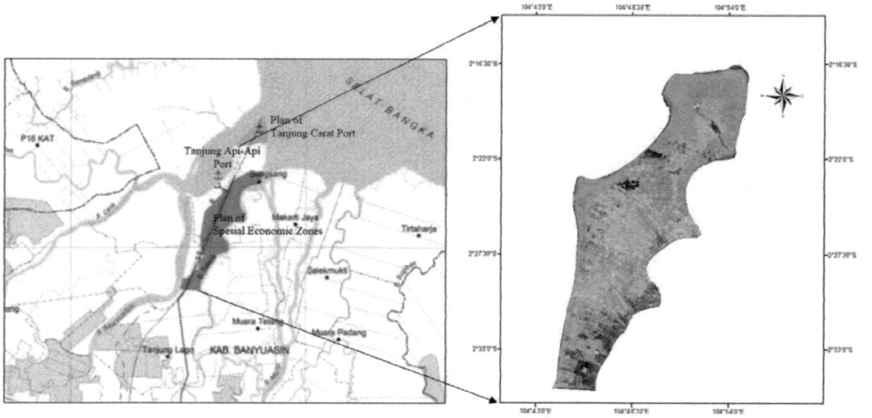

Figure 1. Location of research.

Tanjung Api-Api area has a flat topography, 0 - 2%, and a land surface height of about 1 - 5 m. Based on the geological map of Palembang sheet, Tanjung Api-Api area is a quarterly formation consisting of swamp deposits, which are mud – silt - and sand, blackish gray which contains many plant residues in the form of roots and leaves, remnants of rotten wood pieces, humus, and peat (figure 2).

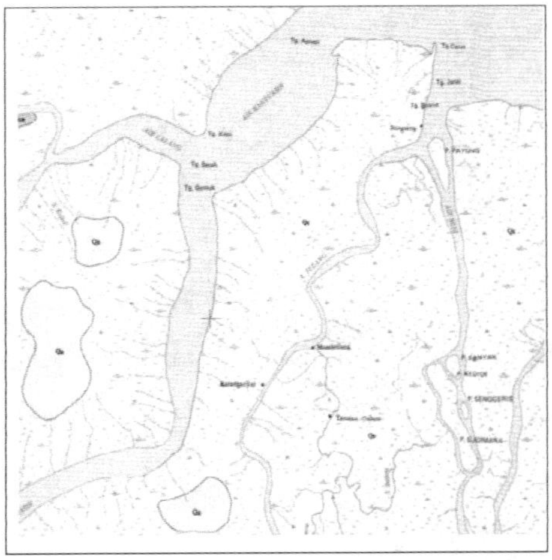

Figure 2. Geological map of Palembang sheet.

Most of the soil in this area consists of soft soil in the depth of more than 20 meters and peat soil with a thickness of fewer than 2 meters. Soil condition of Tanjung Api-Api area is moist to saturated, has a clay, has a large and not well-consolidated coefficient of consolidation and shrinkage limit so that these conditions will trigger land subsidence in case the land is loaded. Tanjung Api-Api area is influenced by tidal conditions because the area is located among the Banyuasin river, Telang river, Bangka strait, and the overall Tanjung Api-Api area which has a swamp ecosystem. Tides occur in one day with a maximum level of 2.39 meters. The results of the research showed that water at Tanjung Api-Api region is the brackish water having the content of sodium (Na) of 1050 mg/liter, chloride (Cl) of 1909,1 mg/liter, pH of 4.55 and corrosive groundwater [6].

2.2. Data analysis

Spatial analysis is done using Quantum Gis version 3.8.1. An analysis is carried out using the AHP (Analysis Hierarchy Process) based on the opinions of people who are experts in the field of geotechnics and the environment to obtain the weight of the factors that cause land subsidence.

In this study, eight thematic maps were used, namely maps of soil type, soft soil thickness, land use or land cover, embankment height, groundwater extraction, topography, geological conditions, and rainfall. By using overlapping techniques and Multi-Criteria Spatial Analysis (MCSA), further zoning of land subsidence will be obtained with two conditions, namely the existing condition and the condition of the planned development of the Tanjung Api-Api area. Figure 3 showed a map of current land use and land use based on the spatial plan of the Banyuasin region.

In this study, three models were used based on variations in land use or land cover, groundwater extraction, and embankment height. Modeling is done based on the weight values of groundwater extraction and embankment height.

Model 1. The potential of land subsidence in case the groundwater and variations in embankment height for each land use or land cover plan are not used.

Model 2. The potential of land subsidence in case the water sources from shallow wells and variations in embankment height for each land use or land cover plan are used.

Model 3. The potential for land subsidence in case the water sources from deep wells or bore wells and variations in embankment height for each land use or land cover plan are used.

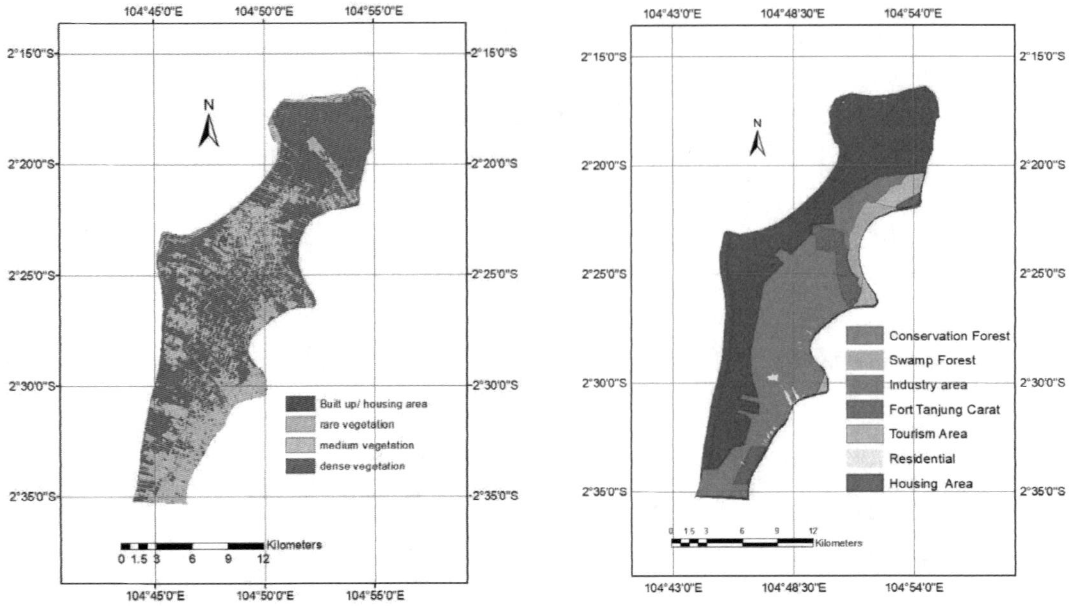

a. The current land use/land cover

b. Plan of land use/land cover based on the Banyuasian district RTRW

Figure 3. Map of land use or land cover at Tanjung Api-Api area.

3. Results and discussion

In general, Tanjung Api-Api area has the same regional characteristics as follows: the type of soil is clay or silt, which thickness is 11m to 20m; geological conditions consist of swamp deposits; the topography is flat; the rainfall is high. Based on these characteristics, the Tanjung Api-Api area has the potential for land subsidence. The potential level of land subsidence depends on the type of land use or land cover, the load of construction, and groundwater extraction.

3.1. Evaluation of factors causing land subsidence using the Analytical Hierarchy Process (AHP) method.

Land subsidence is caused by several factors, namely natural and human factors. Natural factors that cause land subsidence are soil type, soft soil thickness, topography, geological conditions, and rainfall. While the non-natural or human factors that trigger land subsidence are land use or land cover, groundwater extraction, embankment height, and others. To determine the land subsidence potential, weighting and scoring are carried out from these factors using questionnaires obtained from the experts in the geotechnical and environmental. Figure 4 showed the weight of factors causing land subsidence, by using the Analytical Hierarchy Process (AHP).

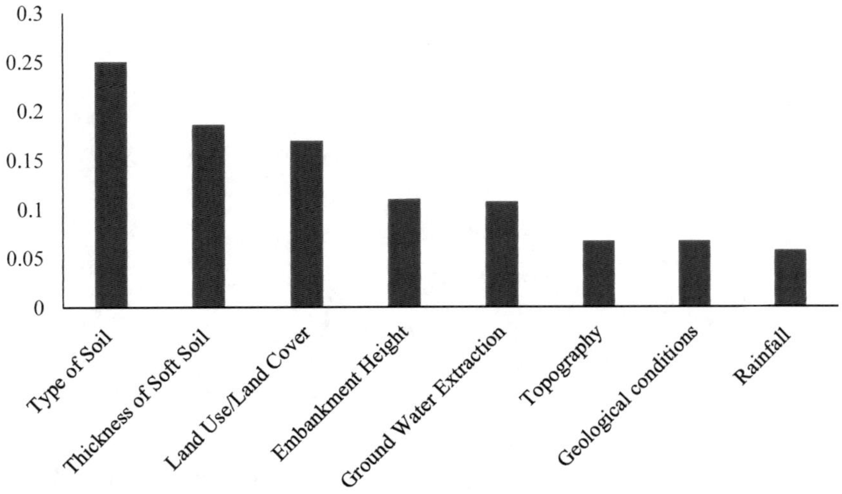

Figure 4. Weights factors causing land subsidence.

Table 1 showed the weighting results and scores for each criterion and sub-criteria of the factors that cause land subsidence.

Table 1. Weight and score of non-natural factors causing land subsidence.

Parameters	Weight	Score	Value
Land Use/ Land Cover			
a. Industrial area	0.169	47.90	7.76
b. Residential/ housing		22.90	3.71
c. Trade and service area		18.70	3.03
d. Agriculture/ plantation		6.80	1.10
e. Bare land		3.60	0.60
Embankment Height			
a. 2 m	0.109	70.20	7.58
b. 1 – 2 m		19.00	2.05
c. < 1 m		10.80	1.17
Ground Water Extraction			
a. Not using groundwater	0.106	13.00	7.738
b. Shallow wells		14.00	1.484
c. Deep wells		73.00	1.378

Figure 4 showed that there are five main factors that cause land subsidence: soil type, soft soil thickness, land use or land cover, embankment height, and groundwater extraction. Parameters that have a high weight value indicate that these parameters are the main factors causing land subsidence, while other parameters are supporting or less essential factors that can lead to land subsidence. The supporting or less essential factors are topography, geological conditions, and rainfall, each of those has a weight value < 0.1. The higher the weight value and score indicate that the area has a high potential for land subsidence. In this study, the area was divided into five zonings based on the highest weight value and the lowest weight value. Table 2 showed the interval score for land subsidence potential.

Table 2. The interval score of land subsidence potential.

The Potential of Land Subsidence	Interval Score
Good	7.55 – 16.64
Low	16.65 – 25.73
Medium	25.74 – 34.82
High	34.83 – 43.91
Very high	43.92 – 53.00

3.2. Existing conditions of land subsidence potential
Figure 5 showed that the condition of the existing area has a low potential for land subsidence (S value between 18.90 to 22.01). This fact is because most of the land use or land cover in the Tanjung Api-Api area consists of mangroves, plantations or agriculture, dryland agriculture, and shrubs, all of which have little weight on land subsidence compared to residential and industrial areas [28]. Currently, the housing area has a low potential for land subsidence because the density of housing in the Tanjung Api-Api area is very low, so that the loading of infrastructure and extraction of groundwater is low, besides that many communities still use houses on stilts.

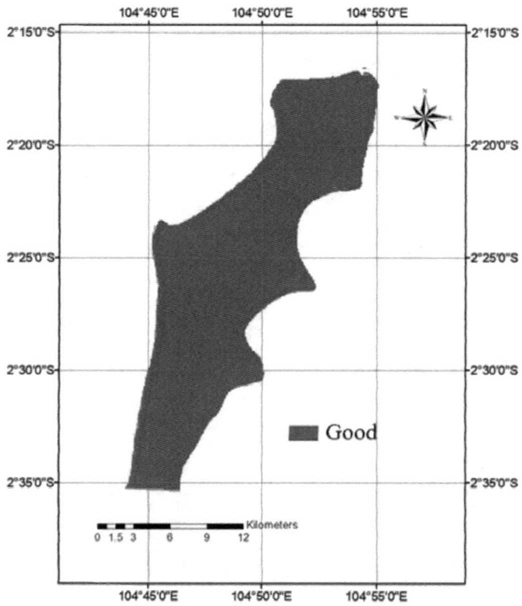

Figure 5. Existing Condition of Land Subsidence Potential.

3.3. Land subsidence potential due to conversion to built-up area
The construction of Special Economic Zones in Tanjung Api-Api will cause land conversion from plantation or agricultural areas to industrial estates and housing areas. Based on the District Spatial Planning Banyuasin for 2010-2030, the Tanjung Api-Api area will be developed into several strategic

IOP Conf. Series: Materials Science and Engineering **933** (2020) 012054 doi:10.1088/1757-899X/933/1/012054

areas of South Sumatra Province, one of which is the Special Economic Zone. The construction of industrial estates and housing areas will lead to groundwater extraction and land reclamation. The spatial model is divided into three models based on the source of groundwater extraction and embankment height. The results of spatial analysis are shown in figure 5.

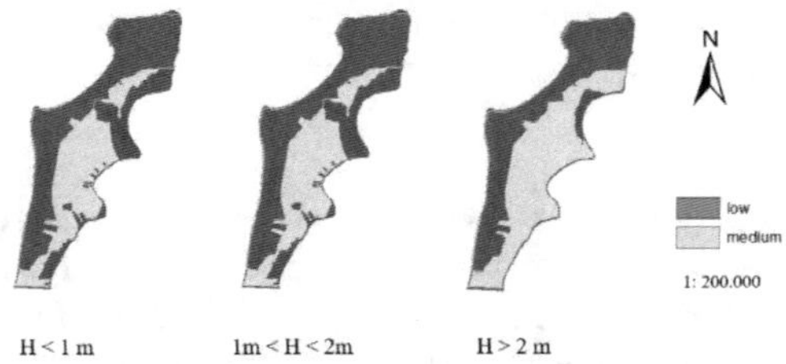

Model 1. The potential of land subsidence in case groundwater was used.

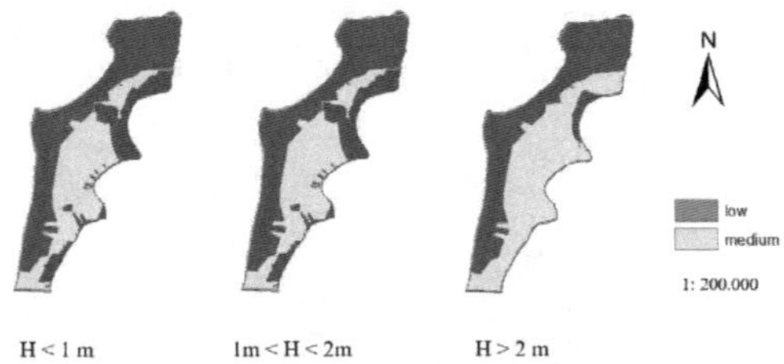

Model 2. The potential of land subsidence in case shallow wells was used.

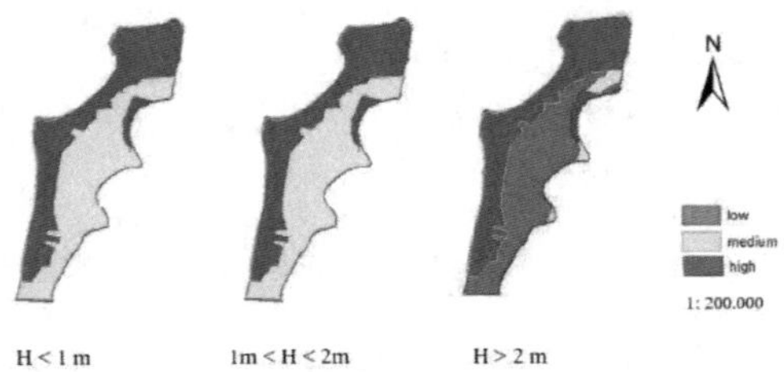

Model 3. The potential of land subsidence in case deep wells was used.

Figure 6. The potential of land subsidence based on variations of height of embankment and groundwater extraction.

Spatial analysis showed that the results based on models 1 and 2 have the same potential pattern of land subsidence. The conversion of lowland functions into industrial estates caused a medium-level potential of land subsidence, while mangrove forests and tourism areas in residential areas have a low-level potential of land subsidence. The low potential for land subsidence in residential areas is due to the absence of groundwater exploitation. The absence of groundwater exploitation is because the community obtains water from other sources such as rainwater and because the building load is still low due to the embankment height of < 2m. The increase of embankment height (h > 2m) causes an increase in load which must be borne by the subsoil; as a result, the potential for soil subsidence is medium.

Model 3 showed that industrial estates and residential areas have high land subsidence potential due to the embankment height of > 2m and the presence of deep well water extraction. Height of embankment of > 2m will cause additional loads, while the use of deep wells or bore wells can cause emptying of cavities in the soil so that there will be a compaction of soil grains that causes soil subsidence. The occurrence of high potential for land subsidence is because of the increase of the load that must be borne by the subgrade. In addition, the occurrence of high potential for land subsidence because the compressible soil will be more easily deformed due to loading and changes in ground level [29-33].

The high potential of land subsidence in industrial estates and residential areas is also caused by a large number of activities in the area resulting in an increase in groundwater extraction and also large-scale stockpiling of lowlands to avoid flooding and tides.

4. Conclusions

Based on the spatial analysis performed, the conversion of lowlands area into industrial estates and housing area potentially cause the land subsidence, compared to the potential of land subsidence occurrence in agriculture, plantation, and bare land areas. Groundwater extraction using deep wells and land reclamation with embankment height more than 2 m triggers high land subsidence in the Tanjung Api-Api area.

Acknowledgements

We would like to thank The Program Bantuan Seminar International Dalam Negeri, Andalas University that has supported this research.

References

[1] Andriani, Ibrahim E, Putranto D D A and Affandi A K 2018 E 3S *Web of Conferences* 68
[2] Fulton A 2006 *Land Subsidence: What is It and Why is It an Important Aspect of Groundwater Management Agriculture and Natural Resources* (University of California)
[3] Abidin Z H, Andreas H. Gumilar I, Gammal M, Fukuda Y and Deguchi T 2009 *FIG Regional Conf. Hanoi Vietnam 19-22 October 2009* pp 1-16
[4] Ismanto A, Wirasatriya A. Helmi M, Hartoko A and Prayogi 2009 *J. Ilmu Kelautan* **14** 4 21-8
[5] Yuwono D B 2013 *J. Teknik* **34** 3 pp 188-195
[6] Hermawan and Sugiyanto 2004 *Bulletin of Enviromental Geology 14* **1** 21-6
[7] San I C 2010 *J. Rekayasa Sriwijaya 19* **1** 7-14
[8] Rahmadi R, Suryadi F X, Susanto R H and Schultz B 2011 *Proc. Int. Seminar-workshop on Integrated Lowland Development and Management* (Palembang, Indonesia) vol 15 no 15 pp 1-20
[9] Aditiya A, Takeuchi W and Aoki Y 2017 *IOP Conf. Series: Earth and Environmental Science* p 98
[10] Mochammad M and Saepuloh A 2017 *IOP Conf. Series: Earth and Environmental Science* p 71
[11] Saharom N S, Diana S C and Kusyala D 2018 *IOP Conf. Series: Earth and Environmental Science* p 152
[12] Andreas H, Abidin H Z, Gumilar I, Sidiq T, Sarsito D A and Pradipta D 2018 *Natural Hazards - Risk Assessment and Vulnerability Reduction* (Intechopen) Chapter 3 pp 39-56
[13] Yang Q, Yinghai K, Dongyi Z, Beibei C, Huili G, Mingyuan L, Lin Z and Xiaojuan L 2018 *Remote Sens* (10) pp 2 - 20
[14] Eddy S, Iskandar I, Ridho M R and Mulyana A 2017 *Biodiversitas* 18 (4) pp 1538-45
[15] Adeyemi A, Botai J, Ramoelo A, Merwel F V D and Tsela P 2015 *South African J. of Geomatics* **4** 351-68
[16] Murakami S, Yasuhara Kand Murata F 2000 *Groundwater Updates* pp 79-84
[17] Putra D P E, Setianto A, Keokhampui K and Fukuoka H 2011 *Mitteilungen zur Ingenieurgeologie und Hydrogeologie-Festschrift zu* pp 39-50.
[18] Sahu P and Sikdar P K 2011 *J Earth Syst. Sci* 120 vol 3 pp 435–446
[19] Zhu L Chen Y, Gong H L, Liu C and Wang R 2013 *Proc. Int. Conf on Modelling and Simulation*

vol 1.(Adelaide: Australia) pp 3267-73
[20] Gumilar I, Abidin H Z, Sidiq T P, Andreas H, Maiyudi R, Gamal M and Fukuda Y 2013 *Indonesian J. of Geospatial 2* vol 2
[21] Udoh J C and Udofia E P 2014 *Current World Enviroment 9* vol 1 pp 1-6
[22] Pradhan B, Abokharima M H, Jebur M N and Tehrany M S 2014 *Nat Hazards* 73 pp 1019-42
[23] Ashraf H and Cawood F 2015 *South African J. of Geomatics 4* vol 3
[24] Liu J, Wang H and Yan X 2015 *Proc. IAHS* vol 372 pp 543–53
[25] Zaid S M, Saleh A A and Al- Mobarak N M 2015 *Middle- East J. of Scientific Research 23* vol 5 pp 902-14
[26] Qu J, Yang L, Ren K, Yu F and Zhou J 2017 *AMSE J. 78* vol 1 pp 83-99
[27] Bhattarai R and Kondoh A 2017 *Advances in Remote Sensing* vol 6 pp 132-46
[28] Andriani, Ibrahim E, Putranto D D A and Affandi A K 2018 *Eco. Env. & Cons. 24* vol 2 pp 651-7
[29] Thoang T T and Giao P H 2015 *Engineering Geology* 199 pp 107-24
[30] Kouda A, Nagata K and Sato T 2015 *Proc. IAHS* 372 pp 95–9
[31] Nguyen H S, Tashiro M, Inagakic M, Yamada S and Noda T 2015 *Soils and Foundation* 55 vol 5 pp 1044–57
[32] Li Y, Gong H, Zhu L and Li X 2017 *Water* vol 9 (1) p 64
[33] Antonellini M, Giambastiani B M S and Greggio N 2019 *Catena* vol 172 pp 76-86

IOP Conf. Series: Materials Science and Engineering **933** (2020) 012055 doi:10.1088/1757-899X/933/1/012055

Spatio-temporal shoreline changes analysis based on Landsat satellite images in Lahewa Beach of North Nias due to tectonics 2005

A Fauzia[1], S Syamsidik[2] and S Syahreza[3]

[1] Civil Engineering Postgraduate Program, Department of Civil Engineering, Universitas Syiah Kuala, Banda Aceh, Indonesia
[2] Department of Civil Engineering, Universitas Syiah Kuala, Banda Aceh, Indonesia
[3] Physics Department, Universitas Syiah Kuala, Banda Aceh, Indonesia

Corresponding e-mail: syamsidik@tdmrc.org

Abstract. The issue of accretion and erosion is one of the most comprehensive challenges in coastal morphology area. North Nias Regency is one part of Nias Island, located in the Megathrust Zone, which unfolds on Indo-Australian and Eurasian Plate subduction zones having a potential to interfere with the coastal region. Therefore, in order to detect the morphological shoreline changes, a time-efficient method of remote sensing is required. The purpose of this research is to detect the horizontal deformation of shoreline in the coastal area of North Nias Regency. This study uses Landsat 5 (TM), Landsat 7 (ETM +) and Landsat 8 (OLI) data. Also, the satellite imagery is utilized which consists of multitemporal data of pre-seismic (2000, 2003, and 2006), co-seismic (2005 and 2006), and post-seismic (2006, 2011, 2018). Furthermore, mapping and DSAS (Digital Shoreline Analysis System) methods are used. Mapping is done by digitizing the shoreline changes that occur by using the Landsat tools. On the other hand, the DSAS analysis is used to calculate and classify the area of erosion and accretion along the coast. The transect is given various colors to distinguish the shoreline alteration differences. The results show that maximum (accretion) and minimum (erosion) EPR values earned in every period in respective order are: pre-seismic with 43.540 m/yr and-80.810 m/yr, co-seismic with 600.840 m/yr and -10.590 m/yr, and post-seismic with 3.420 m/yr and -47.020 m/yr.

1. Introduction

The coastal area is an area that unceasingly exposed to the influence of natural changes and human activities. It has a shoreline that is changing over time as its dynamic condition is. On 28th March 2005, there was a great tectonic earthquake with 8.7 Mw power which caused a vertical and a horizontal shoreline alteration in this area [1]. Horizontally, the beach is characterized by the movement of the shoreline. Erosion and accretion processes can affect daily aspects of human life indirectly such as interfering with infrastructures, cultivations, tourism, and natural resources [2]. Nowadays, the community and the government of North Nias Regency exploit some of its areas as tourism places by building surrounding huts, inns, and infrastructures. These changes will bring an impact on coastal vulnerability to society and ecology [3]. The rate of shoreline changes was calculated in order to design a strategy for planning dynamic conditions and preventing any hazards from the coast [4].

Therefore, in order to detect and analyze the condition of the shoreline morphology, a time-efficient method is required. Historically, the GIS (Geographic Information Systems) and remote sensing applications can be used effectively for detecting the area condition and estimating occurring changes [5-7]. The quantitative and qualitative evaluation of spatial and temporal shoreline changes are solved by using satellite imagery, aerial photographs, and other methods [8]. In this study, we focused on discussing shoreline changes in its lateral way within pre-seismic, co-seismic, and post-seismic periods.

Lahewa Beach is located in the north of Nias coast and contiguous to the Indian Ocean. The length of the Lahewa shoreline is approximately 22365,962 meters. Geographical coordinates of the Lahewa unfold between N 01° 25 '30 "E 97 06' 40" and N 01° 24 '11 "E 97 10' 10" (figure 1). The Geological conditions of the Lahewa coast consist of coral reefs that are visible after the tectonic activity (post-seismic period) in September 2018 (figure 2).

Figure 1. The map of Lahewa Shoreline.

Figure 2. Lahewa coast condition in the post-earthquake period (Aerial photograph, September 2018).

2. Methodology

First, Lahewa Beach was divided into three time periods, which are the period of the year 2000 - 2005 (pre-seismic), the period of the year 2005 - 2006 (co-seismic), and the period of the year 2006 - 2018 (post-seismic). Furthermore, shoreline changes were investigated by using satellite imagery data by Landsat USGS with every photograph having a resolution of 30 meters. The captured Landsat data consist of the sensor of Landsat 4-5 TM (Thematic Mapper), Landsat 7 ETM + (Enhanced Thematic Mapper Plus), and Landsat 8 OLI (Operational Land Imager) (the details can be seen in table 1). All of the data are available on the USGS website (earthexplorer.usgs.gov). This study used the DSAS (Digital Shoreline Analysis System) method which is an add-in toolbox in the Esri ArcGIS software [9]. DSAS is a tool to analyze the shoreline deformation based on satellite imagery [10].

Table 1. Description of the data.

Satellite	Date of Acquisition	Sensor	Resolution
Landsat 4-5	13 March 2005	TM	30 m
	29 December 2006	TM	30 m
	10 February 2011	TM	30 m
Landsat 7	19 May 2003	ETM +	30 m
	13 July 2000	ETM +	30 m
Landsat 8	04 May 2018	OLI	30 m

First, satellite imageries were downloaded from the USGS website. Then, the imageries were extracted by using reflections from several bands obtained. After that, digitizing was conducted by using Landsat tools called Tasseled Cap. This feature generates brightness, greeners, and wetness data. The data format was produced in tiff. Furthermore, tiff data were set as the reference for the digitization process. Each shoreline layer was obtained from different acquisition dates in order to indicate the status of difference [11].

Baseline and shoreline were generated with some basic attribute field. Depending on the shoreline that has been merged, the baseline was stretched along 150 meters from the nearest shoreline. This distance was chosen for the reference to draw the transect lines from the baseline [12]. The formed transect line is a line that perpendicular to the baseline with a specified transect distance [13].

2.1. Classification of erosion and accretion

After the transect line was generated by using the DSAS method, then the transect was calculated by using the equation of EPR (End Point rate) and LRR (Linear Regression Rate) values. The DSAS technique can be used to generate the statistic for several conditions using EPR and LRR [10]. The EPR equation applied in DSAS can be seen in equation (1) as follows:

$$\mathrm{EPR} = \frac{NSM}{Time\ beetwen\ earliest\ and\ recent\ of\ shoreline} \quad (1)$$

The Net Shoreline Movement (NSM) is the distance between the earliest and recent of shoreline in the formed transect in meters [2]. The EPR value is determined by dividing the distance value of the shoreline change (NSM) with the time between the beginning and the end of the year. The LRR is the value produced from a combination of several shorelines and calculated to find the regression equation that fits our data. The linear regression is easily affected by outlier effects disregarding the rate of other statistics [2]. The rate of EPR and LRR Data can be classified as erosion and accretion according to the level of shoreline changes per year [12]. The changes in Lahewa Beach North Nias were categorized into 7 shoreline classifications as in table 2.

Table 2. Shoreline classification.

No.	Shoreline Classification	Rate of Shoreline Change (m/y)
1.	Very High Erosion	>-s
2.	High Erosion	>-s/2 to <-s
3.	Moderate Erosion	>-s/2 to <0
4.	Stable	0
5.	Moderate Accretion	>0 to <s/2
6.	High Accretion	>s/2 to <s
7.	Very High Accretion	>s

3. Results and discussion

Along the Lahewa coast, transect lines were drawn with the gap of 100 meters between the initial transect to the other neighboring transects. Every period was constructed in different transects according to the length of the transect line. This is because the shoreline changes rate had a different rate in each period. Erosion and accretion were found in several periods along the shoreline of Lahewa Beach, Northern Nias.

3.1. Pre-seismic period

Figure 3 shows the level of erosion and accretion that occurred in the pre-seismic period of 2000 – 2005 which are shown as transect lines. Among the 96 transects, the erosion process was occurred in 55 transects (57%). On the other hand, the accretion process happened in 41 transects (43%). The length of the transect is 650 meters long. The tendency of the accretion process is significant in several areas that adjacent to the hill or inland covers as in the east. The erosion process can be seen at the northern shoreline area on the satellite imagery of July 13th, 2000. The maximum and minimum shoreline changes on the EPR is 43.540 m/year and-80.810 m/year in respective order (figure 4).

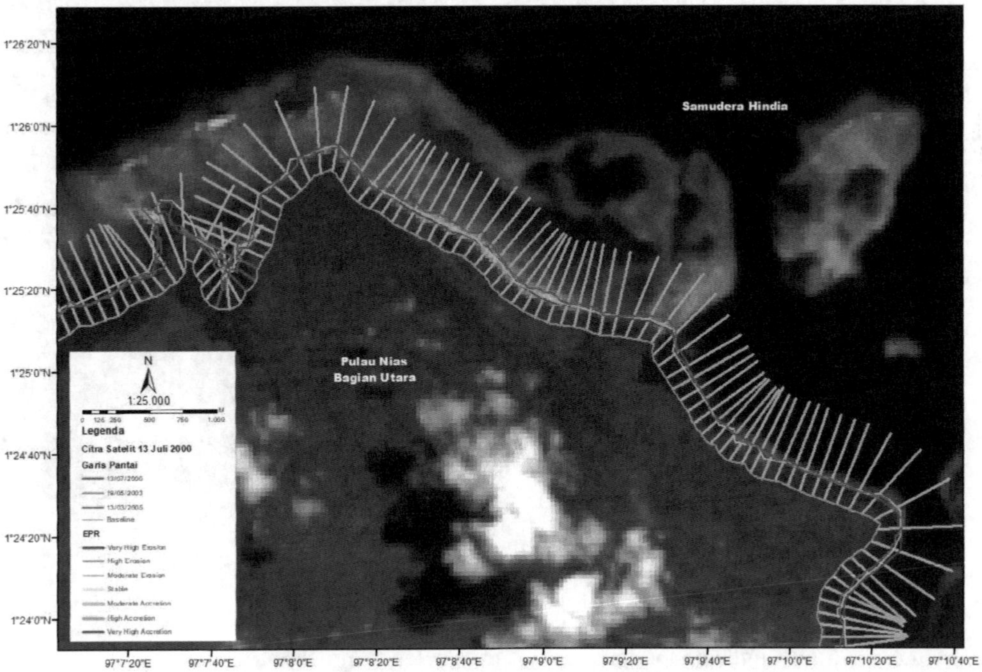

Figure 3. Classification map of erosion and accretion along the shoreline in 2000, 2003, and 2005.

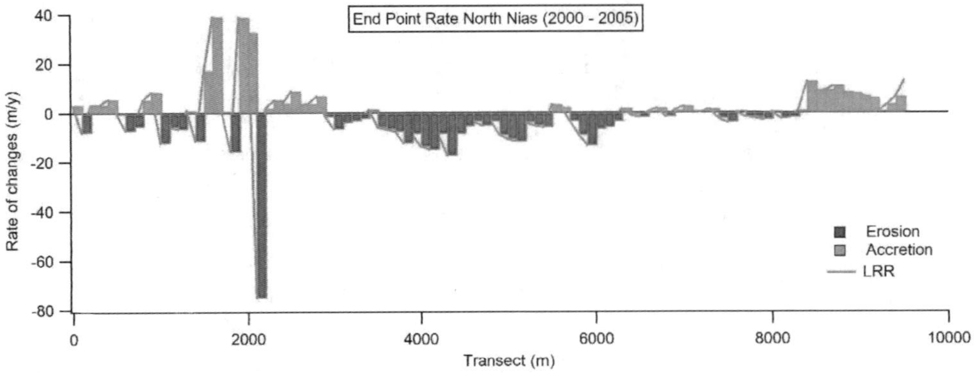

Figure 4. Shoreline changes the rate of Northern Nias in 2000 – 2005.

3.2. Co-seismic period

The shoreline changes rate shown in figure 5 was the most significant which is occurred in the period 2005 – 2006. In this classification, the value was very high, peaking at 600,840 m/yr representing the accretion process on the west coast side. Moreover, in this period, tectonic activity was recorded with notable indications of deformation at several points with the length of the transect reaching 1,300 meters.

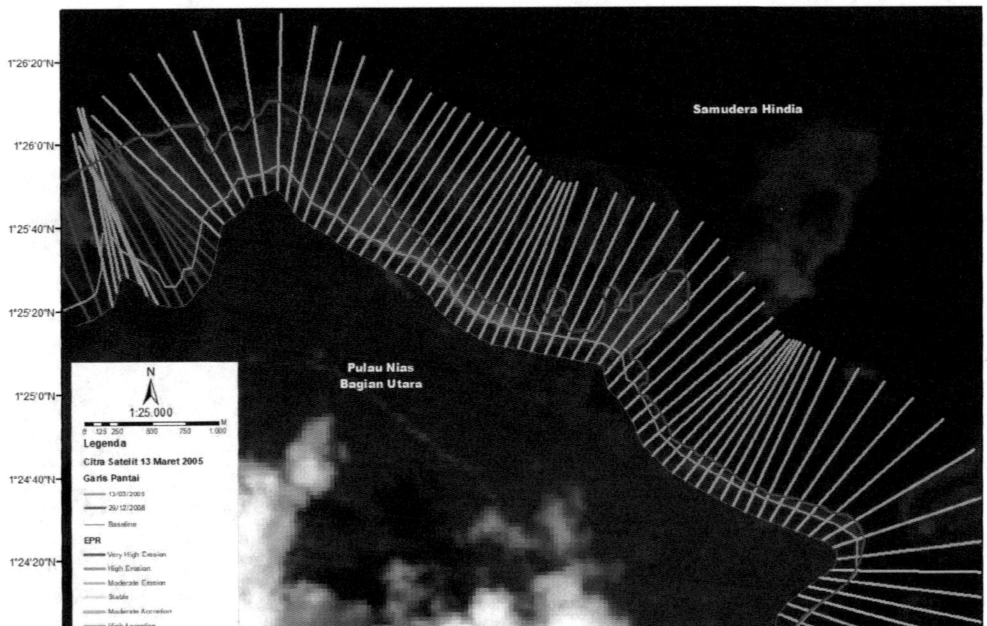

Figure 5. Classification map of erosion and accretion along the shoreline in 2005 – 2006.

The shoreline conditions indicate the tendency of the accretion process to form rapidly and expanding perpendicularly to the shoreline after the tectonic activity. The rate of land loss can only be observed in the southeast area with a minimum value of -10,590 m/yr (figure 6). The satellite imagery was received from the period of co-seismic on March 13, 2005. In this picture, the accretion process was still not apparent.

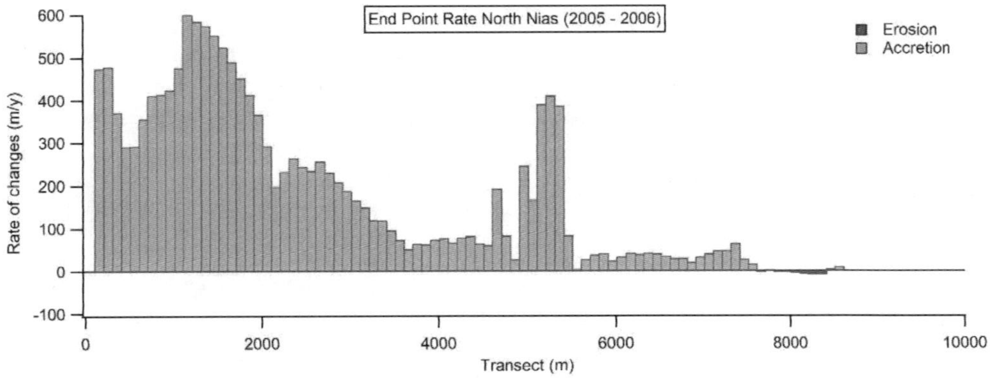

Figure 6. Shoreline changes the rate of Northern Nias in 2005 – 2006.

3.3. Post-seismic period

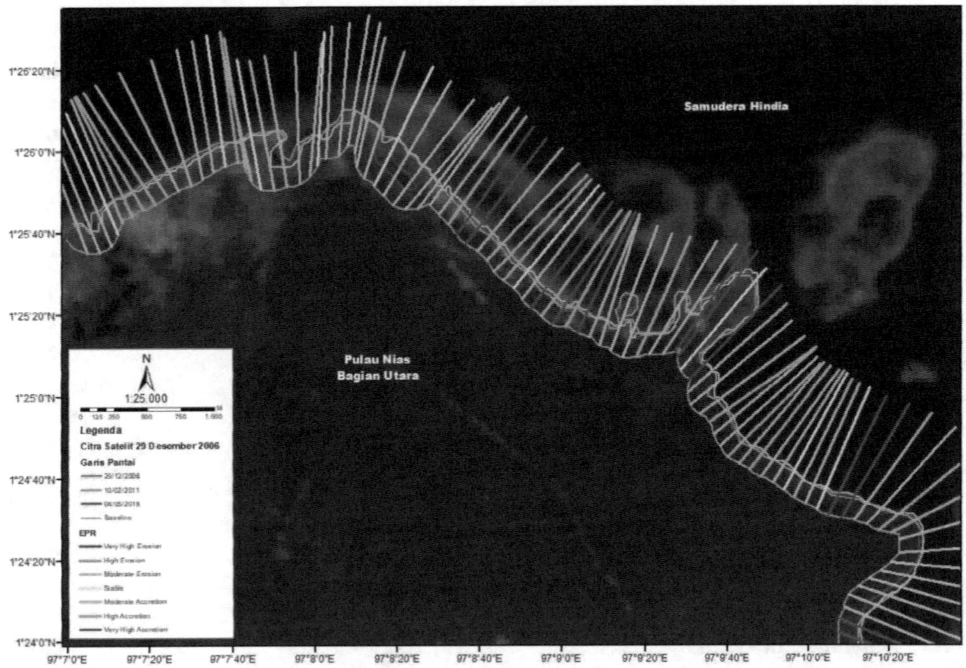

Figure 7. Classification map of erosion and accretion along the shoreline in 2006, 2011, and 2018.

The level of erosion and accretion processes were combined in the period 2006 - 2018 (figure 7) where the changes that occurred are insignificant. After the accretion process, the shoreline had been expanded to hundreds of meters. Also, in this period, the erosion process is unimportant. From the satellite imagery data on December 29, 2006, it can be seen that the significant changes are occurring in the previous period after a year the area being terraformed by a fairly-strong tectonic activity. The lowest rate which represents the erosion process in the co-seismic period is -47,020 m/yr and the highest rate is 3,420 m/yr (figure 8). The length of the transect in the post-seismic period is a thousand meters. During this period, the erosion process was more dominant compared to its counterpart.

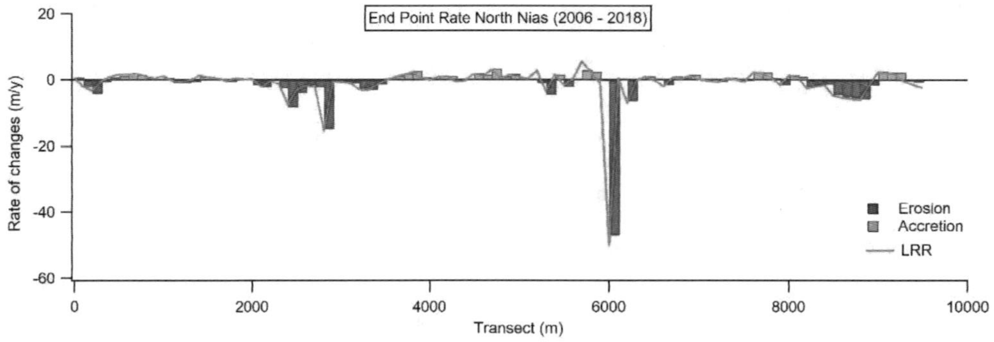

Figure 8. Rate of shoreline changes of North Nias in 2006 – 2018.

Table 3. Statistic of EPR and LRR Northern Nias.

Year		**Pre-seismic**	**Co-seismic**	**Post-seismic**	**Northern Nias**
Total of Transect		**96**	**87**	**96**	**279**
EPR	Average shoreline changes (m/yr)	-1,496	172,718	-0,959	56,90053
	Minimum shoreline changes (m/yr)	-80,810	-10,590	-47,020	-44,223
	Maximum shoreline changes (m/yr)	43,540	600,840	3,420	214,417
	Standard deviation (m/yr)	12,787	179,261	5,429	65,826
	Total transect of erosions	55	7	44	35
	Total transect of accretions	41	80	52	57
LRR	Average shoreline changes (m/yr)	-1,058	-	-0,964	-1,230
	Minimum shoreline changes (m/yr)	-75,060	-	-50,120	-65,465
	Maximum shoreline changes (m/yr)	38,990	-	5,520	24,53
	Standard deviation (m/yr)	12,2082	-	5,721	8,964
	Total transect of erosions	55	-	46	33
	Total transect of accretions	41	-	50	45

The value presented in table 3 came from the statistical analysis of shoreline rate changes of Lahewa, Northern Nias. Negative values indicate the erosion that happened and positive values indicate the accretion that occurred. The highest EPR and LRR values are in the period 2005 – 2006 which are occurred after the tectonic-earthquake activity. Total transects of accretion reached 80 out of 87 total transects. It is also influenced by vertical land movements after the earthquake. The highest value of standard deviation in the EPR came from the period between before and after the tectonic earthquake of the co-seismic period (2000 – 2006). Both conditions between pre-seismic and post-seismic periods are stable where the difference in the rate of erosion and accretion processes is relatively minimal.

4. Conclusion

DSAS and remote sensing methods are capable of detecting horizontal shoreline changes by determining the level of erosion and accretion. Shoreline change is a problem that has to be considered extensively. This is simply because a coastal area is a place where the community daily activities took part; it revolving in activities such as local economies, infrastructure developments, and tourist attractions.

From the DSAS result, every period on Lahewa Beach has a different pace in terms of altering the shoreline. The results showed that the rate of shoreline deformation was greater in the co-seismic period (2005 - 2006) compared to the others. In addition, tectonic activity in the co-seismic period of 2005-2006 brought significant changes to the shoreline. The shoreline position that changes before and after the earthquake made the rate of the changes even smaller. The maximum value of EPR is 600,840 m/yr coming from the co-seismic period and the minimum is coming from the post-seismic period with the rate of -80,810 m/yr. Furthermore, using satellite imagery will provide more accurate data for every time period. Further coastal area management by the government and community is needed to bolster a better regional and municipal development in Lahewa Beach. It is also mandatory for future disaster mitigation and resilience in the area.

References

[1] Briggs R W, Sieh K, Meltzner A J, Natawidjaja D, Galetzka J, Suwargadi B, Hsu Y-J, Simons M, Hananto N, Suprihanto I, Prayudi D, Avouac J-P, Prawirodirdjo L and Bock Y 2006 Deformation and slip along The Sunda Megathrust in The Great 2005 Nias Simeulue Earthquake *Science* 311

[2] Sheeja P S and A G A J 2016 *Application of Digital Shoreline Analysis System in Coastal Erosion Assessment* vol 6 no 6 pp 7876–83

[3] Dhiauddin R and Gemilang W A 2018 Coastline alteration rate of Weh Island, Aceh Province, Indonesia *IOP Conf. Ser.: Earth Environ. Sci.* 216 012010

[4] Dolan R, Fenster M S and Stuart J H 1991 Temporal analysis of shoreline recession and accretion *J. of Coastal Research* **7** 723–44

[5] Hegde A V and Akshaya B J 2015 Shoreline transformation study of Karnataka Coast: Geospatial approach *Aquatic Procedia* **4** 151 – 156

[6] Wen S et al. 2019 Coastal erosion monitoring and hazard degree assessment at Penglai Sandy Coast based on remote sensing *IOP Conf. Ser.: Earth Environ. Sci.* 234 012014

[7] Dong D et al. 2014 Automated techniques for quantification of coastline change rates using Landsat imagery along Caofeidian, China *IOP Conf. Ser.: Earth Environ. Sci.* 17 012103

[8] Selamat S N et al 2017 Extraction of shoreline changes in Selangor coastal area using GIS and remote sensing techniques *J. Phys.: Conf. Ser.* 852 012031

[9] Himmelstoss E A, Henderson R E, Kratzmann M G and Farris A S 2018 Digital Shoreline Analysis System (DSAS) Version 5.0 User Guide (Virginia)

[10] Suhaimi H M et al 2018 Assessment of river bank erosion at Kilim River, Langkawi using geospatial technique *IOP Conf. Ser.: Earth Environ. Sci.* 169 012012

[11] Raj N, Gurugnanam B, Sudhakar V and Francis G P 2019 Estuarine shoreline change analysis along The Ennore river mouth, southeast coast of India, using digital shoreline analysis system *Geodesy and Geodynamics* **10** 205 – 212

[12] Natesan U, Parthasarathy A, Vishunath R., Kumar E J G and Ferrer A V 2015 Monitoring longterm shoreline changes along Tamil Nadu, India using geospatial techniques *Aquatic Procedia* **4** 325 – 32

[13] Dewi S Ratna 2018 Monitoring long-term shoreline changes along the coast of Semarang *IOP Conf. Ser.: Earth Environ. Sci.* 284 012035

Geophysical investigation of road failure the case of the main street of Alue Naga, Banda Aceh, Indonesia

M Syukri[1], A M Taib[2] , Z Fadhli[2] and R Safitri[1]

[1] Physics Department, Universitas Syiah Kuala, Banda Aceh, Indonesia
[2] Geophysical Engineering Program, Earth Sciences Engineering Department, Universitas Syiah Kuala, Banda Aceh, Indonesia

Corresponding e-mail: m.syukri@unsyiah.ac.id

Abstract. A geophysical investigation has been conducted on the road to assess the subsurface soil characteristics. The study investigates the causes of road failure around the body of the damaged roads in the main street of Alue Naga, Syiah Kuala, Banda Aceh to determine the variations of rocks charge ability and resistivity value of the subsurface of the study area. The method employed for this study was 2D Electrical Resistivity and Induced Polarization (IP) method with using Wenner-Schlumberger array. Two profiles covering a distance of 100 meters each were established parallel to the road pavement along the stable and unstable sections of the road. Data were collected along the two profiles using ABEM Terra meter SAS 4000. The observed field data were processed and inverted using 2-D modelling inversion algorithm (RES2DINV Software). The result shows the presence of low resistivity and charge ability value at several parts of both line 1 and line 2. Both lines observed the 0-1.73 msec of charge ability value and <1 ohm.m of resistivity value; those values were interpreted as alluvium. In conclusion, the low resistivity and charge ability value present that the area investigation was a layer of permeable, especially in the central part which was cohesive, and led to the instability and weakness of the load-carrying capacity.

1. Introduction

In developing nations, the number of disappointments of structures such as buildings, dams, bridges and streets has expanded geometrically [1]. Numerous building professions, such as civil, geotechnical and electrical building, have a contention that geophysical assessment was taken significance portion amid the plan prepare of structure and utilities to support data of region projects such as bridges, dams, and interstates [6].

Road failure is one of the common failures of structure in developing countries. It has led to the loss of billions of dollars over decades due to either poorly constructed roads and under-maintained roads. The consequent daily loss of human life and economically significant properties should make road failure an alarming issue [10]. The Alue naga road is the major road that links Alue Naga and Banda Aceh. It has contributed significantly to the social-economic development of society surrounding the area. Its incessant failure has both economic and social effects on the inhabitant of suburban communities because it makes transportation difficult from moving into the region.

The study was conducted to investigate the effect of the geological factors in terms of the nature of the subsoil and the near-surface structures as the potential cause of failures around the body of the main road in Alue Naga Village, Syiah Kuala, Banda Aceh. The study utilized integrated geophysical

IOP Conf. Series: Materials Science and Engineering **933** (2020) 012056 doi:10.1088/1757-899X/933/1/012056

investigations. Geophysical surveys are efficient and cost-effective in providing geotechnical information since both the high speed- and appreciable-accuracy procedures are combined in providing subsurface information area [7].

2. Methodology

2.1. Geology of the study area

The location of the research was carried out at the main entrance from Banda Aceh City to Alue Naga Village, Syiah Kuala District. Geographically, Alue Naga Village is located at coordinates of 5 ° 58'48.02" N - 95 ° 35'25.59" E until 5 ° 59'72.19" N - 95 ° 34'37.66" E with elevations between 0.45 m to +1, 00 m above sea level (asl), with an average of 0.8 m asl. Based on the Aceh Regional Geological map, the measurement locations generally consist of Alluvial (Qh) deposits composed of loose materials such as clay, silt, gravel sand to lumps [2], as in figure 1.

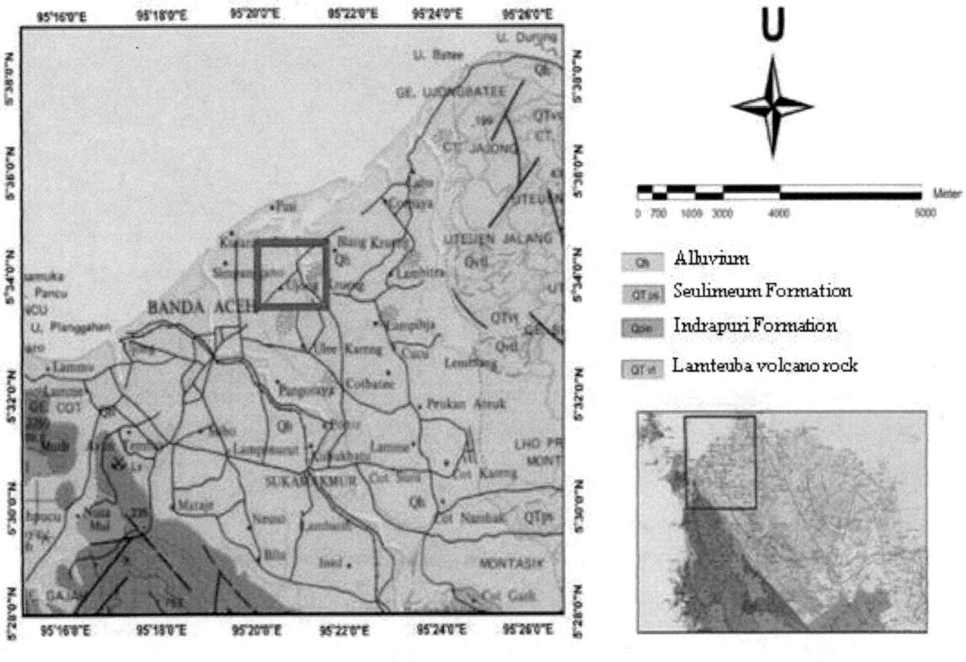

Figure 1 Geology map of study area [2].

2.2. Methodology

2-D resistivity method is an active method that is commonly applied in investigating the electrical properties of the subsurface by measuring the resistivity distribution of the material [3]. In 2-D resistivity method, the electrical current was injected into the ground through a paired electrode (A and B), and the measurement of the potential difference was made by another pair electrode (M and N), as shown in figure 2.

Induced polarization (IP) is a geophysical imaging technique used to identify the electrical charge ability of subsurface materials, such as ore. In this method, the researcher was looking for the portion of the earth where current flow is maintained for a short time after the applied current is terminated. The presence of chargeable material in the ground can often provide an excellent proxy for the distribution of sulphides in the subsurface. Traditionally, [8] charge ability is mapped using induced polarization (IP) surveys. In geotechnical and engineering, the use of IP has been limited mainly for groundwater exploration.

IOP Conf. Series: Materials Science and Engineering **933** (2020) 012056 doi:10.1088/1757-899X/933/1/012056

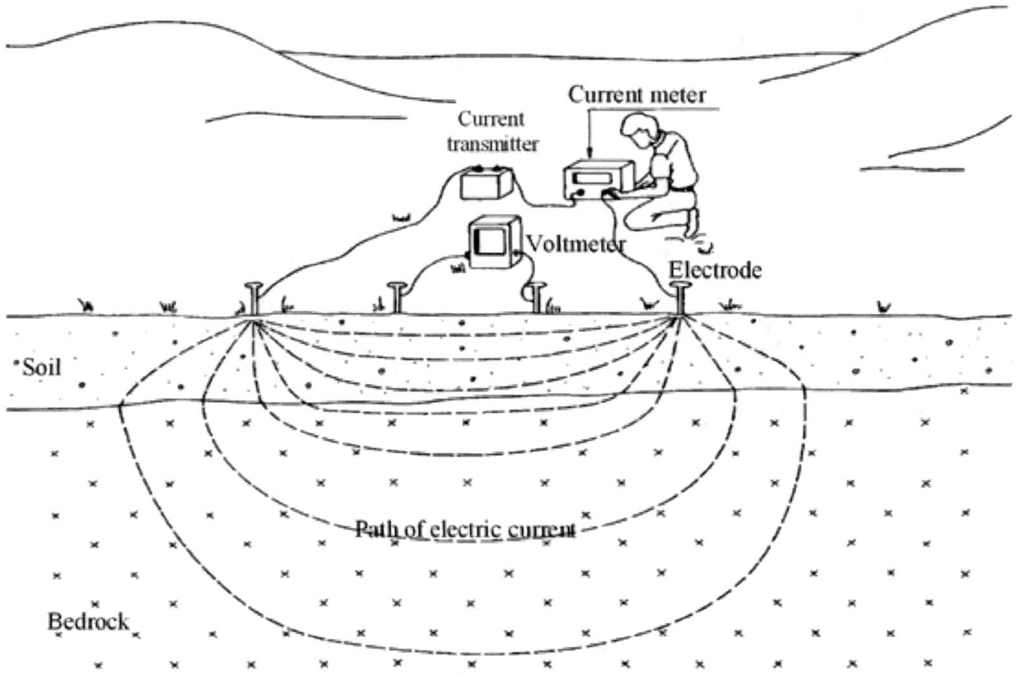

Figure 2. Four points electrodes arrangement with current and potential [4].

Earth materials such as igneous and metamorphic rock have high resistivity value based on the degree of fractures, while sedimentary rock has low resistivity value due to high water content. The resistivity value of rock and soil depends on several factors which are porosity, water saturation and mineral content. Table 1 shows the resistivity values of common rocks and soil types.

Table 1 resistivity value of rock and soil [5].

Material	Resistivity (ohm.m)
Alluvium	10-800
Sand	60-1000
Clay	1-100
Groundwater (fresh)	10-100
Sandstone	8-4 x 10^3
Shale	20-2 x 10^3
Limestone	50-4 x 10^3

Table 2 Chargeability value of rock and soil [8].

Material	Charge ability (m sec)
Volcanic Tuff	300-800
Rock Flakes	50-100
Granite rocks	10-50
Limestone	10-20
Water	0
Alluvium	1-4
Gravel	3-9
Silt	5-20
Sand	3-12
Clay	3-10

In the present study, the 2D resistivity and Ip survey were conducted out simultaneously along the same line. Two lines were used to measure both the resistivity and charge ability of subsurface material in the road failure area. The Wenner-schlumberger configuration array was used [9]. The lines were designed with 2.5 m electrode spacing with a total length of the spread of 100 m respectively. Figure 3 shows 2-D resistivity and IP lines surveys.

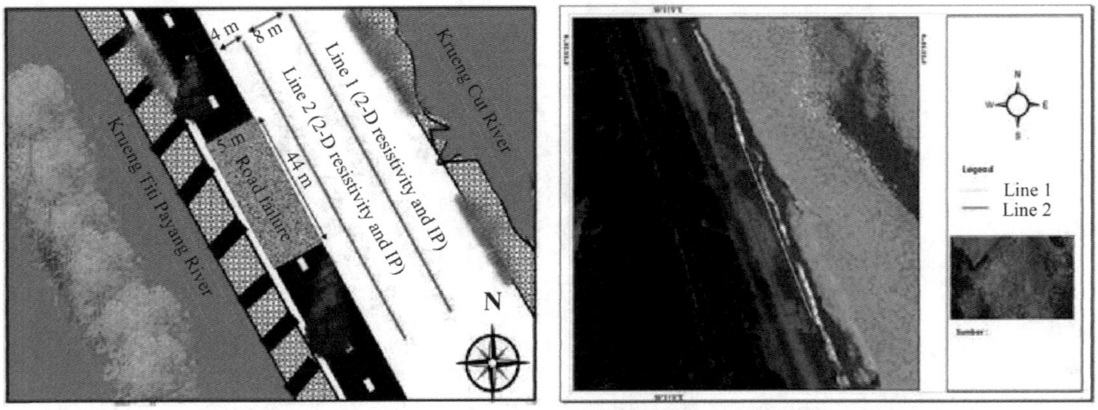

Figure 3. Layout line survey.

The IP-charge ability and resistivity data were generated and recorded automatically by The ABEM SAS4000 Terrameter that employed in a multi-electrode array using ES10-64C electrode selector, smart cable, 41 stainless steel electrodes and jumpers. These data were later extracted and processed using RED2INV software to convert the apparent resistivity data to true resistivity by inversion. The resistivity meter measures apparent resistivity from which pseudo sections were developed and subsequently inverted to true resistivity 2D sections.

3. Results and discussion

The line 1 (2-D resistivity and IP survey) was located in the east of Krueng Raya river (figure 3). The charge ability value of inversion results shown a range of 0 – 200 m.sec with the depth of 21 m approximately (figure 5). The charge ability value is 0 – 1.73 m.sec at the depth 0 - 5 m, approximately, which indicated the presence of low charge ability value as alluvium that contains sand.

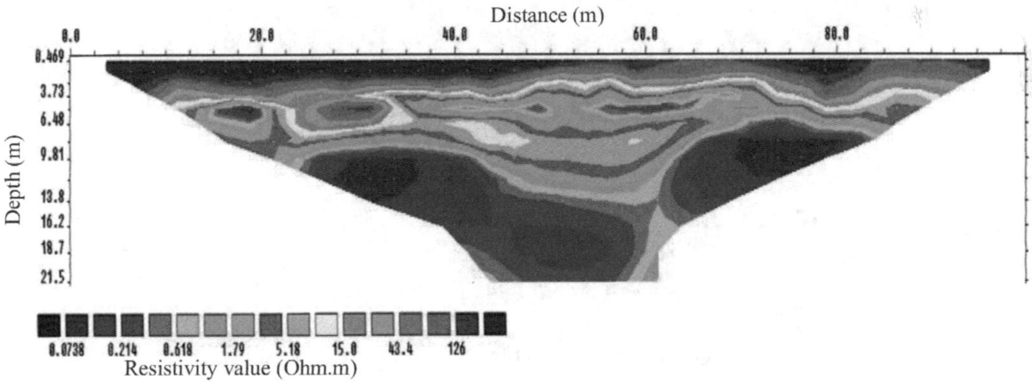

Figure 4. Inversion result of 2-D resistivity section line R1.

The charge ability value of 5.66 - 100 m.sec was indicated as sand clay. At a depth of 10 - 25 m, the charge ability value was more than 100 m.sec, which is indicated as clay. Furthermore, there are three layers that are presented in the line 2 for the inversion result of 2-D resistivity. The first layer is topsoil as sandstone at a depth of 1 - 4 m, approximately with the resistivity value of > 126 ohm.m. The second layer, at a depth of 4 - 10 m, was indicated as sand with a resistivity value of 1.01 – 15 ohm.m. The last layer is Water-intruded clay with a resistivity value of less 1 ohm.m at a depth of 10 – 20 m approximately.

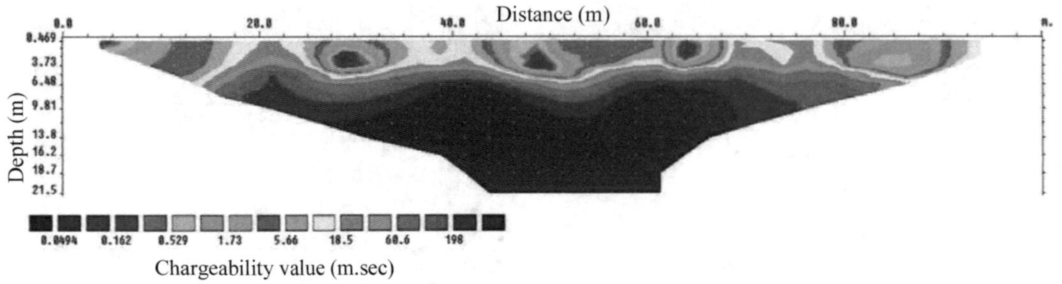

Figure 5. Inversion result of IP section line R1.

The position of the second line (2-D resistivity and IP survey) has a distance of 8 m with line 1 (figure 3). Lane 2 is located exactly on the road failure at a distance of about 30-70 m. The value of the charge ability value on line 2 is 0.01-200 m.sec with a depth of about 25 m. In general, the charge ability value of the second track is a low value (figure 7). A charge ability value of 0-1.73 m.sec at a depth of 0-5 m indicates a low power value as the sand containing river water seepage. While the charge ability value of 5.66-190 m.sec is indicated as the sand which is at a depth of 5 -10 m. The charge ability value > 100 m.sec indicated the clay which is at a depth of 10-25 m. On the other hand, the resistivity value at line 2 yields the results with the same structure and the value of the resistivity range with line 1. The results of the resistivity method at line 2 indicate the presence of three layers for the inversion results from the 2-D resistivity. The first layer is top soil as sandstone at a depth of 1-4 m with a resistivity value of 13-70 ohms. The second layer, which is at a depth of 4-10 m, is shown as sand with resistivity values of 1.01-15 ohm.m. The last layer is an intrusion clay with a resistivity value of less than 1 ohm.m at a depth of about 10-20 m.

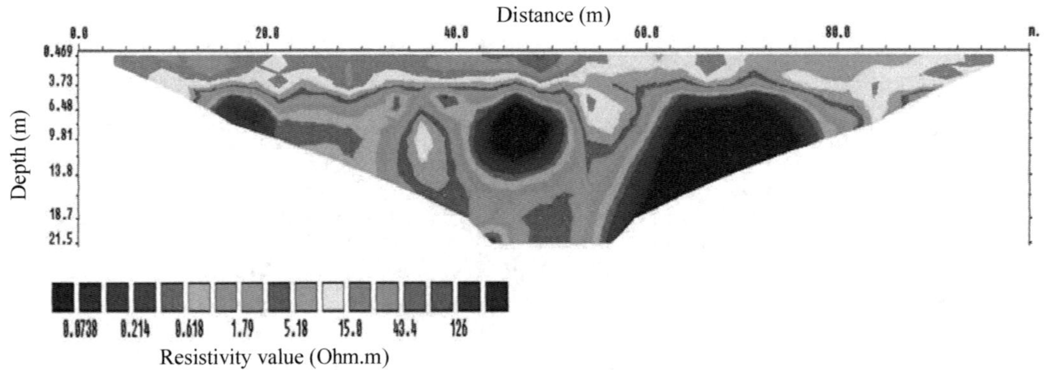

Figure 6. Inversion result of 2-D resistivity section line R2.

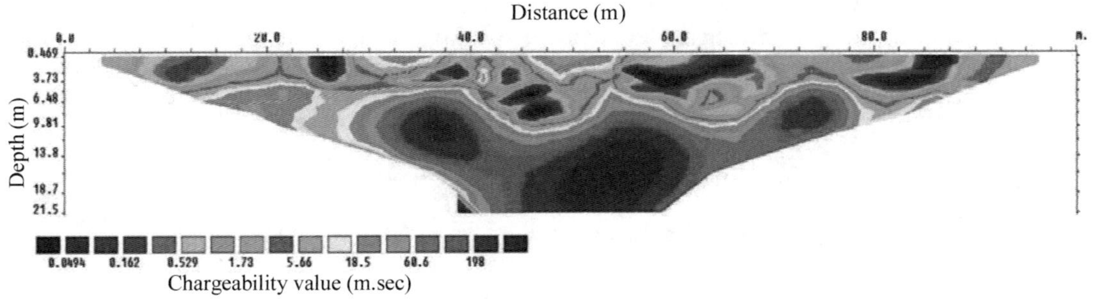

Figure 7. Inversion result of IP section line R2.

4. Conclusion

The indicated resistivity value is low (less than 1 ohm.m), and the chargeability value is also low (below 1 m.sec), which is obtained in several positions on paths 1 and 2. These facts indicate the flow of seepage. Seepage flow is indicated as an underground flow that connects 2 rivers namely the Krueng Cut river and the Krueng Titi Panyang river.

Acknowledgments

The authors would like to thank the Ministry of Education and Culture Indonesia, Syiah Kuala University, for its financial support in the scheme of Decentralized Research Competitive Grants Program PLK 2019, Geophysics team of Universiti Sains Malaysia, Penang (Malaysia), students and staffs of Faculty of Sciences and Faculty of Engineering, Syiah Kuala University, Banda Aceh (Indonesia).

References

[1] Akintorinwa O J and Adeusi F A 2009 Integration of geophysical and geotechnical investigation for a proposed lecture room complex at the Federal University of Technology, Akure, Southwestern, Nigeria *Ozean J. of Applied Sciences* **2** 3 ISSN 1943-2429

[2] Bennett J D, Bridge D McC, Cameron N R, Djunuddin A, Ghazali S A, Jeffery D H, Kartawa W, Keats W, Rock N M S, Thomson S J and Whandoyo R 1981 *Geologic Map of The Banda Aceh Quadrangle, Sumatra* (Geological Research and Development Centre) Systematic Geological Map, Indonesia

[3] Saad R, Muztaza M N, Zakaria M T and Saidin M M 2017 Application of 2d resistivity imaging and seismic refraction tomography to identify Sungai Batu sedimentation depositional origin *J. of Geology and Geophysics* **6** 268

[4] Dahlin T 2001 The development of DC resistivity imaging techniques *Computers & Geosciences* (Elsevier) **27** 9 1019-29

[5] Keller G V and Frischknecht F C 1966 *Electrical Methods in Geophysical Prospecting* (New York: Pergamon Press)

[6] Olusola F O, Victor O and Philips O M 2015 Geotechnical studies of the sub-grade material beneath Akure-Ijare Highway, Ondo State, Southwestern Nigeria *The Int. J. of Science and Technoledge* **3** 11 62

[7] Adenika C I, Ariyibi E A, Awoyemi M O, Adebayo A S, Dasho O A and Olagunju E O 2018 Application of geophysical approach to highway pavement failure: a case study from basement complex terrain southwestern Nigeria *Int. J. of Geo-Engineering* **9** 1 8

[8] Telford W M, Geldart L P, Sheriff R E and Keys D A 1990 *Applied Geophysic* 2th Edition (London: Cambridge University Press)

[9] Sumner J S 1978 *Principle of Induced Polarization for Geophysical Exploration* (Amsterdam: Elsevier Scientific)

IOP Conf. Series: Materials Science and Engineering **933** (2020) 012056 doi:10.1088/1757-899X/933/1/012056

[10] Peter J E, Rafiu A A, Udensi E E, Salako K A, Alhassan U D and Adetona A A 2018 2D Electrical Resistivity Imaging Investigation On Causes Of Road Failure Along Kutigi Street, Minna, North Central, Nigeria *American J. of Innovative Research and Applied Sciences* **6** 5 221-6

Noise mitigation based on noise barrier for railway Makassar – Parepare line

M Hustim[1], I Ramli[2], R Zakaria[1] and E Evianista[1]

[1] Department of Environmental Engineering, Faculty of Engineering, Universitas Hasanuddin, Indonesia
[2] Department of Civil Engineering, Faculty of Engineering, Universitas Hasanuddin, Indonesia

Corresponding e-mail: muraliahustim@ft.unhas.ac.id

Abstract. This research aimed to analyze the noise reduction based on the noise barrier along the Makassar-Parepare railway. The measurement was conducted on Makassar-Parepare line, which consists of 4 types of areas. Six of the observation points are in residential areas, 1 point in commerce area, 1 point in the green open space area, and 1 point in the station area. Each region has 6 measurement points. So that the whole measurement points are 54 measurement points. The noise level measurement was conducted using simple SLM (sound level meter) placed at 3 meters from the side of the road with a height of 1.2 meters from the road surface and omitted the background noise by using windscreen from the SLM. Measurements were operated every 10 minutes for four-time intervals which are 06.00-09.00, 09.00-14.00, and at 14.00-17.00. The research result shows the settlement areas consisting of 36 measurement points, in which 69% has exceeded the quality standard, and 31% are still below the standard. For the green open space area which includes six measurement points, 17% has exceeded the quality standard, and 83% are still below the standard. For the commerce area and stations which consist of 6 measurement points, 100% of the areas are still below the standards. The noise barrier design could reduce the noise level 5.0 until 7.7 dB.

1. Introduction

Population growth in an area will have a linear correspondence to the increase in transportation, where transportation has a critical role to support city development. Transportation in the community is one of the essential factors that support the movement of human life because it cannot be denied that transportation becomes a very important need to be able to carry out all activities. One of the most popular types of transportation in Indonesia is the train because it is not jammed and can get to the destination faster.

According to Decree of the Minister of State for the Environment: KEP-48 / MENLH / 11/1996 [1], it stated that while in the train, passengers will feel the vibrations generated by the train engine. The vibration is a regular movement of objects or media with an alternating motion from a position of balance.

One of the ways to overcome traffic congestion is that the Makassar city government is planning to build the Makassar-Parepare railway, which is where the construction is part of a transportation network system that connects from Makassar to Parepare in South Sulawesi. Train activities that will operate will cause both noise and vibrations as negative impacts on the environment and the community around the railway tracks because the noise generated from the train is quite noisy. The railway track passes through quite dense settlements that cause noise disturbance so that it can disturb the comfort of residents around the railroad tracks.

Based on the noise source, there are also various noise controls based on the source, which is usually a silencer made around the noise source. In the transmission, noise reduction is planned in the form of barriers, and noise control for the recipient is usually done by using PPE (personal protective equipment), for example, earplugs and earmuffs. So, to reduce noise levels, it is necessary to plan the noise controls using barriers in residential areas with the Nomograph method which is used to calculate how high the barriers needed to reduce noise as much as desired. So the results will be subtracted from the total scenario simulation. In this study, a noise management plan will be carried out using a barrier intended for settlement areas on the Makassar-Parepare railway as an action to reduce noise levels and can maintain health and psychology to create a residential environment that comfortable and calm.

The noise from the train also has a double form of steam, sound, and vibration due to the friction of the train wheels and railway tracks which are also made of hard material. The noise that comes from the train engine, horn, and friction between the wheels and brakes, often produce a squeaking sound. Noise from the train is felt by people who live or do some activities around the train station. Therefore, ideally, buildings along the railway line are designed with good acoustics to reduce noise and are designed with good attenuation to minimize vibrations from entering the building [2].

Based on these problems, the purpose of this study is to analyze the noise level in several areas along the Makassar-Parepare railway, make a mapping of noise distribution patterns along the Makassar-Parepare railway, and analyze the noise reduction based on noise barrier along the Makassar-Parepare railway.

2. Methodology

2.1. Data collection

The noise level measurement points on Makassar-Parepare line consist of 4 types of areas, i.e., 6 of the observation points in residential areas, 1 point in the commerce area, 1 point in the green open space area, and 1 point in the station area. Figure 1 shows the four measurement areas. Each region has 6 measurement points. So that the whole measurement point is 54 measurement points. Table 1 provides the distance between each observation point. The selected points considered suitable location condition criteria such as not blocked and the gap between instruments reflecting the sound.

The data collection process was conducted in March until May 2019. The noise level measurement used simple SLM (sound level meter) placed at 3 meters from the side of the road with height 1.2 meters from the road surface and omitted the background noise by using windscreen from the SLM [3]. The measurements were every 10 minutes for four-time intervals, i.e., in the morning between 06.00-09.00 WITA, afternoon measurements between 09.00-14.00 WITA, evening measurements at 14.00-17.00 WITA and night measurements at 17.00-22.00 WITA. Figure 2 shows the measurement method and equipment.

2.2. Methodology of data collection

2.2.1. Noise level. Value of L_{Aeq} and $L_{\mathrm{Aeq,day}}$ (daily equivalent) obtained by calculation of equation (1) and equation (2) [4].

$$L_{Aeq} = L_{50} + 0.43\,(L_1 + L_{50}) \quad (1)$$

Where:

L_{Aeq} = Noise level equivalent (dB),
L_{50} = Noise level 50% (dB), and
L_1 = Noise level 99% (dB).

$$L_{Aeq,day} = 10 \log 10 \left[\left(\frac{1}{tn}\right).\left(10^{\frac{L_{eq}1}{10}}\right) + \cdots + \left(10^{\frac{L_{eq}tn}{10}}\right)\right] \quad (2)$$

Where:

$L_{Aeq,day}$ = Daily noise level equivalent (dB),
$L_{eq}tn$ = Day-n noise level (dB), and
tn = Daily number of measurements.

2.2.2. Noise mapping using surfer 12.0. To find out the pattern of noise spreads that occurs at the school area, noise mapping was created by using the Surfer 12.0 program. This program is needed to create the visual of noise spreads from traffic activities around the school after calculating the noise level at each point of observation, where the distribution is classified in color.

Figure 1. Observation points.

Table 1. Distance between each observation points.

Land use	Observation points	Distance (m)					
		1	2	3	4	5	6
Settlements	Jl. Baco Enni	46	30	25	33	25	41
	Jl.A.Beddu Dg.Mange	54	25	41	25	45	49
	Jl.Pahlawan	35	25	25	83	96	110
	Jl.Pramuka	52	25	39	25	73	55
	Jl.A. Usman	200	145	110	85	25	25
	Jl.H.Ambo Tang	25	28	25	60	42	29
Green open space	Jl. Ceppaga	46	25	45	25	47	34
Commerce	Pasar Ajakkang	51	25	61	25	42	47
Stations	Stasiun Pekkae Soppeng	25	25	61	30	33	44

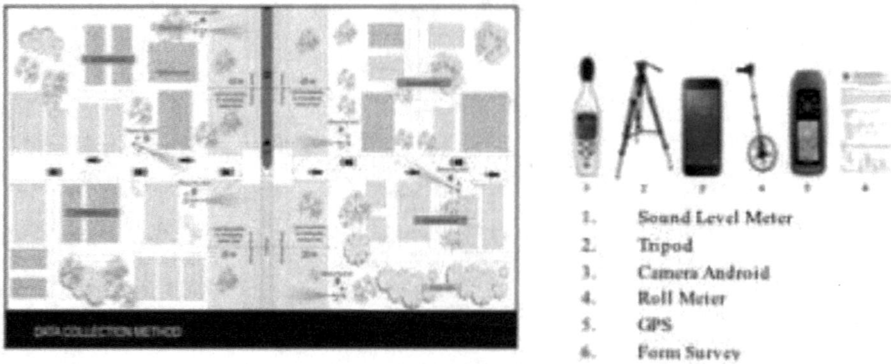

Figure 2. Measurement method.

2.2.3. Noise reduction. Noise reduction planned in the form of a fence (barrier) to reduce noise levels by using the Nomograph method and Noisetools software on the railway project in Barru Regency. Nomograph method was used to calculate how high the barrier needed to reduce noise as much as desired. Nomograph provides an accurate prediction for reducing the sound pressure level. Sound transmission loss is also affected by the frequency. For low frequencies, TL is influenced by the thickness of the wall, whereas for greater frequencies, it is affected by the mass of the wall [5]:

$$TL = (20 \log W) + (20 \log f) - C \tag{3}$$

The noise barrier is the most effective method to reduce the noise caused by vehicle engines (such as on the highway, in railway tracks, and airports) and industrial noise sources without stopping the activity of using this method.

3. Results and discussion

3.1. Results of measurement of noise level analysis in the settlement area

3.1.1. Noise level analysis in the settlement area. The value of noise level resulted from the measurements for the settlement area can be seen in figure 3 until figure 8. Figure 3 shows that point 1 has exceeded the standard by 2 dBA, and point 6 exceeds the standard by 1 dBA, while the other points are still below the standards. Figure 4 shows that the noise level is maximum at 61 dB and minimum at 54 dB. Comparing to the noise quality standard, which is 55 dBA, there are 5 points have exceeded the standard and 1 point is still below the standards. The result in figure 5 shows the noise maximum at 70 dB and minimum at 68 dB. All the point measurements exceeded the standard 55 dB. The result in figure 6 shows the noise maximum at 70 dB and minimum at 68 dB. All the point measurements exceeded the standard 55 dB. The result in figure 7 shows the noise maximum at 52 dB and minimum at 54 dB. All the point measurements are below the standard 55 dB. Figure 8 shows that the noise level maximum at 83 dB and minimum level at 69 dB if compared with the noise quality standard, which is 55 dBA, whole the points exceed the standard.

IOP Conf. Series: Materials Science and Engineering **933** (2020) 012057 doi:10.1088/1757-899X/933/1/012057

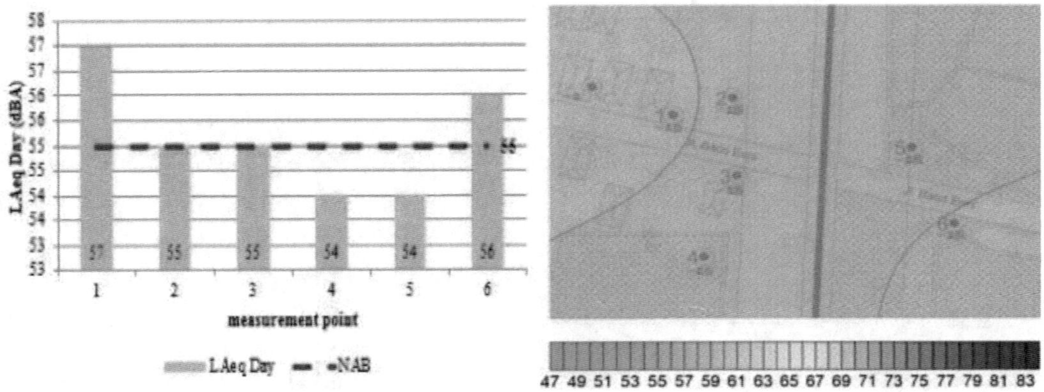

Figure 3. Noise level ($L_{Aeq,Day}$) in settlement area Jl. Baco Enni, Kab. Barru.

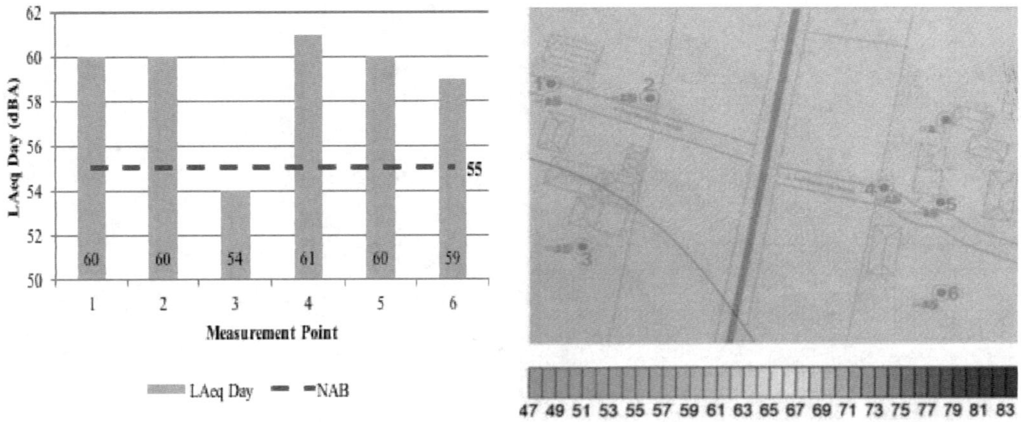

Figure 4. Noise level ($L_{Aeq,Day}$) in settlement area Jl. A. Beddu Dg. Mange, Barru Regency.

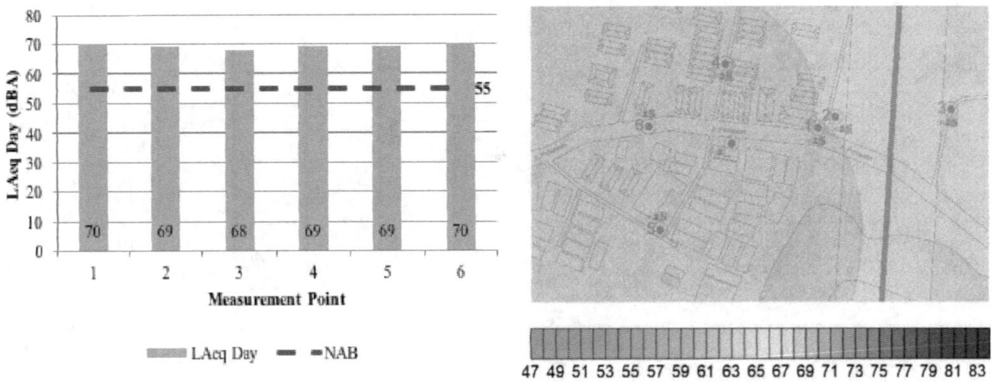

Figure 5. Noise level ($L_{Aeq,Day}$) in settlement area Jl. Pahlawan, Barru Regency.

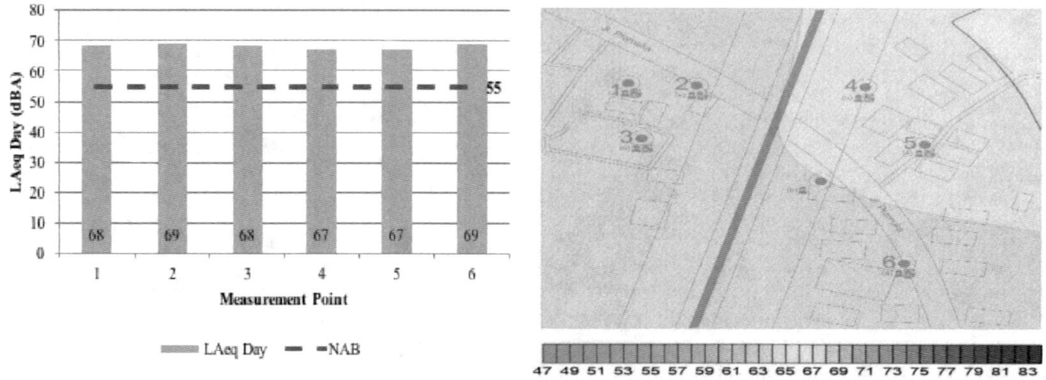

Figure 6. Noise level ($L_{Aeq,Day}$) in settlement area Jl. Pramuka, Barru Regency.

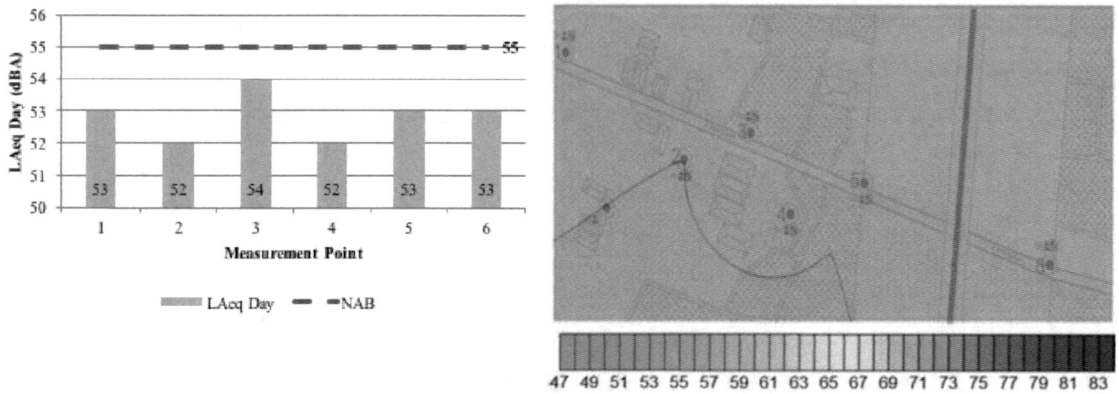

Figure 7. Noise level ($L_{Aeq,Day}$) in settlement area Jl. A. Usman, Barru Regency.

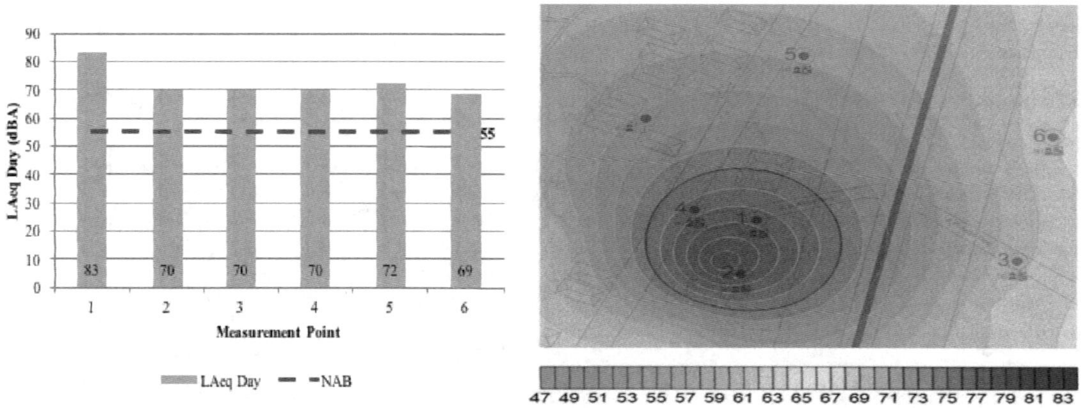

Figure 8. Noise level ($L_{Aeq,Day}$) in settlement area Jl. H. Ambo Tang, Barru Regency.

IOP Conf. Series: Materials Science and Engineering **933** (2020) 012057 doi:10.1088/1757-899X/933/1/012057

3.1.2. Noise level analysis in the green open space area. Compared to the noise quality standard according to pollution Decree of the Minister of State for the Environment: KEP-48 / MENLH / 11/1996, for the green space area of 50 dBA, point 1 has exceeded the standard with 4 dBA. Points 2, 3, 4, and 5 have exceeded the standard with 2 dBA. While point 6 is still below the standards.

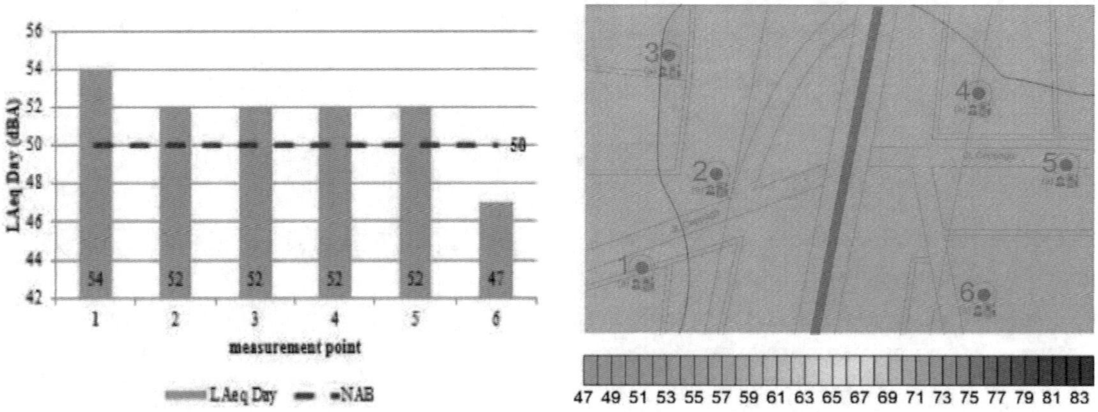

Figure 9. Noise level ($L_{Aeq,Day}$) in green open space Jl. Ceppaga, Barru Regency.

3.1.3. Noise level analysis in the commerce area. The noise level in commerce in figure 10 shows the maximum level is 62 dB if compared with the environmental quality standard of noise pollution Decree of the Minister of State for the Environment: KEP-48 / MENLH / 11/1996 for the commerce area of 70 dBA, all observation points were below the quality standard.

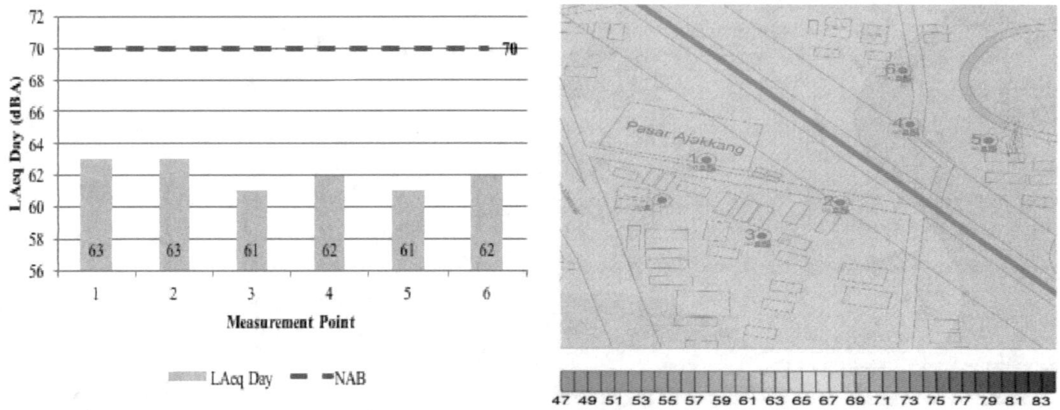

Figure 10. Noise level ($L_{Aeq,Day}$) in commerce area Pasar Ajakkang, Barru Regency.

3.1.4. The noise level was analysis in the station area. Compared with the noise standard according to the RI Health Minister No.718 of 1987 for Zone D for the station area, 60 dBA is recommended and 70 dBA is allowed. Then, all points had exceeded the recommended limit, i.e., points 1,3,5 exceeds 4 dBA and points 2, 4, 6 exceeds 5 dBA, so they have the potential to cause health problems

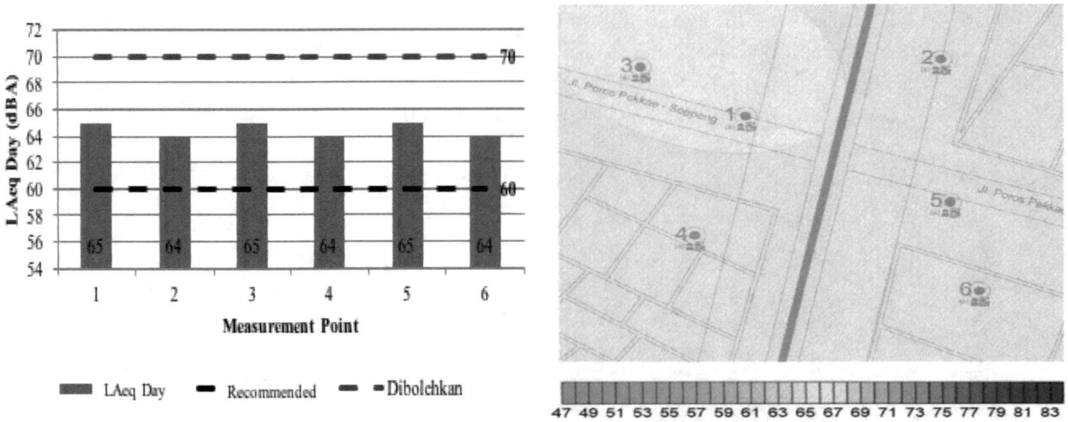

Figure 11. Noise level ($L_{Aeq,Day}$) in station area Jl. Poros Pekkae Soppeng, Barru Regency.

3.2. Noise reduction using the nomograph method

To reduce noise levels due to the construction of the Makassar-parepare railroad, the noise barrier method is used. The reduction by the nomograph method is applied by plotting on the graph. The location plan of the barrier design plan can be seen in figure 12, 13 and 14.

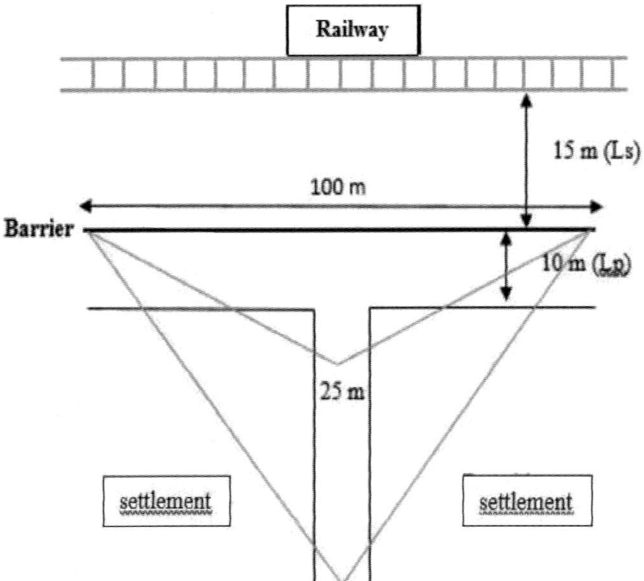

Figure 12. Noise barrier design plan.

The noise level before the barrier is 57 dBA, which if compared to the quality standard according to Decree of the Minister of State for the Environment: KEP-48/MENLH/11/1996 is 55 dBA and it means that it has exceeded the quality standard. The Nomograph graphic plotting results shows that with a distance between the noise source to the receiver of 25 m using a 15 m barrier break, barrier position of 25.3 m, and the angel substandard of 136.87, the result is the attenuation barrier of 7.2 dB which can be seen in table 2.

The result from noise barrier analysis indicates that noise barrier design could reduce noise level 5.0 until 7.7 dB. The noise level reduction, the difference between noise measured before and after the existence of the barrier, on the H. Ambo Tang road with a noise value before the barrier of 83 dB and after the barrier, the barrier attenuation is 5.0 dB. While the minimum difference can be seen in the Baco Enni road with a noise value before the barrier of 57 dB and after the barrier, the barrier attenuation is 7.2 dB in settlement areas. In the commercial area which is located in the Ajakkang road with a noise level before the barrier of 54 dB and after the barrier, the attenuation barrier value is 7.5 dB, and for the Pekkae station area, the noise level value before the barrier is 65 dB and after the existence of barrier is 7.0 dB.

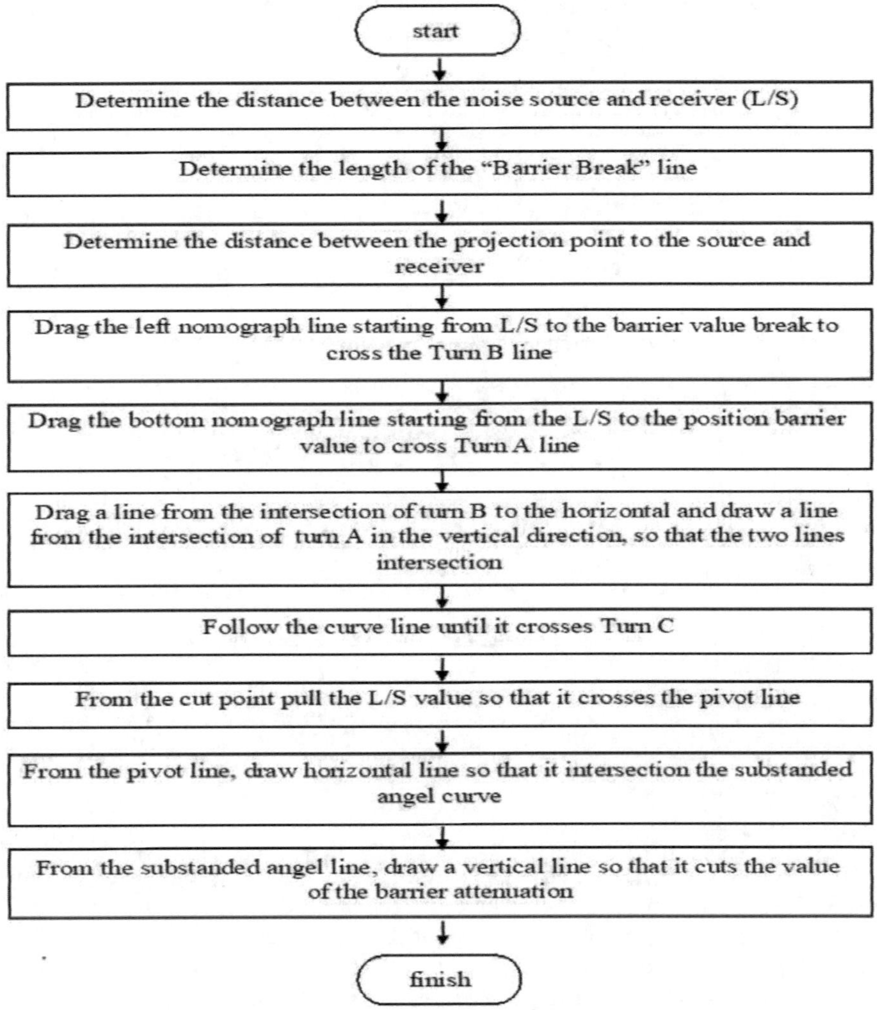

Figure 13. Nomograph method flowchart.

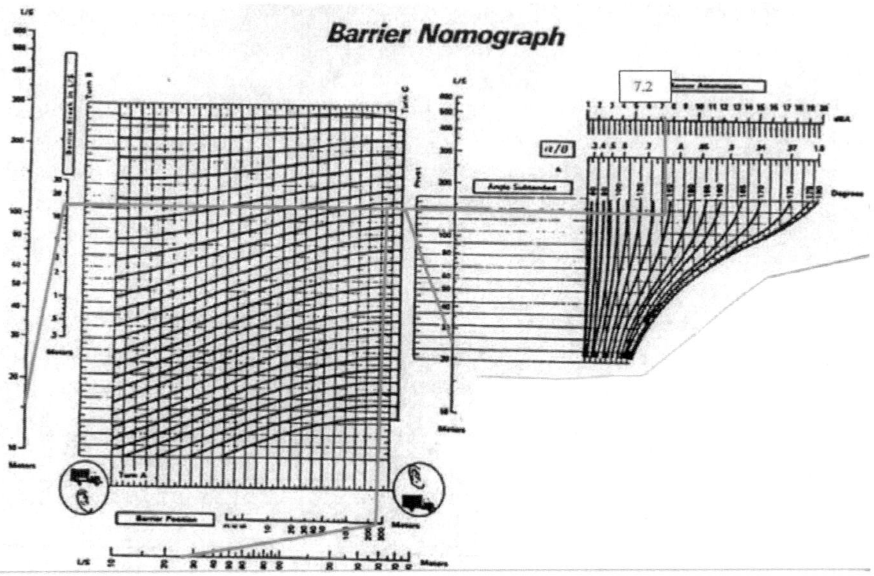

Figure 14. Nomograph graphic plotting.

Table 2. Barrier attenuation values for all location.

Observation points	Distance	$L_{Aeq\ day}$ Measurenment (dB)	Angle subtended	*Barrier Attenuation (dBA)*	NAB
Baco Enni	46	57	136.87°	7.2	55
A.Beddu	54	60	142.78°	7.6	55
Pahlawan	35	70	125.22°	6.0	55
Pramuka	52	68	141.44°	7.7	55
A. Usman	25	53	108.14°	5.0	55
Jl. H. Ambo Tang	25	83	108.14°	5.0	55
Pasar Ajakkang	51	54	140.82°	7.5	70
Stasiun Pekkae	44	65	135.08°	7.0	60

4. Conclusion

Based on the analysis of noise measurement for 4 types of areas around the railway tracks in Barru Regency, the different noise level percentages were obtained. The percentage of noise levels for settlement areas, green open space, and commerce is compared with the environmental quality standard for noise pollution, while the percentage of noise levels for the station area is compared with the noise standard of the Republic of Indonesia Minister of Health No.718 of 1987. For settlement areas consisting of 36 measurement points, 69% have exceeded the quality standard and 31% are still below the standard. For green open space area consists of 6 measurement points, 17% has exceeded the quality standard and 83% are still below the standard. For the commerce area and stations which include 6 measurement points, 100% of the area is still below the standards.

From the results of the noise measurement by using a different type of barrier reduction, different results are obtained. The result from noise barrier analysis indicates that noise barrier design could reduce noise level 5.0 until 7.7 dB.

IOP Conf. Series: Materials Science and Engineering **933** (2020) 012057 doi:10.1088/1757-899X/933/1/012057

References

[1] Kementrian Lingkungan Hidup Republik Indonesia 1996 *Keputusan Menteri Lingkungan Hidup Nomor 48 Tahun 1996 Tentang Baku Mutu Tingkat Kebisingan.*

[2] Mayangsari A P 2013 *Perancangan Barrier Untuk Menurunkan Tingkat Kebisingan Pada Jalur Rel Kereta Api Di Jalan Ambengan Surabaya Dengan Menggunakan Metode Nomograph* (Surabaya: Institut Teknologi Sepuluh November (ITS)).

[3] Ogata Y and Nagakura K 2012 Noise Reduction Effect of Low Barries Installed Adjacent to Rail *QR of RTRI* **53** No.3.

[4] Mediastika C E 2005 *Akustika Bangunan: Prinsip-prinsip dan Penerapannya di Indonesia* (Erlangga).

[5] Agustina S L 2012 *Pengaruh Intensitas Kebisingan Kereta Api Tehadap Gangguan Pendengaran Pada Masyarakat Tegalharjo Yang Tinggal Dipinggiran Rel Kereta Api* (Surakarta: Universitas Sebelas Maret).

IOP Conf. Series: Materials Science and Engineering **933** (2020) 012058 doi:10.1088/1757-899X/933/1/012058

Hospital fire safety improvement effort (study case Rumah Sakit Ibu dan Anak, Banda Aceh)

I Wahyuni[1], N Ismail[1] and I Izziah[2]

[1] Magister of Disaster Science, Universitas Syiah Kuala, Banda Aceh, Indonesia
[2] Architecture Department, Engineering Faculty, Universitas Syiah Kuala, Banda Aceh, Indonesia

Corresponding E-mail: dr.ika.wahyuni@gmail.com

Abstract. Rumah Sakit Ibu dan Anak (RSIA) Aceh Province is a public service agency that equipped well and also becomes a reference hospital. As a reference hospital, RSIA has many patients with a severe and critical condition which are difficult to be evacuated in a fire case. Therefore, the building of RSIA should have good fire safety to minimize risk for patients, hospital personnel, and equipment. The objective of this research is to appraise the safety building of RSIA with Fire Risk Index method. The appraisal was done for 23 hospital rooms by scoring 17 index parameters. Each parameter score then multiplied with respective weight and sum up to get a Fire Risk Index. Moreover, this research also assessed the hospital fire management with deep interview approach for several key persons. Next, the knowledge of hospital personnel and patients were also revealed with the questionnaire survey. The result found that RSIA has an intermediate level of building fire safety with an average index of 3.2. For fire management, it can be concluded that there is a system of fire management already available which is managed and commanded by Health and Safety unit. In addition, the equipment related to fire protection has been maintained regularly with support from the Disaster Mitigation Agency of Banda Aceh municipality. Related to the knowledge of hospital personnel, it revealed that some items need to be improved such as emergency code, the use of fire extinguisher, and emergency signal. For patients, the knowledge of using electrical equipment which has the potential to create fire should be improved. Lastly, patients also need to be informed about fire mitigation and evacuation.

1. Introduction

Fire safety is an important aspect of a Hospital day to day operation. Therefore, there are a few regulations that were issued to mitigate and prevent fire danger, such as Minister of Health decree No. 66/2016, which states that a hospital must carry on fire safety management to minimize risk on hospital resources, patients, visitors, and neighborhood environment [1].

Based on data that was issued by the Indonesian National Disaster Mitigation Agency [2], there were 2.048 fire cases from 2008 until 2018 and caused a loss of up to 43 billion Rupiah. On the other hand, Fire Fighting and Rescue Department of Banda Aceh Municipality recorded 287 fire cases from 2014 until 2017. Moreover, the Disaster Mitigation Agency of Aceh Province recorded 237 disaster events in range January until May 2019, which is dominated by the fire with 105 cases. These figures made fire is the most non-natural disaster which happens more frequently than others. Based on the facts [2], [3], and [4], fire mitigation and prevention actions are required and must be implemented in particular for public service buildings like the hospital.

Hospital is a high risk from fire case due to many people are inside the building and could cause live loss. A high number of people could be patients with various conditions and caused evacuation efforts are difficult to be done. In addition, damage and property loss are also taking into account due to the availability of costly medical and non-medical equipment.

Rumah Sakit Ibu dan Anak (RSIA) Banda Aceh is a public service agency and become a reference hospital for another hospital within Aceh Province. RSIA is equipped with good medical equipment and has been qualified as a B category for minimum standard medical service. As a reference hospital, the condition of patients could be in severe or even in a critical stage which is difficult to be evacuated in a fire case. Therefore RSIA must have a good fire safety, reliable management system and personnel ability to keep patients safe from fire. Although RSIA never gets a fire incident, the frequent use of mechanical and electrical devices has a high probability of a fire accident. A fire accident can be ignited from many activities among those: the use of equipment with high power consumption (i.e. radiology device), electrical generator for power backup, short circuit, gas stove in the kitchen, pressure tube, chemical material usage in laboratory or pharmacy, and so forth.

From preliminary observation, it was found that RSIA already has some active and passive fire protection tools such as fire extinguisher, smoke detector, hydrant, etc. However, it is not known how those tools affect the safety of RSIA building from fire. The existence of those tools, of course, must be enhanced by management and resources to gain an optimum safety level. But, from previous research that was conducted by [5] about analysis of policy implementation towards hospital disaster plan on earthquake and tsunami at RSIA, it found that several aspects of disaster management had not been accomplished, namely: disaster mitigation team was not ready, no standard operating procedure available and the lack of hospital personnel awareness priority for rescuing patients.

Based on those facts and problems, this research aimed to appraise the safety building of RSIA in a fire case and also looking for a strategy to get the right approach in improving building fire safety to minimize loss and saving lives of people in RSIA.

2. Methodology

This research is a combination of qualitative and quantitative methods (mixed methods). The quantitative method was used to appraise the fire risk index of RSIA building and to measure the knowledge of hospital personnel and patients about fire safety. The qualitative method took part in the key person deep interview activity which explores strategy to improve fire safety.

Fire Risk Index (FRI) is a method to appraise the safety level of a building for a fire case. This method was developed by the Fire Safety Department of Lund University Sweden in the year 2000. This method can be applied to all building types such as residential, commercial, government, public service, etc. The result of this method at the end is a score index within range 0-5. The higher score index shows the higher fire safety of a building and vice versa the lower index indicates less fire safety [6].

The fire safety appraisal using FRI method is carried out by assessing 17 parameters, there are: building wall material, fire suppression, fire service, building compartment, structure separation, doors, windows, facade, attic, adjacent building, smoke control system, detection system, signal system, escape routes, structure load-bearing, maintenance and information, and ventilation. Each parameter has a respective weight as in table 1.

Table 1. FRI Parameter weight.

No.	Parameter (P)	Weight (w)
1	Building wall material	0.0576
2	Fire suppression	0.0668
3	Fire service	0.0681

No.	Parameter (P)	Weight (w)
4	Compartment	0.0666
5	Structure separation	0.0675
6	Doors	0.0698
7	Windows	0.0473
8	Facade	0.0492
9	Attic	0.0515
10	Adjacent building	0.0396
11	Smoke control system	0.0609
12	Detection system	0.063
13	Signal system	0.0512
14	Escape routes	0.062
15	Structure load-bearing	0.063
16	Maintenance and information	0.0601
17	Ventilation	0.0558

Each parameter is rated from 0 to 5 and multiplied by the respective weight. The final index is a summation of those multiplication results and can be formulated with equation 1.

$$S = \sum_{i=1}^{17} w_i x_i \tag{1}$$

Where S = FRI index, w_i = weight of i_{th} parameter, x_i = Score of i_{th} parameter.

To measure the knowledge of fire safety for hospital staff and patients, the questionnaire instrument was used. The questionnaire for both groups is different. The knowledge level of correspondent was determined by a total true answer in percent. Instrument validity and reliability were tested in the current site at RSIA because there are specific questions in the instrument which cannot be applied for other hospitals due to the discrepancy of facilities and equipment. Bivariate Pearson correlation was used in the instrument validity testing. This validity method correlates each item score to the total score. In the process it will filter each question which has a significant correlation with the total score, it means the corresponding question is important to support the result validity.

The population number of hospital staff is 473 which is the total number of RSIA employees. Meanwhile, the population number of patients is 137, which was determined from the daily average number of patients. The sampling number was determined using Slovin formula as seen in equation 2. Using a 10% error margin for both groups, gave 83 hospital staff and 58 patients as sampling correspondent.

$$n = \frac{N}{1+Ne^2} \tag{2}$$

Where n = Sampling number, N = Population number, e = Margin error.

The key person-deep interview approach was carried out to explore more detail about hospital management and equipment related to fire safety. The find out from FRI and questionnaire will be consulted with the key person in order to formulate the strategy to improve fire safety at RSIA.

3. Results and discussion

The appraisal of fire safety was done for 23 rooms in RSIA. Each parameter was assessed and rated with the help of a local building consultant. The result shows that some parameters already have maximum score such as building wall material, structure separation, signal system, load-bearing capacity. On the other hand, the lower score can be found mostly in parameter fire suppression, attic and smoke control systems. Other parameters score varies from low to high. The spread of the score for each room can be seen in figure 1.

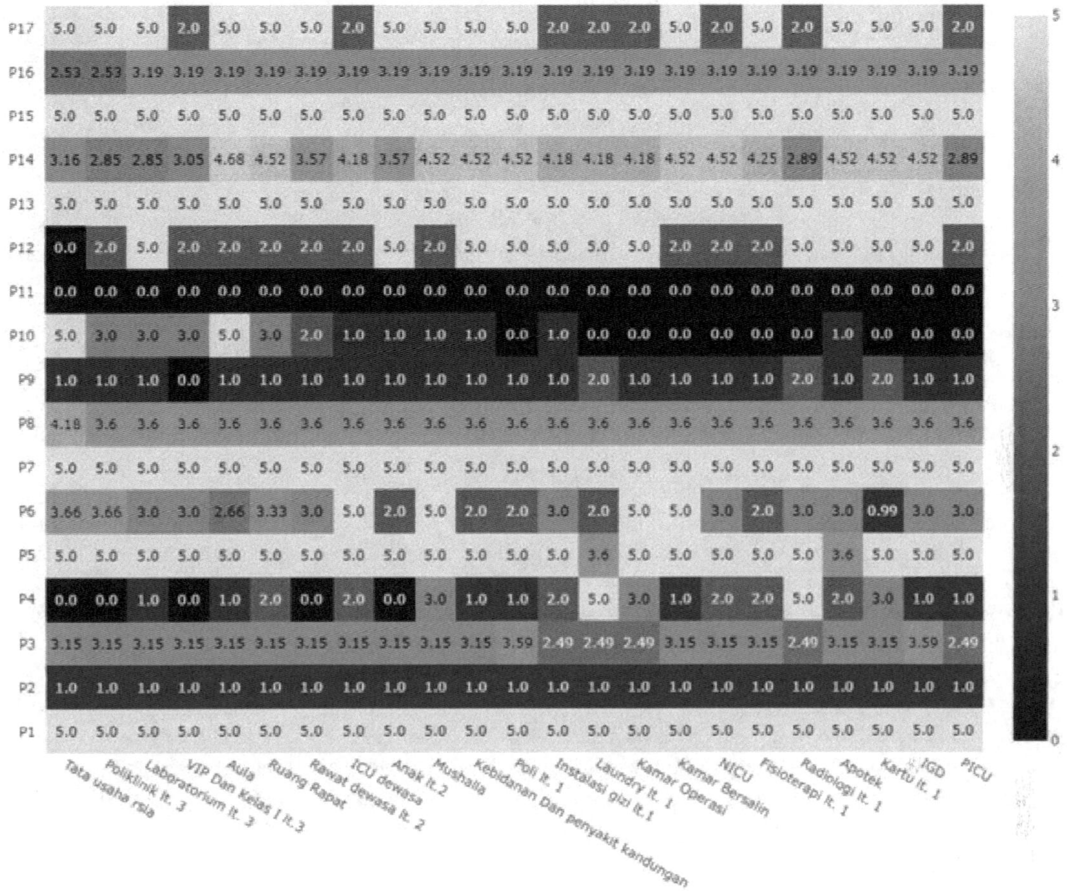

Figure 1. Score value for 23 rooms in RSIA.

After scoring each room, then each score was multiplied by the corresponding weight. table 2 is the final result of FRI for each room. The FRI mean value is 3.2. It can be concluded that RSIA has an intermediate level of fire safety. Mushalla/Prayer room has the highest score with 3.41 and PICU got the lowest one at 2.78.

Table 2. Fire Risk Index for 23 rooms in RSIA.

Room	Fire Risk Index	Room	Fire Risk Index
Administration office in RSIA	3.05	First-floor nutrition installation	3.16
Third-floor polyclinic	3.05	First-floor laundry	3.21
Third-floor laboratory	3.30	Operation room	3.32

Room	Fire Risk Index	Room	Fire Risk Index
VIP room and third-floor first-class room	2.84	Birthing room	3.24
Hall	3.28	NICU	3.00
Meeting room	3.30	First-floor physiotherapy room	3.08
Second-floor adult room	3.05	First-floor radiology	3.29
Adult ICU	3.15	Pharmacy	3.30
Second-floor children room	3.13	First-floor card room	3.33
Prayer room	3.41	Emergency room	3.31
Obstetry and gynecology	3.25	PICU	2.78
First-floor poly	3.25		

For fire management, it can be concluded that there is a system of fire management already available which is managed and commanded by Health and Safety unit. In addition, the equipment related to fire protection has been maintained regularly with support from the Disaster Mitigation Agency of Banda Aceh municipality.

Related to the knowledge of hospital personnel, it revealed that both hospital staff and patients mostly have an intermediate level of fire safety knowledge with a percentage of 74.7% and 61.4%. Only 12.5% of hospital staff have good knowledge of fire safety and the rest 13.25% is bad. On the other hand, 17.54% of patients have a good knowledge of fire safety and 21.05% is bad. From the analysis of the survey result, some items need to be improved such as emergency code, the use of fire extinguisher, and emergency signal. For patients, the knowledge of using electrical equipment which has the potential to cause a fire should be improved. Lastly, patients also need to be informed about fire mitigation and evacuation.

4. Conclusion

RSIA has intermediate fire safety with a mean index of 3.2. The highest index is 3.41, which belongs to the prayer room; the lowest one is 2.8, which belongs to the Pediatric Intensive Care Unit (PICU). There are some parameters which are rated as the lowest score and considered to be improved. Those parameters are fire suppression, attic, and smoke control system. The knowledge level of hospital personnel regarding fire safety needs to be improved because the knowledge of 74.7% of the hospital staff was categorized at the intermediate level, 12.5% was categorized as good and 13.25% was categorized as bad. On the other hand, for the knowledge of patients for fire safety, 61.4% is at an intermediate level, 17.54% is good and 21.05% is bad. Some aspects required to be improved for hospital staff are emergency code, the use of fire extinguisher, and emergency signal. For patients, the knowledge of using electrical equipment that could cause a fire should be improved. Also, the patients need to be informed about fire mitigation and evacuation. For fire management, it can be concluded that the system of fire management at RSIA is already available and managed by the health and safety unit. In addition, the equipment related to fire protection is maintained regularly.

Acknowledgments
The researcher would like to acknowledge Mr. Nazli as theses supervisor and Mrs. Izziah as co-supervisor. Medical Support Unit, Health and Safety unit of RSIA including all staff which had contributed to this research.

References

[1] Keputusan Menteri Kesehatan Nomor 66 tahun 2016 tentang *Kesehatan dan Keselamatan Kerja di Rumah Sakit*
[2] Badan Penanggulangan Bencana Aceh (BPBA) 2019 *Data Kejadian Bulan Januari-Mei 2019 Provinsi Aceh*
[3] Data Informasi Bencana Indonesia (DIBI) http://dibi.bnpb.go.id/dibi diakses tanggal 22 April 2019
[4] Dinas Pemadam dan Penyelamatan Kota Banda Aceh 2018 *Data Kejadian Tahun 2014-2017* Kota Banda Aceh.
[5] Kasmawati 2016 Analisis Implementasi Kebijakan Hospital Disaster Plan di BLUD Rumah Sakit Ibu dan Anak Provinsi Aceh Menghadapi Bencana Gempa Bumi dan Tsunami *Jurnal Kedokteran Syiah Kuala.* **16** 75-81
[6] Larsson D 2000 Developing the Structure of a Fire Index Method for Multistorey Apartment Buildings *Hazard Analysis Technique for System Safety* (New Jersey:John Willey & Sons Inc.) Department of Fire Safety Engineering Lund University Sweden Ericson C A 2005

AUTHOR INDEX